ATOMIC PHYSICS 5

1969—Atomic Physics 1

Proceedings of the First International Conference on
Atomic Physics, June 3-7, 1968, in New York City
V. W. Hughes, Conference Chairman
B. Bederson, V. W. Cohen, and F. M. J. Pichanik, Editors

1971—Atomic Physics 2

Proceedings of the Second International Conference on
Atomic Physics, July 21-24, 1970, Oxford, England
G. K. Woodgate, Conference Chairman
P. G. H. Sandars, Editor

1973—Atomic Physics 3

Proceedings of the Third International Conference on
Atomic Physics, August 7-11, 1972, Boulder, Colorado
S. J. Smith and G. K. Walters, Conference Chairmen and Editors

1975—Atomic Physics 4

Proceedings of the Fourth International Conference on
Atomic Physics, July 22-26, 1974, Heidelberg, Germany
G. zu Putlitz, Conference Chairman, E. W. Weber and A. Winnacker, Editors

1977—Atomic Physics 5

Proceedings of the Fifth International Conference on
Atomic Physics, July 26-30, 1976, Berkeley, California
Richard Marrus, Conference Chairman, Michael Prior and Howard Shugart, Editors

A Continuation Order Plan is available for this series. A continuation order will bring delivery of each new volume immediately upon publication. Volumes are billed only upon actual shipment. For further information please contact the publisher.

ATOMIC PHYSICS 5

Editors
RICHARD MARRUS
Conference Chairman

MICHAEL PRIOR

and

HOWARD SHUGART

University of California, Berkeley

PLENUM PRESS • NEW YORK AND LONDON

The Library of Congress cataloged the first volume of this title as follows:

International Conference on Atomic Physics.

    Atomic Physics; proceedings. 1st—
1968—
New York [etc.] Plenum Press.
    v. illus. 26 cm. biennial.

    1. Nuclear physics—Congresses. I. Title.
QC173.I 53   539.7   72-176581

Library of Congress Catalog Card Number 72-176581
ISBN-13: 978-1-4613-4204-5   e-ISBN-13: 978-1-4613-4202-1
DOI: 10.1007/978-1-4613-4202-1
Proceedings of the Fifth International Conference
on Atomic Physics, held in Berkeley, California,
July 26—30, 1976

© 1977 Plenum Press, New York
Softcover reprint of the hardcover 1st edition 1977
A Division of Plenum Publishing Corporation
227 West 17th Street, New York, N.Y. 10011

Victor William Cohen (1911-1974)

Conference Chairman

R. Marrus, University of California, Berkeley, California

Organizing Committee

S. Bashkin	University of Arizona, Tucson, Arizona
P. Bender	Joint Institute for Laboratory Astrophysics, Boulder, Colorado
J. Brossel	École Normale Supérieure, Paris, France
U. Fano	The University of Chicago, Chicago, Illinois
S. Feneuille	Laboratoire Aimé Cotton, Orsay, France
V.W. Hughes	Yale University, New Haven, Connecticut
A. Kastler,	École Normale Supérieure, Paris, France
I. Lindgren	Chalmers University of Technology, Göteborg, Sweden
H. Narumi	Hiroshima University, Hiroshima, Japan
E.E. Nikitin	Institute of Chemical Physics, Moscow, USSR
F.M. Pipkin	Harvard University, Cambridge, Massachusetts
I.I. Rabi	Columbia University, New York, New York
A.L. Schawlow	Stanford University, Stanford, California
H.A. Shugart	University of California, Berkeley, California
S.J. Smith	Joint Institute for Laboratory Astrophysics, Boulder, Colorado
G.K. Woodgate	Clarendon Laboratory, Oxford, England
G. zuPutlitz	Universität Heidelberg, Heidelberg, W. Germany

Program Committee

A. Dalgarno	Center for Astrophysics, Cambridge, Massachusetts
W. Happer	Columbia Radiation Laboratory, New York, New York
V. Hughes	Yale University, New Haven, Connecticut
D. Kleppner	Massachusetts Institute of Technology, Cambridge, Massachusetts
R. Marrus	University of California, Berkeley, California, Chairman

Local Committee

P. Bucksbaum

S. Chu

E.D. Commins

R.S. Conti

B. Davis

S. Davis

H. Gould

T. Hadeishi

J.M. Hardin

A. Huq

R.D. Knight

V.W. Lam

J. Lynch

D. MacDonald

R. Marrus

P. Mohr

D. Neuffer

A. Ozmen

M. Prior

L. Schipper

H.A. Shugart

E.C. Wang

P.G. Yarnold

Sponsoring Agencies

Energy Research and Development Agency (ERDA)

National Science Foundation (NSF)

Office of Naval Research (ONR)

International Union for Pure and Applied Physics (IUPAP)

University of California at Berkeley (UCB)

Conferences Hosts

University of California, Berkeley; Albert Bowker, Chancellor

Lawrence Berkeley Laboratory; Andrew Sessler, Director

We gratefully acknowledge a donation from the International Business Machines Corporation

PREFACE

The Fifth International Conference on Atomic Physics was held July 26-30, 1976 in Berkeley, California. Invited talks were solicited which were representative of the most important developments since the fourth conference held in Heidelberg, Germany in 1974. In this volume, we have collected the manuscripts of the invited speakers, in the belief that they represent a guide to contemporary research in atomic physics. Experimental work on such topics as the search for parity violation, spectroscopy and collision processes of fast, highly-stripped heavy ions, exotic atoms, high-Rydberg states, laser spectroscopy, photoelectron spectroscopy, and others are described. The work described in these manuscripts is a clear measure of the continued vitality of our field.

One unhappy event since the last conference was the passing of Dr. Victor William (Bill) Cohen (1911-1974) of Brookhaven National Laboratory. Bill was one of the scientists who recognized early the need for personal communication among atomic physicists and was the prime mover in establishing the present international conference series. Everyone who has enjoyed the stimulation of these conferences is indebted to Bill Cohen, and we dedicate this volume of the proceedings to his memory.

Richard Marrus
University of California
Berkeley, 1976

Contents

Parity Violation Effects Induced by Neutral
Currents in Atoms   1
    C. Bouchiat, M.A. Bouchiat, and L. Pottier

An Optical Rotation Experiment to Search for
Neutral Current Effects in Atomic Bismuth 23
    Norval Fortson

Search for Parity Non-Conserving Optical
Rotation in Atomic Bismuth 27
    P.E.G. Baird, M.W.S.M. Brimicombe,
    G.J. Roberts, P.G.H. Sandars, and
    D.N. Stacey

Atomic Physics Tests of Quantum
Electrodynamics 37
    Peter J. Mohr

Hadronic Atoms 63
    Chien-Shiung Wu

Detection of π-μ Coulomb Bound States 95
    R. Coombes, R. Flexer, A. Hall,
    R. Kennelly, J. Kirkby, R. Piccioni,
    D. Porat, M. Schwartz, R. Spitzer,
    J. Toraskar, S. Wiesner, B. Budick,
    and J.W. Kast

Review of Precision Positronium Experiments . . . 103
    Allen P. Mills, Jr., Stephen Berko,
    and Karl F. Canter

Hyperfine Structure of Stored Ions —
Results for 2s ^{3}He$^+$ 125
    Michael H. Prior and Edmond C. Wang

Multiphotonic High Resolution Spectroscopy 147
    B. Cagnac

xi

Laser Spectroscopy and Predissociation of
Molecules 167
      J.C. Lehmann

Time-Resolved Laser Spectroscopy:
Quantum Beats and Superradiance 179
      S. Haroche, C. Fabre, M. Gross,
      and P. Pillet

Atomic Beam Experiments at the Isolde
Facility at CERN 201
      Curt Ekström and Ingvar Lindgren

High Resolution Laser Spectroscopy of
Radioactive Sodium Isotopes 215
      H.T. Duong, P. Jacquinot, P. Juncar,
      S. Liberman, J. Pinard, and J.L. Vialle
      G. Huber, R. Klapisch, C. Thibault

Nuclear Excitation by Electron Transition 227
      M. Morita

Hyperfine and Isotope Shift Measurements Far-
Off Stability by Optical Pumping 239
      E.W. Otten

Highly-Excited Atoms 269
      Daniel Kleppner

Anisotropy and Time Dependence in Atomic
Collisions 283
      Joseph Macek

R-Matrix Theory of Atomic and Molecular
Processes 293
      Philip G. Burke

Electron Correlation in Atoms From
Photoelectron Spectroscopy 313
      D.A. Shirley, S.T. Lee, S. Süzer,
      R.L. Martin, E. Matthias, and
      R.A. Rosenberg

Polarized Electrons 325
      M.S. Lubell

Role of Impurities in Magnetically Confined
High-Temperature Plasmas 375
      C.F. Barnett

Diagnostic Problems of Large Tokamaks 391
    Kenneth M. Young

Relativistic Magnetic-Dipole Transitions
in Atoms, Ions, and Psions 415
    Joseph Sucher

The Electric and Magnetic Dipole Moments
of the Neutron 453
    Norman F. Ramsey

Spectroscopy of Highly-Ionized Atoms
Produced by a Low-Inductance Vacuum Spark 473
    U. Feldman and G.A. Doschek

Influence of Electron Capture on X-ray
Production in Heavy-Ion Collisions 493
    Hans D. Betz, F. Bell, and
    E. Spindler

Ion-Induced Continuum X-ray Emission 509
    F.W. Saris and Th. P. Hoogkamer

The Ionization of Inner Shells of Atoms
With Account of Outer Shell Rearrangement 537
    M. Ya Amusia

Index 567

PARITY VIOLATION EFFECTS INDUCED BY NEUTRAL CURRENTS IN ATOMS

C. BOUCHIAT, M.A. BOUCHIAT and L. POTTIER

Laboratoire de Physique de l'Ecole Normale Supérieure

24, rue Lhomond, 75231 PARIS CEDEX 05 - France

INTRODUCTION

The problem of space reflexion symmetry in atomic physics is a long story which began with the Wigner explanation of the Laporte rule. The symmetry was not questioned until the discovery of parity violation in weak interactions, following the brilliant analysis of Lee and Yang [1]. In 1958, Zel'dovich [2] gave the first discussion of the possible effects in Atomic Physics of a weak electron-nucleon interaction induced by neutral currents. In particular, he estimated the rotation of the plane of polarization of visible light propagating in optically inactive matter to be of the order of 10^{-13} radian/meter, and, at that time, his conclusion was that such an effect, obviously, could not be observed. A similar investigation, somewhat more detailed, was performed later by F.C. Michel [3]; the predicted effects concerning mainly the Hydrogen atom, were also not very encouraging.

In 1971, G.'t Hooft presented the first proof that spontaneously broken gauge theories (S.B.G.T.) were renormalizable. The unified model of weak and electromagnetic interactions proposed earlier by Weinberg [5] and Salam [6] was a prototype of such a theory and immediately became the favourite candidate for a field theory of weak interactions. The most evident implication of the Weinberg-Salam theory was the existence of weak neutral currents involving leptons and hadrons. So began a very active search for muonless neutrino interactions in high energy neutrino scattering on nuclei. Positive evidence for neutral currents was first announced by the Gargamelle collaboration [7], rapidly confirmed by the two Fermilab experiments [8][9].

1

In 1973, stimulated by these developments in weak interaction physics, there was a renewed interest for the problem of parity violation in atomic physics. The hopes of observing the effects of neutral currents in atoms appeared much brighter ([10]), when it was realized that parity mixing amplitudes induced by neutral currents in atoms follow a Z^3 law, and are thus greatly enhanced (by factors of the order of $10^6 - 10^7$) in heavy atoms. Furthermore, the dye laser technology was opening a new field of experiments in atomic spectroscopy, allowing, in particular, the study of highly forbidden radiative transitions where a weak electron–nucleus interaction has the best chance to show up. In the last three years, various tests of parity violation induced by neutral currents in electronic atoms and muonic atoms have been proposed ([10-16]) and some of them have led to experiments now under completion. The level of accuracy at which the gauge models of weak interactions can be tested will certainly be reached in the near future.

The informations provided by these experiments are eagerly sought for by weak interaction theorists. In the renormalizable gauge theories of weak interactions, the observed events –muonless neutrino scattering on nuclei– are associated with the exchange of a heavy neutral vector boson Z_0 between the neutrino and the nuclei as illustrated in the diagram of fig. 1-a. This occurs in contradistinction with previously well-known weak interactions with charged currents, illustrated in the diagram of fig. 1-b : a heavy vector boson is also exchanged but it bears the electric charge necessary for ensuring total charge conservation :

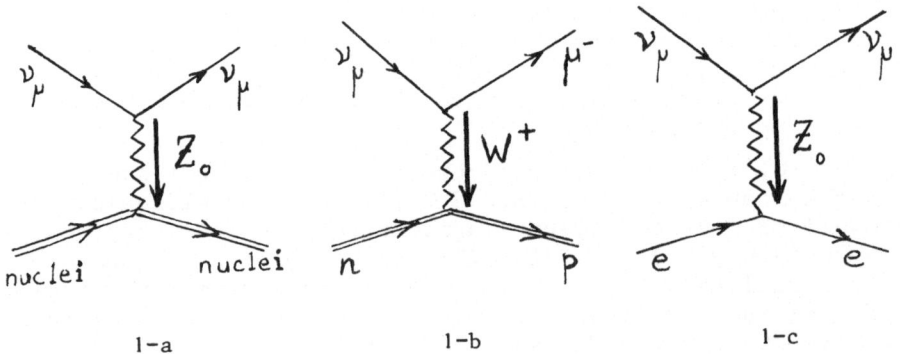

1-a   1-b   1-c

Figure 1 :   *Diagrams illustrating weak-interactions with neutral currents (a and c) and with charged currents (b)*

Neutral current interaction has also been observed in the scattering of neutrino ν_μ, on electrons (three events in Gargamelle), So we already know that the neutral boson, if it exists, interacts with neutrinos, nuclei and almost certainly with electrons.

A positive result, demonstrating the existence of neutral currents in Atomic Physics, with roughly the expected size, would confirm the existence of electronic neutral currents and the "current-current" picture for neutral currents interactions suggested by the intermediate boson hypothesis : interactions such that (1-a) and (1-c) imply the existence of an electron-nucleus neutral current induced weak interaction (schemed on fig. 2) in any theory in which the neutral current interaction is produced by the exchange of a neutral intermediate boson.

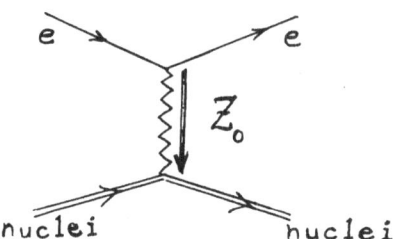

Figure 2 : *Scheme of the electron-nucleus neutral-current weak interaction under search in Atomic Physics*

Furthermore there is presently no direct evidence for parity-violation in weak neutral current interaction. In the Weinberg-Salam model the electron-nucleus interaction violates P-reflection symmetry almost maximally. The observation of any sort of parity violation would immediately rule out the simplest version of another type of weak interaction model, the vector gauge model, where neutral currents interactions do conserve parity.

It would be also the first time that a weak interaction would be detected through an interference effect with the electromagnetic interaction and that the absolute sign of the weak interaction coupling constant could be measured. Note this sign is a measurable quantity specifically predicted in any SBGT model.

Furthermore experiments performed on different atoms should allow to test the isospin structure of weak-neutral currents in the low energy limit.

I. THE PARITY VIOLATING ELECTRON–NUCLEUS POTENTIAL INDUCED BY NEUTRAL CURRENTS *

The parity violating electron–nucleon interaction will be described by a Hamiltonian density of the current-current type. It can be considered as the limit of any neutral current interaction mediated by a neutral vector boson, for infinite boson mass. We shall restrict ourselves to vector and axial vector currents :

$$
H_{p.v} = \frac{G_F}{\sqrt{2}} \int d^3x \; \overline{e}(x) \; \gamma^{\mu}\gamma_5 e(x) \left(C_{Vp} \; \overline{p}(x)\gamma_{\mu} \; p(x) + C_{Vn} \; \overline{n}(x)\gamma_{\mu} \; n(x) \right)
$$

$$
+ \frac{G_F}{\sqrt{2}} \int d^3x \; \overline{e}(x)\gamma^{\mu} \; e(x) \left(C_{Ap} \; \overline{p}(x) \; \gamma_{\mu}\gamma_5 p(x) + C_{An} \; \overline{n}(x) \; \gamma_{\mu}\gamma_5 n(x) \right)
$$

$$
(1)
$$

At low energy the nucleons can be treated as point-like particles; $e(x)$, $n(x)$ and $p(x)$ are the Dirac fields associated with the electron, neutron and proton. Each term appearing in the sum is the product of an electronic current by an hadronic current, a vector current (γ_{μ}) being always associated with an axial vector one ($\gamma_{\mu}\gamma_5$). We are interested only in processes where the nucleus does not suffer any transition. It can be shown that only the time component ($\mu = 0$) of the vector nucleonic current and the spatial component ($\mu \neq 0$) of the axial-vector nucleonic one give contributions. The real constants C_{Ap}, C_{An}, C_{Vp}, C_{Vn} are the weak axial-vector and vector charges of the proton and neutron.

The Weinberg–Salam model makes definite predictions about the constants C_A and C_V :

$$
C_{Vp} = -\frac{1}{2} + 2 \sin^2\theta_w \qquad\qquad C_{Ap} = \lambda \left(2 \sin^2\theta_w - \frac{1}{2}\right)
$$

$$
C_{Vn} = \frac{1}{2} \qquad\qquad\qquad\qquad\qquad C_{An} = - C_{Ap}
$$

$$
(2)
$$

θ_w is the now familiar weak interaction angle, which, from the analysis of neutrino experiments is restricted to the following values :

$$
0.3 \lesssim \sin^2\theta_w \lesssim 0.4 \tag{3}
$$

$\lambda \simeq 1.25$ is the ratio between the vector and axial vector coupling constants in neutron β decay.

Although the latest neutrino data on neutral currents seem to favour the Weinberg–Salam model (recent analysis ([17]) seems to ex-

---

* *A weak short range electron-electron interaction has been shown to play a negligible role* ([10]).

clude a pure V, or pure A, hadronic currents) the knowledge of the coupling constants C_{Vp} , C_{Vn} , etc... is crucial to discriminate between various gauge models of weak interactions. As we have already said, the simplest version of the vector-like model predicts :

$$C_{Vp} = C_{Vn} = C_{Ap} = C_{An} = 0 ,$$

values inconsistent with the Weinberg-Salam model.

In order to discuss the physics of neutral currents in atoms, it is convenient to derive from the second quantized Hamiltonian $H_{p.v.}$ a non relativistic electron-nucleus potential, $V_{p.v.}(r)$:

$$V^{p.v.}(r) = -\frac{G_F}{\sqrt{2}} \left\{ \frac{\vec{s}_e \cdot \vec{P}_e}{m_e c} \delta^3(r) \left[Z\, C_{Vp} + (A - Z)\, C_{Vn} \right] \right.$$

$$\left. + \left(\frac{\vec{P}_e}{m_e c} - 2i \frac{\vec{s}_e \wedge \vec{P}_e}{m_e c} \right) \delta^3(r) \left[\vec{S}_p\, C_{Ap} + \vec{S}_n\, C_{An} \right] \right.$$

$$\left. + \text{Hermitian conjugate} \right\}$$

$$\tag{4}$$

r, m_e , $\vec{s}_e$, $\vec{P}_e$ are respectively the position, mass, spin $\vec{s}_e = \frac{1}{2}\vec{\sigma}_e$ and momentum operators of the electron. $\vec{S}_p$ and $\vec{S}_n$ are the $\underline{\text{total}}$ spin operators of the protons and neutrons in the nucleus. If one wishes to include scalar and pseudo-scalar neutral currents (or tensor pseudo-tensor currents), one has simply to add an imaginary part to the constants C_{Vp} and C_{Vn} (or C_{Ap} and C_{An}) [18].

In the first term of $V_{p.v.}$ involving the weak vector charge of the nucleons, the effects of the different nucleons add $\underline{\text{coherently}}$, in the same way as the electric charge. This is not true for the term involving the weak axial-vector charges, the different nucleons, instead, add their contribution as their individual spins. For this reason in high Z atoms the term independent of the nuclear spin associated with the hadronic vector current will give the dominant contribution (by a factor Z), except in quite special cases when it vanishes identically [10].

II. THE MATRIX ELEMENT OF $V^{p.v.}$ BETWEEN VALENCE STATES OF ATOMS. VARIATION WITH ATOMIC NUMBER : THE Z^3 LAW

Under the influence of the parity violating potential, parity mixing occurs between atomic wave functions. In the independent particle model of the atom, $V^{p.v.}$, because of its zero-range nature, will mix only s and p states. The matrix element of $V^{p.v.}$ is given in terms of the radial wave functions $R_{n\ell}(r)$ by :

$$< n \, s_{1/2} \, |v^{p.v.}| \, n'p_{1/2} > \; =$$

$$= \frac{-3i \, \hbar}{8\pi m_e c} \, \frac{G_F}{\sqrt{2}} \left[C_{Vp} \, Z + C_{Vn} \, (A-Z) \right] \, \frac{d}{dr} R_{n1}(0) \cdot R_{n0}(0) \qquad (5)$$

This matrix element presents some similarity with the expression giving the hyperfine splitting of atomic states which is proportional to $|R_{n0}(0)|^2$. There exists in the literature a semi-empirical formula due to Fermi and Segré giving the wave function at the origin $R_{n0}(0)$, which only requires the knowledge of the energy spectrum of the valence s electron. When used to compute hyperfine separations of atoms and ions with one s electron, the formula appears to be remarkably accurate [19], provided relativistic corrections are included properly. In ref. [10], the Fermi-Segré formula has been extended in order to give the starting coefficient of the radial wave function at the origin,

$$\lim r \to 0 \; \frac{R_{n\ell}(r)}{r^\ell} \; , \text{ for any value of the orbital angular momentum } \ell.$$

Let us quote the result :

$$\lim r \to 0 \; \frac{R_{n\ell}(r)}{r^\ell} = \nu_n^{-3/2} \, \frac{2^{\ell+1} \; z^{\ell+1/2}}{(2\ell+1) \, ! \; a_0^{\ell+3/2}} \left[1 + \delta_\ell \, (\epsilon_n, \, Z) \right] \; ,$$
$$(6)$$

with :

$$\delta_\ell(\epsilon, Z) = -\epsilon_n^{3/2} \, \frac{d\mu_{n\ell}}{d\epsilon} - \zeta \, \frac{\ell(\ell+1)(2\ell+1)}{6} + 0(\epsilon_n^2) + 0(\zeta^2) \; . \qquad (7)$$

a_0 is the Bohr radius, $\epsilon_n = -1/\nu_n^2$ is the binding energy in atomic units. $\mu_{n\ell}(\epsilon)$ is the interpolated quantum defect $(\mu_{n\ell}(\epsilon_n) = n - \nu_n)$; the constant ζ is given in terms of the average potential acting on the valence electron by

$$\zeta = \frac{a_0}{z^2 e^2} \, \frac{d}{dr} \left(rV(r) \right) \Big|_{r=0} \; .$$

For $\epsilon_n \ll 1$ (small binding energy) and high Z, $\delta_\ell(\epsilon, Z)$ is at most of the order of 10%.

The accuracy of the formula for $\ell = 0, 1$ has been tested with experiment ($\ell = 0$) or with numerical computation ($\ell = 1$), and found to give the correct result within ten percents.

The remarkable feature of this formula, besides its simplicity, is the fact that it only requires a good knowledge of the energy spectrum and it is independent of the detailed shape of the electronic potential.

We shall not give here the detailed derivation of the formula which can be found in ref. (^{10}), but rather try to give some justifications of the Z-dependence of the wave function near the origin :

$$R_{n\ell} \propto r^{\ell} Z^{\ell + \frac{1}{2}}$$

The derivation of (^{10}) involves the use of what may be called the Coulomb J.W.K.B. approximation. In the limit of <u>small binding</u> <u>energy</u> (high quantum number), the radial wave function of an electron in an unscreened Coulomb potential of charge $Z |e|$ is given, up to a constant factor, in terms of a Bessel function :

$$u_{n\ell}(r) = r R_{n\ell}(r) \simeq C_{\ell} \sqrt{8Zr} J_{2\ell+1} (\sqrt{8Zr}) ,$$

(r is expressed in atomic units). For the case of a screened Coulomb potential, for small binding energies, one tries, for $u_{n\ell}(r)$, a representation of the form :

$$u_{n\ell}(r) = C_{\ell} \left(\frac{d\phi}{dr} \right)^{-\frac{1}{2}} \sqrt{8\phi} J_{2\ell+1} (\sqrt{8\phi}) , \tag{8}$$

where the function $\phi(r)$ obeys a non linear differential equation. This procedure is completely analogous to the ordinary J.W.K.B. approximation, the Bessel function playing the role of the trigonometric functions. In fact, when the argument $\sqrt{8\phi}$ of the Bessel function is large, one can replace the Bessel function by its asymptotic expansion and one recovers the ordinary J.W.K.B. expression. The function ϕ is obtained through an iteration procedure in which the first approximation is obtained by neglecting the derivative of ϕ of order higher than one. To first order ϕ is obtained from a quadrature. For instance, in the case $\ell = 0$, ϕ is given by the simple expression :

$$\sqrt{8\phi} = \int_{o}^{r} (-V(r) + \varepsilon_n)^{\frac{1}{2}} dr. \tag{9}$$

In the vicinity of the nucleus, the screening can be neglected and the potential energy is also much larger than the binding energy ε, so that the above expression reduces to :

$$\phi \simeq Zr .$$

When r is larger than the atomic core radius r_c , the potential is coulombic again but corresponds to a unit electric charge. The first order approximation to ϕ, becomes accurate again, provided r is large but still smaller than the external turning point $r_e \simeq 2/|\varepsilon_n|$. Then ϕ is given by the simple expression :

$$\sqrt{8\phi} \underset{\sim}{\sim} \sqrt{8r} + \text{constant} + 0\left(\frac{1}{r}\right), \tag{10}$$

where the constant can be shown to be equal to the quantum defect times π.

On the other hand, when $r > r_c$, the quantum defect theory yields an exact expression of the normalized radial wave function in terms of the binding energy and of the interpolated quantum defect $\mu_\ell(\epsilon)$. In the wave function thus obtained, all explicit dependence on the nuclear charge has disappeared. The amplitude of this external wave function is of the order of $\nu_n^{-3/2}$; it reflects the simple fact that the spatial extension of the wave function is, in atomic units, of the order of ν_n. The wave function given by the quantum defect theory can be matched in the region $r_c < r \ll 1/|\epsilon_n|$ with the one given by the Coulomb J.W.K.B. approximation. The matching yields the normalization constant C_ℓ under the form of an asymptotic expansion in ϵ_n. C_ℓ is found to be of the order of $\nu_n^{-3/2}$ with no dependence on the nuclear charge Z. It is possible to see this without entering the details of the calculation. One remarks that, in the region $r_c < r \ll 1/|\epsilon_n|$, the argument $\sqrt{8\phi}$ of the Bessel function being approximately equal to $\sqrt{8r}$, the J.W.K.B. Coulomb amplitude is independent of Z and ν_n. From this fact, it follows immediately that the matching will lead to a constant C_ℓ of the order of $\nu_n^{-3/2}$. Going to the limit $r \to 0$ in (8), one gets at once :

$$\lim r \to 0 \quad \frac{R_{n\ell}(r)}{r} \propto \nu_n^{-3/2} \, Z^{\ell+1/2}$$

Using the results given in formula (6) to evaluate the matrix element of $V^{\text{P.V.}}$, one arrives at the final expression :

$$< ns_{1/2} \, |V^{\text{P.V.}}| \, n'p_{1/2} > =$$

$$= - i \, \frac{\hbar G_F \, Z^3 \, [\, 1 + \delta_0(\epsilon_n) + \delta_1(\epsilon_{n'})\,]}{2\pi m_e c \, \sqrt{2} \, a_o^4 \, (\nu_n \, \nu_{n'})^{3/2}} \left[C_{vp} + \left(\frac{A}{Z} - 1\right) C_{vn} \right] \tag{11}$$

From the above expression, it is apparent that the matrix element follows a Z^3 law. Summarizing the above considerations, we may say that one power of Z expresses the <u>coherent effect of the nucleons</u> (the weak vector charge of the nucleus is the sum of the vector charges of the constituents). The second power of Z reflects the fact that the <u>density of the valence electron at the nucleus in-creases</u> like Z, because of the Coulomb attraction of the nucleus; the third power follows from the proportionality of $V^{\text{P.V.}}$ to the <u>velocity of the electron near the nucleus which grows like Z</u>.

The matrix element of $V^{\text{P.V.}}$ expressed in eq. (11) has still to be corrected for relativistic effects. The relativistic correcting factor, K_r, is quite substantial due to the fact that the radial Dirac wave functions of an $s_{1/2}$ and $p_{1/2}$ electron moving in the Coulomb field of a point charge are infinite at the origin. K_r has been obtained in ref. ([10]) in terms of the nuclear radius R and gives rise to <u>a further increase with Z</u> :

$$K_r(Z,R) = \left[\frac{\Gamma(3)}{\Gamma(2\sqrt{1-Z^2\alpha^2}+1)} \left(\frac{2ZR}{a_o} \right)^{\sqrt{1-Z^2\alpha^2}-1} \right]^2 \qquad (12)$$

For example the value of K_r is 2.8 for Cs (Z = 55) and 9.0 for Pb (Z = 82).

Up to now, our discussion has been limited to valence electron states, which give rise to radiative transitions involving visible light. One may think of transitions, in the X ray range, involving internal shells K or L. For heavy atoms the matrix element of $V^{P.V.}$ between K and L states obeys a Z^5 law. However, negative powers of Z associated with the energy denominators and the ratio E_1/M_1 between electric and magnetic dipole amplitudes appear in the computation of the observable quantities and they almost completely cancel the factor Z^5 of the matrix element of $V^{P.V.}$. The same phenomena do not occur with valence states and the observable parity violation effects will reflect the variation with Z of the matrix element of $V^{P.V.}$.

III. PHYSICAL IMPLICATIONS OF PARITY MIXING IN ATOMIC WAVE FUNCTIONS

a) The Question of P and T Violation

The effects of a parity violating e-nucleus interaction are very different, depending on the behaviour on time reflection T of the parity violating potential.

First, let us very briefly discuss the case of T-odd parity violating potential, since this is not the case for the interaction involving vector axial-vector neutral currents (*). Such a potential gives rise to a linear Stark effect on non-degenerate atomic levels. The experiments designed to look for an electric dipole moment of the electron have been reanalyzed in order to set limits on the existence of scalar-pseudo scalar currents ([20-21]). From the existing data, one can exclude a weak electron-nucleon interaction involving the product of a pseudo-scalar electronic current by a hadronic scalar current, with a coupling constant larger than

(*) *It can be checked on expression (1) or (4) that the interaction is P-odd and T-even unless the constants C_V or C_A are pure imaginary : as we already mentioned, this reflects the contribution of scalar-pseudo scalar or tensor-pseudo tensor neutral currents.*

10^{-3} G_F.

A similar analysis, performed through the experiment of Harrison et al. on Thallium fluorine ([22]), which was designed to detect an eventual electric dipole of the proton, gives a similar limit on a pseudo-tensor-tensor interaction ([21-18]).

Let us, now, turn our attention onto the observable effects produced by a short-range P-violating but T-invariant electron-nucleon interaction as predicted by the Weinberg-Salam model (C_A and C_V real constants). It is well known that such an inter-action <u>cannot</u> give rise to a linear Stark shift of non-degenerate atomic <u>levels</u>, or equivalently, to a static electric dipole moment.

b) Abnormal Electric Dipole Amplitude Induced by Neutral Currents Between States of Same Parity

T-invariance forbids static electric dipole moment, but per-mits electric dipole transition moment between states of same par-ity. The basic principle of most experiments proposed so far is the observation of the interference between the abnormal $E_1^{p.v.}$ am-plitude and the normal magnetic dipole amplitude M_1, or the elec-tric dipole amplitude E_1 ind. induced by an external static elec-tric field.

Before going to a description of the possible experimental methods let us give order of magnitude estimates of $E_1^{p.v.}$ and M_1. We define $E_1^{p.v.}$ and M_1 as the matrix elements of the z-component of the electric dipole d_z and of the magnetic dipole operator μ_z divided by c.

The $E_1^{p.v.}$ amplitude between two s-states (or two p-states) is given to first order in $V^{p.v.}$ by an expression of the form :

$$E_1^{p.v.} = \sum_{n''} \left(\frac{< ns|V^{p.v.}|n''p >< n''p|d_z|n's >}{E_n - E_{n''}} \right.$$

$$\left. + \frac{< ns|d_z|n''p >< n''p|V^{p.v.}|n's >}{E_{n'} - E_{n''}} \right) \tag{13}$$

The allowed electric dipole amplitudes $< ns|d_z|n''p >$ are typically of the order of $|e| a_o$. Up to factors of the order of unity, the modulus of the matrix element of $V^{p.v.}$ can be expressed in unit of $1/2\, m_e c^2 \alpha^2$ (Rydberg) as :

$$\frac{|< V^{p.v.} >|}{\frac{1}{2} m_e c^2 \alpha^2} \sim \alpha^2 \frac{G_F m_e^2 c}{\pi\sqrt{2}\, \hbar^3} Z^2 A K_r \overline{C}_V \, ,$$

where we have defined $\overline{C}_v$ as :

$$\overline{C}_v = \frac{Z}{A} C_{vp} + \left(1 - \frac{Z}{A} \right) C_{vn} .$$

Assuming that there is no large cancellation in the sum over the intermediate states, the following expression can be used to get an order of magnitude of $\left| E_1^{p.v.} \right|$:

$$\left| E_1^{p.v.} \right| \sim \alpha^2 \frac{G_F m_e^2 c}{\pi\sqrt{2} \, \hbar^3} Z^2 A \, K_r \, \overline{C}_v \, |e| \, a_o \qquad (14)$$

Using the numerical value $G_F \frac{m_e^2 c}{\hbar^3} \sim 3 \times 10^{-12}$, one obtains finally :

$$\left| E_1^{p.v.} \right| \sim 3.6 \times 10^{-17} Z^2 A \, K_r \, \overline{C}_v \, |e| \, a_o \qquad (15)$$

Taking for instance the case of Bi $(Z = 83, A = 209)$

$$K_r \sim 9, \qquad \left| E_1^{p.v.} \right| \sim 4.7 \times 10^{-10} \, \overline{C}_v \, |e| a_o \qquad (16)$$

By going to heavy atoms, $Z > 50$, a factor of the order of $10^6 - 10^7$ is gained.

For a normal magnetic dipole transition, the magnetic dipole amplitude is of the order of one Bohr magneton over c :

$$|M_1| \sim |\mu_B| / c = \frac{\alpha}{2} |e| a_o .$$

One introduces the dimensionless ratio :

$$\left| \frac{E_1^{p.v.}}{M_1} \right| \sim \frac{\sqrt{2}}{\pi} \alpha \ G_F \frac{m_e^2 c}{\hbar^3} Z^2 A \, K_r \, \overline{C}_v$$

$$\sim 10^{-14} Z^2 A \, K_r \, \overline{C}_v \qquad (17)$$

Taking again the case of Bi one finds :

$$\left| \frac{E_1^{p.v.}}{M_1} \right| \sim 1.3 \times 10^{-7} \, \overline{C}_v \qquad (18)$$

These order of magnitude estimates have been confirmed by explicit computations which have been performed by using essentially two methods. The first one involves an explicit summation over the intermediate states [10]. The matrix elements of $V^{p.v.}$ are those given by the semi-empirical formula. The electric dipole amplitudes $< n's|d_z|n''p >$ are taken either directly from experiments or from phenomenological calculations. The summation is usually dominated by a finite number of states corresponding to low excita-

tion energies. The big advantage of this method is that the important ingredients are all accessible to experimental measurements.
The second method, perhaps mathematically more elegant, uses a
Green function technique [10][23][24][25] to perform the summation
over the intermediate states, but it requires a detailed knowledge
of the potential. One introduces the wave function $\Phi(E, E_n)$ corresponding to an angular momentum $\ell = 1$, defined as :

$$\Phi(E, E_n) = (E - H_1)^{-1} d_z |ns >. \tag{19}$$

H_1 is the one-particle Hamiltonian for p states.

The radial part of $\Phi(E, E_n)$ is the solution of an inhomogeneous
second order differential equation which can be solved by numerical
methods.

Knowing $\phi(E, E_n)$, the electric dipole amplitude E_1 is obtained
by the following expression :

$$E_1 = < \phi(E_n, E_{n'}) |V^{p.v.}|ns > + < n's|V^{p.v.}|\phi(E_{n'}, E_n) > \tag{20}$$

The two methods have been used to compute the 6s-7s transition
of Cesium and they give results which are in very good agreement
with each other [10][25].

IV. EXPERIMENTS

a) Different Types of Experiments

Many experiments have been suggested which could lead to an
experimental demonstration of the existence of parity violating,
time-reversal invariant weak neutral currents in atoms. It is not
possible to review here all the proposals. To our knowledge, there
are presently two types of experiments which have materialized as
experimental projects now in a well advanced stage or near completion. They proceed from two different philosophies and both have
their advantages and their drawbacks.

The first type of experiment is a search for a circular dichroism in highly forbidden magnetic dipole transitions [10], as
for instance the 6S-7S transition of Cesium. Because of the hindrance factor affecting M_1, the circular dichroism is expected to
reach the level of 10^{-4}. The drawback of an extremely small absorption cross-section is at least partially compensated by the large
photon flux that can deliver a tunable laser and by the use of optical fluorescence, a sensitive method of detection. The elimination of the systematic errors does not seem a priori too difficult,
but because of the low intensities involved the difficulties lie
in the signal to noise ratio. The present stage of this experiment
is discussed at length below.

The second type of experiment involves the measurement of the
optical rotatory power of an atomic Bismuth vapour for wavelengths
in the vicinity of a normal magnetic transition [14][15][26][27].
The choice of normal transition is dictated by practical considerations. Assuming a simple Breit-Wigner shape, the rotation of the
plane of polarization ϕ, through unit length, can be written in
terms of the absorption coefficient at the line peak, κ_{max}, as :

$$\phi(\omega) = \frac{\pi}{\lambda} \ \text{Re} \ (n_L - n_R)$$
$$= 2\kappa_{max} \ \frac{\frac{\omega-\omega_0}{\Gamma/2}}{1 + \left(\frac{\omega-\omega_0}{\Gamma/2}\right)^2} \ \text{Im} \left\{ \frac{E_1^{p.v.}}{M_1} \right\} \tag{21}$$

$\omega = 2\pi c/\lambda$ is the frequency of light; ω_0 and Γ are respectively the
frequency and the width of the line. From looking at this formula,
it would seem that one should work with highly forbidden M_1 transition in order to have a ratio δ/κ_{max} as large as possible. However,
in twice forbidden transitions, like the $6S_{1/2} - 7S_{1/2}$ of Cesium, the
angle $\ell\phi$ being proportional to $|E_1^{p.v.} \cdot M_1^*|$ is far too small for any
reasonable length ℓ of traversed matter, and one has to restrict
to normal magnetic transition with $|M_1| \sim |\mu_B|/ c$. A good choice
seems to be the transitions between the ground state of Bi and the
first excited states (all these states belong to the configuration
$6p^3$) in particular the transitions $^4S_{3/2} - ^2D_{3/2}$ (λ = 878 nm) and
$^4S_{3/2} - ^2D_{5/2}$ (λ = 648 nm). Very encouraging results are being obtained
by the Washington University [27] and Oxford University groups.
We shall not enter into the details of these very interesting experiments since they will be the subject of the two following papers
[28][29]. Let us only remark that in such experiments where the
effect is observed in <u>transmission</u>, there is no counting rate problem, but the rotation angle to be observed being of the order of
10^{-7} radian, the problem of systematic errors is certainly the most
serious one. Furthermore, up to now, it is not quite clear how the
physical quantity measured in such experiments can be calibrated
in a proper way to give a result on the coupling constant $\overline{C}_V$.

b) Circular Dichroism in the 6s-7s Transition of Cs

This experiment is designed to measure the $E_1^{p.v.}$ electric dipole
transition amplitude between the $6S_{1/2}$ and $7S_{1/2}$ states of atomic
Cesium, induced by the parity-violating time-reversal invariant interaction. The amplitude $E_{1z}^{p.v.}$ was computed [10] in the framework
of the Weinberg-Salam model with $\sin^2\theta_w = 0.35$:

$$E_1^{p.v.} = -|e| < 6S, m_S=1/2|z|7S, m_S=1/2 >$$
$$= -i \ 1.7 \times 10^{-11} \ |e| a_o \tag{22}$$

The error resulting from uncertainty on the atomic wave function is estimated to be less than 15%.

In the absence of a parity violating interaction, the radiative transition, photon + $6S_{1/2} \to 7S_{1/2}$, is normally of the magnetic dipole type. But because of the orthogonality of 6S and 7S radial wave functions, the magnetic dipole transition moment vanishes in the non-relativistic limit. A non-vanishing transition amplitude is likely to arise from the mechanisms which lead to a semi-quantitative explanation of the deviation of the gyromagnetic ratio of the Cs ground state from that of the free electron $\left(\frac{\Delta g}{g} = \frac{g_{Cs} - g_e}{g_e} \simeq 1.1 \times 10^{-4} \right)$. A magnetic dipole transition moment M_1 of the order of $10^{-4} \mu_B$ is thus expected. The phase and the modulus of M_1 have been measured experimentally [30] :

$$M_1 = < 6S, m_S=1/2 \mid \frac{\mu_z}{c} \mid 7S, m_S=1/2 > = -4.24 \times 10^{-5} \mid \frac{\mu_B}{c} \mid \quad (23)$$

The same phase convention has been used in the expression of $E_1^{p.v.}$ and M_1.

With the above results we can predict a circular dichroism for the resonant frequency of the 6S → 7S transition :

$$P_c = \frac{\sigma_+ - \sigma_-}{\sigma_+ + \sigma_-} = 2 \, Im \left\{ \frac{E_1^{p.v.}}{M_1} \right\} = 2.2 \times 10^{-4} \quad , \quad (24)$$

where σ_+ and σ_- are respectively the excitation cross sections, photon + $6S_{1/2} \to 7S_{1/2}$, for the photon of helicity +1 and −1. P_c is also the circular polarization of the photons emitted in the spontaneous transition $7S_{1/2} \to 6S_{1/2}$. P_c transforms like a helicity and consequently is odd under P and even under T, so that P_c receives contribution from the interaction involving vector and axial vector currents.

Instead of measuring directly the $E_1^{p.v.}$ M_1 interference, it appears more convenient to observe a polarization effect which is characteristic of an interference between $E_1^{p.v.}$ and the electric dipole transition moment $E_1^{ind.}$ induced by a static electric field $\vec{E}_0$. With this technique, it is possible to suppress completely the rather important background associated with the $6S_{1/2} - 7S_{1/2}$ radiative transition induced by collisions :

$$\text{photon} + Cs + Cs \to Cs + Cs^{\star}$$

In ref. [10], it was shown that the interference between the mixed $M_1-E_1^{p.v.}$ amplitude with the d.c. field induced amplitude $E_1^{ind.}$ produces a spin polarization in the final state $\vec{P}_e^{(1)}$ given by the following expression :

$$\vec{P}_e^{(1)} = \frac{8}{3} \frac{F(F+1)}{(2I+1)^2} \left[Re\left\{ \frac{M_1}{E_1^{ind.}} \right\} + \xi_i \ Im\left\{ \frac{E_1^{p.v.}}{E_1^{ind.}} \right\} \right] \hat{k}_i \wedge \hat{E}_o$$

$$(25)$$

F is the hyperfine quantum number of the lower as well as the upper state ($\Delta F = 0$ transition). I denotes the nuclear spin, $\hat{k}_i$ and $\hat{E}_o$ are unit vectors along the incident photon momentum and along the electric field. ξ_i is the circular polarization of the incident photon. The vector $\hat{k}_i \wedge \hat{E}_o$ has the transformation properties of an angular momentum under space and time reflection. The presence in $P_e^{(1)}$ of a term of the form $\xi_i \hat{k}_i \wedge \hat{E}_o$ is a clear indication that the atomic Hamiltonian contains a P-odd, T-even piece.

The value of M_1 quoted previously (eq. 23) is obtained from an experimental determination of the ξ-independent term in formula (25). The principle of the experiment now under completion at the E.N.S. (Paris) consists in the measurement of the difference :

$$\vec{\delta P}_e^{(1)} = \vec{P}_e^{(1)} \ (\xi_i = +1) - \vec{P}_e^{(1)} \ (\xi_i = -1) = \frac{16}{3} \frac{F(F+1)}{(2I+1)^2} \ Im\left\{ \frac{E_1^{p.v.}}{E_1^{ind.}} \right\}$$

$$(26)$$

Simple considerations of invariance predict also the existence of a component of $\vec{P}_e$ along $\hat{k}_i$ when the incident beam is circularly polarized :

$$\vec{P}_e^{(2)} = p(F) \ \xi_i \ \hat{k}_i \qquad (27)$$

where $p(F)$ is a number of the order of 10^{-1} which can be obtained from independent spectroscopic measurements ([31]). The spin polarization $\vec{P}_e^{(2)}$ results from spin-orbit coupling in the P states admixed to the S states under the effect of the d.c. electric field. It is important to note that $\vec{P}_e^{(2)}$ and $\vec{P}_e^{(1)}$ can be unambiguously distinguished, first because of their different directions but also owing to the fact that $\vec{P}_e^{(1)}$ reverses with the electric field while $\vec{P}_e^{(2)}$ is independent of $\vec{E}_o$. In fact, the detection of the known polarization $\vec{P}_e^{(2)}$ offers a very convenient way of calibrating correctly optical signals in terms of spin polarization.

We now outline the experimental method. Cesium atoms, in a saturated vapor (10^{-2} to 10^{-1} torr), are excited by a single mode c.w. laser beam tuned at the frequency of the 6S-7S transition ($\lambda = 5393.5$ Å), in an external d.c. field (1000 volt/cm) perpendicular to the beam. Resonance is detected by monitoring the fluorescence on the allowed 7S-6P$_{1/2}$ transition, in the direction $\hat{k}_f = \hat{k}_i \wedge \hat{E}_o$. The hyperfine structure of the transition is fully resolved in spite of Doppler broadening of the lines. Observed selection rules for the different hyperfine components agree very well with those expected theoretically for free Cs atoms [31]. Al subsequent measurements are performed at the peak frequency of a $\Delta F = 0$ transition.

The apparatus is schematized on Figure 3. The incident circular polarization ξ_i is modulated like $\sin \omega_i t$ by letting the laser beam go through a quarter wave plate rotating at the angular frequency $\omega_i/2$. Then the beam enters through a hole and, slightly off axis, into a spherical mirror interferometer inside which is placed the Cesium cell. Multipassages of the laser beam present several advantages : first, of all the signal $\delta P_e^{(1)}$ is substantially enhanced, since each reflection conserves the photon spin, furthermore different possible causes of systematic asymmetries (coming from spurious magnetic fields, or geometric asymmetry, although none of them has been observed yet), can be compensated in this way.

To detect a spin polarization of the 7S state along $\hat{k}_i \wedge \hat{E}_o$, one measures the intensity of the circularly polarized fluorescence light. The efficiency ξ_f of the circular analyzer is actually modulated, like ξ_i, but at a different frequency ω_f. The $E_1 P \cdot V \cdot - E_1$ ind. interference manifests itself through the apparition of a component of the photodetector signal at the frequencies $\omega_f + \omega_i$ and $\omega_f - \omega_i$, which it is convenient to weight by the inverse of another photodetector component (at $2\omega_i$) characteristic of total fluorescence light. Let us denote by $\mathcal{F}^{(1)}$ (E_o) the corresponding signal :

$$\mathcal{F}^{(1)}(\vec{E}_o) = \frac{\kappa}{2} \overrightarrow{\delta P}_e^{(1)} \cdot \hat{k}_f \quad .$$

κ is a proportionality factor which accounts for all possible sources of depolarization and is eliminated in the calibration procedure. $\mathcal{F}^{(1)}(\vec{E}_o)$ is digitally integrated while the laser frequency is kept resonant for the 6S (F=4) $\to$ 7S (F=4) transition. For elimination of an eventual stray signal resulting from weak intensity modulations of the fluorescence light at ω_i and ω_f one uses the fact that $\delta P_e^{(1)}$ reverses when $\vec{E}_o$ does : one sequentially reverses the d.c. field $\vec{E}_o$ at constant amplitude and subtract the results obtained with two consecutive orientations :

$$\Delta \mathcal{F}^{(1)}(E_o) = \mathcal{F}^{(1)}(\vec{E}_o) - \mathcal{F}^{(1)}(-\vec{E}_o) = \kappa \overrightarrow{\delta P}_e^{(1)} \cdot \hat{k}_f \qquad (28)$$

The mean value of $\mathcal{F}^{(1)}(E_o)$ and standard deviation on the mean, are calculated in real time. The successive data $\mathcal{F}^{(1)}(\vec{E}_o)$ and $\mathcal{F}^{(1)}(-E_o)$ are also registered on magnetic tape for ulterior statistical analysis (detection of spurious noise or slow drift elimination).

Calibration of the sensitivity of the detection is an important step of the measurement. The known polarization $\vec{P}_e^{(2)}$ created in the 7S state along $\hat{k}_i$ (eq. 27) is used. By applying a magnetic

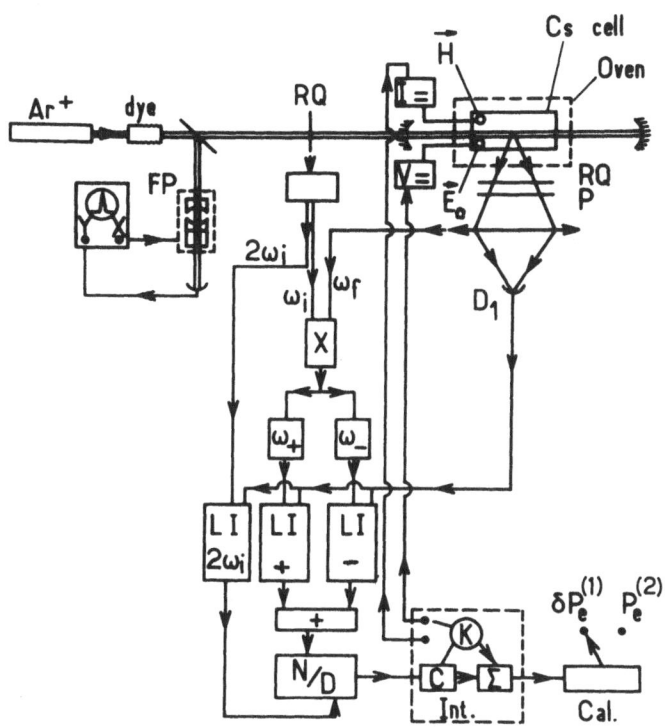

<u>Figure 3</u> : *Schematic view of the apparatus. FP, Fabry-Perot,*
$\overline{RQ}$, quarter wave plates rotating at frequencies $\omega_i/2$ and $\omega_f/2$.
P fixed polarizer. LI_+ , LI_- , $LI(2\omega_i)$ lock-in amplifiers tuned to
$\omega_+=\omega_i+\omega_f$, $\omega_-=\omega_i-\omega_f$ and $2\omega_i$. Selective amplifiers ω_+ and ω_- provid-
ing the reference signals. N/D, ratiometer. Int., digital integra-
tor. Cal., calculator. K, clock monitoring the sequential reversal
of $\vec{E}_0$ (provided by power supply V=) when $\delta\vec{P}_e^{(1)}$ is detected (H=0),
or that of $\vec{H}$ (power supply I=) when calibration is done (E_0 fixed).

field $\vec{H}$ parallel to $\vec{E}_0$, $\vec{P}_e^{(2)}$ acquires a non-zero component along
$\hat{k}_f$ (Hanle effect) :

$$\vec{P}_e^{(2)} \cdot \hat{k}_f = p(F) \frac{H/\Delta H}{1 + (H/\Delta H)} \tag{29}$$

which changes its sign when $\vec{H}$ is reversed. So the calibration is
performed in a simple way ; it consists in the replacement of the
sequential $\vec{E}_0$-reversal at $\vec{H} = 0$, used to measure $\delta\vec{P}_e^{(1)}$, by a sequential $\vec{H}$-reversal at constant $\vec{E}_0$. The same data analysis leads to :

$$\Delta \mathcal{G}^{(2)} = \mathcal{G}^{(2)}(H) - \mathcal{G}^{(2)}(-H) = \kappa p(F) \frac{2H/\Delta H}{1 + (H/\Delta H)^2} \tag{30}$$

Measurements done for different values of H and fitted with the variation predicted by eq. (30) yield ΔH, the characteristic width of the Hanle effect, and the proportionality factor κ.

The preliminary results of the experiment (^{32}), from 10 hours' data analysis, give a measured value of $\delta P_e^{(1)}$, compatible with a true value equal to zero. From the r.m.s. uncertainty, it can be concluded that :

$$|E_1^{p.v.}| < 2.0 \times 10^{-9} \; |e|a_o \quad \text{at 90\% confidence level} \quad (31)$$

To get this number, one has to use a theoretical value (^{10}) of the electric dipole amplitude $E_1^{ind.}$ which is affected by a theoretical uncertainty of the order of 10%.

This result can also be expressed in terms of the asymmetry $|E_1^{p.v.} / M_1|$ (since the ratio $M_1/E_1^{ind.}$ has been measured (^{30}))

$$|E_1^{p.v.} / M_1| < 1.3 \times 10^{-2} \text{ at 90\% confidence level,}$$

or in terms of an upper limit on the average coupling constant :

$$\overline{C}_v = \frac{Z}{A} C_{vp} + \left(1 - \frac{Z}{A} \right) C_{vn}$$

$$|\overline{C}_v| < 44 \quad \text{at 90\% confidence level} \quad (32)$$

As we have already said, the main limitation of this experiment is the signal to noise ratio. However, since the above preliminary result was obtained, recent measurements have shown that the signal over noise was definitely improved by a factor 5.5 (in part due to increase in light intensity) and it is certainly possible to get further improvement. Different stabilization controls are being put in position and will also allow to lengthen the time of averaging. So it should be possible, in a near future, to improve quite substantially the results given by expressions (31) and (32).

c) Stray Magnetic Fields : A Source of Spurious Effects

We shall not give here a complete discussion on the possible sources of systematic errors which may affect the two types of experiments we have just described, but rather concentrate our attention on the spurious effects to be associated with stray magnetic fields. The existence of a component of a d.c. magnetic field along the direction of the photon momentum gives rise to effects which may simulate a parity violation.

A difference δn between the complex indices n_R and n_L for right and left circularly polarized photons (Faraday effect) can be induced by two mechanisms :

a) The transition frequencies relative to a given hyperfine component are slightly displaced for right and left photons (we consider
here the weak field limit where the Zeeman shifts are small compared
to hyperfine splittings).

b) The transition probabilities relative to a given hyperfine component averaged over the initial magnetic quantum numbers are different for right and left photons; the difference being of the order
of $\omega_L/\Delta W$ where ω_L is the Larmor frequency and ΔW the hyperfine splitting. The effect averages to zero for a source with a broad profile
such that the different components are excited with the same light
intensity.

The difference $\delta n^{(a)} = n_R - n_L$ associated with the (a) mechanism is proportional to the derivative of the index $\partial n(\omega)/\partial \omega$. The
variations with frequency of Im $\delta n^{(a)}$ and Re $\delta n^{(a)}$ are reversed
with respect to those of Im n and Re n :

Im $(\delta n^{(a)})$ has a <u>dispersive</u> shape and Re $(\delta n^{(a)})$ an <u>absorptive</u>
one. In the Cesium experiment a small spurious dichroism appears
when the laser frequency is not exactly centered at the line peak.
With a detuning $\Delta \omega = \omega - \omega_o$, one gets Im $\delta n/$Im n $\simeq \omega_L \Delta \omega / \Gamma_D^2$, where
Γ_D is the Doppler width as defined in ref. ([10]). In a typical situa-
tion corresponding to a magnetic field of 10^{-1} Gauss and a detuning
$\Delta \omega = \Gamma_D/10$, one gets for Cesium Im $\delta n^{(a)}/$ Im n $\simeq .5 \ 10^{-5}$, which is
to be compared with an external neutral currents effect of the or-
der of 10^{-4}. In the Bismuth experiments, one takes advantage of the
fact that the true and the spurious effects associated with effect
a) have different symmetry properties with respect to the line center.

The b) mechanism is more troublesome because the associated
$\delta n^{(b)}$ has, in the vicinity of a hyperfine line component, the same
frequency shape as the true parity violating effect. It affects
both the Cesium experiment in which the hyperfine structure is resolved and the Bismuth experiments if the different hyperfine components are not excited with the same intensity. As an example, the
dichroïsm induced by the b) mechanism in the $F = 4 \rightarrow F = 4$ transition in Cesium with a field of 0.1 Gauss is equal to 4.3×10^{-5}. A
result of a similar order of magnitude is expected in the case of
Bismuth so that the stray magnetic field has to be kept below the
level of 10^{-4} Gauss. In fact, the problem is not as severe as it
may look a priori. A large fraction of the $\delta n^{(b)}$ induced by the
magnetic field can be eliminated by a reversal of the direction of
propagation of the incident light, all the rest of the apparatus
being kept fixed. $\delta n^{(b)}$ changes its sign while the true effect
$\delta n^{p.v.}$ remains obviously unchanged.

CONCLUSION

We have tried in this paper to show that the fields of Weak Interactions and Atomic Physics are becoming very close together. In spite of several unnegligible experimental difficulties, it is now reasonable to expect significative information about fundamental properties of weak neutral current interaction between atomic electrons and nuclei to emerge, in a near future, from the search for P without T violation atomic physics experiments.

REFERENCES

[1]   T.D. LEE and C.N. YANG, Phys. Rev. 104, 254 (1956)
[2]   Ya.B. ZEL'DOVICH, Zh. Eksperim. i. Theor. Fiz. 36, 964 (1959)
      (transl. Soviet Phys. JETP, 9, 682 (1959))
[3]   F.C. MICHEL, Phys. Rev. B 138, 408 (1965)
[4]   G.'T HOOFT, Nucl. Phys. B 33, 173 (1971); ibid. B 35, 167 (1971)
[5]   S. WEINBERG, Phys. Rev. Lett. 19, 1264 (1967)
[6]   A. SALAM, "Elementary Particle Physics", ed. by N. Svartholm,
      p. 367 (1968)
[7]   F.J. HASERT et al., Phys. Letters, 46B, 138 (1973)
[8]   A. BENVENUTI et al., Phys. Rev. Letters, 32, 800 (1974)
[9]   S.J. BARISH et al., Phys. Rev. Letters, 33, 448 (1974)
[10] M.A. BOUCHIAT and C.C. BOUCHIAT, Phys. Letters 48B, 111 (1974);
      ibid., J. de Phys. 35, 899 (1974); ibid., J. de Phys. 36, 493
      (1975)
[11] G. FEINBERG and M.Y. CHEN, Phys. Rev. D 10, 190 (1974)
[12] J. BERNABEU et al., Phys. Letters, 50B, 467 (1974)
[13] A.N. MOSKALEV, Zh. E.T.F. Pis. Red. 19, 229, 394 (1974)
      (JETP Lett. 19, 141, 216 (1974))
[14] I.B. KHRIPLOVICH, Zh. Eksperim. i. Theor. Fiz. Pis. Red. 20,
      689 (1974) (Transl. Sov. Phys. JETP Lett. 20, 315 (1974))
[15] P.G.H. SANDARS in Atomic Physics IV, ed. G. Zu Putlitz,
      E.M. Weber, A. WINNACKER, Plenum Press, New York (1975)
[16] R.R. LEWIS and W.L. WILLIAMS, Phys. Letters, 59B, 70 (1975)
[17] V. BRISSON, Communication at the XIe Rencontre de Moriond,
      March 1976
[18] C. BOUCHIAT, Proceedings of Neutrino 75 Conference, Balaton
      (June 1975)
[19] H.G. KUHN, Atomic Spectra, p. 342-346 (Longman), 1969
[20] C. BOUCHIAT, Phys. Letters, 57B, 284 (1975)
[21] E.A. HINDS et al., Phys. Letters, 62B, 97 (1976)
[22] G.E. HARRISON et al., Phys. Rev. Letters, 22, 1263 (1969)
[23] M. BRINICOMBE, C.E. LOVING and P.G.H. SANDARS, J. Phys. B,
      Atom. Mol. Phys. 9L, 237 (1976)
[24] E.M. HENLEY and L. WILETS, University of Washington, Preprint
      RLO 1388713
[25] C.E. LOVING and P.G. SANDARS, J. Phys. B, Atom. Mol. Phys. 8L,
      336 (1975)
[26] D.C. SOREIDE and E.N. FORTSON, Bull. Am. Phys. Soc. 20, 491
      (1975)
[27] D.C. SOREIDE et al., Phys. Rev. Letters, 36, 352 (1976)
[28] P.G.H. SANDARS, Invited paper to the FICAP Conference (this
      book)
[29] E.N. FORTSON, Invited paper to the FICAP Conference (this book)
[30] M.A. BOUCHIAT and L. POTTIER, J. Phys. Lettres 37 L-79 (1976)
[31] M.A. BOUCHIAT and L. POTTIER, J. Phys. Lettres 36 L-189 (1975)
[32] M.A. BOUCHIAT and L. POTTIER, Preprint E.N.S., March 1976, to
      be published in Phys. Letters B, 62B, 327 (1976)

AN OPTICAL ROTATION EXPERIMENT TO SEARCH FOR NEUTRAL CURRENT EFFECTS IN ATOMIC BISMUTH

Norval Fortson

University of Washington

Seattle, Washington 98195

We discuss here the current status of an experiment[1,2] at the University of Washington in which we are searching for a small parity non-conserving (PNC) effect in atoms. As first pointed out by Bouchiat and Bouchiat[3], effects of measurable size are expected in heavy atoms if the neutral weak current interaction[4-6] between electrons and nucleons has a PNC component. In our experiment, the effect of interest is an optical rotation associated with the 8755Å magnetic-dipole absorption line in atomic bismuth vapor. Similar experiments have been proposed by Sandars[7] and Khriplovich[8], and one using the 6476Å line in Bi is underway and is being reported at this conference.[9]

We are able to set an upper limit on a PNC optical rotation which is somewhat smaller than anticipated on the basis of atomic calculations[10,11] using the Weinberg model of neutral currents. Below this limit our present data shows a very small optical rotation feature which we are currently investigating in order to establish whether it is a true PNC effect.

An earlier description of our experiment is given in reference 2. We look for an optical rotation of the form

$$\delta\phi_p = 4\pi\ell\lambda^{-1}(n-1)\,F$$

where ℓ is the path length and n the index of refraction of the absorbing vapor. The PNC fraction $F \equiv \mathrm{Im}(E1/M1)$ is the ratio of opposite parity electric dipole amplitude E1 that is coupled into the magnetic dipole amplitude M1 by PNC forces within the atom. The standard optical convention is used whereby $\delta\phi$ is considered positive for a rotation that appears clockwise when looking into the source.

The rotation $\delta\phi$ should follow the index of refraction dispersion curve about the absorption line. A convenient experimental quantity is $\delta\bar{\phi}_p$ the rotation at the low frequency dispersion peak when ℓ is one absorption length at the line center. For most line shapes, a very good approximation is $\delta\bar{\phi}_p = \frac{F}{2}$.

The $6^4S_{3/2} \rightarrow 6^2D_{3/2}$ absorption line in atomic Bi was selected because it is free of overlapping Bi_2 molecular absorption bands, and can still be reached by tunable lasers. In addition, Bi is a heavy atom where neutral current effects are enhanced (reference 3).

There are by now a number of calculations of the PNC fraction expected for our Bi line. Brimicombe, Sandars, and Loving[10] obtain $F = 3.2 \times 10^{-7}$ and Henley and Wilets[11] find $F = 3.5 \times 10^{-7}$. Both calculations use an independent particle Hartree-Fock model and a Weinberg angle[5] $\sin^2\theta_\omega = 0.35$. Other calculations[12,13] yield values 30% to 50% smaller.

Our apparatus is the same as the one in reference 2. A narrow light beam from the tunable laser (an optical parametric oscillator[14]) passes in turn through a Nicol prism polarizer, a water-filled Faraday cell for modulating the polarization angle, a long alumina tube with a one-meter length of Bi vapor, and a Nicol prism analyzer. The ratio of the light intensities before and after the Nicol analyzer measures the optical rotation. The Bi vapor tube is filled with one atmosphere of He gas to retard Bi migration, and is placed in a fire-brick oven which can heat the tube to above $1550^\circ K$. A solenoid surrounds the tube in the oven to cancel the Faraday rotation that accompanies the Bi absorption line.

For optimum sensitivity, the two Nicol prisms are set with their polarization axes nearly perpendicular; i.e. at close to maximum extinction of the transmitted beam. When the Nicol analyzer is a small angle ϕ from maximum extinction it has a transmission coefficient of $\phi_0^2 + \phi^2$, where ϕ_0^2 is the transmission coefficient at maximum extinction. With our newest calcite Nicols, $\phi_0^2 < 10^{-8}$. Any small optical rotation $\delta\phi$ between the polarizers produces a fractional change $\delta\phi/\phi_0$ in transmitted light at the optimum setting, $\phi = \phi_0$. The Faraday cell is used to rotate the polarization back and forth between $\pm\phi_0$ on successive laser pulses, and the difference in transmission between the two settings provides a signal that is sensitive only to rotation of the light polarization.

The atomic absorption line is split into 10 hyperfine components spread over a total width of $\Delta\nu_0 = 0.3\text{Å}$ (12 GHz). The present tunable laser source has a broader width $\Delta\nu_s$, and thus our source determines the observed absorption line width and line shape. In order to maximize optical rotation in this situation, we operate at such a high Bi density that practically all wavelengths within the atomic line profile are absorbed, causing peak absorption of at least the fraction $\frac{\Delta\nu_0}{\Delta\nu_s}$ of the laser light. In a typical situation, when we operate with an oven temperature of about $1500^\circ K$, our one-meter absorption path is 10 absorption lengths for wavelengths within $\Delta\nu_o$, as determined by a high resolution Fabry-Perot mono-chrameter; and the overall peak absorption of our laser beam with $\Delta\nu_s$ = 40 GHz is measured to be 45%. It is readily seen that any Bi optical rotation would be increased at the dispersion peak in pro-portion to the number of absorption lengths on the atomic line. The exact factor can be found by integrating over the actual observ-ed laser line shape in any given case. We apply this correction to get the final values of $\delta\bar{\phi}_p$.

An axial magnetic field H produces a Faraday rotation $\delta\phi_f$ associated with the absorption line. The peak magnitude of this rotation is about 10^{-4} radians/gauss for the conditions described in the previous paragraph. The current in the coil about the Bi tube is adjusted to make the observed $\delta\phi_f \sim$ o. As a further precaution, beams are sent through the Bi tube in opposite directions without changing H and the rotation signals subtracted. Faraday rotation should disappear whereas a PNC rotation would be reinforced.

In order to look for parity effects, we scan the wavelength of the laser across the absorption line with $\delta\phi_f$ reduced to a minimum. Typically the scan is 4 Å wide (about 4 $\Delta\nu_s$) and the laser sweep takes about 2 minutes per scan. The rotation signals for successive scans are added and displayed on a signal averager. The result is fit to a dispersion curve centered about the observed absorption line center. At the highest Bi densities, with the large number of absorption lengths, the values of F calculated for this line should produce peak rotations >10^{-6} rad.

We have two major sources of measured rotation angle uncertain-ty. One is random fluctuations which seem to be caused by small ran-dom motion of the light beam across the polarizers. This noise amounts to about 10^{-6} rad. in one second, and becomes completely negligible after several minutes of averaging. The other uncer-tainty comes from a wavelength-dependent change in polarization angle which seems to be caused by some interference effects in the Nicol prisms. This variation can amount to 10^{-6} rad./Å and is a major source of systematic variations of the rotation angle.

The result of approximately 1200 scans with different Bi vapor pressures, Nicol orientations, and including both directions of the beam through the apparatus is that a small dispersive feature re-mains that has a peak value reduced to unit path length of:

$$\delta\bar{\phi}_p = (+5 \pm 2) \times 10^{-8} \text{ radians}$$

The indicated error is statistical and represents two standard deviations. In addition there is about a 20% uncertainty in our determination of the optical path length. This feature has an excellent fit to a dispersion curve centered about the atomic line and hence is an effect associated directly with Bi atoms. However, we have not yet made all tests necessary to rule out other systematic effects in Bi vapor (light bending, etc.) which might also be associated with refractive dispersion in our apparatus.

We are now checking to determine whether this feature is indeed a PNC effect. At this stage, the most that can be said is that if any PNC effect exists, it is smaller in Bi than predicted by the Weinberg model of neutral currents plus simple atomic theory.

We plan to install a new laser system and improved optics (spatial filters, etc.) with the aim of increasing our sensitivity by another order of magnitude. We also plan a similar experiment in Tl using the 1.2 micron $6P_{1/2} \rightarrow 6P_{3/2}$ absorption line.

My colleagues in the work reported here are all listed as the co-authors of reference 2. I would like to express my appreciation to them; and also to E. M. Henley for useful discussions about the theory of neutral currents.

References

[1] D. C. Soreide and E. N. Fortson, Bull. Am. Phys. Soc. **20**, 491 (1975).

[2] D. C. Soreide, D. E. Roberts, E. G. Lindahl, L. L. Lewis, G. R. Apperson, and E. N. Fortson, Phys. Rev. Lett. **36**, 352 (1976).

[3] M. A. Bouchiat and C. C. Bouchiat, Phys. Lett. **48B**, 111 (1974), and J. Phys. (Paris) **35**, 899 (1974), and **36**, 493 (1975).

[4] A. Salam and J. Ward, Phys. Lett. **13**, 68 (1964).

[5] S. Weinberg, Phys. Rev. Lett. **19**, 1264 (1967).

[6] F. J. Hasert et al., Phys. Lett. **46B**, 138 (1973); A. Benvenuti et al., Phys. Rev. Lett. **32**, 800 (1974); S. J. Barish et al., Phys. Rev. Lett. **33**, 448 (1974).

[7] P. G. H. Sandars, in Atomic Physics 4, edited by G. zu Putlitz, E. W. Weber, and A. Winnacker (Plenum, New Yor, 1975).

[8] I. B. Khriplovich, Pis'ma Zh. Eksp. Teor. Fiz. **20**, 686 (1974) [JETP lett. **20**, 315 (1974)].

[9] P. G. H. Sandars, in the following paper.

[10] M. W. S. M. Brimicombe, C. E. Loving, P. G. H. Sandars, J. Phys. B. 9, L1 (1976).

[11] E. M. Henley and L. Wilets, To be published, Physical Review A.

[12] P. G. H. Sandars, Private communication.

[13] I. B. Khriplovich, Private communication.

[14] S. E. Harris, Proc. IEEE **57**, 2096 (1969).

SEARCH FOR PARITY NON-CONSERVING OPTICAL ROTATION IN ATOMIC BISMUTH

P.E.G. Baird, M.W.S.M. Brimicombe, G.J. Roberts,

P.G.H. Sandars and D.N. Stacey

Clarendon Laboratory, Parks Road, Oxford OX1 3PU, U.K.

1. INTRODUCTION

Bouchiat and Bouchiat pointed out[1] soon after the discovery of weak neutral currents[2] that their existence might lead to parity non-conserving (PNC) effects of observable magnitude in atoms. One of the most promising possibilities[3] is to look for optical rotation close to an allowed M_1 absorption line in a heavy element. We report here preliminary results from our experiment[4] on the 648 nm $J = 3/2$ $J = 5/2$ transition in bismuth. We find that the optical rotation, if it exists, must be somewhat smaller than had been expected.

One can readily show that the interference between a parity non-conserving E_1 matrix element and the normal M_1 element leads to an optical rotation given by

$$\varphi = \frac{-4\pi L(n - 1)}{\lambda} R$$

where n is the refractive index and L the path length. R is the PNC ratio;

$$R = \text{Im.} \frac{\langle f | E_1 | i \rangle}{\langle f | M_1 | i \rangle} \quad \text{where } | i \rangle \text{ and } | f \rangle$$

are the initial and final states for the transition. φ is defined to be positive for a clockwise rotation when viewed against the light.

φ thus has a dispersive dependence on wavelength. The consequent rapid variation near resonance is an essential feature of our experiment where it is used to separate the PNC angle from other effects. However, in the region close to the resonance there is also absorption which limits the useful path length. A convenient

Table 1.   Calculated values of the PNC ratio R for M_1 transitions
in Bi assuming a Weinberg angle Sin $^2\theta_w$ = 0.35.

	J=3/2→3/2(876nm) $R/10^{-7}$	J=3/2→5/2(648nm) $R/10^{-7}$
Brimicombe, Sandars & Loving; Semi-empirical potentials (ref 5) I	-3.2	-4.3
Semi-empirical potentials (ref 5) II	-2.4	-3.2
Henley and Wilets; Greens function (ref 6)	-3.5	-
Grant, Pyper and Sandars; Multiconfiguration relativistic Hartree Fock potential (ref 7)	-2.4	-3.2
Kriplovich; Semi-empirical potential with 'scaling' adjustment (ref 8)	-1.65	-2.1

measure of sensitivity is the value ϕ_P of the high frequency
peak of the rotation for one absorption length at line centre.
This is given by

$$\phi_P = + \frac{R}{2}$$

We have chosen Bi for our experiment because it is unique in
having an allowed M_1 transition from the ground state, J=3/2→J=5/2
in the visible region where continuous wave tunable dye lasers are
available. Bi has only a single isotope with spin 9/2. Up to the
present we have confined our attention to the well resolved
F = 6 → 7 hyperfine component.

The parity non-conserving ratio R has been calculated by a
number of authors for some of the allowed M_1 transitions in Bi.
The results are set out in table 1.

From these calculations we expect a rotation angle ϕ of
order 1 to 2 x 10^{-7} radians for one absorption length at line
centre.

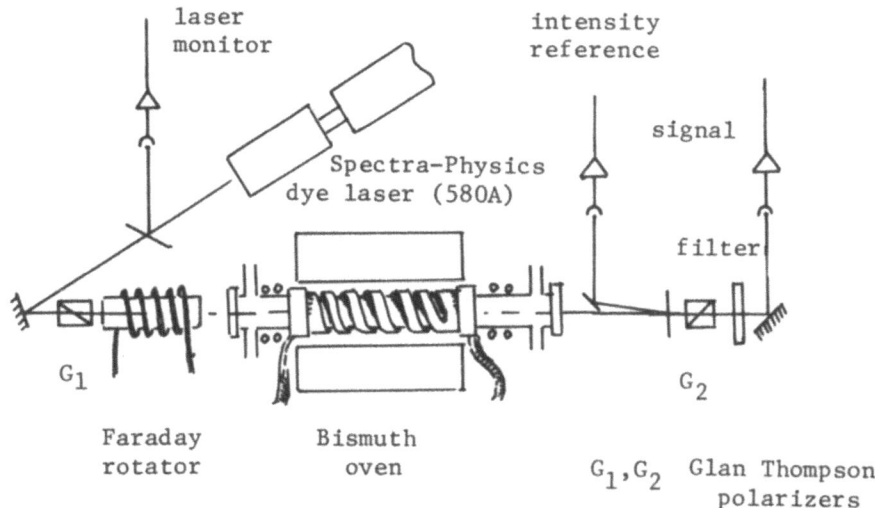

Figure 1   Diagram of apparatus (schematic)

2.   APPARATUS

Our apparatus is illustrated in fig. 1. The Spectra-Physics
dye laser gives a few milliwatts of single mode radiation with a
frequency stability much better than the Doppler width of a single
Bi hyperfine component. The light from the laser is passed through
two crossed Glan-Thompson polarizers which have an extinction ratio
of order 10^{-7}. The detectors are standard low noise photodiodes.
The Bi is contained in an alumina tube which can be heated in a
furnace to 1500 K to give one atomic absorption length. Windows
at either end of the tube are kept at room temperature by water
cooling; coating by the Bi is reduced by use of He buffer gas
at approximately 0.2 atmospheres. The Faraday modulator is an
externally mounted piece of silica to which a magnetic field can
be applied.

The magnetic field produced by the 50 Hz current in the oven

heater is of order 10 Gauss, which causes a Faraday rotation in
the Bi of order 10^{-3} radians. This large signal is readily
detected and is used to monitor both the wave-length of the laser
and the optical depth of Bi in the oven. Indeed the signal is so
large that in order to avoid possible interference the heater
current is switched off for alternate periods of 10 seconds while
the parity channel is open. The width of the Faraday rotation line
agrees with that expected from the normal Doppler width at 1500 K
(1 GHz) together with additional broadening of about 20% due to
the buffer gas.

One potential difficulty in the experiment is the presence of
Bi_2 molecules which have an absorption band overlying the
$J = 3/2 \rightarrow 5/2$ atomic line. At one atomic absorption length the
overall transmission at line centre is only $\sim$ 5%. This causes no
serious problems apart from an undesirable loss in intensity. We
ratio out the major effects of varying absorption in the oven by
normalizing the light transmitted through the second polarizer
against a reference signal, the light which has passed through the
oven but not the analyser (see figure 1).

In order to reduce any Faraday rotation caused by stray DC
magnetic fields on the Bi, the whole oven is placed in a mu-metal
box.

3.   DETECTION SCHEME

For nearly crossed polarizers the transmitted light intensity
has the form

$$I = A(\phi + \phi_R)^2 + B$$

where ϕ is the angle of rotation in the region between the
polarizers and ϕ_R is any residual misalignment between the
polarizers. ϕ is made up of three parts

$$\phi = \phi_{PNC} + \phi_M \cos \omega_M t + \phi_F$$

where ϕ_{PNC} is the PNC optical rotation of interest to us.
$\phi_M \cos \omega_M t$ is the additional time varying angle produced by the
Faraday modulator. ϕ_F is a possible Faraday rotation due to
residual magnetic fields on the Bi.

In order to obtain a signal which is linear in the small angle
ϕ_{PNC} we use phase-sensitive detection methods to pick out that
part of the intensity which varies at frequency ω_M. This is

$$S_M = 2A \phi_M (\phi_{PNC} + \phi_F + \phi_R)$$

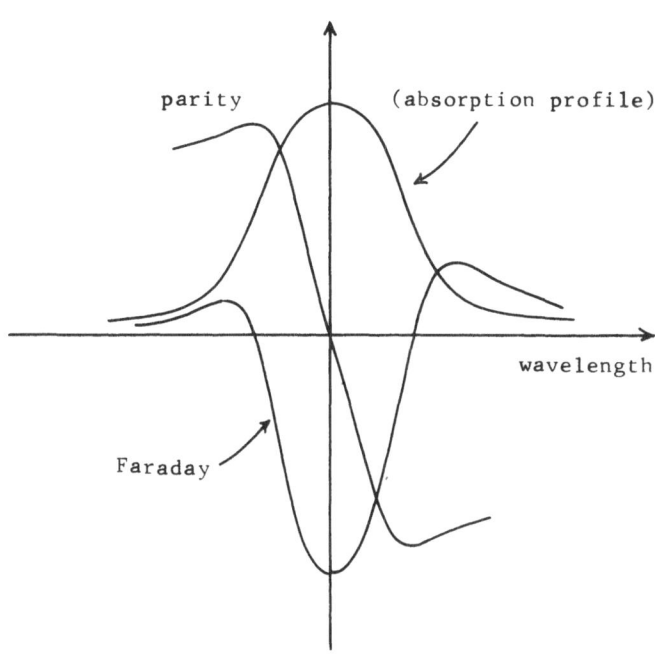

Figure 2   Features of the λ = 648 nm   J = 3/2, F = 6
           $\rightarrow$ J = 5/2, F = 7   transition in Bi

In order to discriminate against the residual angle ϕ_R, which
will in general be very much larger than ϕ_{PNC}, we measure the
change ΔS_M as the laser wave-length is switched from one side
of the atomic line to the other.   Since ϕ_R is largely wave-length
independent $\Delta\phi_R \ll \phi_R$,   whereas $\Delta\phi_{PNC} \approx 2\phi_{PNC}$.

Then
$$\Delta S_M = 2A\phi_M(\Delta\phi_{PNC} + \Delta\phi_F + \Delta\phi_R)$$

where we have assumed that both ϕ_M and the coefficient A are
independent of wavelength.   A more complete analysis can be made
taking into account small changes in these quantities.

ϕ_F is expected to be quite small because the oven is
magnetically shielded and there is DC field compensation.   However,
in order to discriminate against any residual Faraday effects the
two wavelength positions are chosen so that they have the same
Faraday rotation, i.e. $\Delta\phi_F$ = 0 .   This is ensured by a

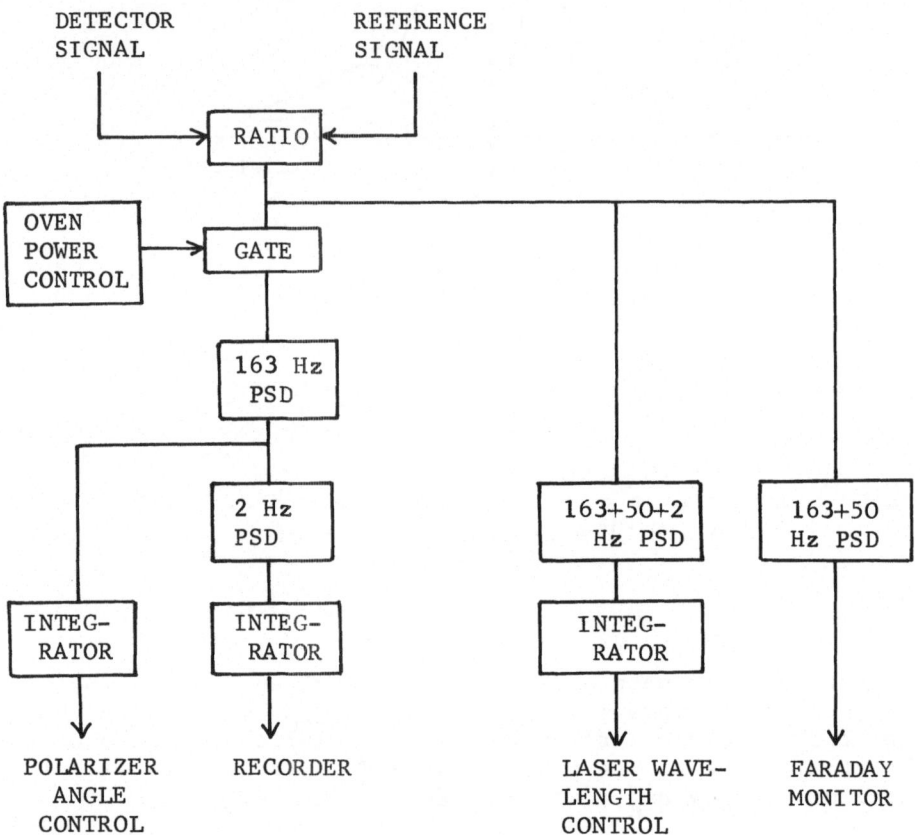

<u>Fig. 3</u> Simplified block diagram of detection system.

wavelength adjusting feed-back loop using the Faraday effect
produced by the large magnetic field from the oven heater.

It will be seen from the Faraday curve in fig. 2 that there
are three regions where $\Delta\phi_F = 0$ can be obtained for a given
wavelength change $\Delta\lambda$. One is close to line centre, the other
two are in the wings. It is convenient to change the wavelength
by jumping between cavity modes of the dye laser; this is done
by changing the spacing of the internal etalon. Cavity modes
are separated by about 400 MHz, and with the buffer gas pressure
used, 3 modes closely match the 1.2 GHz change required to move
from one peak to the other of the dispersion curve. We have used
this 1.2 GHz, 3 mode change throughout.

A block diagram of the detection scheme is given in figure 3.
The Faraday modulation frequency is 163 Hz, the laser wavelength
is modulated at 2 Hz, and the oven heater field has a frequency
of 50 Hz.

4.   EXPERIMENTAL PROCEDURE

The signal to noise in this experiment is extremely satisfactory; with the laser operating well we can readily detect
angles of order 10^{-8} radians in a 4 minute integration period.

Our main difficulty occurs because the residual wavelength
dependent angle $\Delta\phi_R$ is not zero but has a value of order 10^{-7}
radians which varies in magnitude and sign over times of order
hours. We believe that the origin of this effect is as follows.
For any polarizer, there are likely to be small variations in the
effective polarizer angle over the spatial distribution of the
beam. A change in wavelength necessarily involves a change in the
spatial characteristics of the laser beam, with a consequent small
change $\Delta\phi_R$ in the nett residual angle. $\Delta\phi_R$ depends critically
on the details of the spatial distribution of the laser light and
will therefore vary over periods of order hours in which thermal
and other changes are known to make appreciable changes in the
laser operating characteristics.

In order to circumvent this difficulty in these initial
experiments we have adopted a procedure where ΔS_M is measured
in as short a period as possible at a Bi temperature T such
that there is one atomic absorption length, and then at a slightly
lower temperature T^1 when the Bi optical depth and hence $\Delta\phi_{PNC}$
should be appreciably different. The change in ΔS_M from the
one temperature to the other will be proportional to $\Delta\phi_{PNC}$
provided $\Delta\phi_R$ has not changed in the intervening period; this
is checked by repeating the experiment at the first temperature.
In other words, a measurement can be represented as follows:

$$M = \Delta \frac{S_M}{2}^T(1) - \Delta S_M^{T^1} + \Delta \frac{S_M}{2}^T(2)$$

Each such "sandwich" takes some 30 minutes.

The optical path length is monitored at all times through the level of the reference signal. Absolute calibration is achieved in terms of the Faraday effect produced by a known DC magnetic field applied to the Bi through the heater coil.

5. RESULTS

We have taken some 62 measurements switching across line-centre. Our result (normalized to one atomic absorption length at line centre) is

$$\phi_P = (\pm 3.5 \pm 3) \times 10^{-8} \text{ radians}$$

where the statistical error represents two standard deviations. This is appreciably smaller and of the opposite sign to the predicted angle $\phi_P \approx -1.5 \times 10^{-7}$.

We have also made a similar series of measurements at the point on the long-wavelength side of line-centre where two equal Faraday angles are again obtained for the same wavelength change as used in the line centre experiment. Here, where one would expect a vanishingly small optical rotation, we find

$$\phi = (-4.5 \pm 3) \times 10^{-8} \text{ radians}$$

Although these results suggest that $R \lesssim 10^{-7}$ for 648 nm in Bi, there is apparently a systematic effect present giving a rotation of order 5.10^{-8} radians which precludes any more definite statement at this stage. We are currently modifying and improving the apparatus in a number of ways in order to clarify the situation.

References

1. Bouchiat, M.A. and Bouchiat, C.C., Phys. Lett. 48B, 111 (1974).

2. Hasert, F.J. et, Phys. Letts.46B, 138 (1973).
    Benvenuti, A. et al, Phys. Rev. Lett. 32, 800 (1974).
    Barish, S.J. et al, Phys. Rev. Lett. 33, 448 (1974).

3. Kriplovich, I.B., J.E.T.P. Lett. 20, 315 (1974).
    Sandars, P.G.H., Atomic Physics ed. G. zu Putlitz,
    E.W. Weber and A. Winnacker (New York; Plenum Press),
    (1975).
    Soneide, D.C. and Fortson, E.N., Bull.Am.Phys.Soc. 20,
    491 (1975).

4. P.E.G. Baird; M.W.S.M. Brimicombe; G.J. Roberts;
    P.G.H. Sandars and D.N. Stacey, to be published.

5. M.W.S.M. Brimicombe, C.E. Loving, P.G.H. Sandars,
    J.Phys. B. 9, L1 (1976).

6. E.M. Henley and L. Wilets, Private communication.

7. I.P. Grant, Private communication.

8. I.B. Kriplovich, Private communication.

ATOMIC PHYSICS TESTS OF QUANTUM ELECTRODYNAMICS*

Peter J. Mohr

Department of Physics and Lawrence Berkeley Laboratory
University of California
Berkeley, California 94720

I.   INTRODUCTION

Experiments with simple atomic systems, for which the effects of quantum electrodynamics can be accurately predicted, have steadily become more precise and have been extended to systems with a wide variety of constituent particles. The corresponding increased demands on quantum electrodynamics theory have so far been met successfully. At this time, there does not appear to be any serious disagreement between theory and experiment.[1]

In this paper, the tests of quantum electrodynamics derived from bound systems and the free electron and muon magnetic moments will be reviewed. The emphasis will be on the areas in which recent developments in theory or experiment have taken place. In addition, determinations of the fine structure constant from the Josephson effect and the fine structure of helium will be discussed.

II.   JOSEPHSON EFFECT DETERMINATION OF THE FINE STRUCTURE CONSTANT

The most accurate value for the fine structure constant α is presently derived from the Josephson effect determination of e/h. The value of α is related to the value of e/h through a series of precisely known constants:[2]

$$\alpha^{-2} = \left[\left(\frac{1}{4R_\infty} \right) \left(\frac{c\Omega}{\Omega_{NBS}} \right) \left(\frac{\mu'_p}{\mu_B} \right) \left(\frac{2e}{h} \right) \right] \frac{1}{\gamma'_p}$$

The constants in the square brackets have been measured with uncertainties of 5 parts in 10^8 or less. The gyromagnetic ratio of the proton in water γ_p' has been the least well known, and has recently been measured with improved precision by Olsen and Williams[2] at the National Bureau of Standards. Their result

$$\gamma_p' = 2.675\ 131\ 4(11) \times 10^8\ \text{Rad s}^{-1}\ T_{NBS}^{-1} \qquad\qquad 0.42\ \text{ppm}$$

combined with measured values for the remaining constants yields[2]

$$\alpha^{-1} = 137.035\ 987(29) \qquad\qquad 0.21\ \text{ppm}$$

The value of α obtained this way does not depend directly on quantum electrodynamics theory.

III.   FINE STRUCTURE IN HELIUM

The calculation and measurement of the fine structure of the 2^3P levels of helium to a precision of 1 ppm provides a test of the theory of the two-electron system; or assuming the validity of the theory, the theoretical and experimental results can be used to determine an accurate value for the fine structure constant.[3] The theoretical program for such a calculation was outlined by Schwartz in 1964.[4] Precision experiments were undertaken by Hughes and collaborators at Yale. The experimental values they obtained are[5]

$$\nu_{01} = [E(2^3P_0) - E(2^3P_1)]/h = 29\ 616.864(36)\ \text{MHz} \qquad 1.2\ \text{ppm}$$

$$\nu_{12} = [E(2^3P_1) - E(2^3P_2)]/h = 2\ 291.196(5)\ \text{MHz} \qquad 2.2\ \text{ppm}$$

The theory is based on a reduced Hamiltonian H_R obtained from the Bethe-Salpeter equation:[6]

$$H_R = H_0 + H_2 + H_a + H_4 + \ldots$$

The reduced Hamiltonian consists of the nonrelativistic two-electron Pauli-Schrödinger Hamiltonian H_0, the relativistic corrections H_2 of order $\alpha^2 Ry$ associated with the Breit equation, a Hamiltonian associated with the anomalous magnetic moment of the electron H_a, and the remaining corrections H_4 of order $\alpha^4 Ry$. The corrections are evaluated by perturbation theory starting with numerical eigenfunctions of H_0. The leading term is the first order correction due to the spin-dependent part of H_2, which must be evaluated accurately.[4,7] This correction is of order $\alpha^2 Ry$. The first order correction due to H_a contributes in order $\alpha^3 Ry$ and $\alpha^4 Ry$.[4] The first order correction due to H_4 is of order $\alpha^4 Ry$.[8] In addition, H_2 contributes through second order perturbation theory terms of order α^4Ry, and through cross terms between the mass polarization operator and H_2 terms of order $(m/M)\alpha^2$Ry.[9,10,11] There are additional corrections of order $(m/M)\alpha^2$Ry.[12,8] Earlier results are summarized in Ref. 13.

The largest theoretical uncertainty has been in the terms evaluated in second order perturbation theory. Recent work of Lewis, Serafino, and Hughes has reduced the uncertainty of these terms.[11] Their method of evaluation is based on Dalgarno-Lewis variational perturbation theory with up to 455 term Hylleraas trial functions.

Assuming the 1973 recommended value for α,[14] α^{-1} = 137.036 04(11), the theoretical fine structure splittings are[11]

ν_{01} = 29 616.883(43) MHz   1.4 ppm

ν_{12} = 2 291.282(81) MHz   35 ppm

in good agreement with the experimental values. Using the calculated and measured values of ν_{01} to obtain a value for α, one finds

α^{-1} = 137.036 08(13)   0.94 ppm

This result is consistent with other determinations of α (see Section IX).

IV. ELECTRON G-FACTOR ANOMALY

The electron g-factor anomaly $a_e = (g_e - 2)/2$, which is a purely quantum electrodynamical quantity, is the deviation of the g-factor of the electron from its value predicted by the Dirac equation $g_e = 2$.

The experimental value of Wesley and Rich is[15,16]

a_e = 1 159 656.7(3.5) $\times$ 10^{-9} 3 ppm

Work on a new experiment, which is expected to improve the accuracy by up to an order of magnitude, is in progress by Rich and co-workers at Michigan.[17] A recent preliminary value measured in a resonance experiment by Van Dyck, Ekstrom, and Dehmelt at the University of Washington is[18]

a_e = 1 159 655(5) $\times$ 10^{-9} 4 ppm

This value was obtained with a single electron trapped in a magnetic bottle.[19] The precision is comparable to the precision of the

Wesley-Rich measurement.

In view of the prospects for improved accuracy in the experiments, it is appropriate to review the present status of the theory.
The theoretical value for the g-factor anomaly is expressed as a
series in powers of α/π

$$a_e = a_e^{(2)}\frac{\alpha}{\pi} + a_e^{(4)}\left(\frac{\alpha}{\pi}\right)^2 + a_e^{(6)}\left(\frac{\alpha}{\pi}\right)^3 + \cdots$$

where each coefficient $a_e^{(n)}$ is determined by evaluation of the nth
order radiative corrections. The first two coefficients in this
series are known exactly:[20, 21]

$$a_e^{(2)} = \frac{1}{2}$$

$$a_e^{(4)} = \frac{197}{144} + \frac{\pi^2}{12} - \frac{\pi^2}{2}\ln 2 + \frac{3}{4}\zeta(3) = -0.328\ 48$$

The term $a_e^{(6)}$ requires the evaluation of 72 Feynman diagrams.[22]
Recent work has led to improved numerical accuracy in the evaluation of the diagrams and analytic evaluation of many diagrams. The
contribution of graphs with vacuum polarization insertions, such as
the one shown in Fig. 1(a), is known analytically.[23] The contribution of graphs with light-by-light insertions, such as in Fig. 1(b),
is known accurately numerically.[24] Some combinations of the remaining graphs which contain no electron loops, such as in Fig. 1(c),
are now known analytically.[25,26] Combining the exact value of the
analytically known graphs with numerical values for the remaining
graphs, one obtains the following theoretical values for the sixth
order coefficient (the three values correspond to three independent
numerical evaluations):[26]

$$a_e^{(6)} = 1.206(49) \qquad \text{Levine and Wright}[27]$$

$$a_e^{(6)} = 1.188(17) \qquad \text{Cvitanović and Kinoshita}[28]$$

$$a_e^{(6)} = 1.070(39) \qquad \text{Carroll and Yao}[29]$$

Higher order corrections, hadronic effects, muon loops, and weak interaction effects are expected to have the effect of a change of
less than 0.01 in $a_e^{(6)}$.[26]

The sixth order coefficient can be compared with the derived
experimental value, obtained by subtracting the lower order theoretical values (assuming $\alpha^{-1} = 137.035\ 987(29)$) from the experimental

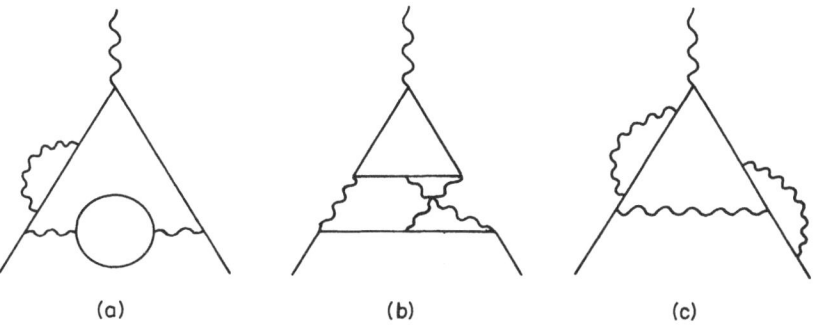

Fig. 1.   Feynman diagrams which contribute to $a_e^{(6)}$.

value of Wesley and Rich:

$$a_e^{(6)} = 1.53(33)$$

The theoretical uncertainty in $a_e^{(6)}$ is approximately an order of magnitude smaller than in the experimentally determined value.

V.   MUON G-FACTOR ANOMALY

A recent measurement of the muon g-factor anomaly by Bailey et al. at CERN has yielded the value[30]

$$a_\mu = 1\ 165\ 895(27) \times 10^{-9} \hspace{3cm} \text{23 ppm}$$

In this experiment, the g-factor anomaly is determined by observation of the precession frequency of muons in a storage ring.[31]

The quantum electrodynamical contributions to the muon anomaly are expressed in a power series in α/π in analogy with the electron anomaly:

$$a_\mu(\text{QED}) = a_\mu^{(2)}\frac{\alpha}{\pi} + a_\mu^{(4)}\left(\frac{\alpha}{\pi}\right)^2 + a_\mu^{(6)}\left(\frac{\alpha}{\pi}\right)^3 + \dots$$

There are two classes of Feynman graphs which contribute to $a_\mu^{(n)}$. One class consists of the graphs obtained by replacing the electrons by muons in the graphs associated with $a_e^{(n)}$. The contribution of these graphs to $a_\mu^{(n)}$ is just $a_e^{(n)}$. The other class consists of mass-dependent graphs with electron loop insertions such as the one shown in Fig. 2:

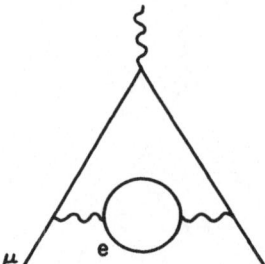

Fig. 2.   Fourth order mass-dependent contribution to $a_\mu^{(4)}$.

$$a_\mu^{(n)} = a_e^{(n)} + \text{graphs with electron loops}$$

In second order, there are no such mass-dependent graphs:

$$a_\mu^{(2)} - a_e^{(2)} = 0$$

In fourth order, there is one mass-dependent graph, which is shown in Fig. 2:[32]

$$a_\mu^{(4)} - a_e^{(4)} = \frac{1}{3} \ln \frac{m_\mu}{m_e} - \frac{25}{36} + \frac{\pi^2}{4} \frac{m_e}{m_\mu} + \ldots = 1.094\ 26$$

In sixth order, all the mass-dependent graphs except those with light-by-light insertions are known analytically.[33]   Recent work by Samuel and Chlouber has improved the numerical evaluation of the graphs with light-by-light insertions.[34]   The combined contribution of analytically known graphs plus light-by-light graphs is

$$a_\mu^{(6)} - a_e^{(6)} = 1.944 + 21.32(5) = 23.26(5)$$

An estimate for the eighth order term, obtained by evaluating the eighth order graphs which are expected to give the dominant contribution, is given by[35]

$$a_\mu^{(8)} \simeq 150(70)$$

In addition to the quantum electrodynamical contributions to a_μ, there is a hadronic contribution which arises from the graph in Fig. 2 with the electron loop replaced by virtual hadronic intermediate states.   The value for this contribution is obtained from measured values for the cross section $\sigma(e^+e^- \to \text{hadrons})$ by means of a dispersion relation.[22]   The result of a recent calculation is[36]

a_μ(hadronic) = 66(10) × 10^{-9}

Weak interaction contributions are expected to be small (a_μ(weak) $\sim$ 2 × 10^{-9}).

The total theoretical value for a_μ is

a_μ = 1 165 918(10) × 10^{-9}

in good agreement with the experimental value. The hadronic contribution is necessary in order to have this agreement.

VI. GROUND STATE FINE STRUCTURE OF POSITRONIUM

Two recent measurements of the ground state fine structure of positronium have been made. Mills and Bearman obtained[37]

$\Delta\nu$(exp) = 203.3870(16) GHz

Egan, Frieze, Hughes, and Yam have determined[38]

$\Delta\nu$(exp) = 203.3849(12) GHz

The current theoretical value is less accurate. There remain terms of order α^2 relative to the main term which have not yet been evaluated. The theoretical expression up to this order is[39]

$$\Delta\nu(\text{th}) = \alpha^2 \text{Ry} \left[\frac{7}{6} - \frac{\alpha}{\pi} \left(\frac{16}{9} + \ln 2 \right) + \frac{1}{2}\alpha^2 \ln \alpha^{-1} + O(\alpha^2) \right]$$

corresponding to a value of

$\Delta\nu$(th) = 203.404 GHz + $O(\alpha^4 \text{Ry})$

Some of the corrections of order α^4Ry have been evaluated. The quantum electrodynamical correction without recoil yields[40]

-1.937α^4Ry = -0.018 GHz

Vacuum polarization corrections to the one-photon annihilation diagram contribute[41]

-0.298 α^4Ry = -0.003 GHz

These contributions, combined with the lower order result, give a tentative theoretical value of

$\Delta\nu$(th) $\simeq$ 203.383 GHz

However, a definitive test of the theory must await evaluation of the remaining terms of relative order α^2.

VII. 2^3S_1 - 2^3P_2 ENERGY SPLITTING IN POSITRONIUM

The first measurement of an excited state energy splitting in positronium has been reported by Mills, Berko, and Canter.[42] For the 2^3S_1 - 2^3P_2 splitting, they obtained the value

$$\Delta E(2^3S_1 - 2^3P_2) = 8628.4(2.8) \text{ MHz}$$

The theoretical value has been calculated to lowest order in α by Ferrell:[43]

$$\Delta E(\alpha^2 Ry) = 8394 \text{ MHz}$$

The corrections of relative order α, which include the leading order radiative corrections, have been calculated by Fulton and Martin:[44]

$$\Delta E(\alpha^3 Ry) = 231 \text{ MHz}$$

The total theoretical result

$$\Delta E(2^3S_1 - 2^3P_2) = 8625 \text{ MHz} + O(\alpha^4 Ry)$$

is in good agreement with the experimental value.

The experiments on positronium mentioned in this section and the preceding section, and the recent measurement of the lifetime of orthopositronium in a powder by Gidley, Marko, and Rich[45] are reviewed by Mills, Berko, and Canter in these proceedings.[46]

VIII.   MUONIUM HYPERFINE INTERVAL

A precision measurement of the muonium hyperfine interval by Casperson et al. at LAMPF has yielded[47]

$$\Delta\nu = 4463.3022(14) \text{ MHz} \qquad\qquad\qquad 0.3 \text{ ppm}$$

This result is consistent with the "world average" of previous measurements quoted by Kobrak et al:[48]

$$\Delta\nu = 4463.3040(18) \text{ MHz} \qquad\qquad\qquad 0.4 \text{ ppm}$$

The theoretical value is given by[22]

$$\Delta \nu = \frac{16}{3}\alpha^2 Ry \frac{\mu_\mu}{\mu_B}\left(1 + \frac{m_e}{m_\mu}\right)^{-3}\left(1 + b + q + \delta_\mu\right)$$

where b is the Breit correction, q is the quantum electrodynamic correction, which includes the anomalous magnetic moment of the electron, and δ_μ is the recoil correction of order m_e/m_μ. The theoretical expression for $\Delta\nu$ can be written in terms of the experimentally determined parameters α and μ_μ/μ_p:[47]

$$\Delta \nu = \alpha^2 \frac{\mu_\mu}{\mu_p}\left[2.632\ 957\ 87 \pm 0.6\ \text{ppm}\right] \times 10^7 \text{ MHz}$$

The error quoted in the square brackets arises from the estimated uncertainty in the quantum electrodynamical contributions of relative order α^3.[49] Equating the theoretical and experimental values, assuming the value for α given in Section II, one obtains[47]

$$\frac{\mu_\mu}{\mu_p} = 3.183\ 329\ 9(25) \qquad\qquad\qquad 0.8 \text{ ppm}$$

and

$$\frac{m_\mu}{m_e} = 206.769\ 27(17) \qquad\qquad\qquad 0.8 \text{ ppm}$$

The value for μ_μ/μ_p obtained this way differs by two standard deviations from the value[50]

$$\frac{\mu_\mu}{\mu_p} = 3.183\ 346\ 7(82) \qquad\qquad\qquad 2.6 \text{ ppm}$$

determined by observation of the muon precession frequency in liquids.

IX.   DERIVED VALUES OF α

It is of interest to compare the values of α derived from comparison of theory and experiment for various precision experiments. The values from the hydrogen fine structure and hydrogen hyperfine structure are discussed in Ref. 14. The results from muonium hyperfine structure, helium fine structure, electron anomalous g-value, and a c Josephson effect are discussed in the preceding sections. The derived values are shown in Fig. 3.

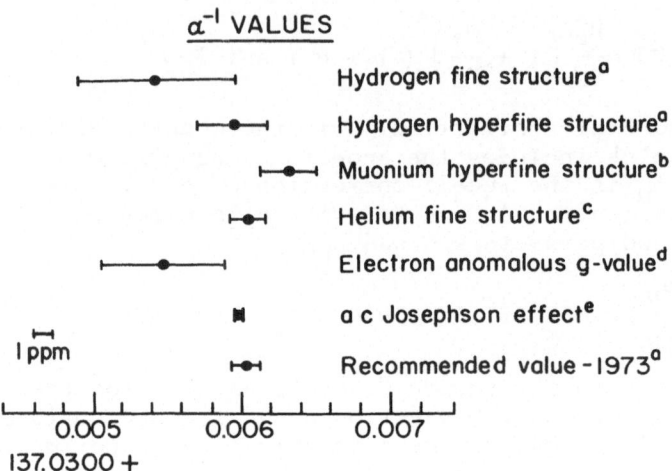

Fig. 3. Values of α derived from various measurements: [a]Ref. 14; [b]Sec. VIII; [c]Sec. III; [d]Sec. IV; [e]Sec. II.

X.   HYPERFINE SPLITTINGS IN ^{3}He$^+$

The theoretical expression for the hyperfine splitting in the $nS_{\frac{1}{2}}$ state of a hydrogen-like system is given by

$$\Delta\nu_n = \Delta\nu_n^F\left(1 + \frac{m}{M}\right)^{-3}\left(1 + b_n + q_n + \delta_n\right)$$

where $\Delta\nu_n^F$ is the Fermi splitting, M is the mass of the nucleus, b_n is the Breit correction, q_n is the quantum electrodynamic correction for a fixed point nucleus, and δ_n represents the remaining corrections including the effect of nuclear size and recoil. In ^{3}He$^+$, δ_n is of the order of 10^{-4} and can be calculated only with limited precision. Uncertainty arising from this term limits the accuracy of a direct comparison of theory and experiment as a test of quantum electrodynamics. The main contribution to δ_n is proportional to $|\psi_n(0)|^2 \propto n^{-3}$ as is the Fermi splitting $\Delta\nu_n^F$, so the dominant part of δ_n is independent of n. Therefore, by comparing theory and experiment for the difference $8\Delta\nu_2 - \Delta\nu_1$, the main contribution of the nuclear structure effects, which is proportional to $\delta_2 - \delta_1$ in the difference, is eliminated. This fact was recognized some time ago and calculations of the necessary differences have been done. The results for the case of ^{3}He$^+$ are given by

$$\Delta\nu_1{}^F\left(1 + \frac{m}{M}\right)^{-3} (b_2 - b_1) = 1.152\ 98\ \text{MHz}$$

$$\Delta\nu_1{}^F\left(1 + \frac{m}{M}\right)^{-3} (q_2 - q_1) = 0.036\ 03\ \text{MHz}$$

$$\Delta\nu_1{}^F\left(1 + \frac{m}{M}\right)^{-3} (\delta_2 - \delta_1) = 0.000\ 80\ \text{MHz}$$

The difference $b_2 - b_1$ includes Breit corrections through order $(Z\alpha)^4$.[51] The difference $q_2 - q_1$ has been calculated by Zwanziger through order $\alpha(Z\alpha)^2$.[52] The difference $\delta_2 - \delta_1$ was first calculated by Sternheim[53] through order $(m/M)(Z\alpha)^2$ and includes corrections calculated by Schwartz.[54]

An improved comparison for ${}^3\text{He}^+$ has been made possible by a recent measurement of $\Delta\nu_2$ by Prior and Wang,[55] which is described by Prior in these proceedings.[56] A precise measurement of $\Delta\nu_1$ has been made by Schuessler, Fortson, and Dehmelt.[57] Both of these measurements were made by an ion storage method. The experimental values are

$$\Delta\nu_2 = 1083.354\ 982\ 5(76)\ \text{MHz} \qquad\qquad 7\ \text{ppb}$$

$$\Delta\nu_1 = 8665.649\ 867(10)\ \text{MHz} \qquad\qquad 1\ \text{ppb}$$

The differences are

Exp: $8\Delta\nu_2 - \Delta\nu_1 = 1.189\ 993(62)\ \text{MHz}$

Th: $8\Delta\nu_2 - \Delta\nu_1 = 1.189\ 80\ \text{MHz}$

Comparison of the experimental and theoretical values confirms the quantum electrodynamical correction to approximately 0.5%. The residual difference between theory and experiment is presumably due to uncalculated contributions to $\delta_2 - \delta_1$, including terms of order $(Z\alpha)^2\delta_1$, and uncalculated contributions to $q_2 - q_1$ of order $\alpha(Z\alpha)^3$.

XI. LAMB SHIFT

The main theoretical contribution to the Lamb shift $S = \Delta E(2S_{1/2}) - \Delta E(2P_{1/2})$ in a hydrogen-like system arises from the lowest order self energy and vacuum polarization corrections corresponding to the Feynman diagrams shown in Fig. 4. The contribution of the self energy to the Lamb shift S_{SE} is given by

(a)   (b)

Fig. 4. Feynman diagrams for the lowest order self energy (a) and vacuum polarization (b) corrections.

$$S_{SE} = \frac{\alpha}{\pi} \frac{(Z\alpha)^4}{6} mc^2 \left[\ln(Z\alpha)^{-2} - \ln \frac{K_0(2,0)}{K_0(2,1)} + \frac{11}{24} + \frac{1}{2} \right.$$

$$+ 3\pi \left(1 + \frac{11}{128} - \frac{1}{2} \ln 2 \right)(Z\alpha) - \frac{3}{4} (Z\alpha)^2 \ln^2(Z\alpha)^{-2}$$

$$\left. + \left(\frac{299}{240} + 4 \ln 2 \right)(Z\alpha)^2 \ln(Z\alpha)^{-2} + (Z\alpha)^2 G_{SE}(Z\alpha) \right]$$

and the contribution of the vacuum polarization to the Lamb shift S_{VP} is

$$S_{VP} = \frac{\alpha}{\pi} \frac{(Z\alpha)^4}{6} mc^2 \left[-\frac{1}{5} + \frac{5}{64} \pi(Z\alpha) - \frac{1}{10} (Z\alpha)^2 \ln(Z\alpha)^{-2} \right.$$

$$\left. + (Z\alpha)^2 G_{VP}(Z\alpha) \right]$$

In each expression, the energy shift is divided into known parts of lowest order in $Z\alpha$ and a function which contains the exact higher order remainder. Calculations of the lower order terms are summarized in Ref. 22. The remainder, contained in $G_{SE}(Z\alpha)$ and $G_{VP}(Z\alpha)$, gives a significant contribution (-0.17 MHz) to the Lamb shift in hydrogen. Erickson and Yennie calculated the main contribution $G_{SE}(0) = -19.08 \pm 5$.[58] Erickson has estimated the value $G_{SE}(\alpha) = -17.1 \pm 0.6$ and has obtained an approximation for $G_{SE}(Z\alpha)$ for a wide range of Z.[59] Recently, the self energy has been evaluated numerically to all orders in $Z\alpha$ for the $2S_{1/2}$ and $2P_{1/2}$ states for Z in the range 10-110.[60] The results, in terms of the function $G_{SE}(Z\alpha)$, are shown in Fig. 5. The value at Z = 1, $G_{SE}(\alpha) = -23.4 \pm 1.2$, is the extrapolated value obtained by fitting the function

$$a + b(Z\alpha)\ln(Z\alpha)^{-2} + c(Z\alpha)$$

to the calculated values at Z = 10, 20, and 30.[60]

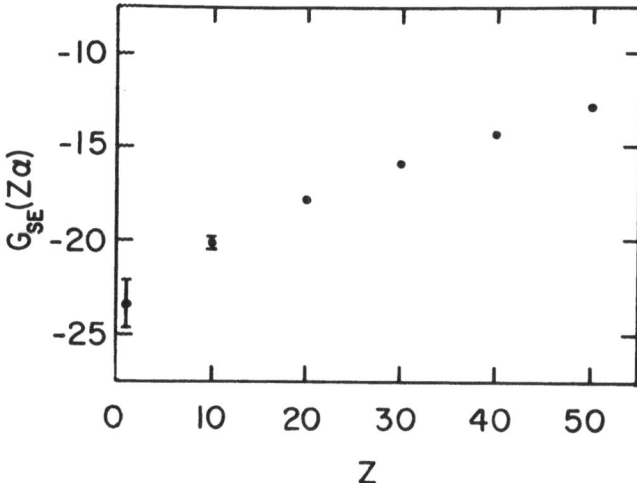

Fig. 5. Calculated values of $G_{SE}(Z\alpha)$ and the extrapolated value at Z = 1, from Ref. 60.

The numerical evaluation is based on the expansion of the coordinate space Green's function in the expression for the self-energy shift as a sum over eigenfunctions of angular momentum.[61] Each term in the sum factorizes into a radial Green's function and a part which contains the functional dependence on the coordinate angles. The radial Green's functions for a Coulomb potential are known in terms of Whittaker functions which can be evaluated to high accuracy. The numerical work consists of evaluating the radial Green's functions, summing the series associated with the angular momentum expansion, and evaluating a three-dimensional integral over the sum. It is necessary to isolate and remove the mass renormalization term. Technical difficulties, such as numerical cancellation for small Z and slow convergence of the angular momentum sum, are dealt with by algebraic rearrangement of the expression, and by the introduction of suitable subtraction terms.

Evaluations of the self-energy contribution to the $1S_{1/2}$-state energy at high Z which take into account electron screening effects and finite nuclear size effects have been done by an independent method based on summation over angular momentum eigenfunctions by Desiderio and Johnson[62] and recently by Cheng and Johnson.[63] The results of Cheng and Johnson for the $1S_{1/2}$-state in a Coulomb poten-

tial and the results of the method described in the preceding paragraph for the $1S_{\frac{1}{2}}$-state[61] are in close agreement for Z in the range
50-130.

Evaluation of the vacuum polarization contribution to the Lamb
shift is facilitated by expanding the vacuum polarization potential
in powers of the external potential. The first term, which is
linear in the external potential, is the Uehling potential[64] which
gives rise to the lowest order terms in S_{VP} and the dominant part
$G_U(Z\alpha)$ of $G_{VP}(Z\alpha)$:[65,60]

$$G_{VP}(Z\alpha) \simeq G_U(Z\alpha) = -\frac{1199}{2100} + \frac{5}{128} \pi(Z\alpha)\ln(Z\alpha)^{-2} + 0.5(Z\alpha) + \ldots$$

There are additional well-known corrections to the Lamb shift
which are reviewed in Ref. 22. They arise from fourth order radiative corrections, reduced mass effects, relativistic recoil corrections, and the effect of nuclear size. The nonrelativistic size
correction

$$S_{NS} = \frac{(Z\alpha)^4}{12}(R/\lambdabar)^2 mc^2$$

depends on the r.m.s. nuclear charge radius $R = \langle r^2 \rangle^{\frac{1}{2}}$ which is
determined by experiments on muonic atoms and by electron scatter-
ing experiments. In the case of hydrogen, only the latter information is presently available.

The result of a recent analysis of electron-proton elastic
scattering data by Borkowski et al. is R = 0.87(2) fm.[66] This value
is larger than the values R = 0.805(11) fm[67] and R = 0.80(2) fm[68]
which have been generally used in evaluating the Lamb shift in
hydrogen.[58,59,22,60] The difference, ΔR = 0.07 fm, produces a
change in the theoretical value for the Lamb shift of 0.02 MHz
which is at the level of precision of the most recent measurement.[69] In order to choose a "best" value for R to evaluate the
Lamb shift, we consider the relevant electron scattering data.
Fig. 6 shows a sample of measured cross sections over a wide range
of the four-momentum transfer squared q^2. Fig. 7 shows data at low
q^2 which is included in the analysis of Ref. 66. A normalization
error of 3-4% in the data in the figures is not included in the
error bars. The data are normalized to the dipole fit which is
the cross section based on the Rosenbluth formula with

$$G_E^D(q^2) = (1 + q^2/18.23 \text{ fm}^{-2})^{-2}$$

$$G_M^D(q^2) = \mu G_E^D(q^2)$$

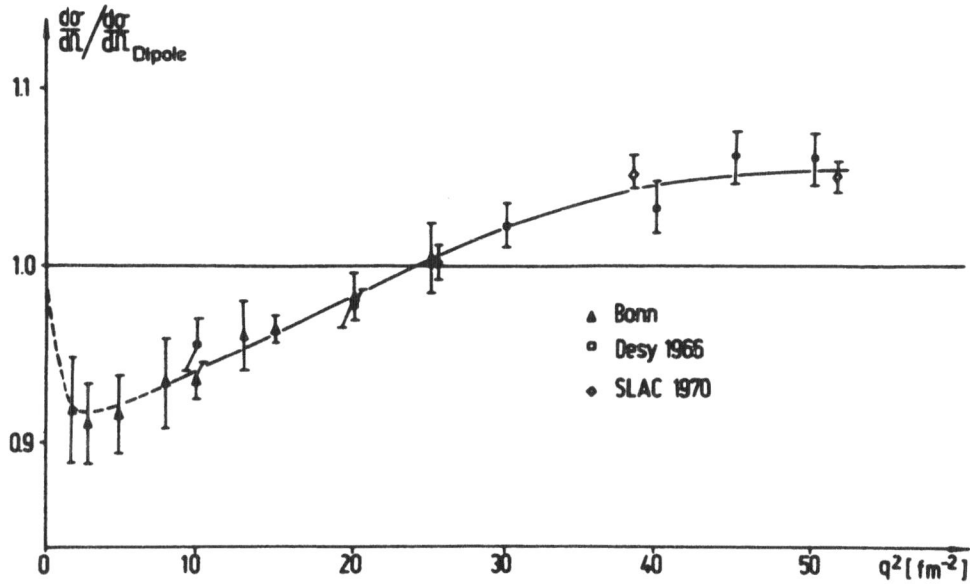

Fig. 6. Elastic electron - proton cross sections normalized to the dipole fit, from Berger et al., Ref. 70. The DESY data have been multiplied by 0.989 and the SLAC data by 0.984.

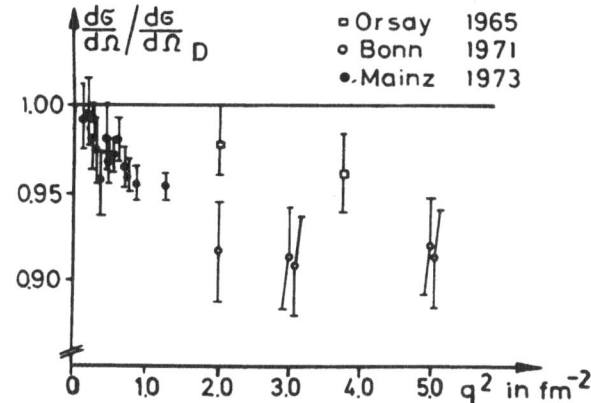

Fig. 7. Low q^2 elastic electron-proton cross sections normalized to the dipole fit, from Borkowski et al., Ref. 71.

substituted for the electric and magnetic form factors $G_E(q^2)$ and $G_M(q^2)$ of the proton; μ is the magnetic moment of the proton. The proton r.m.s. charge radius is related to the electric form factor by

$$R^2 = -6 \left. \frac{dG_E(q^2)}{dq^2} \right|_{q^2 = 0}$$

The dipole form factor $G_E^D(q^2)$ gives R = 0.81 fm. If the approximation of one photon exchange together with the scaling relation $G_M(q^2) = \mu G_E(q^2)$ describes the scattering, then the cross section is proportional to $G_E^2(q^2)$, and deviation of the slope of the data from zero at $q^2 = 0$ signifies deviation of the value of R from the dipole value. The data in Figs. 6 and 7 indicate a systematic deviation of the cross section from the dipole fit and appear to approach $q^2 = 0$ with a negative slope. The apparent slope of the data is consistent with the value R = 0.87 fm, but not with R ≃ 0.80 fm which would require a small positive slope. Hence, we shall tentatively assume R = 0.87(2) fm in evaluating the Lamb shift in hydrogen. Additional experimental information on the proton charge radius would be useful.

The contributions to the Lamb shift are listed in Table I. The theoretical value, with an estimated uncertainty for uncalculated terms, is[22,60]

$$S_{TH} = 1057.888(13) \text{ MHz}$$

where R = 0.87(2) fm[66] has been assumed.

TABLE I.   Contributions to the Lamb shift in hydrogen

CORRECTION	ORDER (mc^2)	VALUE (MHz)
Self energy	$\alpha(Z\alpha)^4[\ln(Z\alpha)^{-2}, 1, Z\alpha, \ldots]$	1085.812
Vacuum polarization	$\alpha(Z\alpha)^4[1, Z\alpha, \ldots]$	-26.897
Fourth order	$\alpha^2(Z\alpha)^4$	0.101
Reduced mass	$\alpha(Z\alpha)^4 m/M[\ln(Z\alpha)^{-2}, 1]$	-1.636
Relativistic recoil	$(Z\alpha)^5 m/M[\ln(Z\alpha)^{-2}, 1]$	0.359
Nuclear size	$(Z\alpha)^4(R/\lambdabar)^2$	0.148
TOTAL		1057.888

TABLE II. Comparison of theory and experiment
for the Lamb shift (Z ⩾ 3)

ION	THEORY	EXPERIMENT	REF.
Li^{2+}	62 737.5(6.6) MHz	62 765(21) MHz	76
		62 790(70) MHz	77
		63 031(327) MHz	78
C^{5+}	781.99(21) GHz	780.1(8.0) GHz	79
O^{7+}	2196.21(92) GHz	2215.6(7.5) GHz	80
		2202.7(11.0) GHz	81
F^{8+}	3343.1(1.6) GHz	3339(35) GHz	74
Ar^{17+}	38.250(25) THz	38.3(1.2) THz	73

There has been a substantial improvement in the measurement of
the Lamb shift in hydrogen. Lundeen and Pipkin, using the separated
oscillatory field method to produce a resonance narrower than the
natural linewidth, have obtained[69]

$$S_{EXP} = 1057.893(20) \text{ MHz}$$

which agrees with the theoretical value. For discussion of the
experiment, see Ref. 72.

Measurements of the Lamb shift in high Z hydrogen-like ions
have been extended to Ar^{17+} by Gould and Marrus who have reported
a preliminary value at this conference.[73] A result for F^{8+} has
been reported by Kugel et al.[74] The theoretical and experimental
values are listed in Table II; the theoretical values are from
Ref. 75. There is generally good agreement between theory and
experiment. Measurements at high Z provide a test of strong field
binding effects in the Lamb shift.

XII. FINE STRUCTURE IN MUONIC HELIUM

The first measurement of the fine structure $E(2P_{3/2}) - E(2S_{1/2})$
in $(\mu^4He)^+$ has been reported by Bertin et al.[82] The experiment was
described in the previous conference of this series.[83] The measured
value is

$$\Delta E(2P_{3/2} - 2S_{1/2}) = 1527.4(0.9) \text{ meV}$$

TABLE III. Theoretical contributions to the fine structure
$E(2P_{3/2}) - E(2S_{1/2})$ in muonic helium (in meV)

Fine structure	145.7
Vacuum polarization, order α	1666.1
Vacuum polarization, order α^2	11.6
Self energy, muon vac. pol.	-10.7 ± 1
Finite nuclear size	$-103.1 \langle r^2 \rangle$ fm^{-2}
Nuclear polarization	3.1 ± 0.6
TOTAL	$1815.8 \pm 1.2 - 103.1 \langle r^2 \rangle$ fm^{-2}

Early theoretical studies of low-Z muonic atoms were done by
Di Giacomo,[84] and by Campani.[85] Two conflicting estimates have been
made for the nuclear polarization correction. Bernabéu and
Jarlskog [86] estimated 3.1 ± 0.6 meV with a calculation based primarily on measured photoabsorption cross sections as input data.
Henley, Krejs, and Wilets[87] estimated a correction of 7.0 ± 1.5 meV,
based on a harmonic oscillator model for the helium nucleus, in a-
greement with an earlier calculation of Joachain.[88] Bernabéu and
Jarlskog,[89] in an analysis of the two results, point out that the
harmonic oscillator model predicts a nuclear electric dipole polarizability which disagrees with the value deduced from measured
photoabsorption cross sections; hence, the smaller value appears
more likely to be correct. An independent calculation by Rinker
confirms the value 3.1 meV.[90]

Theoretical contributions to the energy separation are listed
in Table III. The values in that table are taken from the compila-
tion of Borie,[91] with the exception of the vacuum polarization correction of order α which is taken from Rinker,[90] and the nuclear
polarization correction which is taken from Bernabéu and Jarlskog.[86]
It is interesting to note that the electron vacuum polarization
contribution to the fine structure is an order of magnitude larger
than the Dirac fine structure splitting. The splitting is sensitive
to the nuclear size, so that it is convenient to parameterize the
corresponding energy shift in terms of the r.m.s. nuclear charge
radius. The self energy and muon vacuum polarization terms are
point nucleus values with an estimated uncertainty due to finite
nuclear size corrections. Relativistic recoil corrections, which
are estimated to be less than 0.5 meV,[90,91] are not included.

Assuming the value $\langle r^2 \rangle^{\frac{1}{2}} = 1.650(25)$ fm for the nuclear radius,
based on a weighted average of electron scattering results,[82] the
theoretical value for the fine structure is

$$\Delta E(2P_{3/2} - 2S_{1/2}) = 1535(9) \text{ meV}$$

which is in good agreement with the experimental value. The error
is mainly due to uncertainty in the nuclear radius. If the theory
is assumed to be correct, one can use the measured energy splitting
to deduce the value

$$\langle r^2 \rangle^{\frac{1}{2}} = 1.673(4) \text{ fm}$$

for the nuclear charge radius.

XIII. HIGH Z MUONIC ATOMS

Accurate measurements of the x rays emitted in transitions between large n states in high Z muonic atoms provide a test of
quantum electrodynamic corrections to the energy levels. In particular, the muon levels are sensitive to the effect of vacuum
polarization, which is tested to better than 1% with present-day
experimental precision. Recent experiments by Tauscher et al.[92]
and by Dixit et al.[93] have yielded results for muon transition
energies which are in good agreement with theory, in contrast to
the results of earlier experiments which disagreed with theory.[94,95]

A recent improvement in the theory has been made by the evaluation of the effect of nuclear size on the higher order vacuum
polarization correction. This effect has been calculated by
Arafune,[96] by Brown, Cahn, and McLerran,[97] and by Gyulassy.[98] For
the $5g_{9/2} - 4f_{7/2}$ transition in muonic lead, the finite nuclear size
correction decreases the magnitude of the higher order vacuum polarization correction by 6 eV from the Coulomb value of about -50 eV to
-44 eV. A calculation of the higher order vacuum polarization correction which takes finite nuclear size into account has been done
by Rinker and Wilets.[99]

There has been interest in a correction of order $\alpha^2 (Z\alpha)^2$ corresponding to the Delbrück-like diagrams shown in Fig. 8. It was suggested that the correction could be large ($\simeq$ -35 eV) for the
$5g_{9/2} - 4f_{7/2}$ transition in muonic lead.[100] However, Wilets and
Rinker estimated a range of 1 to 3 eV for the correction.[101] The

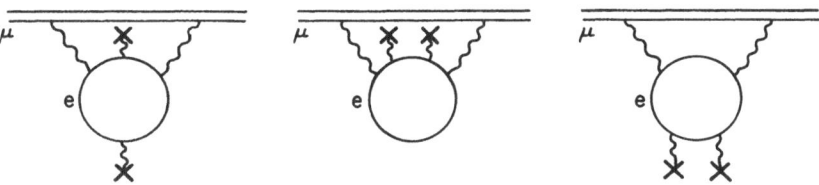

Fig. 8. Feynman diagrams for order $\alpha^2 (Z\alpha)^2$ Delbrück-like corrections to muon energy levels.

TABLE IV. Muonic Atom Energy Level Differences (eV)

Transition	Point Nucleus	Finite Size	Vacuum Polarization				Self En.	Rel. Rec.	Nuc. Pol.	Elec. Scr.	Total
			I	II	III	IV					
$_{56}$Ba											
$4f_{5/2}$-$3d_{3/2}$	439 069±1	-146±8	2436	-21±2	17	1	9±3	3	7	-18±1	441 357±9
$4f_{7/2}$-$3d_{5/2}$	431 654±1	-55±5	2328	-20±2	16	1	-8±2	3	7	-18±1	433 908±6
$5g_{7/2}$-$4f_{5/2}$	200 544±1	0	761	-9±1	5	0	2±1	1	0	-31±2	201 273±3
$5g_{9/2}$-$4f_{7/2}$	199 194±1	0	747	-9±1	5	0	-2±1	1	0	-31±2	199 905±3
$_{80}$Hg											
$5g_{7/2}$-$4f_{5/2}$	414 182±1	-8±1	2047	-42±2	14	1	7±2	2	3	-78±4	416 128±5
$5g_{9/2}$-$4f_{7/2}$	408 465±1	-2	1972	-40±2	14	1	-6±2	2	3	-79±4	410 330±5
$_{81}^{203}$Tl											
$5g_{7/2}$-$4f_{5/2}$	424 850±1	-9±1	2117	-44±2	15	1	7±2	2	4	-79±4	426 864±5
$5g_{9/2}$-$4f_{7/2}$	418 837±1	-3	2039	-43±2	14	1	-7±2	2	4	-81±4	420 763±5
$_{82}$Pb											
$5g_{7/2}$-$4f_{5/2}$	435 666±1	-10±1	2189	-46±2	15	1	7±2	2	4	-81±4	437 747±5
$5g_{9/2}$-$4f_{7/2}$	429 344±1	-4	2106	-45±2	15	1	-7±2	2	4	-83±4	431 333±5

I. $\alpha(Z\alpha)$ Uehling term.   II. $\alpha(Z\alpha)^3$ + finite size.   III. $\alpha^2(Z\alpha)$.   IV. $\alpha^2(Z\alpha)^2$ Delbrück term.

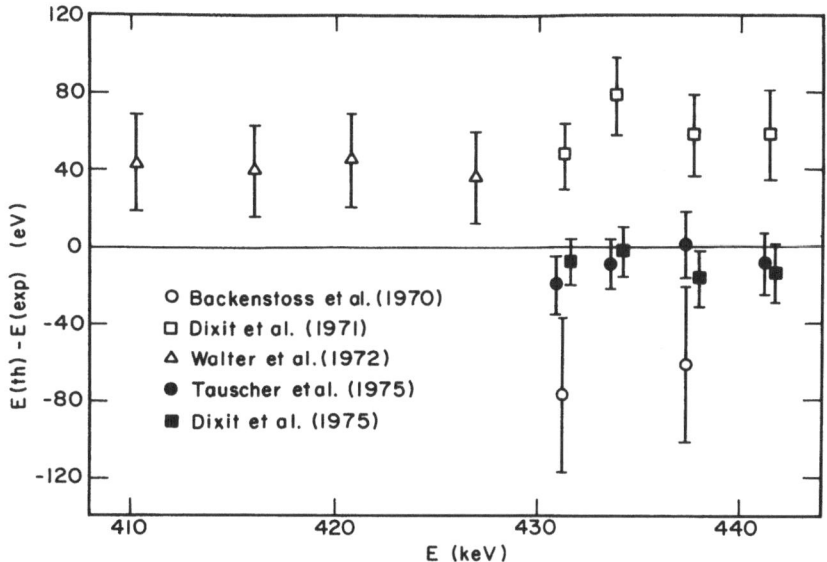

Fig. 9.   Comparison of theory and experiment for muonic atom transition energies.   The experimental values are from Backenstoss et al.,[105] Dixit et al.,[94] Walter et al.,[95] Tauscher et al.,[92] and Dixit et al.[93]

small value has been confirmed by Fujimoto;[102] Borie has recently reported a value of 1 eV for the correction.[103]

The contributions to the muon energy level differences for a selection of transitions are listed in Table IV.   The point nucleus contribution is the reduced-mass Dirac energy separation.   The value employed here for the mass of the muon is based on the muonium hyperfine structure determination described in Section VIII.   The remaining numerical values are from Table 2 of the review by Watson and Sundaresan,[104] with the following exceptions.   The vacuum polarization correction in column II includes a finite size correction based on the formula of Arafune[96] (which predicts a 5 eV correction in the $5g_{9/2} - 4f_{7/2}$ transition in lead).   The $\alpha^2 (Z\alpha)^2$ term is based on results in Refs. 101-103.   Theory and experiment for muonic atom transitions with energy > 410 keV are compared in Fig. 9.   We note that the most recent experimental results would disagree with theory without the higher order vacuum polarization correction.

ACKNOWLEGMENT

Helpful discussion with S. J. Brodsky, V. W. Hughes, J. D. Jackson, and M. H. Prior is gratefully acknowledged.

REFERENCES

*Work supported by the United States Energy Research and Development Administration.

1. For recent reviews of quantum electrodynamics see Refs. 3, 14, 22, 31; S. J. Brodsky, "The Impact of Quantum Electrodynamics," presented to LAMPF Users Group, Nov. 1975. For a discussion of the high-energy tests, see R. Hofstadter, Proceedings of the 1975 International Symposium on Lepton and Photon Interactions at High Energies, W. T. Kirk, ed. (SLAC, Stanford, 1975), p. 869.

2. P. T. Olsen and E. R. Williams, Atomic Masses and Fundamental Constants, 5, J. H. Sanders and A. H. Wapstra, eds. (Plenum Press, New York, 1976), p. 538.

3. For reviews see M. L. Lewis, Atomic Physics 4, G. zu Putlitz, E. W. Weber, and A. Winnacker, eds. (Plenum Press, New York, 1975), p. 105; V. W. Hughes, Atomic Physics 3, S. J. Smith and G. K. Walters, eds. (Plenum Press, New York, 1973), p. 1; N. M. Kroll, ibid., p. 33.

4. C. Schwartz, Phys. Rev. 134, A1181 (1964).

5. A. Kponou, V. W. Hughes, C. E. Johnson, S. A. Lewis, and F. M. J. Pichanick, Phys. Rev. Lett. 26, 1613 (1971); S. A. Lewis, F. M. J. Pichanick, and V. W. Hughes, Phys. Rev. A 2, 86 (1970).

6. M. Douglas and N. M. Kroll, Ann. Phys. (N. Y.) 82, 89 (1974).

7. B. Schiff, C. L. Pekeris, and H. Lifson, Phys. Rev. 137, A1672 (1965).

8. J. Daley, Ph.D. Thesis, University of California, Berkeley (1971), unpublished.

9. L. Hambro, Phys. Rev. A 5, 2027 (1972); Phys. Rev. A 6, 865 (1972); Phys. Rev. A 7, 479 (1973).

10. M. L. Lewis and P. H. Serafino, Bull. Am. Phys. Soc. 18, 1510 (1973).

11. M. L. Lewis, P. H. Serafino, and V. W. Hughes, Phys. Lett., to be published.

12. M. Douglas, Phys. Rev. A 6, 1929 (1972).

13. J. Daley, M. Douglas, L. Hambro, and N. M. Kroll, Phys. Rev. Lett. 29, 12 (1972).

14. E. R. Cohen and B. N. Taylor, J. Phys. Chem. Ref. Data 2, 663 (1973).

15. J. C. Wesley and A. Rich, Phys. Rev. A 4, 1341 (1971).

16. S. Granger and G. W. Ford, Phys. Rev. Lett. 28, 1479 (1972).

17. D. Newman, E. Sweetman, and A. Rich, Atomic Masses and Fundamental Constants, 5, J. H. Sanders and A. H. Wapstra, eds. (Plenum Press, New York, 1976), p. 506.

18. R. van Dyck, Jr., P. Ekstrom, and H. Dehmelt, Bull. Am. Phys. Soc. 21, 818 (1976).

19. H. G. Dehmelt, Atomic Masses and Fundamental Constants, 5, J. H. Sanders and A. H. Wapstra, eds. (Plenum Press, New York, 1976), p. 499.

20. J. Schwinger, Phys. Rev. 73, 416 (1948).
21. R. Karplus and N. M. Kroll, Phys. Rev. 77, 536 (1950); C. M. Sommerfield, Phys. Rev. 107, 328 (1957); A. Petermann, Helv. Phys. Acta 30, 407 (1957); Nucl. Phys. 3, 689 (1957).
22. B. E. Lautrup, A. Peterman, and E. de Rafael, Phys. Reports 3, 193 (1972).
23. R. Barbieri and E. Remiddi, Nucl. Phys. B90, 233 (1975); J. A. Mignaco and E. Remiddi, Nuovo Cim. 60A, 519 (1969).
24. J. Calmet and A. Peterman, Phys. Lett. 47B, 369 (1973); C. Chang and M. J. Levine, unpublished, quoted in Ref. 26.
25. R. Barbieri and E. Remiddi, Atomic Masses and Fundamental Constants, 5, J. H. Sanders and A. H. Wapstra, eds. (Plenum Press, New York, 1976), p. 519, and references therein.
26. M. J. Levine, R. C. Perisho, and R. Roskies, Phys. Rev. D 13, 997 (1976), and references therein.
27. M. J. Levine and J. Wright, Phys. Rev. D 8, 3171 (1973).
28. P. Cvitanović, quoted in Ref. 26.
29. R. Carroll and Y. P. Yao, Phys. Lett. 48B, 125 (1974); R. Carroll, Phys. Rev. D 12, 2344 (1975).
30. J. Bailey, K. Borer, F. Combley, H. Drumm, C. Eck, F. J. M. Farley, J. H. Field, W. Flegel, P. M. Hattersley, F. Krienen, F. Lange, G. Petrucci, E. Picasso, H. I. Pizer, O. Runolfsson, R. W. Williams, and S. Wojcicki, Phys. Lett. 55B, 420 (1975).
31. For discussions of the experiment, see F. H. Combley, Proceedings of the 1975 International Symposium on Lepton and Photon Interactions at High Energies, W. T. Kirk, ed. (SLAC, Stanford, 1975), p. 913; V. W. Hughes, High Energy Physics and Nuclear Structure-1975, D. E. Nagle, et al., eds. (AIP, New York, 1975), p. 515.
32. H. Suura and E. Wichmann, Phys. Rev. 105, 1930 (1957); A. Petermann, Phys. Rev. 105, 1931 (1957); H. H. Elend, Phys. Lett. 20, 682 (1966); 21, 720 (1966); G. W. Erickson and H. T. Liu, UCD-CNL-81 report (1968).
33. See Ref. 23.
34. M. A. Samuel and C. Chlouber, Phys. Rev. Lett. 36, 442 (1976).
35. B. Lautrup, Phys. Lett. 38B, 408 (1972); B. Lautrup and E. de Rafael, Nucl. Phys. B70, 317 (1974); M. A. Samuel, Phys. Rev. D 9, 2913 (1974); J. Calmet and A. Peterman, Phys. Lett. 56B, 383 (1975); and Ref. 34.
36. V. Barger, W. F. Long, and M. G. Olsson, Phys. Lett. 60B, 89 (1975).
37. A. P. Mills, Jr. and G. H. Bearman, Phys. Rev. Lett. 34, 246 (1975).
38. P. O. Egan, W. E. Frieze, V. W. Hughes, and M. H. Yam, Phys. Lett. 54A, 412 (1975).
39. R. Karplus and A. Klein, Phys. Rev. 87, 848 (1952); T. Fulton, D. A. Owen, and W. W. Repko, Phys. Rev. A 4, 1802 (1971), Phys. Rev. Lett. 24, 1035 (1970), 25, 782(E) (1970); D. A. Owen, Phys. Rev. Lett. 30, 887 (1973); R. Barbieri, P. Christillin,

and E. Remiddi, Phys. Rev. A 8, 2266 (1973), Phys. Lett. 43B, 411 (1973).

40.   T. Fulton, Phys. Rev. A 7, 377 (1973).

41.   R. Barbieri, P. Christillin, and E. Remiddi, Phys. Rev. A 8, 2266 (1973); M. A. Samuel, Phys. Rev. A 10, 1450 (1974).

42.   A. P. Mills, Jr., S. Berko, and K. F. Canter, Phys. Rev. Lett. 34, 1541 (1975).

43.   R. A. Ferrell, Phys. Rev. 84, 858 (1951); see also Ref. 44.

44.   T. Fulton and P. C. Martin, Phys. Rev. 95, 811 (1954).

45.   D. W. Gidley, K. A. Marko, and A. Rich, Phys. Rev. Lett. 36, 395 (1976); see also the discussion by G. W. Ford, L. M. Sander, and T. A. Witten, Phys. Rev. Lett. 36, 1269 (1976).

46.   A. P. Mills, Jr., S. Berko, and K. F. Canter, in these proceedings.

47.   D. E. Casperson, T. W. Crane, V. W. Hughes, P. A. Souder, R. D. Stambaugh, P. A. Thompson, H. Orth, G. zu Putlitz, H. F. Kaspar, H. W. Reist, and A. B. Denison, Phys. Lett. 59B, 397 (1975).

48.   H. G. E. Kobrak, R. A. Swanson, D. Favart, W. Kells, A. Magnon, P. M. McIntyre, J. Roehrig, D. Y. Stowell, V. L. Telegdi, and M. Eckhause, Phys. Lett. 43B, 526 (1973).

49.   S. J. Brodsky and G. W. Erickson, Phys. Rev. 148, 26 (1966).

50.   K. M. Crowe, J. F. Hague, J. E. Rothberg, A. Schenck, D. L. Williams, R. W. Williams, and K. K. Young, Phys. Rev. D 5, 2145 (1972).

51.   G. Breit, Phys. Rev. 35, 1447 (1930).

52.   D. E. Zwanziger, Phys. Rev. 121, 1128 (1961); we have numerically recomputed the integrals obtained in that paper to the accuracy necessary for the present comparison. The coefficient of the $\alpha(Z\alpha)^2 \ln(Z\alpha)^{-2}$ term has been confirmed independently by: A. J. Layzer, Nuovo Cim. 33, 1538 (1964); S. J. Brodsky and G. W. Erickson, Phys. Rev. 148, 26 (1966).

53.   M. M. Sternheim, Phys. Rev. 130, 211 (1963); H. Grotch and D. R. Yennie, Rev. Mod. Phys. 41, 350 (1969).

54.   C. Schwartz, Ann. Phys. (N. Y.) 2, 156 (1959); M. Douglas, Phys. Rev. A 11, 1527 (1975).

55.   M. H. Prior and E. C. Wang, Phys. Rev. Lett. 35, 29 (1975); M. H. Prior, private communication.

56.   M. H. Prior, in these proceedings.

57.   H. A. Schuessler, E. N. Fortson, and H. G. Dehmelt, Phys. Rev. 187, 5 (1969).

58.   G. W. Erickson and D. R. Yennie, Ann. Phys. (N. Y.) 35, 271, 447 (1965).

59.   G. W. Erickson, Phys. Rev. Lett. 27, 780 (1971); see also Ref. 14.

60.   P. J. Mohr, Phys. Rev. Lett. 34, 1050 (1975).

61.   P. J. Mohr, Ann. Phys. (N. Y.) 88, 26, 52 (1974).

62.   A. M. Desiderio and W. R. Johnson, Phys. Rev. A 3, 1267 (1971).

63.   K. T. Cheng and W. R. Johnson, paper contributed to this conference, and private communication.

64. E. A. Uehling, Phys. Rev. 48, 55 (1935).
65. E. H. Wichmann and N. M. Kroll, Phys. Rev. 101, 843 (1956).
66. F. Borkowski, G. G. Simon, V. H. Walther, and R. D. Wendling, Z. Physik A 275, 29 (1975).
67. L. N. Hand, D. G. Miller, and R. Wilson, Rev. Mod. Phys. 35, 335 (1963).
68. R. Hofstadter and H. R. Collard in Nuclear Radii, Group I, Vol. 2 of the Landolt-Börnstein new series, H. Schopper, ed., (Springer-Verlag, Berlin, 1967), p. 21.
69. S. R. Lundeen and F. M. Pipkin, Phys. Rev. Lett. 34, 1368 (1975).
70. Ch. Berger, V. Burkert, G. Knop, B. Langenbeck, and K. Rith, Phys. Lett. 35B, 87 (1971).
71. F. Borkowski, P. Peuser, G. G. Simon, V. H. Walther, and R. D. Wendling, Nucl. Phys. A222, 269 (1974).
72. S. R. Lundeen, Atomic Masses and Fundamental Constants, 5, J. H. Sanders and A. H. Wapstra, eds. (Plenum Press, New York, 1976), p. 571; F. M. Pipkin, Atomic Physics 4, G. zu Putlitz, E. W. Weber, and A. Winnacker, eds. (Plenum Press, New York, 1975), p. 119.
73. H. Gould and R. Marrus, paper contributed to this conference, and private communication.
74. H. W. Kugel, M. Leventhal, D. E. Murnick, C. K. N. Patel, and O. R. Wood, II, Phys. Rev. Lett. 35, 647 (1975).
75. P. J. Mohr, Beam-Foil Spectroscopy, I. A. Sellin and D. J. Pegg, eds. (Plenum Press, New York, 1976), p. 89.
76. M. Leventhal, Phys. Rev. A 11, 427 (1975).
77. D. Dietrich, P. Lebow, R. deZafra, H. Metcalf, Bull. Am. Phys. Soc. 21, 625 (1976).
78. C. Y. Fan, M. Garcia-Munoz, and I. A. Sellin, Phys. Rev. 161, 6 (1967).
79. H. W. Kugel, M. Leventhal, and D. E. Murnick, Phys. Rev. A 6, 1306 (1972).
80. G. P. Lawrence, C. Y. Fan, and S. Bashkin, Phys. Rev. Lett. 28, 1612 (1972).
81. M. Leventhal, D. E. Murnick, and H. W. Kugel, Phys. Rev. Lett. 28, 1609 (1972).
82. A. Bertin, G. Carboni, J. Duclos, U. Gastaldi, G. Gorini, G. Neri, J. Picard, O. Pitzurra, A. Placci, E. Polacco, G. Torelli, A. Vitale, and E. Zavattini, Phys. Lett. 55B, 411 (1975).
83. A. Bertin, G. Carboni, J. Duclos, U. Gastaldi, G. Gorini, G. Neri, J. Picard, O. Pitzurra, A. Placci, E. Polacco, G. Torelli, A. Vitale, E. Zavattini, and K. Ziock, Atomic Physics 4, G. zu Putlitz, E. W. Weber, and A. Winnacker, eds. (Plenum Press, New York, 1975), p. 141.
84. A. Di Giacomo, Nucl. Phys. B11, 411 (1969).
85. E. Campani, Lett. al Nuovo Cim. 4, 982 (1970).
86. J. Bernabéu and C. Jarlskog, Nucl. Phys. B75, 59 (1974).
87. E. M. Henley, F. R. Krejs, and L. Wilets, Nucl. Phys. A256, 349 (1976).

88. C. Joachain, Nucl. Phys. $\underline{25}$, 317 (1961).
89. J. Bernabéu and C. Jarlskog, Phys. Lett. $\underline{60B}$, 197 (1976).
90. G. A. Rinker, Phys. Rev. A $\underline{14}$, 18 (1976).
91. E. Borie, Z. Physik A $\underline{275}$, 347 (1975).
92. L. Tauscher, G. Backenstoss, K. Fransson, H. Koch, A. Nilsson, and J. De Raedt, Phys. Rev. Lett. $\underline{35}$, 410 (1975).
93. M. S. Dixit, A. L. Carter, E. P. Hincks, D. Kessler, J. S. Wadden, C. K. Hargrove, R. J. McKee, H. Mes, and H. L. Anderson, Phys. Rev. Lett. $\underline{35}$, 1633 (1975).
94. M. S. Dixit, H. L. Anderson, C. K. Hargrove, R. J. McKee, D. Kessler, H. Mes, and A. C. Thompson, Phys. Rev. Lett. $\underline{27}$, 878 (1971).
95. H. K. Walter, J. H. Vuilleumier, H. Backe, F. Boehm, R. Engfer, A. H. v. Gunten, R. Link, R. Michaelsen, C. Petitjean, L. Schellenberg, H. Schneuwly, W. U. Schröder, and A. Zehnder, Phys. Lett. $\underline{40B}$, 197 (1972).
96. J. Arafune, Phys. Rev. Lett. $\underline{32}$, 560 (1974).
97. L. S. Brown, R. N. Cahn, and L. D. McLerran, Phys. Rev. Lett. $\underline{32}$, 562 (1974); L. S. Brown, R. N. Cahn, and L. D. McLerran, Phys. Rev. D $\underline{12}$, 609 (1975).
98. M. Gyulassy, Phys. Rev. Lett. $\underline{32}$, 1393 (1974); M. Gyulassy, Nucl. Phys. $\underline{A244}$, 497 (1975).
99. G. A. Rinker, Jr. and L. Wilets, Phys. Rev. A $\underline{12}$, 748 (1975).
100. M. Chen, Phys. Rev. Lett. $\underline{34}$, 341 (1975).
101. L. Wilets and G. A. Rinker, Jr., Phys. Rev. Lett. $\underline{34}$, 339 (1975).
102. D. H. Fujimoto, Phys. Rev. Lett. $\underline{35}$, 341 (1975).
103. E. Borie, Bull. Am. Phys. Soc. $\underline{21}$, 625 (1976).
104. P. J. S. Watson and M. K. Sundaresan, Can. J. Phys. $\underline{52}$, 2037 (1974).
105. G. Backenstoss, S. Charalambus, H. Daniel, Ch. Von der Malsburg, G. Poelz, H. P. Povel, H. Schmitt, and L. Tauscher, Phys. Lett. $\underline{31B}$, 233 (1970).

HADRONIC ATOMS

Chien-Shiung Wu

Department of Physics, Columbia University

New York, New York 10027

The title of this talk is Hadronic Atoms. Any negatively charged leptons or hadrons, other than the conventional electrons can be bound in the Coulomb field of an atomic nucleus. These atomic systems created from negatively charged particles from high energy accelerators have been named the Exotic Atoms which are essentially hydrogen-like atoms except for their unusually large energy scales, drastically reduced orbital radii and their transient existence, hence the adjective: Exotic. At present, five different types of exotic atoms have been observed by capturing of the slowed negative muons (μ^-), pions (π^-), kaons (k^-), sigma hyperons (Σ^-) and anti-protons ($\bar{p}$) into atomic orbits (see Table I). The μ^- is a lepton which interacts rather weakly through electromagnetic and weak interactions with the nucleus. The muon in a muonic atom is able to reach the lowest orbit (the 1S orbit) and spent a considerable fraction of its life (2.6 μ sec) inside the nucleus. On the other hand, the π^-, k^-, Σ^-, and $\bar{p}$ are hadrons which interact strongly with the nucleus. They are generally captured by the nucleus before ever reaching the lower orbits. To single out the exotic atoms formed by the hadrons (the strongly interacting particles), we call them the Hadronic Atoms.[1]

If the central Coulomb field is not confined to that of atomic nucleus, then there are more members in the exotic club such as the positronium (e^+, e^-), the muonium (e^-, μ^+), the anti-protonium ($\bar{p}$, p), the charmonium ($\bar{c}$, c) and the pion-muon atoms (π^+, μ^-). These form another interesting and exciting class of hydrogen-like Atoms and some of them will be discussed by other speakers at this conference.

LONG LIVED NEGATIVE PARTICLE SUITABLE FOR EXOTIC ATOMS

Particle	Spin	Mass	Lifetime (Sec)	Radius (1S) fm.	
e^-	1/2	0.511	∞	52,917/Z	
μ^-	1/2	105.6	2.20×10^{-6}	256/Z	
π^-	0	139.6	2.60×10^{-8}	194/Z	⎫ Hadronic
K^-	0	493.8	1.24×10^{-8}	54.7/Z	⎬
$\bar{P}$	1/2	938.3	∞	28.8/Z	⎬ Atoms
Σ^-	1/2	1197.4	1.5×10^{-10}	22.6/Z	⎭
Ξ^-	1/2	1321.3	1.7×10^{-10}	20.5/Z	⎫ To be
Ω^-	3/2	1672.5	1.3×10^{-10}	16.2/Z	⎬ Observed
$\bar{D}$	1	1875.6	∞	14.4/Z	⎭

POSITRONIUM, MUONIUM, ANTI—PROTONIUM, CHARMONIUM, π–μ ATOMS, ETC.

TABLE I

The pionic atoms were first observed in the early fifties
from the conventional cyclotrons ($\sim$ 500 Mev p). The rest of
the hadronic atoms were not observed until 1967 with proton
beams of energies $\geq$ 6 Gev. In fact, the first observation of
the X-rays from the kaomic atoms was reported from this laboratory by Wiegand and Mack.[2]

The great importance that the kaonic atoms might have for
investigation for the nuclear surface was first enthusiastically
pointed out and stressed by Wilkinson[3] as early as in 1960 at
the Rutherford Jubilee International Conference and then again
at the Tokyo International Conference on Nuclear Structure in
1967. Because of the very strong interaction between the kaon
and the nucleon, if the kaon penetrates just a little deep
into the nucleus, there will be so much absorption that sharp
and discrete kaonic levels do not exist. Therefore, the kaon
orbits will have appreciable overlap with the nucleus only on
the nuclear surface (Fig. 1). Regretably, we are still far
from reaching that goal, inspite of long and hard work in
attacking the problem. The main cause for the lack of progress
is probably due to our inadequate knowledge in how to handle
strong interactions properly, particularly, there is the
presence of a quasi-bound state, known as the Y* which is only
27 Mev below the threshold (Fig. 2). Its presence reduces the
certainty in the analysis of the k$^-$-proton scattering amplitude and also makes it highly energy dependent as can be seen
from (Fig. 3). It is obvious we have to gain further knowledge on the k$^-$-nucleon interaction and also to understand
how they are influenced by nuclear effects. These developments and reasonings have led to the current trends on the
experimental investigation of hadronic atoms to veer back to
systematic studies of light nuclei.

However, the hadronic atoms could be used not only as
probes for the nuclear matter distribution. The field of
hadronic atoms encompasses a tremendously broad scope from
chemical effects, atomic physics, nuclear structure, quantum-
electrodynamics to elementary particles. Many interesting
phenomena such as the dynamic E2 excitations in nuclei and
the hyperfine effects have been observed and precision
determinations on the QED tests and the basic properties of
particles have been obtained. It is exciting for a young
and exotic field such as the hadronic atoms to have already
accomplished so much in the past decade.

A. The Dynamic E2 Mixing:

If the exhorted hydrogen-like simplicity of the exotic
atoms, which makes its atomic properties easily calculable

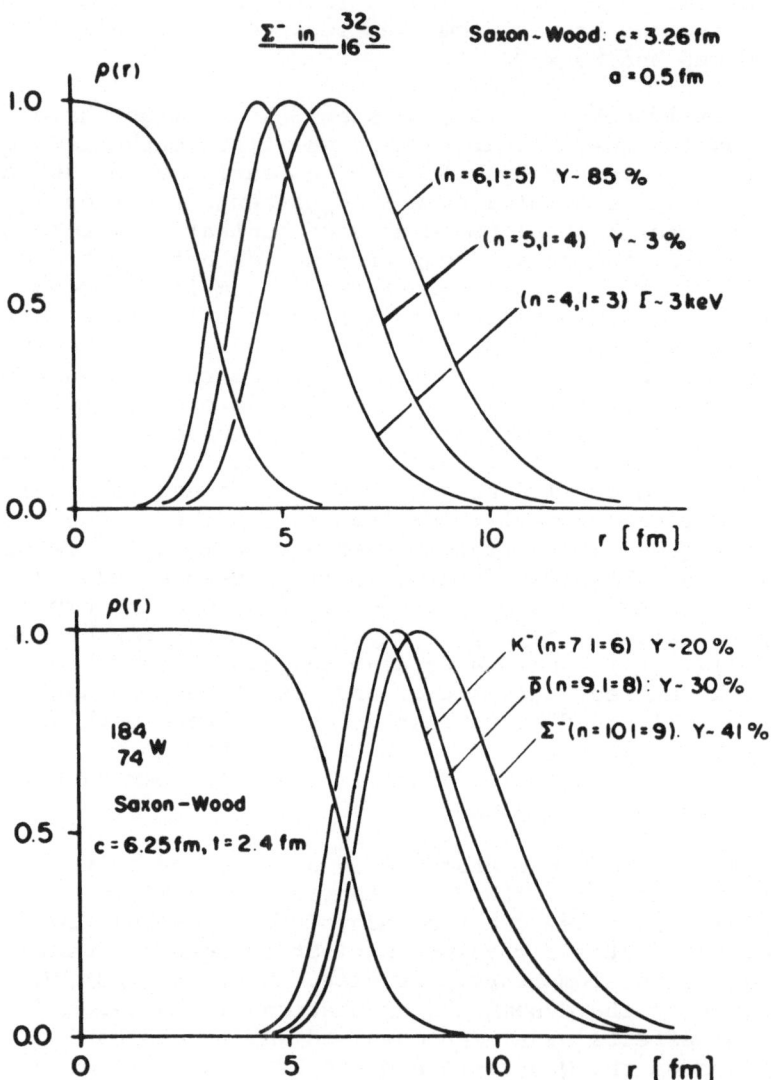

Fig. 1. Showing the overlap integrals $|\phi(r)|^2 r^2 \rho_N(r) n_c$
$\ell = n_c-1$ of Σ^- in $_{16}S$ and also K^-, $\bar{p}$, and
Σ^- in $_{74}^{184}W$ (Ref. 3).

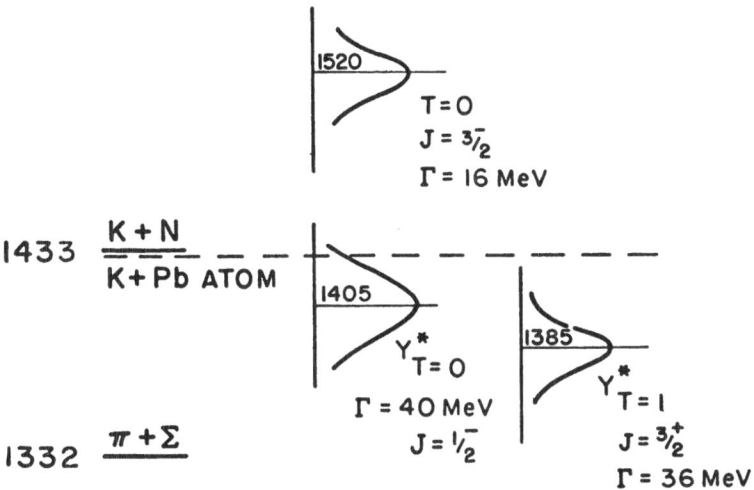

Fig. 2. Excitation functions for the low energy strangeness,
S = -1, two particle channels. In the general region of
1433 Mev. There are the Y_0^* (1405); Y_1^* (1385); and (1520)
resonances.

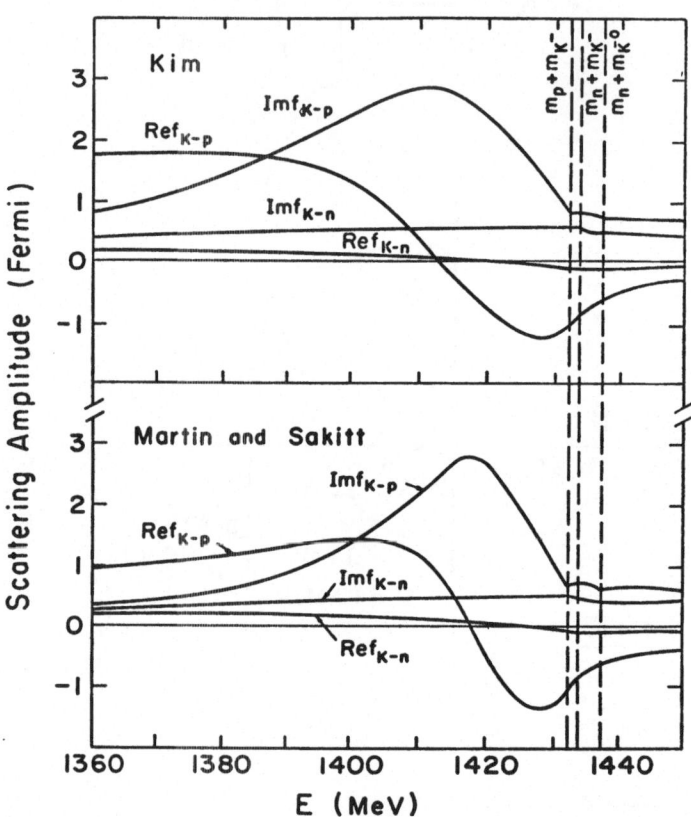

Fig. 3.   The energy dependence of the free k⁻-nucleon scattering
amplitudes for the fits of Kim and of Martin and Sakett.
(Ref. 1).

in contrast to ordinary atoms, were its only virtue; then our interests for its usefulness would be rather limited. Fortunately, because of the atomic interaction energies involved in exotic atoms are usually large enough to be comparable to nuclear energies, a host of interesting dynamic E2 mixing between the atomic and nuclear states has been discovered and yielded unique information on nuclear structure. The interaction between the atomic and the nuclear system illustrates beautifully the manifestation of the quantum mechanical coupling of two otherwise independent systems.

1. Owing to the high Z value and the large quadrupole transition moment of ^{238}U, the electric quadrupole matrix element in its hadronic atoms could be as large as several kev. Since the energy of the first excited 2^+ state is only 44.7 kev, a significant amount of the 2^+ state is mixed in. The actual amount of mixing, of course, depends on the individual atomic level; therefore, the transition energies between such levels are shifted appreciably. The Columbia and Yale group[4] has measured and explained satisfactorily the large energy shifts of the lower atomic transitions of the k^- and $\bar{p}$ in ^{238}U as the dynamic mixing of the first excited quadrupole state with the concerned atomic states. The following two figures (Fig. 4 and Fig. 5) show comparisons of the experimental circular X-ray transition energy with theoretical transition energy with and without corrections due to dynamic E2 mixing in K^--U and $\bar{p}$-U respectively. The comparisons with calculated E2 corrections are very satisfactory.

2. Another kind of E2 dynamic mixing is due to nuclear resonance effect which was theoretically proposed by M. Leon[5] and observed in pionic ^{112}Cd at LASL[6] and in kaonic ^{98}Mo at LBL[7]. This effect occurs when the energy of a nuclear excited state nearly equals a hadronic atom deexcitation energy. Thus, the hadronic atom deexcites by exciting the nucleus as illustrated in Figure 6. The strong absorption of the hadrons from the atomic states due to this dynamic E2 resonance mixing should be easy to observe as it weakens one or two hadronic X-ray line intensities relative to the intensities from other isotopes of the same element (Fig. 7). In Kaonic ^{98}Mo atom, the calculated energy of the $n = 6 \rightarrow n = 4$ transition 798.2 kev is comparable to the energy of a nuclear $0^+ \rightarrow 2^+$ transition 787.4 kev. While in the isotope ^{92}Mo, a corresponding 2^+ level is located much higher, at 1540 kev. The calculated intensity ratio of the 6-5 transition in these two nuclei is $\dfrac{I(n = 6' \rightarrow 5)_{98}}{I(n = 6 \rightarrow 5)_{92}} = \dfrac{0.19}{1.00}$ where 6' state is a mixture

E2 MIXING

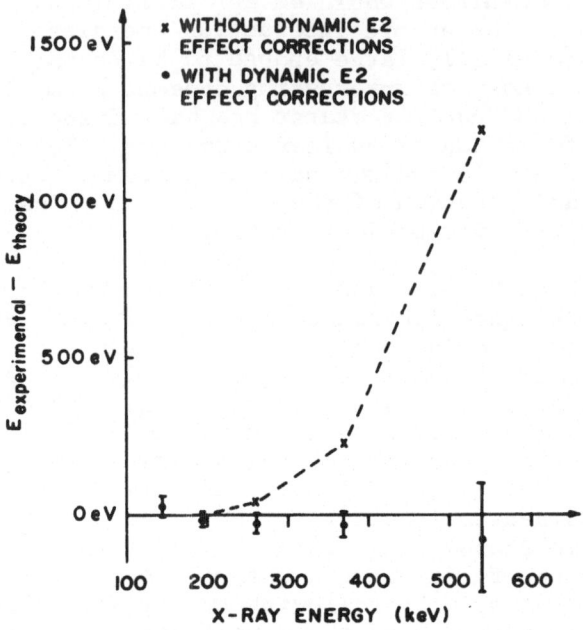

Fig. 4.   Comparison of the experimental circular X-ray transition
          energy with theoretical transition energy, with and without
          corrections due to the dynamic E2 mixing.   The computation
          assumes m_{K^-} = 493.657 MeV.

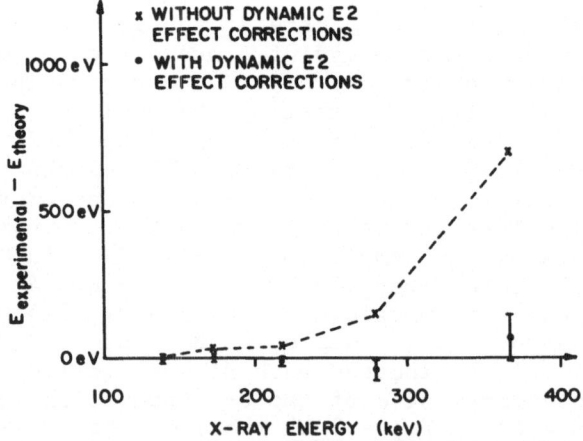

Fig. 5.   Comparison of the experimental circular X-ray transition
          energy with theoretical transition energy, with and without
          corrections due to the E2 dynamic mixing.   The mass of the
          antiproton is assumed to be 938.179 MeV.

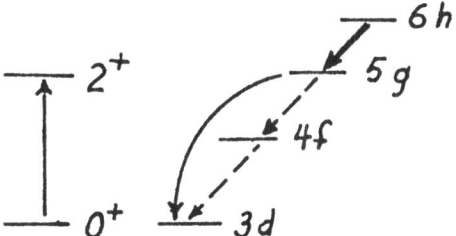

Fig. 6. Nuclear and atomic energy levels relevant to the mixing of the (5g, 0^+) and (3d, 2^+) levels in pionic ^{112}Cd.[6]

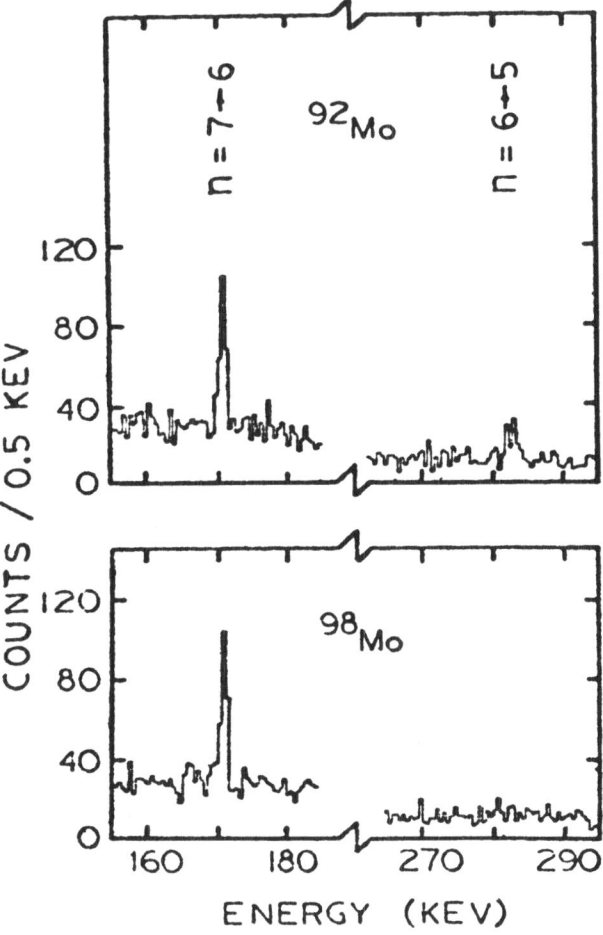

Fig. 7. The relative intensities of the k^-Mo X-rays in ^{98}Mo and ^{92}Mo. Note the drastic reduction of the 6→5 transition in ^{98}Mo.[7]

of the n = 6 state and n = 4 state. The observed ratio was $\dfrac{0.16}{1.00}$
in excellent agreement with the theoretical calculation.

   In a post deadline paper to this conference, J.N. Bradbury
et al.[8] from LAMPF reported a recent observation of the nuclear
resonance effect in pionic ^{110}Pd atom. The significance of
this investigation is to verify the prediction by Ericson
et al.[9] That the p-wave pion–nucleus interaction changes from
attractive to repulsive as Z increases beyond about 36. This
unusual behavior cannot be tested directly by measuring pionic
atom energy shifts because absorption in the 3d state causes
the 3d → 2p X-ray line to disappear for Z > 30. However,
as suggested by M. Leon, the nuclear resonance effect in pionic
^{110}Pd offers an opportunity, not otherwise available, to verify
for this unusual behavior. In the recent LAMPF experiment,[8]
they compared X-ray intensities from a pair of separated isotope (A = 108, 110) palladium targets place simultaneously in
a moderated π^- beam. The observed attenuation for A = 110 is
(20 ± 5)% (preliminary value), which implies very clearly
that the π^- nucleus interaction in the 3p state is indeed
repulsive as predicted by Ericson et al.

B. Hyperfine structure of pionic X-ray lines:

   Although the overlap of the hadrons with the nucleus is
negligible in the outer region, the effect of the electromagnetic properties of the nucleus may be detectable in the
hyperfine structure of the X-ray spectrum. For example, the
spectroscopic quadrupole moment of a deformed nucleus such
as ^{165}Ho has been directly determined in this manner at SIN.[11]
Here is a figure (Fig. 8) showing the resolved hyperfine
structure of the 5g → 4f X-ray transition in π ^{165}Ho. The
quadrupole moment thus determined is Q = 3.47 ± 0.116. This
value of the spectroscopic quadrupole moment is more precise
than that determined from any other methods. However, one
must take into account the strong interaction effects.

C. The fundamental properties of the elementary particles:

   In the region inside of the K-electron orbit but still
outside of the range of the strong interaction of the nucleus,
the hadron finds itself in an unscreened Coulomb field as in a
hydrogen-like atom. The characteristic electric dipole transitions are that between the circular orbits [(n, ℓ = n−1) →
(n−1, ℓ−1)]. Here both the shielding effect due to the
electrons and the strong interaction due to the nucleus are
negligible. The transition energies can be calculated to very

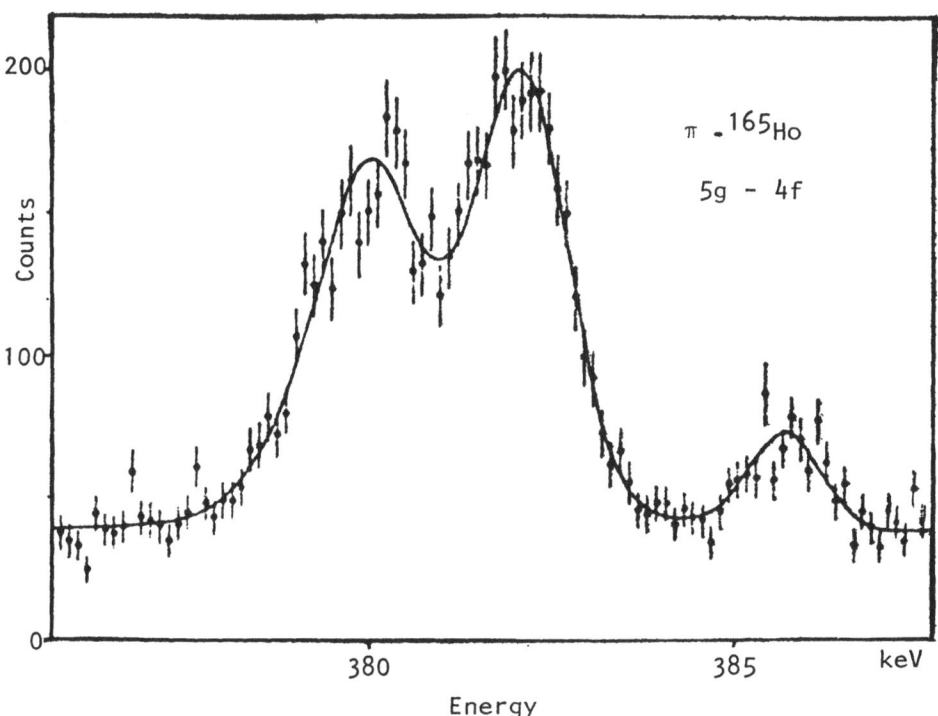

Fig. 8.   The resolved hyperfine splitting of the pionic X-ray
5g→4f in ^{165}Ho.[8]

high accuracy and precision. For this reason, it has been
applied to determine the fundamental properties of the elementary particles and to test certain aspects of quantum
electrodynamics.

The precision determination of the properties of an elementary particle is important, not only for its intrinsic
value, but also for studying small effects such as the vacuum
polarization effects which could perturb the binding energy
of the particle in its atomic orbit. Fortunately, the
existence of exotic atoms provides us with a particularly
beautiful means of making such measurements.

The bound state energy eigen-values of an hydrogenic atom
are given approximately by:

$$E_n = -\frac{1}{2} \frac{(Z\alpha)^2}{n^2} \; mc^2, \tag{1}$$

where m is the mass of the orbiting particle and Z is the
nuclear charge. The fine structure splitting may be expressed
as

$$E_{FS} = (\mu_D + 2\mu_{anom}) \frac{(Z\alpha)^4}{2n^3} \frac{mc^2}{\ell(\ell+1)}, \tag{2}$$

where μ_D is the Dirac moment, and μ_{anom} is the anomalous part
of the magnetic moment.

Inspection of Eq. (1) and (2), the ratio of the fine
structure splitting to the binding energy is proportional to:

$$\frac{E_{FS}}{E_n} \propto (\mu_D + 2\mu_{anom}) \frac{(Z\alpha^2)}{n\ell(\ell+1)} \tag{3}$$

The splitting is greatest for the largest Z value and corresponding smallest $n \rightarrow (n-1)$ transition available. The toll ex-
acted for working with high-Z exotic atoms is the magnitude of
the radiative corrections to the energy levels since $Z\alpha$ is
not a negligible quantity. These corrections, however, are
fairly well known, and the methods of computation are established.

An extensive program of measuring the mass of K^-, $\bar{p}$ and Σ^- and magnetic moments of $\bar{p}$ and Σ^- were carried out by Columbia and Yale group[12,14,17] at BNL in the last few years. Our results will be compared with previously reported values on these fundamental properties.

For precision mass determination, there are certain cautions of which one must exercise:

a) On the experimental side: One should aim at both the highest degree of accuracy and precision. The energy calibration is generally done by the simultaneous recording of the prompt X-rays and the delayed calibration sources. Even then, small systematic errors in these calibration energies would directly affect the mass measurement. The detection system is preferably controlled by a computer system so that the stability of which gives less than a fraction of a channel shift in several thousand channels during the entire running period.

b) On the theoretical side: The transition energies are generally calculated with highest orders of corrections. In general, using the Klein-Gordon Eq. of a point nucleus for a boson and the Dirac Eq. for a fermion, expanded to $(Z\alpha)^6$, and corrected for the reduced mass as well as the relativistic recoil $B^2/2M_A$, where B is the binding energy of the level. We denote these transition energies as zeroth order energies (see Table II).

The vacuum polarization term $\alpha(Z\alpha)$, including the finite size effect and the iterated term is calculated by perturbation theory and the $\alpha^2(Z\alpha)$, $\alpha(Z\alpha)^3$, $\alpha(Z\alpha)^5$, and $\alpha(Z\alpha)^7$ term are also calculated. The basic theoretical works on these corrections were carried out many years ago. The theoretical transition energies $E_{calc.}$ are the sums of all these terms and are listed in the last column of Table II.

c) The data analysis:

The line contaminations due to parallel transitions between noncircular orbits.

The observed spectral lines also contain, besides the main transitions between the circular orbits which are $(n, \ell = n-1) \rightarrow (n-1, \ell = n-2)$, some unresolved weak transitions due to parallel non-circular transitions $(n, \ell = n-2) \rightarrow (n-1, \ell = n-3$ etc$)$. In order to correct for these weak contaminants, the intensity ratios and the energy differences between the parallel weak

TABLE II

Transition	Zeroth order (keV)	$\alpha(Z\alpha)$ (eV)	$\alpha(Z\alpha)$ iter. (eV)	$\alpha(Z\alpha)$ finite size (eV)	$\alpha^2(Z\alpha)$ (eV)	$\alpha(Z\alpha)^3$ (eV)	$\alpha(Z\alpha)^{5.7}$ (eV)	Finite size high order (eV)	Electron screen (eV)	Nuclear pol. (eV)	E_{calc} (keV)
13 → 12	90.697	284.0	0.5	0.4	2.0	−7.7	−1.1	0.5	−51.8	0	90.924
12 → 11	116.575	419.5	0.8	0.8	2.9	−10.6	−1.5	0.8	−44.9	0	116.943
11 → 10	153.328	632.8	1.4	1.5	4.3	−14.7	−2.1	1.2	−38.2	1	153.916
10 → 9	207.340	980.0	2.3	3.1	6.7	−20.7	−3.0	2.1	−31.9	2	208.280
9 → 8	290.081	1569.5	3.9	7.1	10.9	−29.8	−4.3	3.6	−25.9	5	291.621
8 → 7	423.579	2625.5	7.2	18.2	18.6	−44.5	−6.4	5.9	−20.4	18	426.201

non-circular transitions and the main circular transition must be evaluated from a computer program which can be written based on some well known cascading transition processes and corrected for effects due to strong interactions. Figure 9 illustrates the situation; the experimentally measured transition is the sum of transition T_0, T_1, T_2,-- with energies E_0, E_1, E_2-- and strengths F_0, F_1, F_2-- proportional to the population of the levels. Also, in the case of the $10 \rightarrow 9$ transition, and unresolved $\Delta n=2$ transition ($13 \rightarrow 11$) contaminates the $\Delta n=1$ line. Since the final results to be determined are the energies of the circular transitions only, it is necessary to correct the experimentally measured energies for these contaminants. In order to do this, a cascade calculation must be performed to predict the relative populations of the circular and the non-circular levels. Our experimental tests verified that the cascade calculation for the prediction of the relative populations of levels can be relied upon to an accuracy of $\pm$ 20%.

To correct for the non-circular transitions, for each transition in each set of data, a composite Gaussian representing the main circular transition, with satellite Guassians of the same width, have been fitted to the data. The satellite peaks had relative intensities and separations fixed as previously calculated. The channel members obtained for the main circular transition peaks then correspond to the circular transition energies.

The results:

i) The mass of k⁻; m_{k^-}:[12]

Figure 10 shows the X-ray spectrum of k⁻Pb (upper) and the corresponding routed calibration spectrum of ^{75}Se and ^{198}Au (lower). The transition energies of 5 transitions from $13 \rightarrow 12 \rightarrow 11 \rightarrow 10 \rightarrow 9 \rightarrow 8$ were determined. Since the atomic transition energies are, to first order, proportional to the mass of the atomic particles, the k⁻ mass can be determined by plotting the energy difference $E_{exp} - E_{calc}$ as a function of energy. The slope of the curve gives the scale factor $\Delta m/m$ and can be found by least-square fitting using all five transitions. Figure 11 shows the curve of ($E_{exp} - E_{calc}$) versus E. The best fitted m_{k^-} is:

$$m_{k^-} = 493.657 \pm 0.020 \text{ Mev (Columbia and Yale)}^{[12]}$$

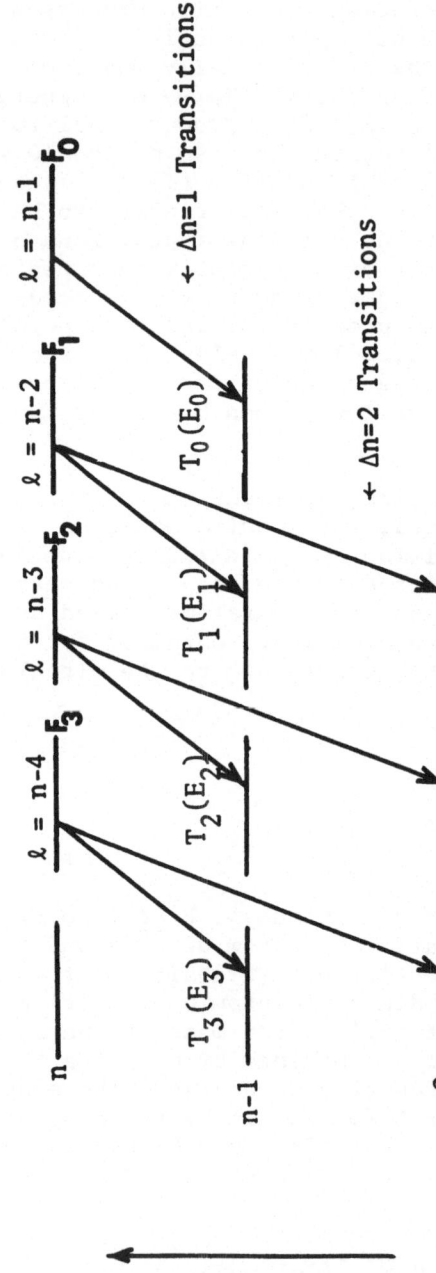

Fig. 9. Illustration of the main circular and the parallel non-circular transition, where $T_0(E_0)$ is circular transition with strength F_0 and $T_1(E_1)$, $T_2(E_2)$, --- are the non-circular transitions with strengths F_1, F_2, ---respectively.

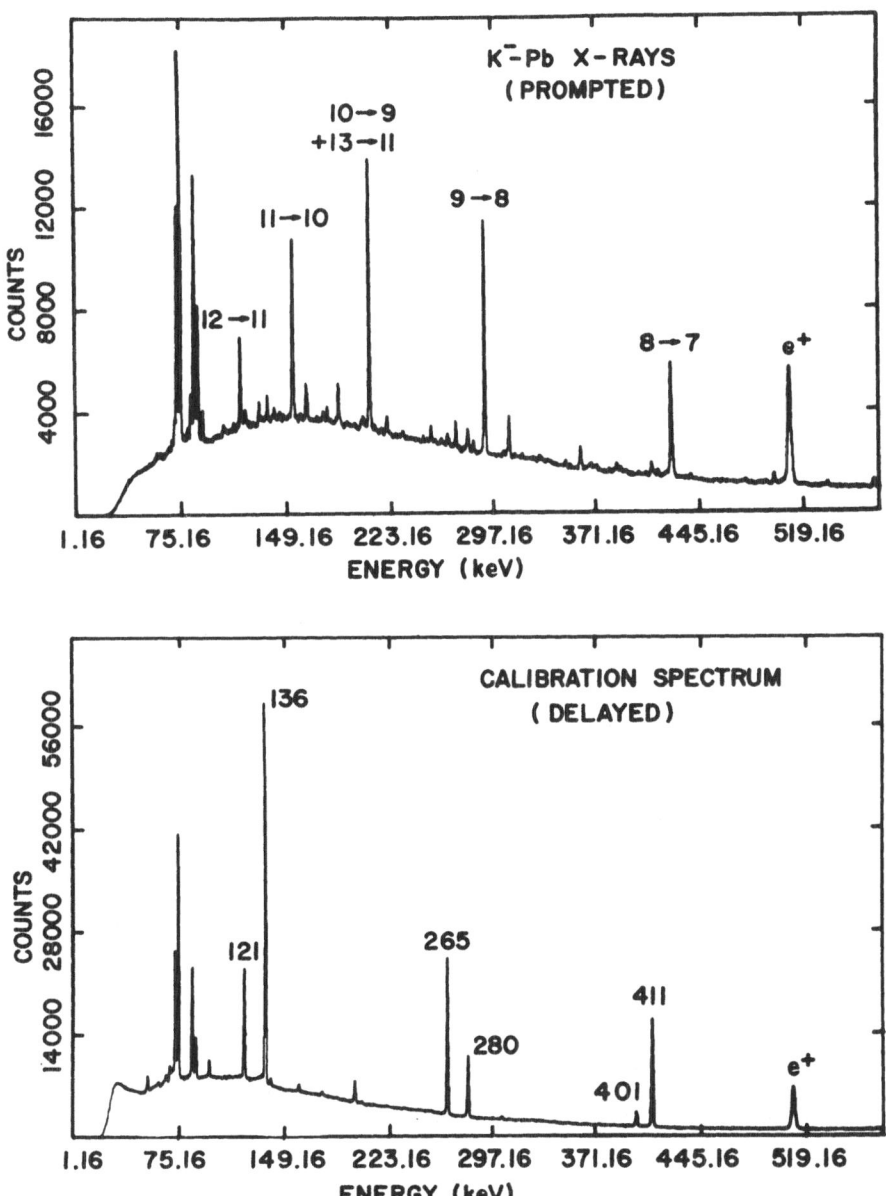

Fig. 10. The X-ray spectrum of k⁻Pb (upper). The corresponding routed calibration spectrum of ^{75}Se and [198]Au (lower).

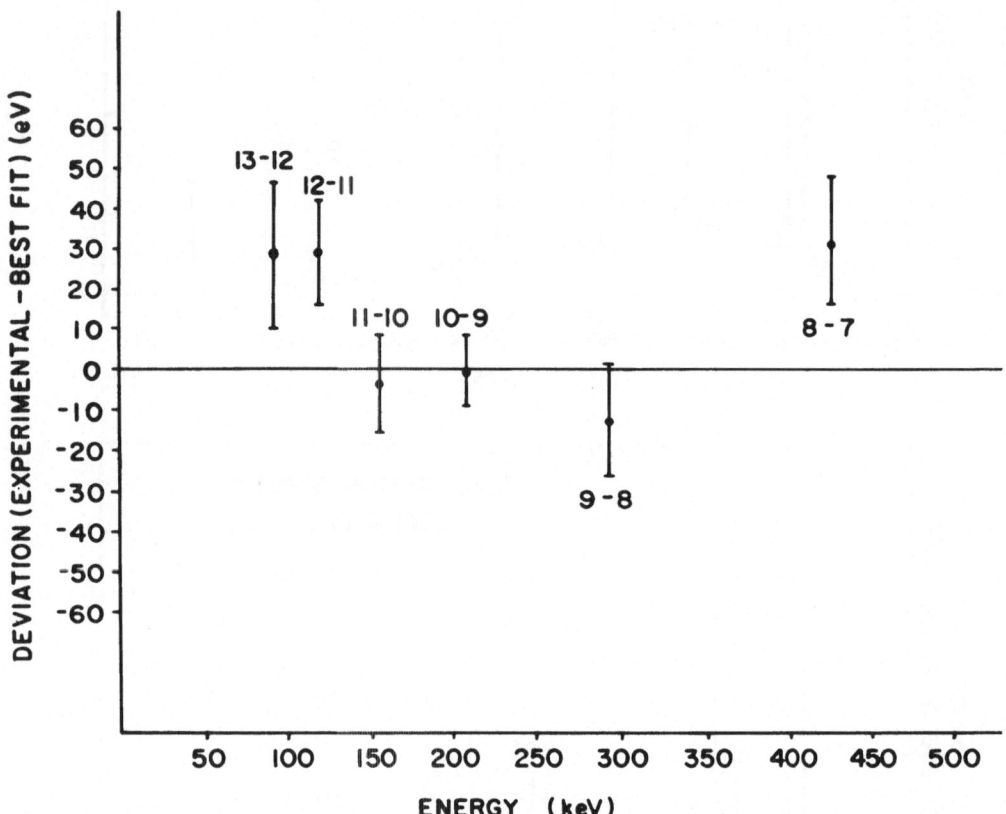

Fig. 11.   The results of the energy differences ($E_{expt.} - E_{calc.}$)
versus E for the six transitions in the k-pb atom.
The 8→7 transition is not used in the k^- mass
determination.   The best-fit is m_{k^-} = 493:657 $\pm$ 0.020
Mev with χ^2 = 6.1 for four degrees of freedom.

This is to be compared with CERN'S value

$$m_{k^-} = 493.688 \pm 0.030 \text{ Mev (CERN)}[13]$$

Both measurements agree within the errors and the average is:

$$m_{k^-} = 493.667 \pm 0.017 \text{ Mev}$$

an error less than 3.5×10^{-5}

ii)   The mass of Σ^{-}[11]; m_{Σ^-}[14]:

When negative kaons are stopped in a target and interact with the nucleons to produce Σ^-.

$$K^- + N \longrightarrow Y + \pi$$

Where Y is a Σ^- or Λ-hyperon and N is a nucleon. Those Σ^- which escape from the nucleus can be slowed and captured by the coulomb field of a nucleus to form excited states of Σ^- atoms. Therefore, k^- and Σ^- X-ray spectra always appear side by side in the same spectrum except the intensity of the Σ^- spectrum is only a few percent of that of k^- spectrum.

We have measured the Σ^--Pb X-rays from $15\rightarrow14\rightarrow13\rightarrow12\rightarrow11\rightarrow10$ transitions (see Fig. 12) and deduced a mass value:

$$m_{\Sigma^-} = 1197.24 \pm 0.14 \text{ Mev (Columbia and Yale)}[14]$$

This is in excellent agreement with measurements done in different methods.

$$m_{\Sigma^-} = 1197.36 \pm 0.10 \text{ Mev (Emulsion method) Bohm et al.}[15]$$

$$m_{\Sigma^-} = 1197.43 \pm 0.11 \text{ Mev (bubble-chamber method) Schmitt}[16]$$

iii)   The mass of $\bar{p}$[14]; $m_{\bar{p}}$.[17]

The CPT theorem predicts the equivalence of mass and lifetime of particle and anti-particle and also predicts that the magnetic moment of particle and anti-particle are equal in magnitude and opposite in sign. It is clearly of interest to

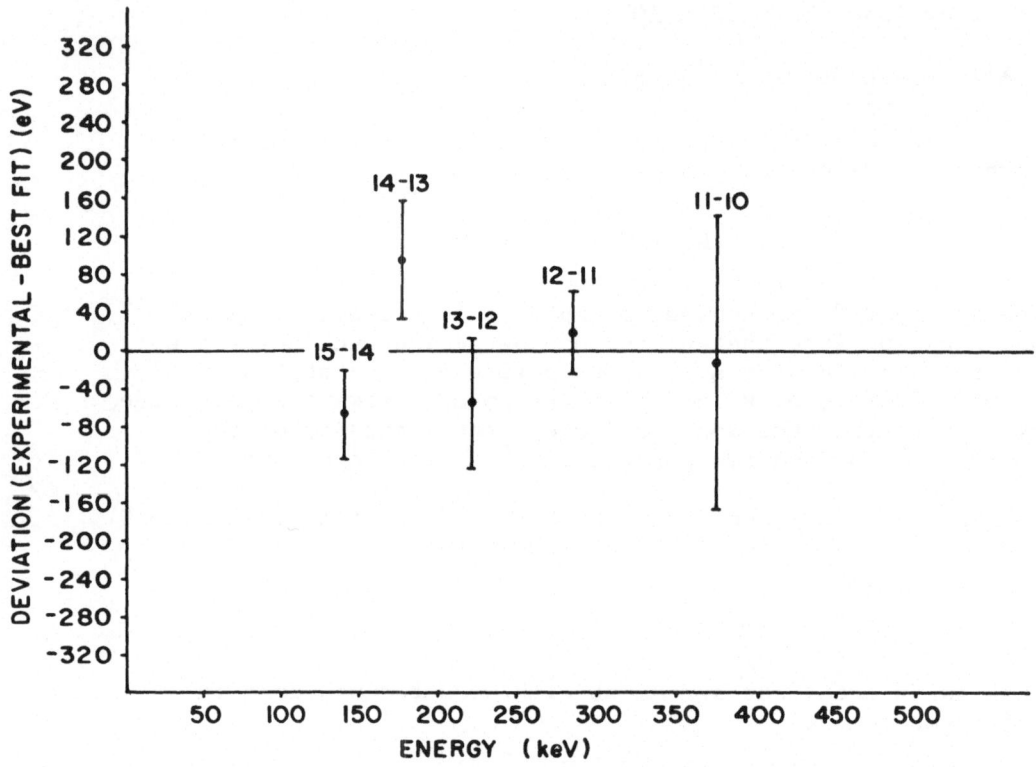

Fig. 12. Deviations between experimental energies and theoretical energies for the best-fit Σ^- mass.

make precise measurements of such basic quantities for the anti-
proton, not only for the intrinsic values, but also providing
another rigid test of the CPT theorem.

In Figure 13, the deviations between experimental energies
and theoretical energies for the best-fit anti-proton mass are
shown. The best fit value of $m_{\bar{p}}$ is

$$m_{\bar{p}} = 938.179 \pm 0.058 \text{ Mev}$$

With $\chi^2 = 2.9$ for five degrees of freedom. We compare this value
to the recently tabulated value of proton mass

$$m_{\bar{p}} = 938.2796 \pm 0.0027 \text{ Mev}$$

The anti-proton mass falls $\underset{\sim}{} 1.7$ standard deviations away from
the proton mass. The probability of such an occurence being due
to a statistical fluctuation is about 10%; this is not sufficiently small to indicate a real evidence for discrepancy.
Furthermore, there are two aspects of our analysis which should
be viewed with some reservation. One is that our absolute energy
scale is derived from calibration source energies determined
in independent experiments. Small systematic errors in these
energies could affect our mass determination. The other is that
the assumption of a simple statistical starting population may
lead to an overestimation of the non-circular contaminants,
which would result in a lower mass. However, our value of the
anti-proton magnetic moment shows excellent agreement with the
proton magnetic moment. Since the magnetic moment is derived
from the measured fine structure splitting instead of the
absolute line energies, any errors introduced by the above
mentioned possibilities would be greatly reduced.

iv) The magnetic moment of anti-proton; $\mu_{\bar{p}}$.[17]

The fine structure splitting and the relative intensities
of its components of a given $\Delta n=1$ transition are shown in
Figure 14. The experimental difficulty is to have sufficient
resolution in order to resolve the fine structure components.
Figure 15 shows the clearly resolved the fine structure splitting in the $1\dot{1} \rightarrow 10$ transition in $\bar{p}$ U.

The value of $\mu_{\bar{p}}$ obtained in our measurements is:

$$\mu_{\bar{p}} = -2.791 \pm 0.021 \; \mu_n \text{ (Columbia and Yale)}$$

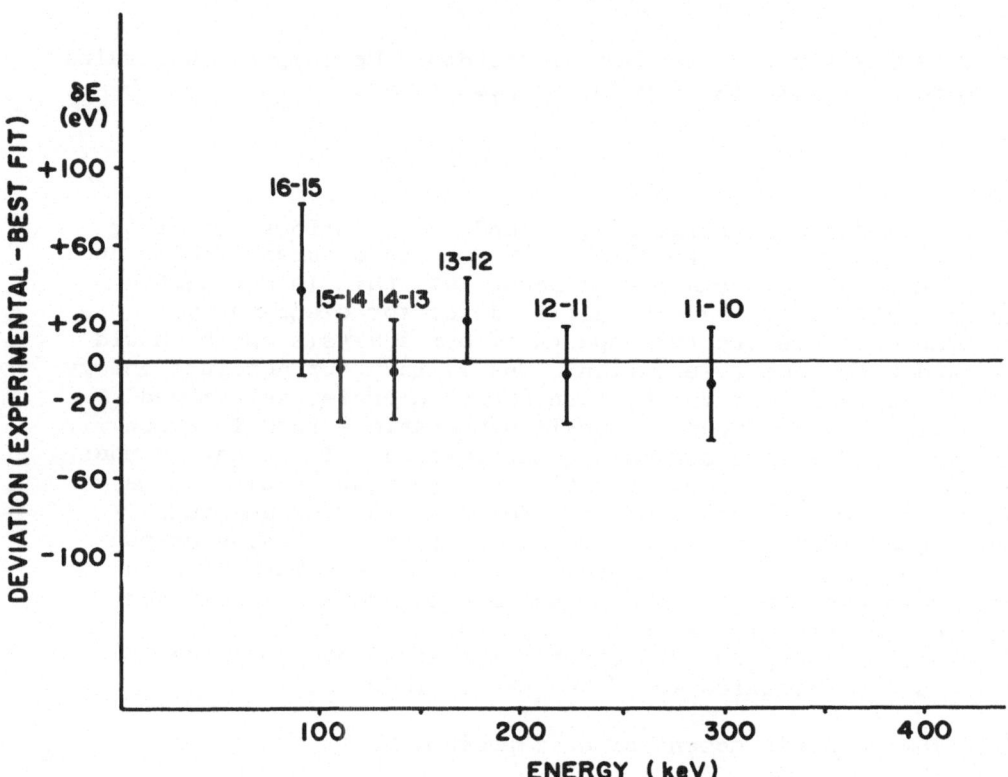

Fig. 13.   Deviations of experimental from theoretical energies for
the best-fit anti-proton mass.

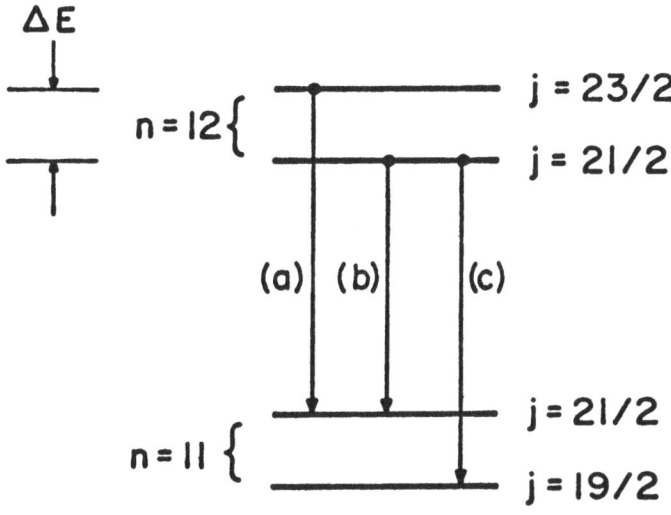

$$\Delta E = (g_0 + 2g_1) \frac{(\alpha Z)^4 mc^2}{2n^3 \, \ell(\ell+1)}$$

$$\mu = (g_0 + g_1) \, \mu_N$$

Fig. 14.   Relative intensities of the fine structure components
of a given Δn=1 transition.

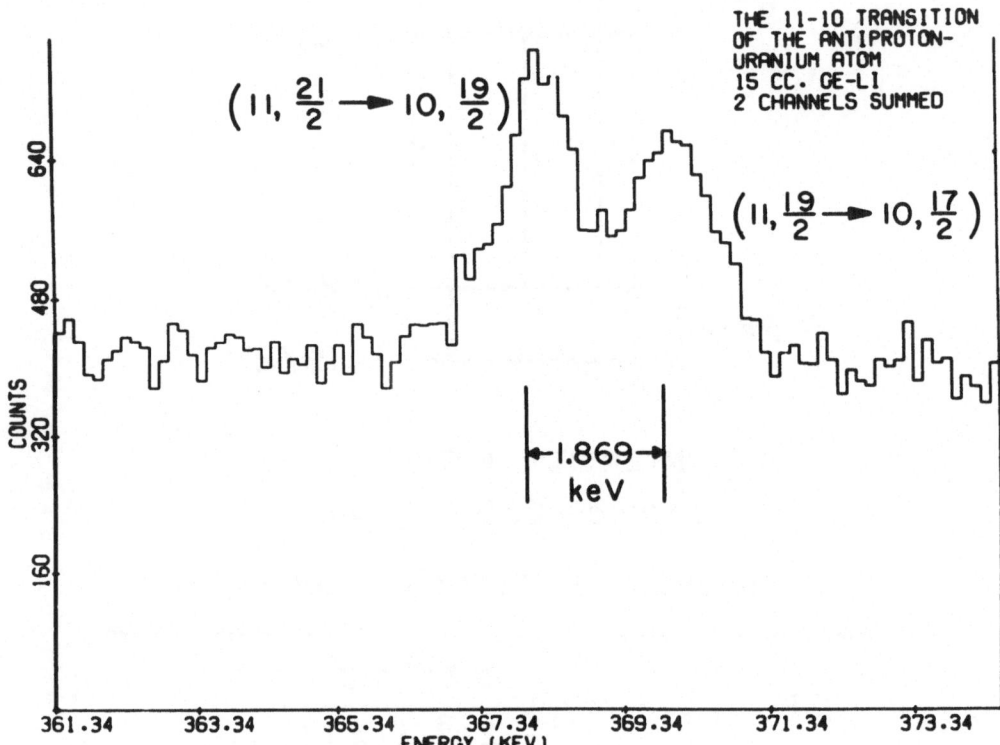

Fig. 15.   The 11→10 transition in p̄U.

This agrees well with the value of proton magnetic moment

$$\mu_{\bar{p}} = 2.793 \ \mu_n \ (\text{error } 1.1 \times 10^{-6} \mu_n)$$

and with a previous determination of the $\mu_{\bar{p}}$.

$$\mu_{\bar{p}} = -2.819 \pm 0.056 \ \mu_n \ (\text{Roberts et al})[19]$$

v) The magnetic moment of Σ^{-11}; μ_{Σ^-}.[14]

The magnetic moment of the Σ^- is difficult to measure. First, its low intensity X-ray spectrum is nestled among the K^- X-ray spectrum. Furthermore, its fine structure splitting is only of ~ 300 ev at most to be compared with the detector re-solution of ~ 1.2 Kev. Figure 16 shows the Σ^- 12→11 transi-tion in natural lead.

In order to extract the magnetic moment of Σ^- from the unresolved fine structure, one has to do some fancy curve fittings. The observed curve has been fitted to a composite line structure consisting of the circular transition doublet plus a non-circular transition doublet with an intensity ratio and energy difference calculated by cascade calculations. The intensity ratio of the doublet components are fixed equal to that appropriate to a statistical level population. The width of each single line was given by the calibration spectrum. The position and the height of the line structure as free pa-rameters. These fits were made as a function of g_1 for each set of data separately; the χ^2 function so obtained were plotted against g_1 (Σ^- magnetons) as shown in Figure 17. The best values for g_1 are:

$$g_1 = \begin{cases} -0.79 \ \begin{array}{l} + 0.52 \\ - 0.36 \end{array} \\ \\ +1.83 \ \begin{array}{l} + 0.36 \\ - 0.51 \end{array} \end{cases}$$

We cannot distinguish between these two solutions on the basis of the χ^2 test.

The magnetic moment of Σ^- thus determined from the Σ^- 12→11 transition in Pb is then

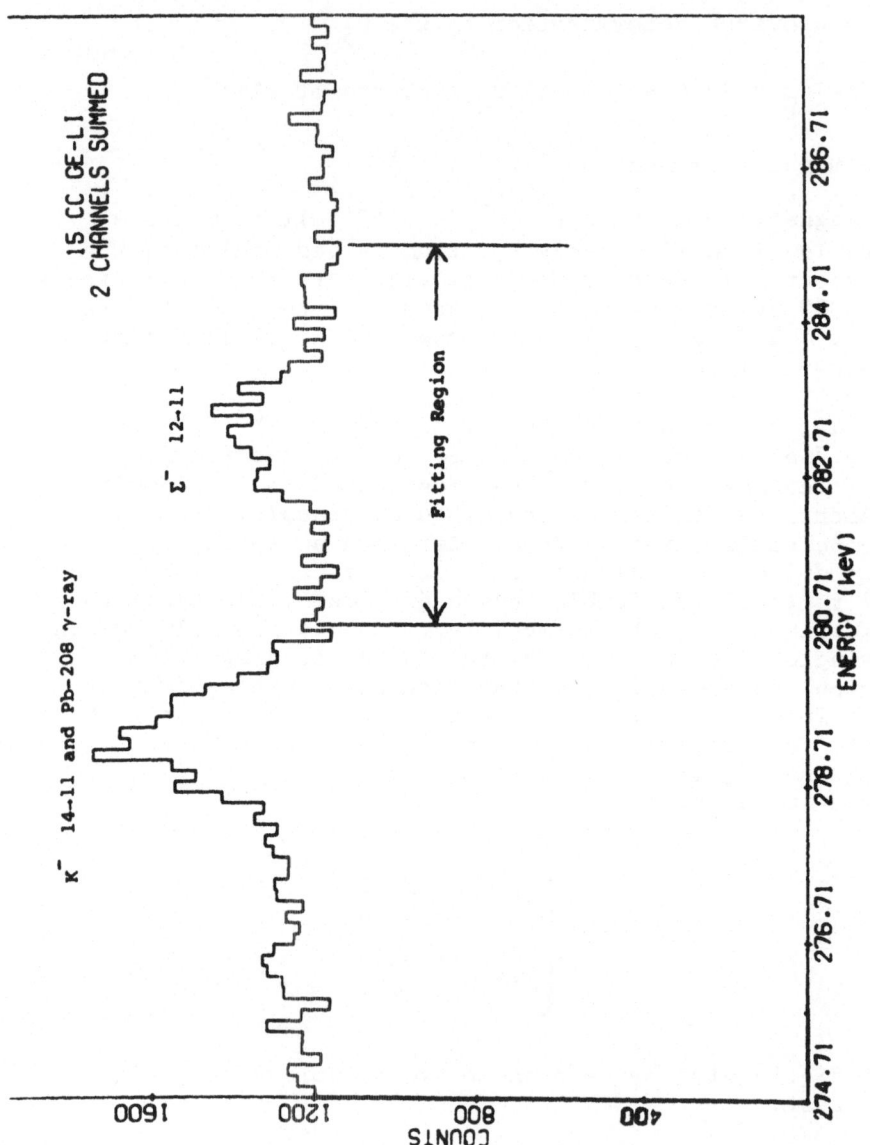

Fig. 16. The Σ^- 12→11 transition in natural lead.

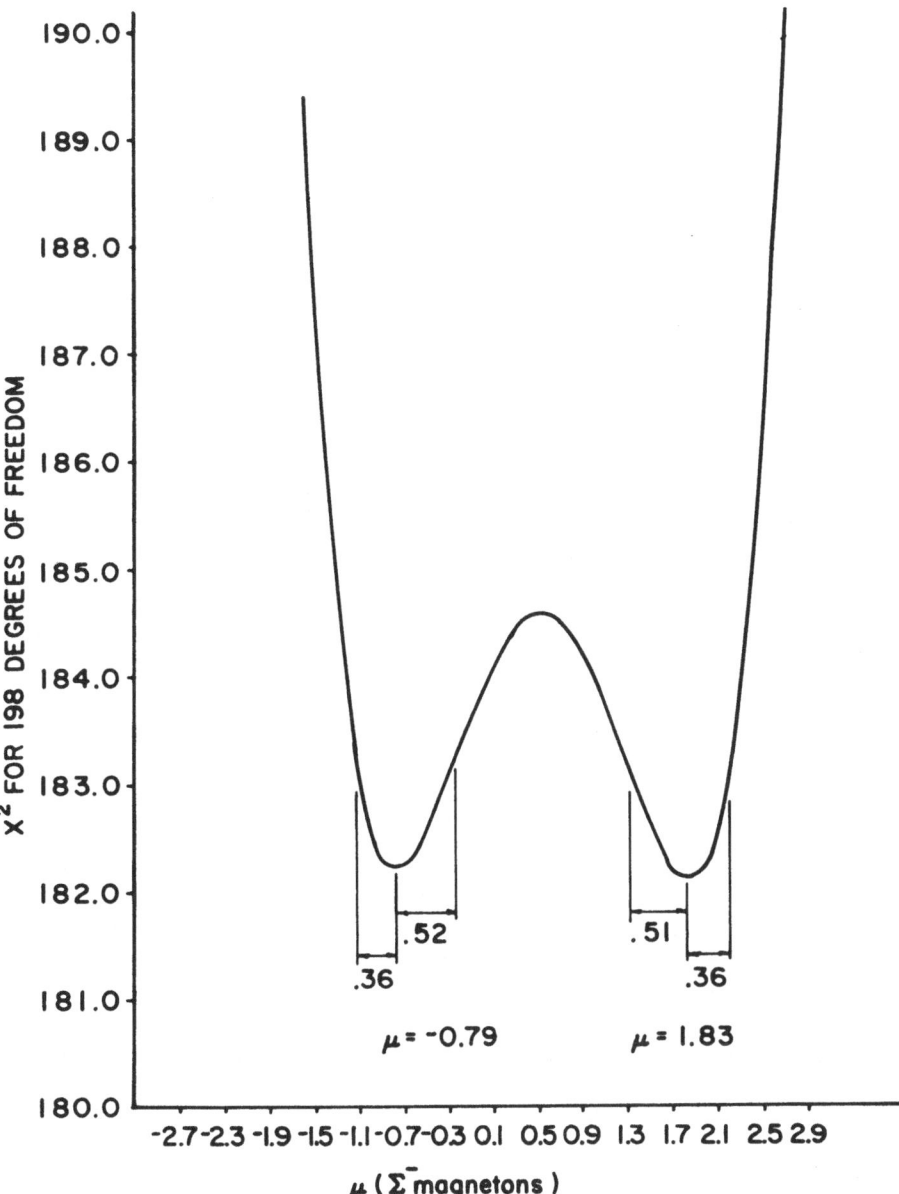

Fig. 17.   Plot of χ^2 versus anomalous moment indicating the best-fit
           values for the moment.

$$\mu_\Sigma = \begin{cases} -1.79 \; {}^{+\,0.52}_{-\,0.36} \; \Sigma^- \text{ magnetons} \\ \\ +0.83 \; {}^{+\,0.36}_{-\,0.51} \; \Sigma^- \text{ magnetons} \end{cases}$$

or equivalently

$$\mu_{\Sigma^-} = \begin{cases} -1.40 \; {}^{+\,0.41}_{-\,0.28} \; \text{nuclear magnetons} \\ \\ \hfill \text{(Columbia and Yale)}[14] \\ \\ 0.65 \; {}^{+\,0.28}_{-\,0.40} \; \text{nuclear magnetons} \end{cases}$$

This result is in good agreement with another determination[19] of the Σ^- magnetic moment by the same technique, which gives

$$\mu_{\Sigma^-} = -1.48 \pm 0.37 \text{ n.m. (Roberts et al)}[19]$$

the predictions of the SU(3) model for the magnetic moment of Σ^- relative to other members of the baryon octet are:

(i) $\mu_{\Sigma^-} = -0.88$ n.m.

(ii) $\mu_{\Sigma^-} = \mu_{\Xi^-}$

This measurement does not have the precision to rule out the first SU(3) prediction (from our χ^2 curve, the probability that $\mu_{\Sigma^-} = -0.88$ n.m. is $\sim$ 23%). The currently accepted value for $\mu_{\Xi^-} = (-1.93 \pm 0.75)$ n.m. is larger than our result for μ_{Σ^-} but not inconsistent with it.

D.   An ingenious approach[20] in the application of the curved crystal spectrometer for the study of pionic-X-rays.

The curved crystal spectrometer under ideal conditions permits a relative accuracy of 10^{-6} which is nearly a factor of 10 better in resolution over that of the solid state detector spectrometer. However, because of its small luminosity, in order that this possibility could be realized, it is necessary that the radiation being measured be intense enough. In the past, the pionic atoms are formed by exposing a target,

whose pionic atoms are being sought, in an extracted beam of
slow pions from a conventional cyclotron. Due to unavoidable
losses in the pion transport system, the intensity of the
pion beam is reduced to 10^6/sec and the maximum counting
rate of the diffracted pionic X-ray line in a crystal spectrometer is only about a few counts per hour.

Recently, an ingenious and bold attempt was made by
looking directly at the intense X-ray source generated in the
pion production target itself by utilizing the characteristic
directionality properties of a curved crystal spectrometer
(5 meter) at the Getchina Laboratory of Leningrad (Fig. 18).
The flux of the external proton beam (1.2 Gev) used was only
8×10^{11} p/sec. The counting rate was greatly increased
by at least, two orders of magnitude. Figure 19 shows the
diffraction curves of the 4f → 3d of π -Ti X-rays taken
at the right and left position of the spectrometer. The counting time for each point is only 20 min. and they were able
to gather sufficient data in 10 hrs. to attain an accuracy of 12
parts per million. A summary of the most recent determinations
of the π mass is listed in Table III. The agreements between
them are excellent.

TABLE III

π atomic X-ray used	Energy (kev)	Method
Ti, Ca, 4f → 3d	139,566 ± 10	Berkeley Crystal spectrometer (1973)[21]
I₁ 5g → 4f	139,569 ± 6	GERN Ge(Li) (1973)[13]
Ti, Ca, 4f → 3d	139,565.7 ± 1.7	Leningrad Crystal spectrometer (1976)[20]

Using the new $m_{\pi^-} = 139{,}565.7 \pm 1.7$ Kev; $m_{\mu^-} = 105{,}659.48 \pm 0.35$
Kev[18] and combined with the newly determined μ^+ momentum from $\pi\pm$

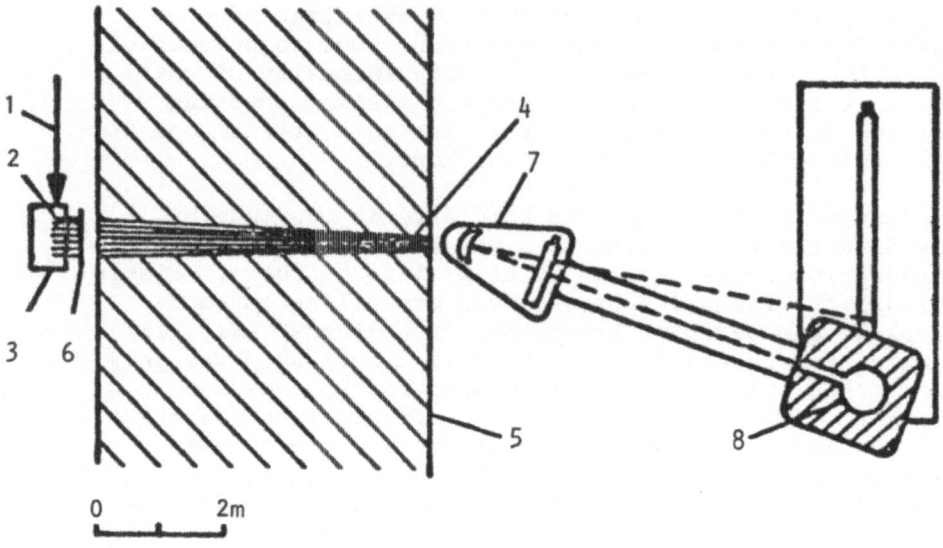

Fig. 18.   Experimental set-up of crystal spectrometer in the
            intense pionic X-ray beam at the Getchina Leningrad
            proton-accelerator.[17]

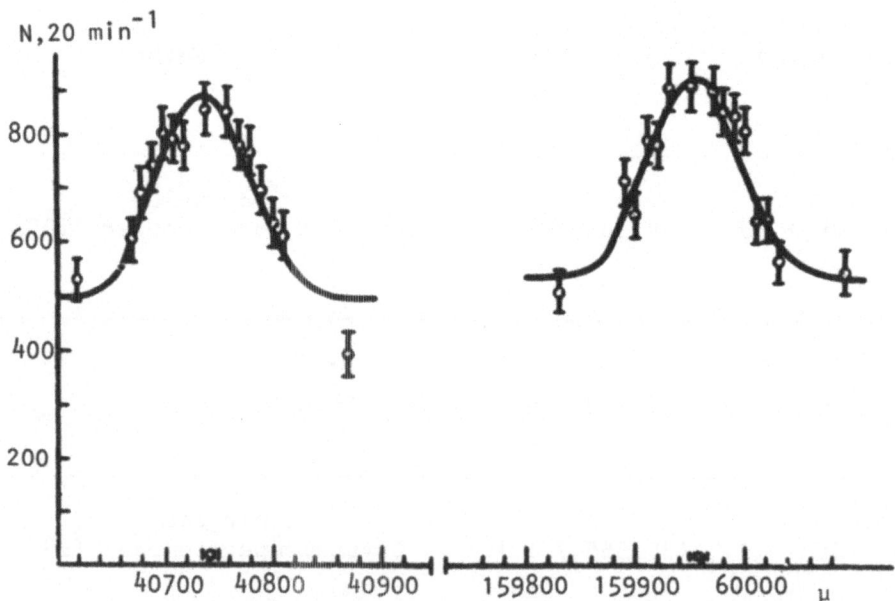

Fig. 19.   The diffraction curves of the 4f→3d of π^- Ti X-rays taken
            at the right and left positions of the spectrometer.

decay at rest $\quad p_\mu = 29787 \pm 5$ kev/c,[22] one obtains

$$m_\nu^2 = 0.047 \pm 0.392 \text{ mev}^2/c^4.$$

$$m_\nu < 0.66 \text{ mev}/c^2 \text{ with a 90\% probability}$$

The revival of the application of the curved crystal spectrometer with this ingenious modification will certainly usher in a lively and significant period in the study of π X-rays in the immediate future. Its eventual application to the high flux k⁻ beams will be eagerly anticipated.

REFERENCES

1. R. Seki and C.E. Wiegand, *Ann. Rev. Nucl. Sci.*, 25, 241 (1975); L. Tauscher, Proceedings of the Conference on High-Energy Physics and Nuclear Structure, Sante Fe 1975, p. 541, AIP Conference Proceedings No. 26; G. Backenstoss, *Ann. Rev. Nucl. Sci.*, 20, 467 (1970).

2. C.E. Wiegand and D.A. Mack, *Phys. Rev. Lett.*, 18, 685 (1967).

3. D.E. Wilkinson, Proceedings of Rutherford Jubliee International Conference, Ed. J.B. Birks, Academic Press 1960; Proceedings of International Conference on Nuclear Structure, Tokyo, Japan 1967. Physical Society of Japan.

4. M.Y. Chen et al., *Nuclear Phys.*, A254, 413 (1975).

5. M. Leon, *Phys. Lett.*, B50, 425 (1974).

6. J.N. Bradbury et al., *Phys. Rev. Lett.*, 34, 303 (1975); *Phys. Rev. Lett.*, 34, 1064 (1975).

7. G.L. Godfrey et al., *Phys. Lett.*, 61B, 45 (1976).

8. J.N. Bradbury et al., Abstracts, FICAP Conference, Berkeley (1976) p. 440.

9. M. Ericson et al., *Phys. Rev. Lett.*, 22, 1189 (1969).

10. M. Leon, *Phys. Lett.*, 53B, 141 (1974).

11. P. Ebersold et al., _Phys. Lett._, 53B, 48 (1974).

12. S.C. Cheng et al., _Nuclear Phys._, A254, 381 (1975).

13. G. Backenstoss et al., _Phys. Lett._, 43B, 539 (1973).

14. G. Dugan, _Nuclear Phys._, A254, 396, (1975).

15. G. Bohm et al., _Nuclear Phys._, B48, 1 (1972).

16. P. Schmitt, _Phys. Rev._, 140, B1328 (1965).

17. E. Hu et al., _Nuclear Phys._, A254, 403 (1975).

18. Particle Data Group, _Rev. Mod. Phys._, 48, S21 (1976).

19. B.L. Roberts et al., _Phys. Rev. Lett._, 33, 1181 (1974).

20. V.I. Marushenenko et al., _J.E.T.P. Letter._ 23, 80 (1976).

21. R.E. Shafer, _Phys. Rev. D_ 8, 2313 (1973).

22. M. Daum et al., _Phys. Lett._, 60B, 380 (1976).

DETECTION OF π-μ COULOMB BOUND STATES[*]

R. Coombes, R. Flexer, A. Hall, R. Kennelly, J. Kirkby,
R. Piccioni, D. Porat, M. Schwartz, R. Spitzer,
J. Toraskar, S. Wiesner

Stanford University, Stanford, California

B. Budick and J. W. Kast

New York University, New York, N.Y.

Abstract

We have observed atoms consisting of a pion and a muon produced in the decay $K_L^O \rightarrow (\pi\mu)_{ATOM}\nu$. This represents the first observations of an atom composed of two unstable particles and of an atomic decay of an elementary particle.

- -

We report herewith the detection of hydrogen-like atoms consisting of a negative (or positive) pion and a positive (or negative) muon in a Coulomb bound state. These pi-mu atoms are formed when the π and μ from the decay $K_L^O \rightarrow \pi\mu\nu$ have sufficiently small relative momentum to bind. We have observed these atoms, produced at relativistic velocities, in the course of an experimental program at the Brookhaven A.G.S.

The basic properties of these atoms are calculable by the formalism used to describe the hydrogen atom. The reduced mass of the system is 60.2 MeV/c^2, its Bohr radius is 4.5×10^{-11} cm and the binding energy of the $1S_{\frac{1}{2}}$ state is 1.6 KeV. To our knowledge, the first calculation of the branching ratio

$$R = \frac{K_L^O \rightarrow (\pi\mu)_{Atom} + \nu}{K_L^O \rightarrow all}$$

was carried out by Nemenov[1], who found that $R \sim 10^{-7}$, with the precise value depending upon the form factors of K_L^O decay. We will present our results on R in a subsequent paper; only the evidence related to the detection of these atoms is discussed herein.

---

[*] Supported by the National Science Foundation

The prime motivations for the experiment are two fold. Firstly, the value R is proportional to the square of the π-μ wave function at very small distances and so an anomaly in its value may be indicative of an anomaly in the π-μ interaction. Secondly, by passing the atoms through a magnetic field at high velocity (γ~10) the 2S states should be depopulated through stark mixing with the 2P states and consequent decay to the 1S states. The extent of this depopulation will be highly dependent upon the vacuum polarization shift (Lamb shift) of the 2S states relative to the 2P states and may, if measured with some accuracy lead to a determination of the pion charge radius.

The K_L^O particles which give rise to our "atomic beam" are produced by a 30 GeV proton beam striking a 10 cm Beryllium target (see Figure 1). A large vacuum tank and a connecting evacuated beam channel lead out to the detection equipment. A 4 ft steel collimator prevents any direct line of sight from the detector system to the target. This is to prevent background particles, in particular K_L^O's, from approaching the neighborhood of our detectors.

Those K_L^O's which decay within the shaded area in the vacuum tank give rise to decay products which may, if properly oriented in their direction of motion, travel down the channel. In order to remove charged particles, we have interposed two magnets along this channel. The first of these, labeled the "sweeping magnet" bends horizontally and has an integrated field strength of 8 kilogauss meters. The second magnet, (originally intended to induce transitions between the 2S and 1S states of these atoms) is called the "transition magnet" and bends vertically with an integrated field strength of 36 kilogauss-meters. Those charged particles which survive have very high momenta or are given a significant deflection before entering the detector region.

We have then a beam consisting largely of γ rays (resulting from π^O's which are in turn the products of kaon decays), highly energetic pions and muons, and occasional atoms. The momentum spectrum of the atoms coming down the channel has no appreciable contribution above about 5 GeV/c.

To dissociate the atoms and make their detection possible, we interpose a thin aluminum foil just before the end of the vacuum channel (see Fig. 2). Ionization of an atom takes place through a series of sequential transitions through the states having highest angular momentum for any given principal quantum number. We have calculated the thickness of foil required to break up a 1S atom to be .010" of Aluminum. In the course of the experiment data was taken with foil thickness of 0.030" and 0.250" of aluminum.

The pion and the muon, now uncoupled, exit the foil at the same velocity (with momenta in the ratio of their rest masses) and in

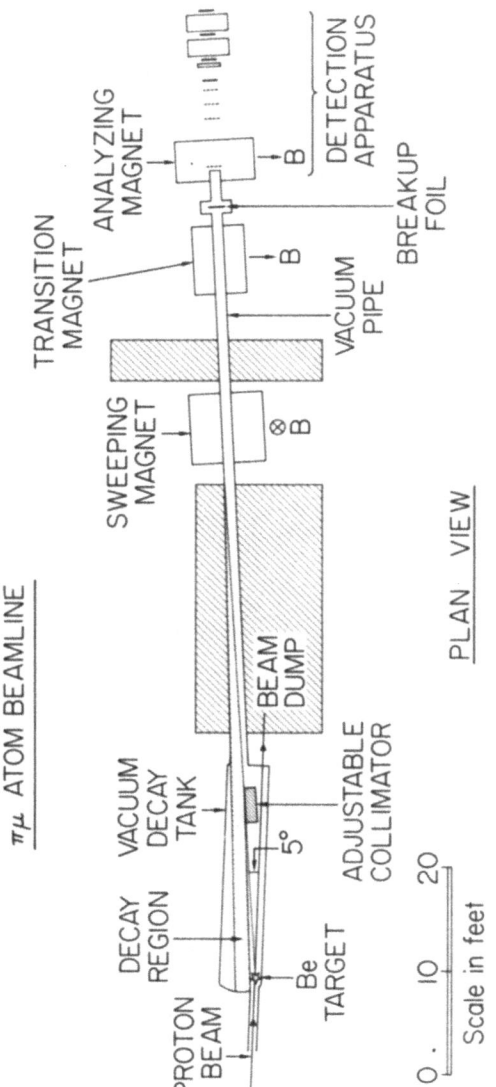

Fig. 1   Experimental Arrangement at the A.G.S.

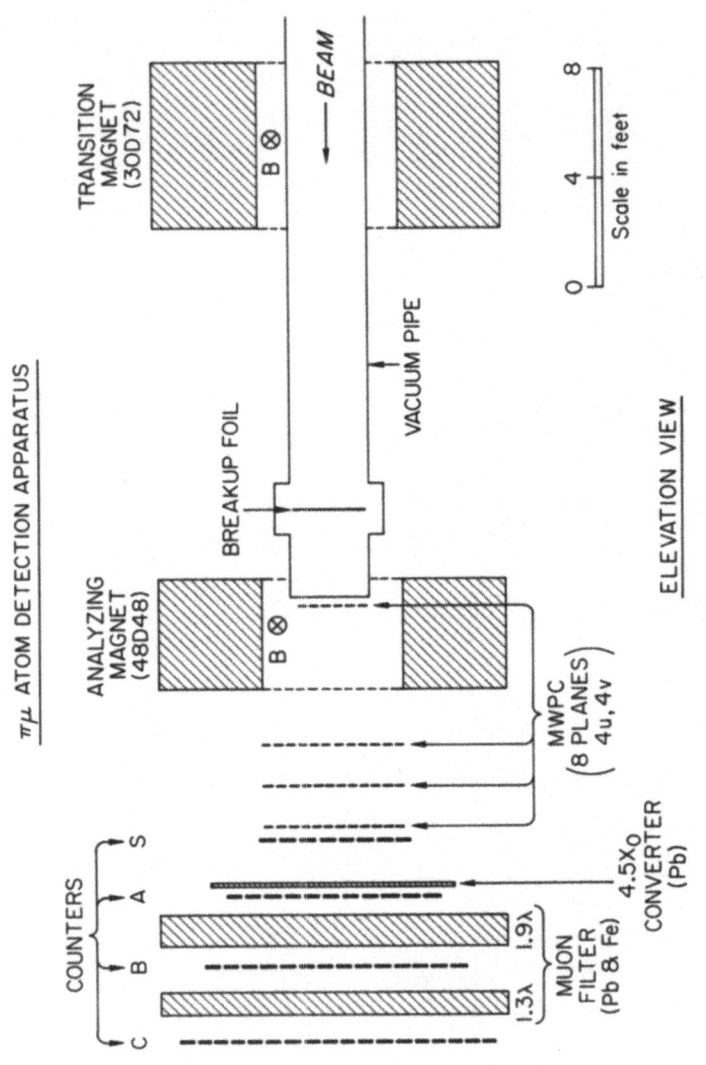

Fig. 2   Detection Apparatus

almost perfect spatial coincidence. The opening angle between them
at a typical atomic momentum of 3 GeV/c, neglecting the multiple
scattering in the foil, should be less than 0.5 milliradians. The
projected multiple scattering of each particle in a .030" Aluminum
foil is about (1.3/p (GeV)) milliradians. Thus the angle between
pion and muon upon emerging from a 0.030" foil should be about two
milliradians. The angle between them in the case of a 0.250" Aluminum foil is about 5 milliradians.

We next introduce these two coincident particles into a horizontal field which serves to separate them vertically. We terminate
the vacuum channel with a thin mylar window where the separation between the pion and muon is about a centimeter for a typical atom.
Just beyond the window we place a multiwire proportional chamber
made of two planes (planes 1 and 2) to allow the reconstruction of
the vertical (x) and horizontal (y) coordinate of each of the particles. Each of these planes is constructed of a set of wires inclined at 60° to the vertical. At the point where the pion and the
muon traverse these planes they are directly above one another and
separated by a vertical distance Δ which is closely correlated to
the sum of their momenta.

After leaving the analyzing magnet the pion and the muon continue
through a series of three further pairs of proportional chambers,
each constructed of wires at ±60° to the vertical. In each of these
planes the x and y coordinates of each track can be localized to
about ± 1mm. Following the last of these chambers, we have, in sequence: a bank of 11 counters (S bank), a sheet of 1" thick lead to
induce showering of electrons, a bank of 15 counters (A bank), a
lead and steel wall embodying 1.9 mean free paths of absorber, another
bank of 19 counters (B bank), a wall comprising 1.3 free paths of
absorber and a final bank of 23 counters (C bank). The absorber re-
moves muons below a momentum of 0.9 GeV/c and about 90% of the pions.
The first crude indication that an event of interest has passed
through the detector comes when we obtain a trigger indicating simultaneous counts in two S counters, two non-adjacent A counters, one
or more B counters and one or more C counters. We next examine planes
1 and 2 to determine rapidly whether two tracks passed directly above
one another within the experimental resolution and with Δ lying between .8 and 3.5 centimeters. We then remove, through the use of our
on-line computer, all events in which more than four tracks passed
through the first plane. The residual events are logged for further
study. The information recorded includes the timing of all counters,
the pulse height on each of the A counters and the positions of the
tracks as they pass through the eight planes.

We carry forth the analysis of the data by subjecting each event
to a sequence of tests, each of which must be passed before it can
be considered a valid candidate for a π-μ atom. The geometrical

characteristics of these tests have been determined through a study of the e^+-e^- pairs which are created by γ rays impinging on the foil and the muons which come down the vacuum channel when the sweeping and transition magnets are turned off. The tests are as follows:

1.   All counters involved in a trigger must be time coincident within ±2 nanoseconds after correction for flight times of the various particles.

2.   The four counters which define the muon track must lie on a straight line within the limits of Coulomb scattering in the absorber. Only one track may penetrate to the C bank.

3.   The pulse height on each of the A counters must be less than 2.5 times that produced by a minimum ionizing particle.

4.   Each of the tracks must have a momentum not less than 0.9 GeV/c.

5.   After the two tracks are reconstructed back through the magnet, we can determine the x and y projections of their apparent separation and the apparent angle between them as they left the foil.

The cuts are as follows:

a)   The vertical separation at the foil must be less than 0.45".

b)   The horizontal separation at the foil must be less than 0.20".

c)   The measured vertical angle between the two tracks as they leave the foil must be less than 0.025 radians.

d)   The measured horizontal angle between the two tracks as they leave the foil must be less than 0.004 radians.

6.   Our study of the e^+-e^- pairs indicates that the vertical spacing, Δ, between the two tracks in planes 1 and 2 is predictable to a wire spacing given the momenta of the two particles. We reject all candidates which do not conform to this constraint within ± 2 wire spacings.

7.   By studying the e^+-e^- pairs we have ascertained that we can project our tracks back to the vicinity of the collimator with a horizontal spatial resolution of ± 1.0". We insist then that all of our tracks of interest point back to 9" wide fiducial region near the collimator, missing both the collimator itself and the walls of the vacuum channel.

8.   Finally, we insist that the sum of the pion and muon momenta be no more than 5 GeV/c.

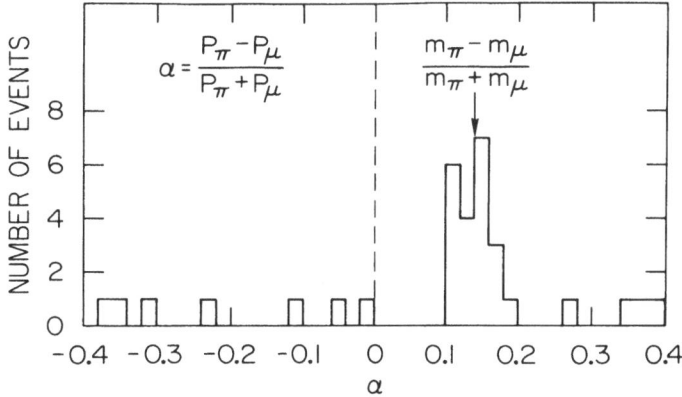

Fig. 3. A Plot of the Parameter α Indicating the Detection of π-μ
         Atoms.

     Having subjected all of the recorded data to these tests, we
arrive at a residue of 33 events. For each of these events we plot
(in Fig. 3) the parameter $\alpha = P_\pi - P_\mu / P_\pi + P_\mu$ where P_π is the pion momentum and P_μ is the muon momentum. A study of this parameter through
an examination of e^+-e^- pairs indicates that the acceptance of our
apparatus, modified by the abovementioned tests, is flat within 30%
from $\alpha = -0.4$ to $\alpha = +0.4$. None of our acceptance tests bias us
toward one or another sign of α. Hence, any bump in this plot would
indicate a strong correlation between pion and muon momenta; in
particular the atoms should be characterized by a value of
$\alpha = m_\pi - m_\mu / m_\pi + m_\mu = 0.14$. The data shows a clear peak at the predicted
point containing a total of 21 events with an estimated background
of 3 events. The width of the peak is consistent with that expected
from measurement errors.

     We conclude that we have observed Coulomb bound states of pions
and muons.

     We would like to express our appreciation to the Brookhaven
A.G.S. staff and to L. Birkwood, J. Bjorken, S. Drell, G. Donaldson,
M. Faessler, M.K. Gaillard, D. Hitlin, B. Kincaid, D. Ouimette,
M. Prasad, C. Rasmussen, A.M. Rushton, J. Tannis, A. Tilghman,
J.D. Walecka, S. Wojcicki and C. Zupancic for valuable assistance.

References

1.   L.L. Nemenov, Sov. Journ. of Nuc. Phys. 16, 67 (1973).

REVIEW OF PRECISION POSITRONIUM EXPERIMENTS

Allen P. Mills, Jr.,[*] Stephan Berko[+] and Karl F. Canter[+]

[*] Bell Laboratories, Murray Hill, N. J. 07974
[+] Physics Department, Brandeis University,
Waltham, Mass. 02154[**]

INTRODUCTION

In the years since its discovery by Martin Deutsch, [1] positronium has been studied by many workers not only because of the intrinsic interest generated by this exotic atom containing antimatter, but also because in some ways positronium being the lightest of all atoms is also in principle the simplest to interpret theoretically. The measurement of the structure of simple atoms provides one of the best opportunities for testing the predictions of quantum electrodynamics, the theory which most accurately accounts for the interactions of electrically charged particles and electromagnetic radiation. For example, atomic hydrogen is the simplest nuclear atom, and its ground state hyperfine splitting is one of the most accurately measured quantities [2]. It is unfortunate, however, that theoretical uncertainties stemming from the finite size, mass, and polarizability of the proton have so far precluded a meaningful comparison with the measurement to better than a few parts per million. In an attempt to be free of such nuclear effects, experimenters have persued the study of the non-nuclear atom positronium which is describable to a very good approximation using only the electron, positron, and photon fields. Compared to the stable atoms, however, positronium presents challenging experimental difficulties attributable in part to its short annihilation decay times and also to the impossibility of producing macroscopic quantities of positronium for observation.

Since there are a number of reviews covering the past work on positronium [3,4], the present discussion will concentrate principally on the recent developments, with particular emphasis on the

discoveries which have allowed measurements to be made on the excited
states of positronium for the first time. An instructive review by
M. A. Stroscio [5] on the theory of positronium still appears to be
up to date as of this writing. An experimentally oriented review
is to be found in the δ Volume (4) of <u>Adventures in Experimental
Physics</u> edited by Bogdan Maglich (1974) [6].

Barring discussion of atomic collision processes, the quantities
of interest in the study of positronium include the annihilation
decay rates, g factors, total mass and binding energy for each state
of positronium; the energy level separations and partial widths for
each pair of levels; and the possible mixing of these levels by
externally applied electric and magnetic fields. Presently measured
with some precision are the annihilation rates and energy splitting
of the ground state which will be discussed in the next section.
The last section will concern the first excited state of positronium
and the measurement of a fine structure splitting in that state.

THE POSITRONIUM GROUND STATE

Formation of Positronium

In 1951 Martin Deutsch discovered that positronium in its
lowest energy state (binding energy $\frac{1}{2}$ Rydberg, orbital angular
momentum zero) is formed when energetic positrons from a radioactive source stop in a gas [1]. Positronium was also later shown
to be formed in liquids [7] and in some insulating amorphous [8]
and crystaline solids [9]. The positronium inside such dense
environments is grossly changed by the interactions with other
atoms, and we shall not discuss it further here. Paulin and
Ambrosino [10] discovered that positronium could also be formed
by stopping positrons in insulating powders of MgO, Al_2O_3, and
SiO_2. The positronium spends most of its time in the voids
between the powder grains. In 1971 Curry and Schawlow [11] found
that they could make some positronium in vacuum by scattering
high energy positrons off a foil coated with MgO powder. Finally,
positronium has been shown to be formed with nearly 100% yield by
letting positrons of ~10 eV energy impinge upon a clean solid
target in vacuum [12]. Although positronium quickly reaches
thermal energies in polyatomic gases and in solids, it may remain
significantly non thermal for much of its lifetime in the noble
gases and in powders. The energy of the positronium formed by
MgO coated foils was measured to be 0.3 eV [11], but the kinetic
energy of positronium emerging from a clean solid target is as yet
unknown.

The Hyperfine Interval

Since the electron and positron both have spin $\frac{1}{2}$, the lowest
state with principle quantum number n=1 is split into a singlet
state and a triplet state. The singlet state lies below the triplet
state by about α^2Ry owing to the interaction of the magnetic moments
of the two particles, the virtual single quantum annihilation of
the pair in the triplet state, and various radiative corrections.
The current theoretical value of this energy difference is [13]

$$\Delta\nu_{th} = \frac{\alpha^4 mc^2}{2h} [\frac{7}{6} - (\frac{16}{9} + \ln2)\frac{\alpha}{\pi} + \frac{1}{2}\alpha^2\ln\alpha^{-1} + \mathcal{O}(\alpha^2)] \tag{1}$$

$$= 203.4040 \pm .0003 \text{ GHz} \quad .$$

While the 1.5 ppm uncertainty of this value stems principally from
the error in the Sommerfeld fine structure constant $\alpha = e^2/\hbar c =$
1/137.03604 ± .7 ppm [14], the neglected terms in the expansion may
well represent a much larger uncertainty in $\Delta\nu_{th}$ since $\alpha^2\Delta\nu =$
0.0108 GHz. Although superficially only a "light isotope" of
hydrogen, positronium differs substantially in several details from
hydrogen: a) given the equal masses of its constituents, positronium
cannot be handled theoretically to any detailed degree by the Dirac
equation, but requires the use of the relativistic two-body equation
(Bethe-Salpeter, Gell-Mann, Low); b) because of the large positron
vs. proton magnetic moment, the distinction between the easily
separable fine vs. hyper-fine structure disappears, the spin-orbit
vs. the spin-spin interaction being of the same order of magnitude;
c) due to the particle-antiparticle system, virtual annihilation
diagrams become important in the positronium fine structure
splitting, already to order Ryα^2.

All precision measurements of $\Delta\nu$ to date have followed the
pioneering experiment of Martin Deutsch and S. C. Brown in 1952
[15-22]. These experiments are all based on the annihilation
selection rules requiring spin-triplet S states to annihilate into
3γ's, spin-singlet states into 2γ's. Positronium is formed by
stopping positrons from a radioactive source in a buffer gas which
is contained in a microwave cavity and located in a uniform magnetic
induction B. The triplet ground state is split into two Zeeman
sublevels by the magnetic induction, and transitions between the
m=0 and the m=±1 states may be induced by the rf magnetic field
perpendicular to B. The Breit-Rabi formula gives the resonant
value of the rf frequency f_{01} in terms of B and $\Delta\nu$:

$$f_{01} = \frac{1}{2}\Delta\nu[(1 + x^2)^{\frac{1}{2}} - 1] \tag{2}$$

where $x = 2\mu_B g' B/h\Delta\nu$ and g' is the bound state electron g factor in
positronium [23]. If the nuclear magnetic resonance of the protons
in a spherical sample of water (proton magnetic moment μ_p') is used

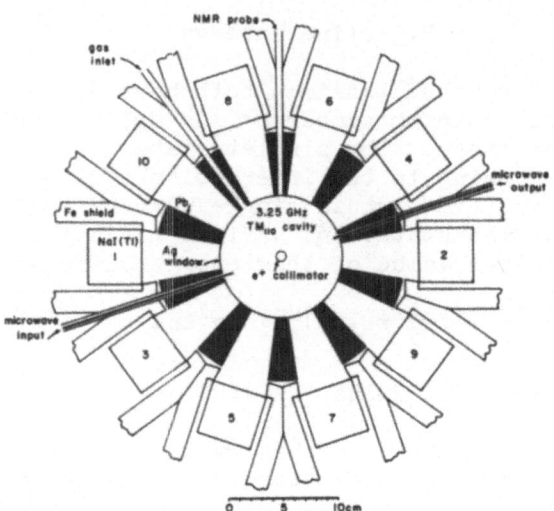

Figure 1., Arrangement of ten NaI(Tl) counters about a microwave
cavity used to detect the positronium Zeeman resonance. The
counters allow five 2γ configurations at 180° and twenty 3γ
configurations. The magnetic field at the center of the cavity is
measured using the removable NMR probe.

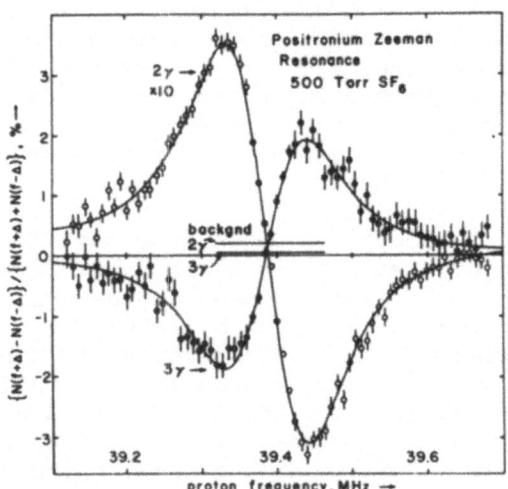

Figure 2. Positronium Zeeman resonance signals obtained using a
500 Torr SF$_6$ buffer gas. Each data point is obtained from a pair
of counting rates measured at proton frequencies f ± Δ, where Δ
is 36 kHz, the signal being the difference divided by the sum.
The fitted curves are first difference Lorentzians with constant
backgrounds. The magnetic field is measured using proton nuclear
magnetic resonance in an oil sample: 39.4 MHz corresponds to
9.2540 kG. (From Ref. 21.)

to measure B we may write

$$\Delta\nu = (g'/g)^2(\mu_e/\mu_p')^2(f_p^2/f_{01}) - f_{01} \tag{3}$$

where f_p is the measured proton resonance frequency in the water, and we use $g'/g = 1 - 11.1\text{x}10^{-6}$ and $(\mu_e/\mu_p')^2 = 433263.56 \pm .06$ [2]. Because of the increased annihilation rate of the mixed m=0 state in a magnetic field, the resonance condition of B and f_{01} causes an increase in the 2γ annihilation yield and a concomitant decrease in the 3γ yield. Since rather large rf fields are required, the experiments use a tuned microwave cavity; and since it is difficult to change the tuning while holding the rf power level constant, a resonance curve is obtained by holding the rf frequency fixed and observing the changes in the annihilation yields as a function of B. The value of f_p at the peak of such a resonance is substituted into Eq. 3 to obtain an experimental value for $\Delta\nu$.

The natural line width (full width at half maximum) of the Zeeman transition is given approximately by [9]

$$\Delta B/B = \Gamma(1^1S_0 \rightarrow 2\gamma)/4\pi\Delta\nu = 0.312\% \tag{4}$$

for small B, where $\Gamma(1^1S_0 \rightarrow 2\gamma)$ is the annihilation rate of the singlet ground state. To achieve a $\Delta\nu$ precision of 5 ppm the Zeeman line center must be measured to 2.5 ppm or about 10^{-3} of the line width. Given a signal with an amplitude S on the order of S=1% thus dictates that enormous amounts of data be collected, the total number of counts required being $N \simeq (10^{-3})^2 S^{-2} \simeq 10^{10}$. The most recent experiments to measure $\Delta\nu$ used multiple counter systems (see Fig. 1) to obtain the necessary high counting rate [21,22]. Typical Zeeman resonances (essentially the first derivatives) are shown in Fig. 2.

Besides obtaining the large amounts of data needed, it is also necessary to make an extrapolation to zero gas density $\Delta\nu(0)$ (See Fig. 3) to remove the effects of the pressure shift first observed by Theriot, Beers and Hughes [18]. Measurements [22] and calculations [24] of the pressure shifts in the rare gases have recently been reported. The experimental $\Delta\nu(0)$ results are shown in Table I. The discrepancy between these values and the $\Delta\nu_{th}$ of Eq. 1 indicates the importance of calculating all the $\mathcal{O}(\alpha^2)$ corrections. The present trend towards higher precision measurements seems to yield a factor of ten increase in accuracy per ten years, and it is desirable that further improvements be made. An eventual measurement of $\Delta\nu$ at the 1 ppm level would present a formidable challenge to theorists, but would in principle provide one of the best determinations of the fine structure constant and a stringent test of QED.

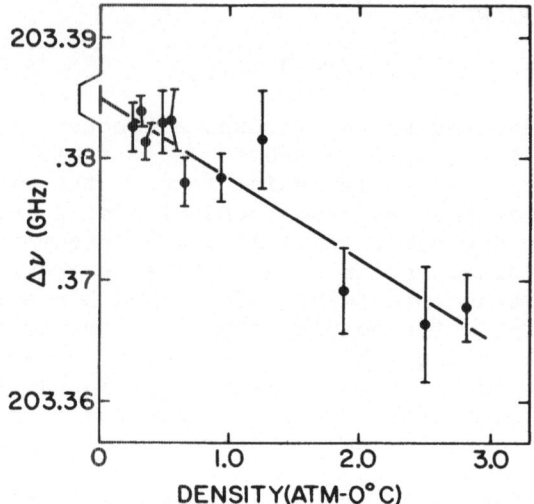

Figure 3. Plot of Δν vs. N_2 gas density. The solid line is the best fit to a linear density dependence and the bar at zero density is the extrapolated value for Δν(0). (From Ref. 22.)

The Annihilation Rates

The predominant annihilation mode for singlet positronium is via 2γ's; 3γ annihilation is prohibited only by the invariance of the electromagnetic interaction under charge conjugation, C. The 2γ rate is calculated to be [25]

$$\Gamma(1^1S_0 \rightarrow 2\gamma) = \frac{\alpha^5 mc^2}{2\hbar}[1-\alpha(5/\pi-\pi/4)] = 7.9854 \times 10^9 \, \text{sec}^{-1} \qquad (5)$$

This rate has been determined by Theriot, Beers and Hughes [19] by extrapolating their Zeeman resonance line width to zero power to obtain a value for the natural line width. They find

$$\Gamma(1^1S_0 \rightarrow 2\gamma) = (7.99 \pm .11) \times 10^9 \text{sec}^{-1} \quad .$$

The forbidden decay $1^1S_0 \rightarrow 3\gamma$ occurs in less than 10^{-5} of the annihilations [26]. Known models of such a decay based on a C nonconserving interaction yield predictions proportional to an unknown coupling constant λ divided by a cut-off mass M raised to a high power. Using λ=1 and M=$2m_e$, the positronium mass, leads to

Authors	Ref.	Date	$\Delta\nu$(GHz)
Deutsch and Brown	[15]	1952	203.2(3)
Weinstein, Deutsch, and Brown	[16]	1954	203.35(5)
Weinstein, Deutsch, and Brown	[16]	1955	203.38(4)
Hughes, Marder, and Wu	[17]	1957	203.33(4)
Theriot, Beers, and Hughes	[18]	1967	203.403(12)
Carlson, Hughes, Lewis, and Lindgren	[20]	1972	203.396(5)
Mills and Bearman	[21]	1975	203.3870(16)
Egan, Frieze, Hughes, and Yam	[22]	1975	203.3849(12)
Theory	[13]		203.404

Table I. Experimental determinations of the positronium hyperfine interval $\Delta\nu$.

predicted rates $\Gamma(1^1S_0 \to 3\gamma) \approx 10^{-6}\Gamma(1^1S_0 \to 2\gamma)$. More believable predictions of $^1S_0 \to 3\gamma$ rates based on virtual transitions to K^0-K^{-0} states which do violate time reversal invariance [27] to a small extent must yield astronomically small rates because of the light mass of positronium compared to the K^0 mesons. Even smaller rates are expected from weak interaction models because of the enormous mass of the hypothetical W vector boson. The rate for the allowed decay into 4γ has been estimated to be $\Gamma(1^1S_0 \to 4\gamma) = 0.35 \times 10^{-6}\Gamma(1^1S_0 \to 2\gamma)$ [28], but so far no one has looked for this decay mode.

Triplet positronium annihilates into 3γ's principally, annihilation into 1γ and 2γ being prohibited by conservation of 4-momentum and angular momentum respectively. The calculated 3γ rate [5] shown in Table II is

$$\Gamma(1^3S_1 \to 3\gamma) = \frac{2}{9\pi} \frac{\alpha^6 mc^2}{\hbar}(\pi^2 - 9)[1 + \frac{\alpha}{\pi}(1.86 \pm .45)] \qquad (6)$$

$$= (7.242 \pm .008) \times 10^6 \text{sec}^{-1}$$

The error bar on this value arises from the difficulty in integration of some of the diagrams involved in the radiative corrections. The rate has been measured in gases by Beers and Hughes [29] and by Coleman and Griffith [30] as shown in Table II. The small discrepancy may reflect the combined error from the calculation plus possible unknown systematic errors associated with

3γ Annihilation Rate ($10^6 sec^{-1}$)		Ref.
7.242 ± 0.008	Theory	[5]
7.275 ± 0.015	Expt in gas	[29]
7.262 ± 0.015	Expt in gas	[30]
7.104 ± 0.006	Expt in SiO_2 powder	[31]

Table II. Theoretical and experimental determinations of the annihilation rate of triplet positronium. The experimental values quoted are zero density extrapolations.

the required large extrapolation to zero gas pressure shown in Fig. 4.

In an attempt to avoid this large extrapolation, Gidley, Marko, and Rich [31] measured $\Gamma(1^3S_1 \rightarrow 3\gamma)$ for the positronium formed in SiO_2 powder. Using powders with estimated grain sizes of 35A° and 70A° they extrapolate to zero powder density (see Fig. 5) to obtain the surprising result listed in Table II, a result which is nearly 2% less than the vacuum decay rate of Eq. 6 and the other experimental values. This result is unusual because most positronium interactions with matter involve some overlap with electrons from other atoms which ordinarily would lead to an increased annihilation rate. Ford, Sander and Witten [32] have considered positronium van der Waals interactions with the grain surfaces and Stark shifts due to the intergrain electric fields, but they consider unlikely the presence of electric fields strong enough ($5 \times 10^7 V/cm$) to explain a reduced triplet annihilation rate of the size observed. However, such electric fields might possibly arise from the ionization trail formed by the stopping positron or from the grain to grain charge differences acquired in the presence of the radioactive source over a period of time. Further experimentation to investigate these effects is required.

The triplet state is forbidden to decay into 4γ and this annihilation mode has been searched for by Rich and Marko [33]. They find $\Gamma(^3S_1 \rightarrow 4\gamma) < 10^{-5}\Gamma(^3S_1 \rightarrow 3\gamma)$ which is consistent with C conservation. A decay $^3S_1 \rightarrow 4\gamma$ might arise by a C violating emission of 3γ's followed by the emission of one extra photon at an ordinary C conserving vertex; or such a decay could be described by a point interaction [34]. As for the decay $^1S_0 \rightarrow 3\gamma$, the light mass of positronium makes an upper limit on $^3S_1 \rightarrow 4\gamma$ difficult to interpret in terms of known possibilities.

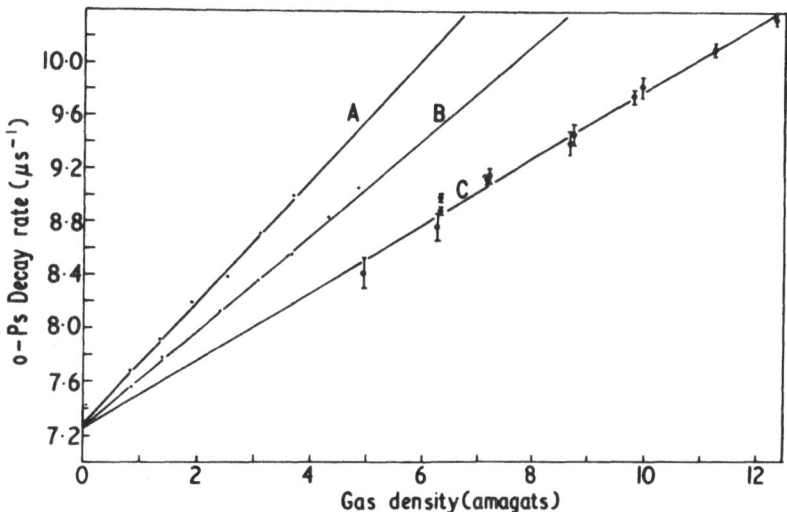

Figure 4. Decay rates λ of orthopositronium in freon and argon as
a function of gas density (A) freon-12, (B) arcton-12 and (C) argon.
(From Ref. 30.)

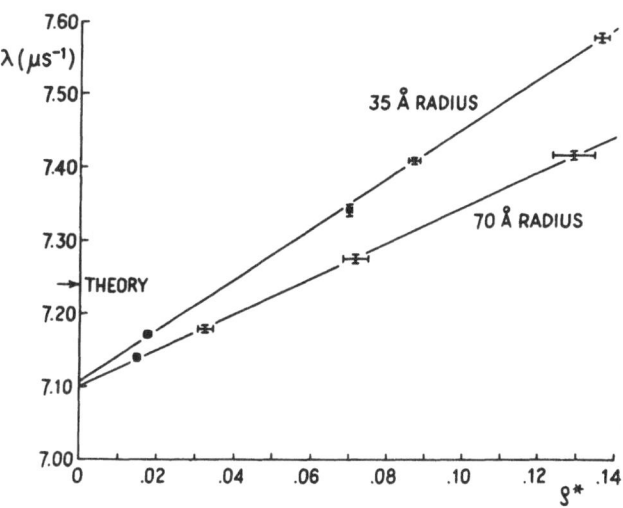

Figure 5. Decay rate λ of orthopositronium in degassed SiO_2 powder
of two different particle sizes as a function of powder density ρ,
where $\rho = \rho_{solid} \cdot \rho^*/(1+\rho^*)$ and $\rho_{solid} = 2.20$ gm cm^{-3}. (From Ref. 31.)

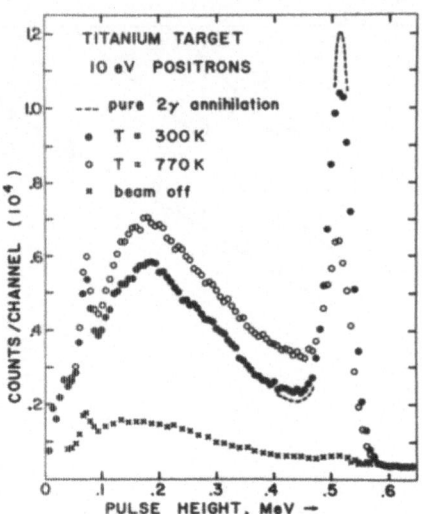

Figure 6. NaI(Tl) pulse height spectra of annihilation γ's for
10 eV positrons incident on a Ti target in vacuum. The "pure 2γ"
spectrum corresponds to no positronium formation, the 300K spectrum
to ∿25% positronium formation, and the 770K spectrum ("clean"
target) to ∿85% positronium formation. (From Ref. 12.)

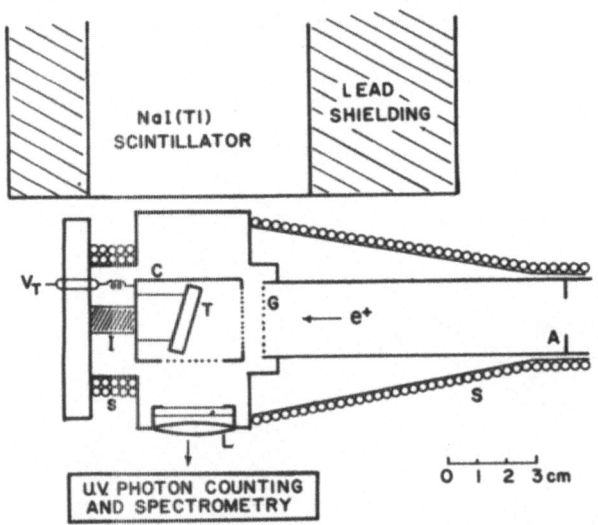

Figure 7. Target chamber for observing the annihilation of slow
positrons (e⁺) incident upon a solid target, T. The positron
beam is guided by the solenoid, S, through the aperture, A, and
grid, G, into the target cage, C. T can be heated to 900K. The
quartz window and lense, L, is used for observing Lyman α radiation
from n=2 positronium. (From Ref. 12.).

THE POSITRONIUM EXCITED STATES

Production of n=2 Positronium

Soon after the discovery of positronium, Deutsch and Kendall [35] attempted to form positronium in its first excited state (principle quantum number n=2) in a gas by optical excitation using a very intense light source with a wavelength near 2430A°, the predicted wavelength of the positronium Lyman α photon from the 2P→1S transition. Other workers have recently reported experiments to excite ground state positronium to the n=2 state using the method of Deutsch and Kendall [36,37]. However, owing to the dense gas and high magnetic fields used, it would be difficult to make measurements on any excited state positronium formed in this way, especially since it would be formed in the short lived P states. Many attempts to observe the Lyman α radiation from n=2 positronium which might occur when positronium is formed in a gas have been unsuccessful [38].

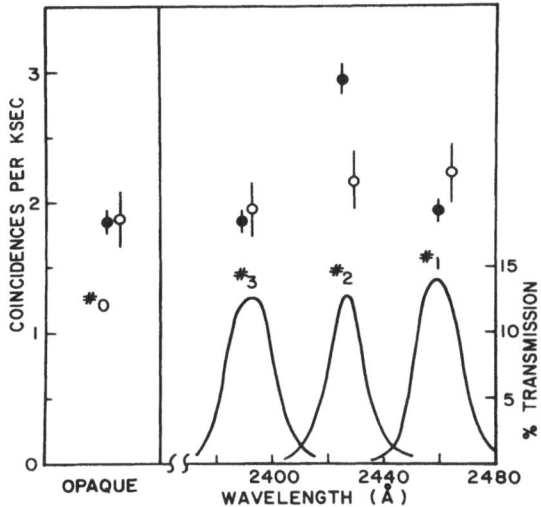

Figure 8. UV photon-annihilation γ coincidence counting rate using different interference filters (#1, #2, #3) in front of the UV photomultiplier. Using the #2 filter which spans the predicted 2430A° Lyman α line a significant increase in the count rate of $(1.04 \pm .13) \times 10^{-3} sec^{-1}$ is observed for 25 eV positrons (solid circles). The data with 400 eV incident energy positrons (open circles) shows a rate which is independent of the filter used. (From Ref. 39.)

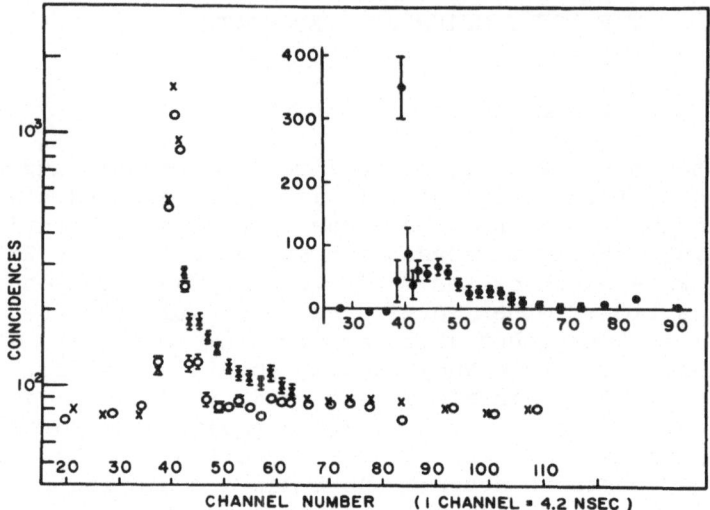

Figure 9. Time spectra of the coincidences between the UV photon detector (start) and the NaI(Tl) γ detectors (stop) as obtained through the #2 (See Fig. 8) "Lyman α" filter (crosses). The inset (solid circles) shows the difference between the #2 spectrum and the average off-line filter (#0, #1, #3) spectrum (open circles). The delayed component difference is about three times greater in intensity than the prompt component difference, in agreement with the 3 to 1 relative abundance of triplet vs. singlet states expected. (From Ref. 39.)

        The first successful observation of the positronium Lyman α radiation was based on recent experiments that showed that n=1 positronium is formed with high probability [12] (see Fig. 6) when low energy (∼10 eV) positrons strike a clean solid target in vacuum far from the radioactive source and moderator which produces the low energy positron beam [39] (see Fig. 7). The positronium so formed has a long lifetime which is affected by the walls surrounding the target. It is concluded therefore that the positronium emerges into the vacuum after formation at the surface. A search for uv photon emission (Fig. 8) preceding the annihilation of the positronium revealed [40] that a small fraction (∼10^{-3}) of the positronium was being formed in the 2P states and emitting light with a wavelength of (2430 ± 30)A°. The relative abundance of triplet and singlet 2P states was found to be about 3 to 1 as expected (Fig. 9). Based on the reasonable assumption that 2S states were also present, an experiment to induce 2S to 2P transitions was set up. The successful completion of this experiment established conclusively the formation of positronium in the 2S and 2P states and resulted in the first

measurement of a fine structure interval in an excited state of
positronium [40].

Measurement of Fine Structure

The importance of the experimental determination of the n=2
Ps fine structure, which includes Lamb shift-like terms, is well
known [42]. In constrast to hydrogen, the positronium 2S and 2P
levels are nondegenerate already in order α^2Ry. The fine structure
of these levels has been computed through order α^3Ry by Fulton and
Martin [43], using the Bethe-Salpeter equation.

An energy level diagram of the n=1 and n=2 states of
positronium is shown in Fig. 10. Of the 15 possible transitions
between the six n=2 energy levels in zero field, the three electric
dipole transitions of the long lived triplet states are the easiest
to detect. These three fine structure intervals are predicted to
lie in the common microwave bands ($2^3S_1 \rightarrow 2^3P_2$, 8625 MHz; $2^3S_1 \rightarrow 2^3P_1$,
13010 MHz; $2^3S_1 \rightarrow 2^3P_0$, 18496 MHz). From the several accessible
fine structure intervals the $2^3S_1 - 2^3P_2$ splitting was chosen for
study, since it was predicted to lie in the X band and it involves
the long lived 2^3S_1 state (10^{-6} sec. for 3γ annihilations – see
term diagram insert of Fig. 2). Fulton and Martin obtain theoretically:

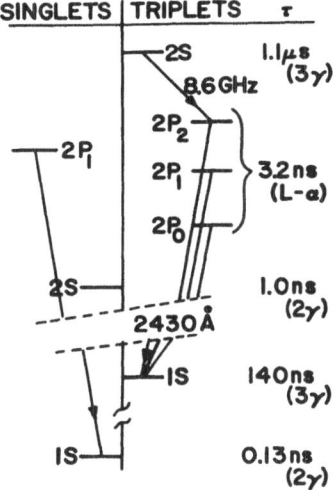

Figure 10. Term diagram of the n=1 and n=2 states of positronium.

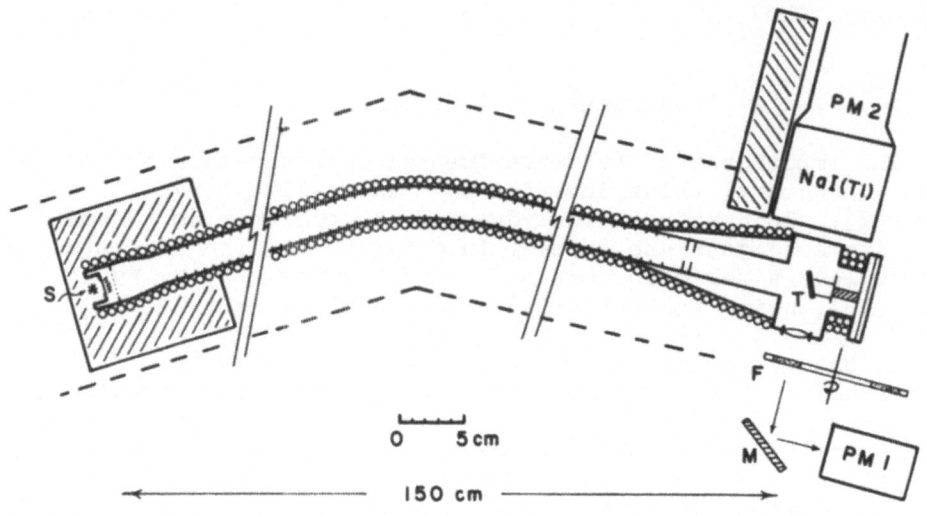

Figure 11. Slow positron beam apparatus: S, ^{58}Co source;
T, target; F, optical filter wheel; M, mirror; PM1, UV photon
counter. (From Ref. 39.)

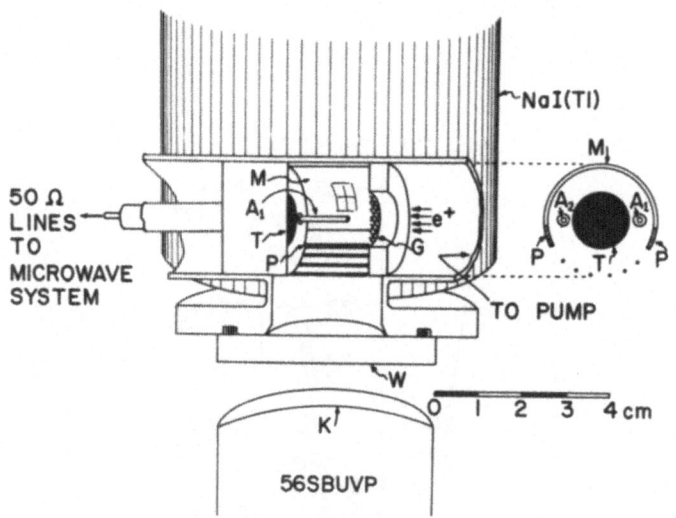

Figure 12. Positron target chamber and microwave cavity.
G, 95%-transmission tungsten grid; T, copper target; M, aluminized
Suprasil quartz mirror; W, Suprasil quartz window; K, CsTe
photocathode; P, support posts; A_1, input antenna; A_2, output
antenna; NaI(Tl), annihilation detector. (From Ref. 40.)

$$\Delta\nu(2^3S_1 - 2^3P_2) = \frac{23}{480}\alpha^2 Ry \{1 + 3.766\alpha\} = 8625.14 \text{ MHz} \quad , \quad (7)$$

and estimate that the neglected terms of order α^4 Ry may amount to several MHz. Present uncertainty in α produces only a 1.5 ppm uncertainty in $\Delta\nu$.

The 2^3S_1 state has a 3γ annihilation rate Γ_s 1/8 of the 1^3S_1 rate to lowest order and should have an extremely small rate for optical transitions to the ground state. The 2P states on the other hand have little probability for annihilation [5] and a rate Γ_p for spontaneous transitions to the 1S ground state by Lyman α emission of

$$\Gamma_p = \Gamma(2P \to 1S) = (\tfrac{2}{3})^8 \; \Gamma(1^1S_0 \to 2\gamma) = (3.18 \text{ nsec})^{-1}$$

to lowest order. The 2^3S_1 state is thus metastable compared to the 2P states. Applying microwave electric fields at the appropriate frequency should cause transitions $2^3S_1 \to 2^3P$ followed by Lyman α emission and finally annihilation in the ground state.

The measurement is thus based on the observation of an enhanced Lyman α emission rate when $2^3S_1 \to 2^3P_2$ transitions are induced by an rf electric field at the proper frequency. The experiment was made possible by the production of Ps in vacuum, using a low energy positron beam, at essentially zero field ($B \approx 50g$), thus curcumventing the problems associated with gas collisions and minimizing Zeeman splitting and motional Stark shifts [35].

A beam of slow positrons (≈ 30 eV) is produced in an apparatus shown in Fig. 11 used in connection with the Lyman α experiment [40], and is guided into a cylindrical rf cavity as shown in Fig. 12. One side of the cavity has been replaced by parallel wires to allow the Lyman α photons (2430A$^\circ$) to be detected by a UV sensitive solar blind photomultiplier. The slow positrons collide with the Cu end face of the cavity and form Ps atoms leaving the surface. Approximately 10^{-3} to 10^{-4} of the incident positrons form Ps in the n=2 state leading to the 2P $\to$ 1S 2430A$^\circ$ photon emission, with the subsequent annihilation of the ground state ($1^1S_0 \to 2\gamma$, $1^3S_1 \to 3\gamma$). Using two NaI(T) detectors place above and below the target chamber to count the annihilations γ's, the time delay spectrum between the Lyman α photons and the subsequent annihilation γ's is recorded, thus registering 2P $\to$ 1S $\to 2\gamma$ or 2P $\to$ 1S $\to 3\gamma$ events.

Figure 13 shows a lifetime spectrum obtained using the Lyman α photon as a start pulse and the annihilation γ's to give the stop signal. The long lifetime component characteristic of ground state Ps occurs when a 2^3P state decays by Lyman α emission to the 1^3S_1

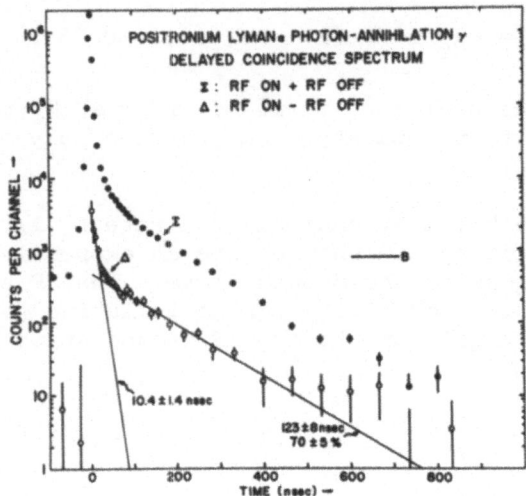

Figure 13. Lifetime spectrum showing the time at which annihilation occurs following the detection of a Lyman α photon signal at T=0. The increased count rate from microwave induced transitions $2^3S_1 \rightarrow 2^3P_2$ is clearly visible in the difference spectrum Δ, RF on - RF off. B is the background which has been subtracted from the sum spectrum Σ, RF on + RF off. (From Ref. 40.)

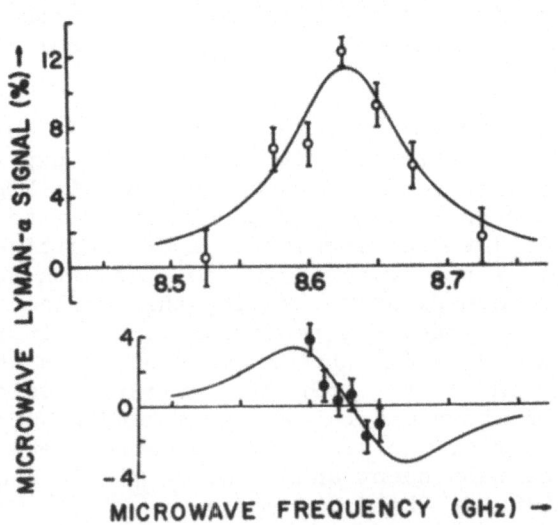

Figure 14. The observed Lyman-α signal S (open circles) and logarithmic first-difference signal S´ (solid circles) as a function of microwave frequency [see Eq. 8]. (From Ref. 40.)

state which then annihilates via 3γ's. Any 2^3S_1 states initially
present would ordinarily annihilate without the emission of a uv
photon and hence are not detected in the delayed coincidence
signal. However, if microwaves at the appropriate frequency are
used to induce transitions $2^3S_1 \rightarrow 2^3P_2$, the extra Lyman α photons
from $2^3P_2 \rightarrow 1^3S_1$ will result in an increase in the intensity of
the detected delayed component. This increased signal is shown in
Fig. 13 as the difference between the microwaves on minus the
microwaves off lifetime spectra. That the long component of
∿120 nsec is shorter than the vacuum 1^3S_1 lifetime (∿140 nsec) is
attributed to wall collisions. Indeed, coating the walls of the
microwave cavity with MgO smoke increases the long lifetime
significantly. There is evidence that Ps is excluded from MgO
powder surfaces [11]. In addition, from the observation of the
long lifetime of 1^3S_1 Ps in MgO powder [10] one would expect a
very small annihilation probability for Ps collisions with MgO
powder. Thus the observation of an increase in the long lifetime
demonstrates that the n=2 Ps is not stuck to the emitting surface.
The intermediate lifetime component (∿10 nsec) is not fully
explained as yet, but could result from the presence of a high
velocity n=2 Ps component.

Figure 14 shows the "microwave Lyman α signal" defined as
$S(f) = \{N_{on}(f)-N_{off}\}/N_{off}$ where N_{on} (N_{off}) is the counting rate
in the time delayed component of the UV photon-annihilation γ time
spectrum, with rf on (off). Also plotted is $S'(f)$

$$S'(f) = \{N(f+\Delta) - N(f-\Delta)\}/\{N(f+\Delta) + N(f-\Delta)\}$$

$$= \{S(f+\Delta) - S(f-\Delta)\}/\{2+S(f+\Delta) + S(f-\Delta)\} \quad , \qquad (8)$$

obtained in separate runs with $\Delta = 30$ MHz. A Lorentzian line shape
$S(f) = \frac{1}{4}A\delta^2\{(f-f_0)^2 + \frac{1}{4}\delta^2\}^{-1}$ was fitted simultaneously to both
sets of data yielding $A = (11.4 \pm 0.6)\%$, $f_0 = (8628.4 \pm 2.8)$MHz, and
$\delta = (102 \pm 12)$MHz with $\chi^2/\nu = 12.1/10$.

The width δ is twice the ≈50 MHz natural linewidth ($\Gamma_{2P}/2\pi$)
of the 2^3P_2 state; this is attributed to power broadening and, to
a smaller extent, to the effects of Ps collisions with the cavity
walls. If all substates of the n=2 level are equally populated
during production, and if there are no geometrical differences
in the detection efficiency between the originally produced 2P
decays and the 2S→2P induced Lyman α decays, one predicts a
maximum effect of 33% for $S(f_0)$. Subtracting the background due to
accidentals in the time spectrum from N(f) results in an observed
effect of $A = (14.1 \pm 1.0)\%$. The change of A with power and the
power broadening of the width agrees with an order of magnitude
computation which includes the natural decay rates, if an additional
decay rate of ≈6×10^7sec^{-1} (possibly accounted for by wall collisions),

is assumed for the 2S state. The details of Ps - wall collisions
are too little understood at present for a more detailed model
computation. These uncertainties, however, influence mainly the
width and magnitude of S(f), and would affect f_0, the resonance
frequency, only via a possible asymmetry of the actual line shape,
an effect subject to further more precise experimentation.

An rf effect A = (0.68 ± 0.09)% is also observed in
the singles rate of the UV photomultiplier. The small size of
the effect is due to the high background of γ induced counts in
the UV photomultiplier and is in agreement with an estimate
based on the observed delayed coincidence rf effect. It is
interesting to note that this constitutes the first observation of
Ps which does not rely on the detection of annihilation γ's.

The center frequency f_0 = (8628.4 ± 2.8)MHz obtained from
the data of Fig. 14 is to be compared with the predicted value
given by Eq. 7 only after a correction is made to account for
the small Zeeman and motional Stark shifts due to the 54G rms
magnetic induction B in the cavity (produced by the slow β^+ beam
solenoid). Estimating that the Ps kinetic energy ranges between
0 eV and 1 eV implies that the corrected theoretical value of
Eq. 7 ranges between 8620 MHz and 8624 MHz, compared to the
experimental value of (8628.4 ± 2.8)MHz. The $\alpha^3 Ry$ radiative
correction ("Lamb shift") in Eq. 7 corresponds to 231 MHz; thus
this experiment already confirms this value to a few percent
accuracy and may indicate the need for higher order corrections.

CONCLUSIONS

Great strides have been made in the study of the properties
of positronium in the quarter century since its discovery [1].
The ground state hyperfine interval is now known to a precision of
5 ppm [21,22]; the triplet decay rate is being measured [29-31]
with accuracies approaching one part in 10^3; the Lyman α radiation
from the first excited state of positronium has been observed [40],
and a fine structure interval in this n=2 level has been measured
[41] with a precision of three parts in 10^4, confirming the Lamb
shift-like radiative corrections [43] to a few percent. It is to
be hoped that further work making use of recent advances in
positronium production [10,11,12] will allow new aspects of the
exotic positronium atom to be investigated and new precision
measurements of its known properties to be made.

REFERENCES

[1]   M. Deutsch, Phys. Rev. 82, 455; 83, 866 (1951).

[2]   B. N. Taylor, W. H. Parker, and D. N. Langenberg, Rev. Mod.
      Phys. 41, 375 (1969), pp. 449-451.

[3]   For an early review of positronium physics see S. De Benedetti
      and H. C. Corben, Ann. Rev. Nucl. Sci. 4, 191 (1954). See also
      V. W. Hughes, Atomic Physics, Vol. 3 (Plenum Publ., New York,
      1969, 1973), and references therein.

[4]   V. W. Hughes, Phys. 1973, Plenarvortr. Physikertag. 37th
      (Physik Verlag, Weinheim, Germany, 1973), pp. 123-155 (in
      English).

[5]   M. A. Stroscio, Physics Reports (Phys. Letters C) 22C,
      215 (1975).

[6]   For a review of the history of the discovery of positronium,
      see "Discovery of Positronium" Adventures in Experimental
      Physics, Vol. 4, issue δ, (B. Maglic, ed., published by
      World Science Education, Princeton, N.J.), 64 (1975).

[7]   R. E. Bell and R. L. Graham, Phys. Rev. 90, 644 (1953); see
      also S. Berko and F. L. Hereford, Revs. Mod. Phys. 28,
      308 (1956).

[8]   L. A. Page and M. Heinberg, Phys. Rev. 102, 1545 (1956);
      V. L. Telegdi, J. C. Sens, D. D. Yovanovitch and S. D. Warshaw,
      Phys. Rev. 104, 8 7 (1956).

[9]   A. Greenberger, A. P. Mills, A. Thompson and S. Berko,
      Phys. Letters 32A, 72 (1970).

[10] R. Paulin and G. Ambrosino, Journal de Phys. 29, 263 (1968).

[11] S. M. Curry and A. L. Schawlow, Phys. Letters 37A, 5 (1971).

[12] K. F. Canter, A. P. Mills, Jr., and S. Berko, Phys. Rev.
      Letters 33, 7 (1974).

[13] T. Fulton, D. A. Owen and W. W. Repko, Phys. Rev. A4, 1802
      (1971); D. A. Owen, Phys. Rev. Letters 30, 887 (1973. See
      also Ref. 5.

[14] E. R. Cohen and B. N. Taylor, J. Phys. Chem. Ref. Data 2,
      663 (1973).

[15] M. Deutsch and S. C. Brown, Phys. Rev. 85, 1047 (1952).

[16] R. Weinstein, M. Deutsch and S. Brown, Phys. Rev. 94, 758 (1954); 98, 223 (1955).

[17] V. W. Hughes, S. Marder and C. S. Wu, Phys. Rev. 106, 934 (1957).

[18] E. D. Theriot, Jr., R. H. Beers and V. W. Hughes, Phys. Rev. Letters 18, 767 (1967).

[19] E. D. Theriot, Jr., R. H. Beers, V. W. Hughes and K. O. H. Ziock, Phys. Rev. A2, 707 (1970).

[20] E. R. Carlson, V. W. Hughes, M. L. Lewis, and I. Lindgren, Phys. Rev. Letters 29, 1059 (1972).

[22] A. P. Mills, Jr. and G. H. Bearman, Phys. Rev. Letters 34, 246 (1975).

[22] P. O. Egan, W. E. Frieze, V. W. Hughes, and M. H. Yam, Phys. Letters 54A, 412 (1975).

[23] H. Grotch and R. A. Hegstrom, Phys. Rev. A4, 59 (1971).

[24] G. H. Bearman and A. P. Mills, Jr., Phys. Letters 56A, 350 (1976).

[25] I. Harris and L. M. Brown, Phys. Rev. 105, 1656 (1957).

[26] A. P. Mills, Jr. and S. Berko, Phys. Rev. Letters 18, 420 (1967).

[27] J. H. Christenson, J. W. Cronin, V. L. Fitch and R. Turlay, Phys. Rev. Letters 13, 138 (1964).

[28] Gerard C. McCoyd, PhD Thesis, St. John's University, N.Y. (1965) (unpublished).

[29] R. H. Beers and V. W. Hughes, Bull. Am. Phys. Soc. 13, 633 (1968); V. W. Hughes, Physics 1973, Plenarvortr. Physikertag, 37th (Physik Verlag, Weinheim, Germany, 1973), pp. 123-155.

[30] P. G. Coleman and T. C. Griffith, J. Phys. B6, 2155 (1973).

[31] D. W. Gidley, K. A. Marko and A. Rich, Phys. Rev. Letters 36, 395 (1976).

[32] G. W. Ford, L. M. Sander and T. A. Witten, Phys. Rev. Letters 36, 1269 (1976).

[33] K. Marko and A. Rich, Phys. Rev. Letters 33, 980 (1974).

[34] H. Mani and A. Rich, Phys. Rev. D4, 122 (1971).

[35] H. W. Kendall, PhD Thesis Massachusetts Institute of Technology (1954), (unpublished).

[36] S. L. Varghese, E. S. Ensberg, V. W. Hughes and I. Lindgren, Phys. Letters 43A, 415 (1974).

[37] S. L. McCall, Bull. Am. Phys. Soc. 18, 1512 (1973).

[38] W. R. Bennett, Jr., V. W. Hughes and C. S. Wu, Bull. Am. Phys. Soc. 1, 68 (1956); V. W. Hughes, J. Appl. Phys. 28, 16 (1957); R. L. Brock and J. R. Streib, Phys. Rev. 109, 399 (1958); W. R. Bennett, Jr., W. Thomas, V. W. Hughes and C. S. Wu, Bull. Am. Phys. Soc. 6, 49 (1961); B. G. Duff and F. F. Heymann, Proc. Roy. Soc., Ser. A 272, 363 (1963); L. W. Fagg, Nucl. Instrum. Methods 85, 53 (1970); M. Leventhal, Proc. Nat. Acad. Sci. 66, 6 (1970); S. M. Curry, PhD Thesis Stanford University, 1972 (unpublished); J. F. Kielkopf and P. J. Ouseph, Bull. Am. Phys. Soc. 19, 592 (1974); A. J. Dahm and T. G. Eck, Phys. Letters 49A, 267 (1974).

[39] P. G. Coleman, T. C. Griffith, and G. R. Heyland, Proc. Roy. Soc., Ser. A 331, 561 (1973); K. F. Canter, P. G. Coleman, T. C. Griffith, and G. R. Heyland, J. Phys. B: Proc. Phys. Soc., London 5, L167 (1972); W. C. Keever, B. Jaduszliwer, and D. A. L. Paul, in Atomic Physics 3, edited by S. J. Smith and G. K. Walters (Plenum, New York, 1973), p. 561; S. Pendyala, P. W. Zitzewitz, J. W. McGowan, and P. H. R. Orth, Phys. Lett. 43A, 298 (1973); D. G. Costello, D. E. Groce, D. F. Herring, and J. W. McGowan, Phys. Rev. B 5, 1433 (1972).

[40] K. F. Canter, A. P. Mills, Jr., and S. Berko, Phys. Rev. Letters 34, 177 (1975).

[41] A. P. Mills, Jr., S. Berko and K. F. Canter, Phys. Rev. Letters 34, 1541 (1975).

[42] See for example, V. W. Hughes, in Atomic Physics, edited by V. W. Hughes, B. Bederson, V. W. Cohen, and F. M. J. Pichanick (Plenham, New York, 1969), p. 15.

[43]   T. Fulton and P. C. Martin, Phys. Rev. 95, 811 (1954).

**   Research at Brandeis University supported in part by grants
from the National Science Foundation.

HYPERFINE STRUCTURE OF STORED IONS - RESULTS FOR 2s ^{2}He+ *

Michael H. Prior and Edmond C. Wang

Department of Physics and Lawrence Berkeley Laboratory
University of California
Berkeley, California 94720

INTRODUCTION

The study of ions and their interactions by means of storage devices which hold them for long periods encompasses a broad range of experimental effort. In the area of collision studies, treated by G. Dunn[1] at the 1974 International Conference on Atomic Physics (ICAP), it includes electron-ion recombination, spin and charge exchange, photo-dissociation, and ion-molecule reactions among others. Another broad area, the original emphasis of the technique, is the radio-frequency spectroscopy of stored ions. This area was pioneered by H. Dehmelt and his collaborators and was reported on at the first ICAP in 1968[2]. The range of possible experiments here includes the traditional domain of rf spectroscopy of atomic and molecular species; e.g. magnetic moments, hyperfine structure and fine structure. New possibilities include the use of lasers as optical pumping sources and, beyond the rf frequency range, one expects two-photon laser spectroscopy on stored ions to emerge[3]. Closely allied to the above, but distinguished by their uniqueness and high degree of refinement, are the experiments of the U. Washington group to measure the anomalous magnetic moment of a single electron[4].

A less explored but potentially large area for study is the measurement of the lifetimes of metastable excited ionic states[3,5].

Of course, one must note the very large effort currently underway world wide to understand and manipulate the behavior of high density plasmas in any of the many controlled thermonuclear reactor (CTR) experiments. The CTR devices differ markedly from the ion traps under discussion here; there is, however, some overlap in the sense that some of the processes of interest[6] to the CTR program for diagnostic and other reasons (e.g., charge exchange, recombination and decay rates of long-lived excited ionic states) are amenable to study with small scale ion traps.

This report lies in the area of radio-frequency spectroscopy and describes the measurement by Dr. E.C. Wang and the author, of the hyperfine structure (hfs) of the metastable 2s state of the ^{3}He$^+$ ion. A brief report of our early results has appeared previously[7].

HFS MEASUREMENTS ON STORED IONS

By way of placing our work in perspective, a brief discussion of the advantages and drawbacks of the ion-storage technique as applied to the study of the hfs of ions, as well as a summary of previous ion-storage hfs measurements follows. For detailed discussion of many of the topics, the reader is referred to the reviews by Dehmelt[8].

One of the strong points of ion-storage methods for the study of hfs is the long observation times available to induce a hyperfine transition. Ions have been stored without loss for periods of many hours (see e.g., Ref. 1) so that, in principle, resonance line widths of unprecedented narrowness (say $\simeq 10^{-4}$ Hz) might be imagined for experiments on ionic ground states. The real questions of the feasibility and utility of an experiment with such a narrow line (requiring perhaps several days to sweep over) detract from its obvious appeal as a tour-de-force. In fact, other considerations often limit the line width and precision of a measurement to less than that allowed on the basis of the ion storage time alone. The quadratic Doppler effect caused by the ion velocity distribution (finite ion temperature) is an example. None-the-less valuable measurements have been made on lines whose widths are only a few Hz[9,10]; this is far narrower than typical atomic beam magnetic resonances linewidths (generally > 1 kHz).

The fact that stored ions are unperturbed by collision with background gas atoms is, of course, responsible for their long storage times and is achieved by the use of standard ultra high vacuum techniques to reduce the background pressure. Ion-ion collisions take place typically at rates of .1 to 10^3 sec^{-1} depending on ion density and temperature, but, because of the long range coulomb force, the wave function overlap is very small and usually results in no significant perturbation to the internal properties of the ions.

Finally, one might include as an advantage of the ion storage technique, the non-destructive detection of the number of stored ions by resonant excitation of their motion. Scattering of resonant optical photons[3,10] is also a useful non-destructive probe of the number of stored ions and can serve to monitor Zeeman sublevels as well.

On the negative side, the number of stored ions is not large being typically 10^6 to 10^8 for ground state ions and perhaps 10 to 1000 times less for excited ionic states (e.g. 2s ^{3}He). Thus the signal to noise ratio of resonances is low and may typically require an hour of integration to achieve a value of 10/1. Monitoring of substate populations via scattering of resonant optical photons may well improve this situation for favorable cases in the near future.

A further weakness is the fact that to date no generally applicable technique for hyperfine state selection exists. Each ion requires its own special solution and one does not have a counterpart for ions of the atomic beam magnetic resonance method which has had such wide application to virtually any paramagnetic atom or molecule. Tunable lasers may somewhat rescue this state of affairs by allowing the technique of optical pumping to assume this role, at least for a subset of favorable ions.

Finally, the radio frequency spectrum may well be perturbed by the confining fields and the ion motion (e.g. Stark, Zeeman and Doppler effects). This is usually not a serious problem as one has considerable flexibility in selection of the type of fields and control of their magnitude. Corrections can be made with some confidence; and, if necessary, extrapolation can be made to field free values.

In the past 14 years or so, since the pioneering work of H. Dehmelt and co-workers[11], two types of ion trap have been used extensively and their properties are well described in Dehmelt's reviews[8]. These are the static field Penning trap and the dynamic field Paul or rf quadrupole trap. Both use a cylindrically symmetric electric potential of the form

$$\Phi(r,z) = U \cdot \frac{2z^2 - r^2}{2z_o^2 + r_o^2}$$

whose equipotentials are the familiar hyperbolas of revolution. U is the potential applied between the cap electrodes separated by $2z_o$ and the ring electrode whose inside radius is r_o; typically z_o, r_o are a few centimeters or less. In the case of the Penning trap (PT), U is constant, usually in the range of 1 to 100 volts, and; in addition, there is present a vector potential $\bar{A} = \bar{r} \times \hat{z} H$ producing a constant magnetic field of magnitude H ($\simeq$ 100 to 10,000 gauss) along the z-axis. Ion motion consists of simultaneous harmonic oscillation parallel to the z-axis, cyclotron motion about the magnetic field lines and a magnetron or $\bar{E} \times \bar{H}$ drift of the cyclotron orbit center about the trap axis.

The rf quadrupole trap (RFQT) uses an oscillating electric potential, i.e., $U = U_o \cos\Omega t$, with Ω typically $\simeq$ .1 → 1.5 MHz, and $U_o \simeq$ 100 to 1000 V and no magnetic field. Ion motion is determined by solution of Mathieu equations and consists of a small amplitude high frequency oscillation at or near Ω and lower frequency larger amplitude harmonic motion in an effective potential ψ of the form

$$\psi = \frac{1}{2}k \cdot (\langle r \rangle^2 + 4\langle z \rangle^2),$$

where $\langle r \rangle$ and $\langle z \rangle$ are r and z coordinates averaged over the high frequency motion.

This brief description of the RFQT and PT schemes will allow some comparison with the purely electrostatic trap design we have used in our experiment. The electrostatic trap was first described by K.H. Kingdon[12] in 1923. More recently, R.G. Herb[13] and associates have developed this configuration as an ionization gauge and, as such, it is known as an "Orbitron." The device consists of a negatively charged wire inside a closed, coaxial cylinder; it has not been applied to the rf spectroscopy of ions previously and it will be discussed in somewhat more detail in a following section.

A summary of published ion-storage hfs measurements is presented in Table I. The entrees are arranged chronologically and the definition of ion-storage is broadened somewhat to include the ion-beam experiment of Novick and Commins[14] on 2s ^{3}He$^+$. (The drifting ions in the beam can be regarded as inertially "confined" during their transit time through the apparatus). This is done because, for many years their experiment stood as the only precise measurement of hfs in a free ion and it forms the point of departure for the work we have done recently on 2s ^{3}He$^+$.

Things to note from Table I are the large values of $Q = \nu/\Gamma$, the ratio of resonance frequency to line width, achieved and the different state selection and analysis schemes used. One notes that all the work except that on 2s ^{3}He$^+$ has been carried out with rf quadrupole traps.

Table I. Hfs Measurements on Stored Ions

Ion	State	hfs freq. (MHz)	Q	Type of Trap	State Selection/ Analysis	Authors (Year)
^{3}He$^+$	2s	1083.3···	1.0×10^4	ion-beam	resonant μwave quenching	Novick, Commins[a] (58)
H$_2^+$	K=1,2; v=4→8	3.9··· →1248.5···	2.4×10^6	RFQT	polarized photo-dissociation	Jefferts[b] (68,69)
^{3}He$^+$	1s	8665.6···	9.0×10^8	RFQT	spin exchange/ spin dependent charge exchange	Scheussler, Fortson, Dehmelt[c] (69)
^{199}Hg$^+$	g.s. $^2S_{\frac{1}{2}}$	40507.4···	5.0×10^9	RFQT	optical pumping/ resonant fluores- cence	Major, Werth[d] (73)
^{3}He$^+$	2s	1083.3···	1.6×10^6	electro- static	resonant μwave quenching	Prior, Wang[e] (75,76)

[a]Ref 14; [b]Ref 15; [c]Ref. 9; [d]Ref. 10; [e]Ref. 7 and this report.

A motivation for the work summarized in Table I has been (H_2^+ and $^3He^+$) to make precision measurements in simple systems where good tests of theory seem possible. In addition, for H_2^+ there was motivation to determine the hfs spectrum of this simplest molecule for astrophysical reasons. For the heavy ion $^{199}Hg^+$, Major and Werth[10] accomplished the first optical pumping of stored ions and the motivation here was, in large measure, the desire to make progress toward the realization of a new time standard. This goal has been often discussed (see e.g., Ref. 16) as a possibility for stored-ion rf spectroscopy, but has yet to be achieved. The low signal to noise ratio of the resonances is the major impediment.

RATIONALE FOR AN IMPROVED 2s $^3He^+$ HFS VALUE

One of the most precisely known quantities in atomic physics is the ground state hfs, $\Delta\nu_1$, in the hydrogen atom. With a fractional uncertainty of $\pm 1.4 \times 10^{-12}$, the experimental value[17] stands as a strong challenge to theory in the one-electron atom. Unfortunately, theory[18] is blocked by uncertainty in the nuclear size and polarizability contributions (-34.6±5.0 ppm) to the nuclear correction $\delta_1(H)$. Thus there are many interesting QED corrections to simple theory which are of the same size or smaller than the uncertainty in $\delta_1(H)$ which cannot be tested by direct comparison with the experimental value. The motivation to extend theory at this level or beyond is thus small.

It is possible to sharply reduce the importance of nuclear corrections, if one has available an additional precision hfs measurement in an excited state. For practical reasons this is restricted to $\Delta\nu_2$, the metastable 2s state hfs. One forms the difference, $D_{21} \equiv 8\Delta\nu_2 - \Delta\nu_1$, which is much less sensitive to nuclear structure (whose leading terms scale like n^{-3}) than $\Delta\nu_1$ or $\Delta\nu_2$ separately. The situation is improved for the QED terms as well. The coefficient of the $(\alpha/\pi)(Z\alpha)^2$ correction term is known exactly for D_{21} (it is -5.5515)[19,20], whereas in the expression for $\Delta\nu_1$ it has an estimated uncertainty of 27% (18.36 ± 5)[21]. It is evidently easier to calculate the QED term differences which contribute to D_{21}, thanks to cancellation of the more difficult state independent terms.

The situation in $^3He^+$ is analogous to that in H except that the nuclear size correction is considerably larger and is less precisely known; two estimates have been made by Sessler & Foley[22] which yielded -183 ppm and -146 ppm depending on the nuclear wave functions used.

The experimental value of D_{21} for $^3He^+$ prior to our work was 1.1901(16) MHz (based on Refs. 9 and 14) with the uncertainty due almost exclusively to that in the $\Delta\nu_2$ measurement of Novick and

Commins. This is to be compared with a theoretical value of
1.1898(5) MHz where the uncertainty is an estimate of uncalculated
terms. The size of these terms could be revealed by a more precise
value of $\Delta\nu_2$, and this was the motivation for our work.

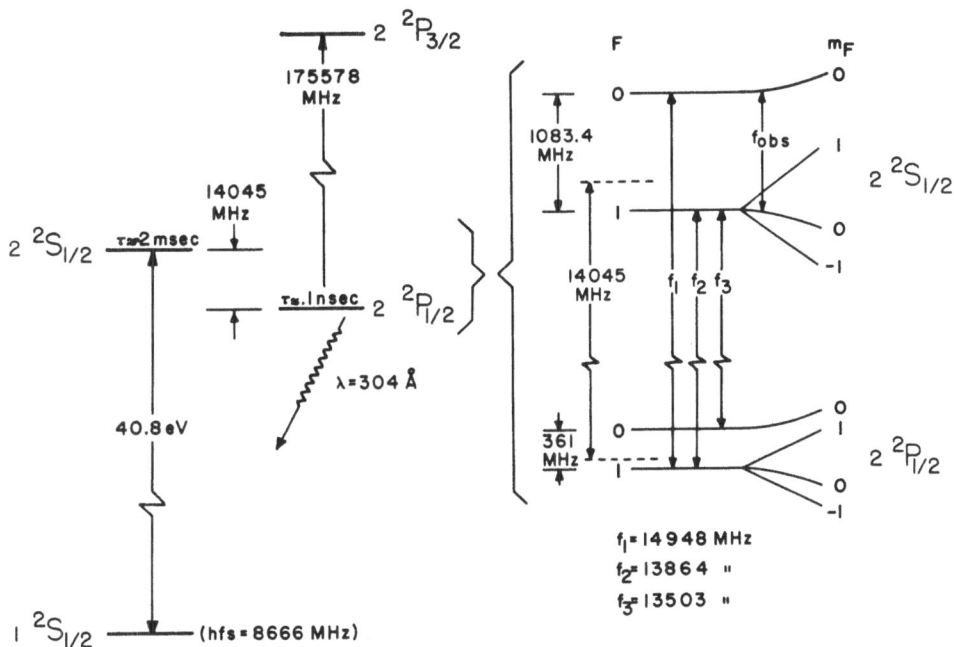

Fig. 1   Energy levels in $^3\text{He}^+$.

METHOD OF MEASUREMENT

Fig. 1 shows the energy level structure of $^3\text{He}^+$ in the n = 1
and 2 states. Our method consists of creating 2s ions inside an
ion-storage device (ion trap) by electron impact on ^{3}He gas at low
pressure (about 4.0×10^{-6} torr). While the ions are confined, we
preferentially remove those in either the F = 0 or F = 1 hyperfine
states by application of a microwave power pulse tuned near the
hyperfine split Lamb shift transitions f_1 or f_2, f_3. (The 1600MHz
width of the $2^2P_{\frac{1}{2}}$ states compared to their 360 MHz hfs splitting
causes the f_2 and f_3 resonances to be unresolved.) Once in one of
the $2^2P_{\frac{1}{2}}$ states, an ion decays with a lifetime of 10^{-10} sec to the
1s ground state by emitting a 304 Å photon. Population of the
depleted 2s hyperfine level can be restored by transfer from the

undepleted level by means of the $\Delta F = 1$, $\Delta m_F = 0$ hyperfine transition marked f_{obs} in Fig. 1. This is done after the microwave
state selection by a suitably polarized oscillating magnetic field
pulse set near the hfs frequency. A second microwave pulse is then
applied and photon detectors and associated electronics count the
number of ensuing 304 Å photons. Counts collected versus frequency applied during the middle hfs transition period yield a
resonance curve, ideally at the unperturbed hfs frequency, $\Delta\nu_2$.
This is the same state-selection and resonance detection scheme
used in the experiment of Novick & Commins[14], the difference being
the use of ion-storage rather than an ion-beam; this requires time-
like rather than spatial separation of functions. By storing the
ions we achieve longer measurement times and have achieved line
widths less than 1kHz (FWHM) compared to 100kHz in the work of
Novick and Commins. The precision of a resonance line center determination can be roughly estimated as the line width divided by
the signal-to-noise ratio. Our signal-to-noise ratio is not as
good as that achieved in the beam experiment so we do not gain the
full factor of 100 or so indicated by the reduction in line width;
in fact, we have achieved about a factor of 20 improvement.

THE ELECTROSTATIC ION TRAP

It was our intention to obtain an experimental configuration
which would produce resonances at essentially the field free hyperfine frequency in order to avoid the need for large corrections
and extrapolations. For this reason, we did not choose to use a
Penning type ion trap such as that used previously to measure the
2s lifetime in ^{4}He$^+$ (Ref 5). The magnetic field needed to operate
the trap would have required a large Zeeman effect correction. The
large amplitude electric fields associated with radio-frequency
quadrupole ion traps were considered prohibitive because of the
associated Stark quenching of the 2s state. For these reasons we
adopted a purely electrostatic confinement scheme. Fig. 2 shows
a cross section view of our device. It is a closed cylinder with a
central rod maintained at a negative potential with respect to the
grounded walls. ^{3}He$^+$ ions are created by impact with electrons
emitted from a filament located outside the bottom end of the cylinder. The electrons move roughly parallel to the rod at a distance of a few centimeters. Ions which have sufficient angular
momentum orbit about the rod in the attractive field and oscillate
along its length in the axial well produced by the grounded cylinder ends. In addition, the structure forms a coaxial cavity resonant in the TE_{011} mode near $\Delta\nu_2$ with a Q of about 1000.

The 304Å photon detectors are CuBe electron multipliers shielded
from metastable neutrals (e.g. ^{3}He 2^1S_0, 2^3S_1) by 800 Å Aluminium
foils. The foils have about 55% transmission at 300 Å.

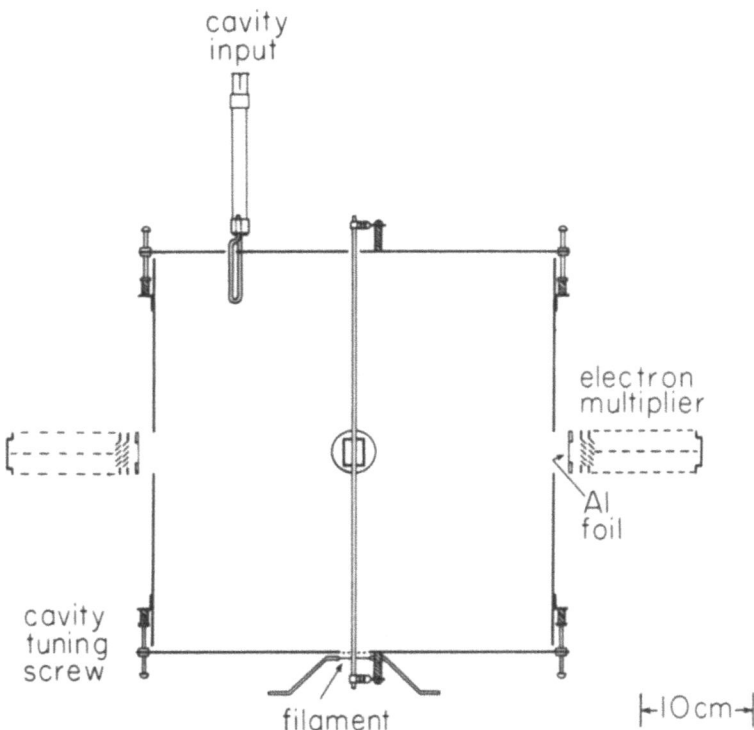

cavity
input

electron
multiplier

Al
foil

cavity
tuning
screw

filament

|←10cm→|

Fig. 2 Sketch of the electrostatic ion trap/rf cavity and photon detectors. The rod is maintained at a negative potential with
respect to the closed cylinder during ion confinement. The rectangular shape shown behind the rod center is the microwave horn used
to induce 2s to $2p_{1/2}$ transitions.

Fig. 3 shows some of the equipotential surfaces and electric field lines for the static trapping field. They are calculable from the Fourier-Bessel series:

$$\Phi(r,z) = \sum_{n=1,3,5\cdots}^{\infty}[A_n I_o(k_n r) + B_n K_o(k_n r)]\cos k_n z,$$

$$k_n = \pi n/z_{max},$$

$$A_n = -B_n K_o(k_n r_o)/I_o(k_n r_o),$$

$$B_n = (-1)^{\frac{n-1}{2}} \frac{4V_R}{\pi n} \frac{I_o(k_n r_o)}{I_o(k_n r_o)K_o(k_n r_i) - K_o(k_n r_o)I_o(k_n r_i)},$$

with I_o and K_o the zeroth order modified Bessel functions and V_R the potential on the rod.

This type of ion trap has a very simple electrode structure and serves well for the present work where long storage times are not important and the possibility of making the structure a simple rf cavity was advantageous. It is possible, however, to refine the idea of electrostatic ion confinement to allow fairly long confinement times and harmonic motion parallel to the z-axis. In particular consider the cylindrically symmetric potential

$$\Phi'(r,z) = \frac{k}{2}(z^2 - \frac{r^2}{2} + B \ln r) + C.$$

For $B = 0$, this potential describes the electric field used in Penning traps and the equipotentials are hyperbolas of revolution about the z-axis. The $\ln r$ term is the potential of a long charged wire and $\Phi'(r,z)$ is a solution of Laplace's equation for arbitrary B. The effective potential $U(r,z)$ acting on an ion of charge q, mass m and having angular momentum L about the z-axis is then,

$$U(r,z) = q\Phi'(r,z) + \frac{L^2}{2mr^2}$$

or,

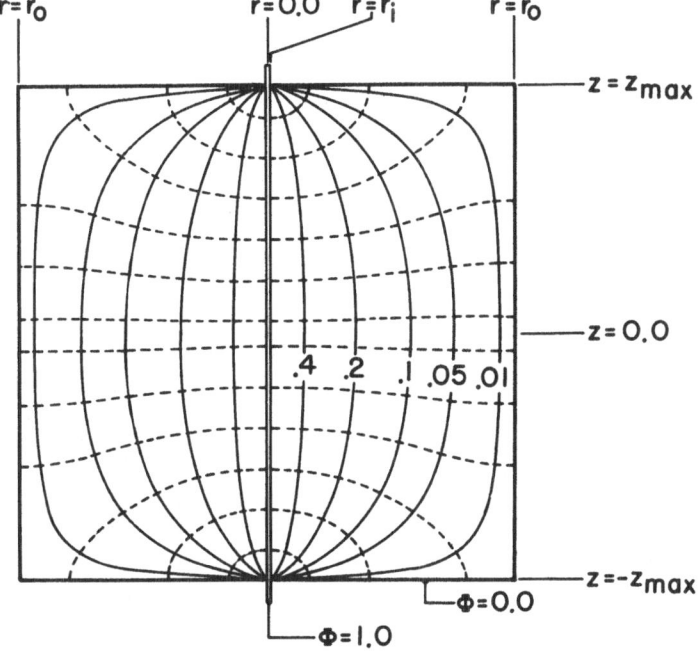

Fig. 3 Potential and field lines for the cylindrical electro-static ion trap.

$$U(r,z) = \psi(r) + \tfrac{1}{2}kqz^2 + C,$$

where

$$\psi(r) = q\frac{k}{2}(B\ln r - \frac{r^2}{2}) + \frac{L^2}{2mr^2}.$$

Thus for $0 < L^2 < (B^2mq/8)$ and qk, B > 0 there exist potential wells in the r and z directions which allow ion confinement. Furthermore, the z component of ion motion will be harmonic with frequency $\omega_z = (qk/m)^{\frac{1}{2}}$.

The author and R. Knight have constructed and partially tested a trap whose surfaces conform to equipotentials of $\Phi'(r,z)$; it has confined N_2^+ ions for several seconds at pressures of a few times 10^{-9} torr and shows promise of longer confinement times at lower pressures.

DATA COLLECTION AND MEASUREMENT PROCEDURE

The data collection scheme and apparatus is shown in Fig. 4. In analogy to atomic beam nomenclature (A, C and B-magnets) we denote the three sequential time intervals as t_A, t_C and t_B. Counts received from the detectors during the B period are stored in a multi-channel scalar (MCS) whose channel address controls the frequency of a synthesizer from which the power to drive the hyperfine transition is derived. Repetitive scans of the resonance are made with about 1000 data cycles at each MCS address during each scan.

The resonance curves were expected and found to fit the Rabi form

$$S(\nu) = AL(\nu)\sin^2[\pi t_c bL(\nu)^{-\frac{1}{2}}] + C,$$

with

$$L(\nu) = b^2/[b^2 + (\nu - f)^2],$$

where f is the line center, A and C are amplitude and base line parameters and b is the magnetic dipole transition matrix element.

Figure (5) shows a series of resonance curves taken at varying values of t_c. The curves are least squares computer fits to the

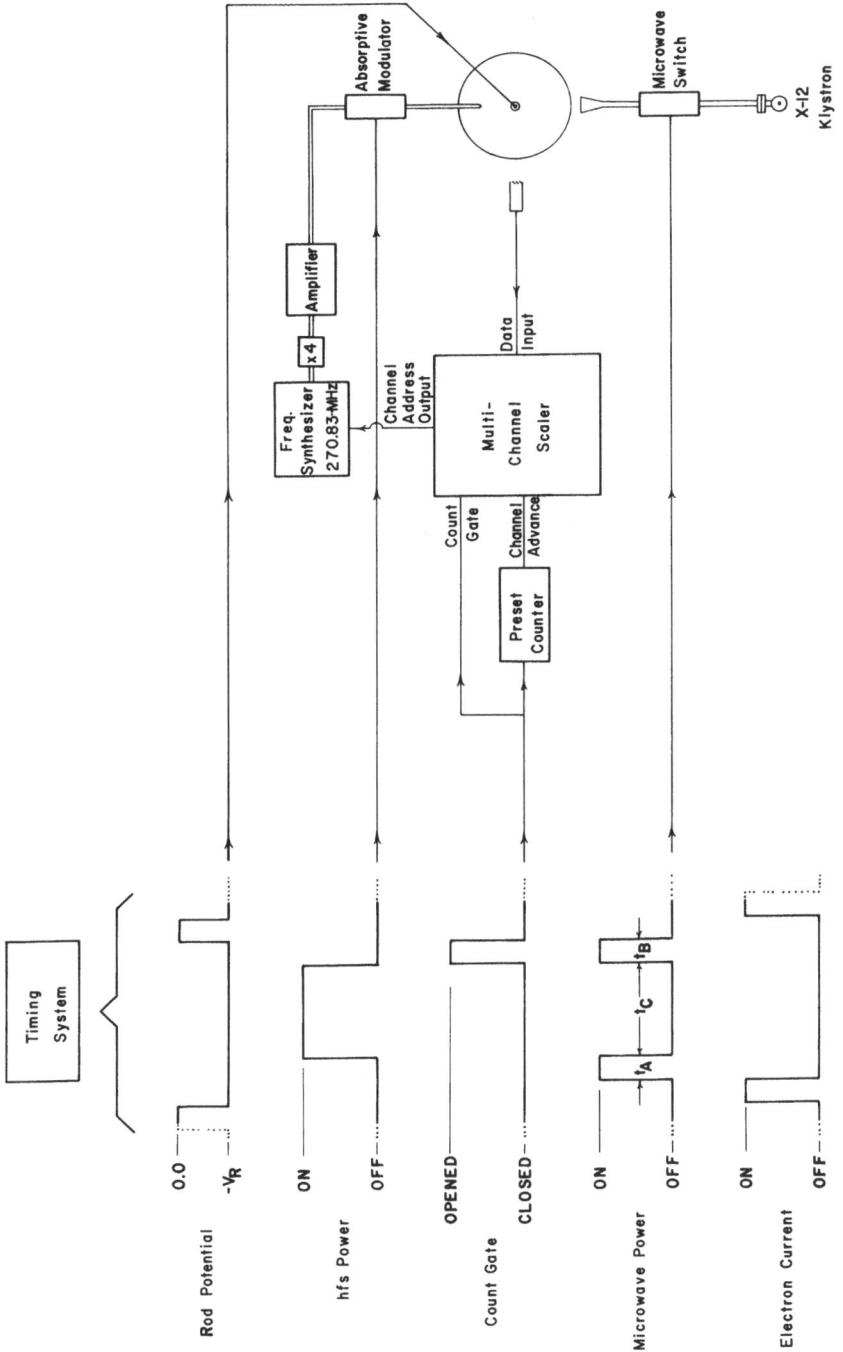

Fig. 4 Sketch of the timing and data collection scheme.

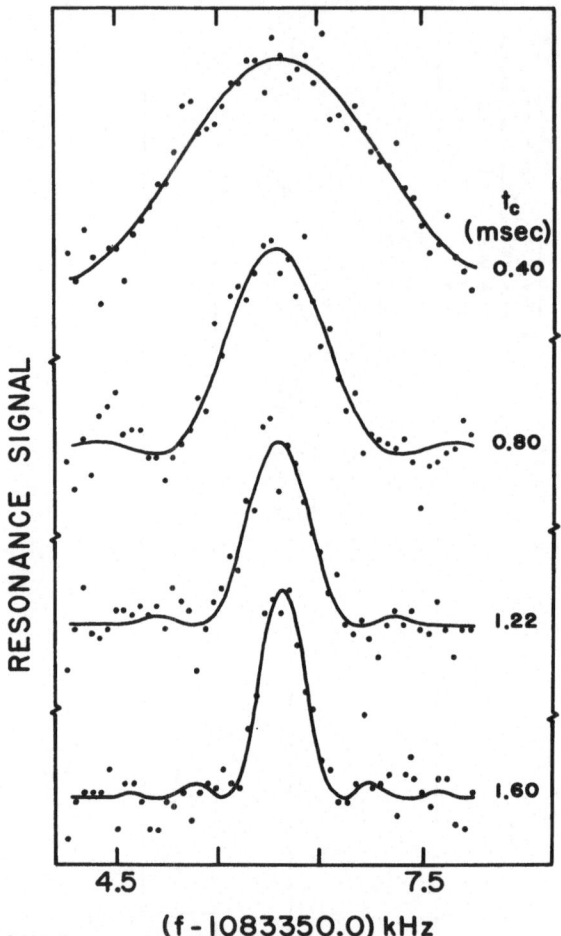

Fig. 5 Resonance curves taken with differing values of t_c.
The solid lines are computer fits to the data. The resonance amplitude is typically 20% of the baseline.

data using the Rabi line shape and, though the side bumps don't show here, other data taken at higher values of rf field show them quite plainly.

The $\Delta F = 1$, $\Delta m_F = 0$ transition has a weak quadratic magnetic field dependence given by

$$f(MHz) = \Delta v_2 + (3.615 \times 10^{-3})H^2$$

where H is in gauss. We use three sets of orthogonal Helmholtz coils to allow arbitrary adjustment of the net field about zero. In fact, we use the observed line centers plotted versus magnet coil current to establish the minimum resonant frequency which we take as our primary measurement of Δv_2. Various corrections to values obtained this way are then applied to achieve a final value for Δv_2.

Fig. 6 shows an example of the measurement procedure. Each resonance curve required about one hour to accumulate and thus one determination of the minimum frequency (nominal zero field hfs) took about six hours. Our new result is based on 36 such determinations.

To eliminate the Stark effect on the hfs, we accumulated data at three rod potentials and extrapolated the mean values of the results to zero rod potential. Fig. 7 shows this procedure.

RESULTS AND DISCUSSION

Table II contains a summary of our data and the various corrections applied to achieve our final result:

$$\Delta v_2 = 1083.354\ 982\ 5(76)\ \text{MHz}$$

Taken with the Δv_1 value of Scheussler et al[8],

$$\Delta v_1 = 8665.649\ 867(10)\ \text{MHz},$$

we obtain,

$$D_{21} = 1.189993(62)\ \text{MHz}$$

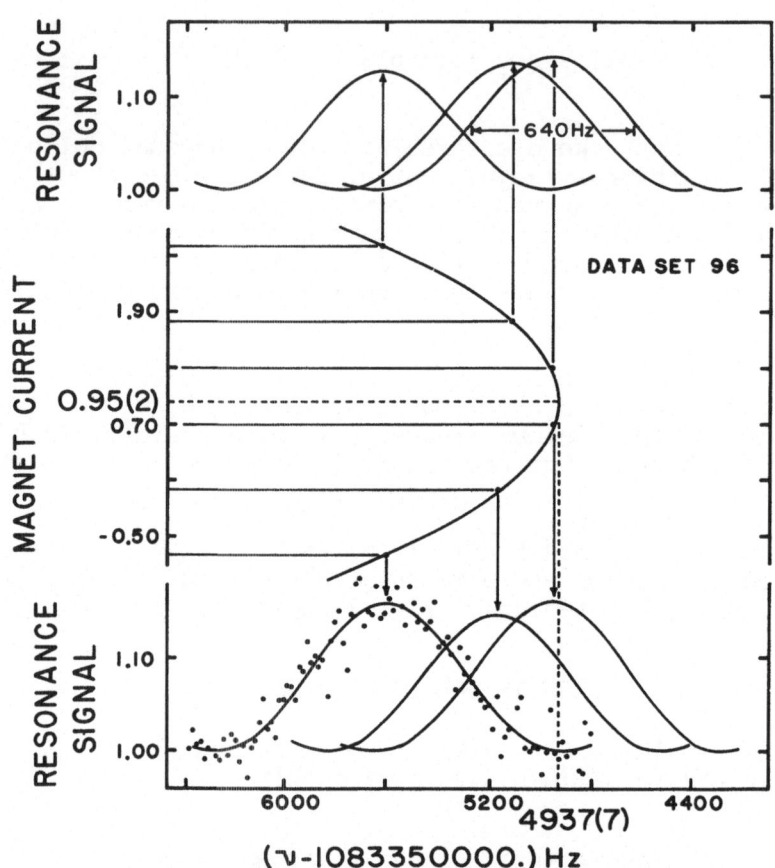

Fig. 6 Variation of resonance line center with Helmholtz magnet current. The solid curves are computer fits. The minimum frequency is the uncorrected hyperfine frequency. Our result is based
on 36 such determinations.

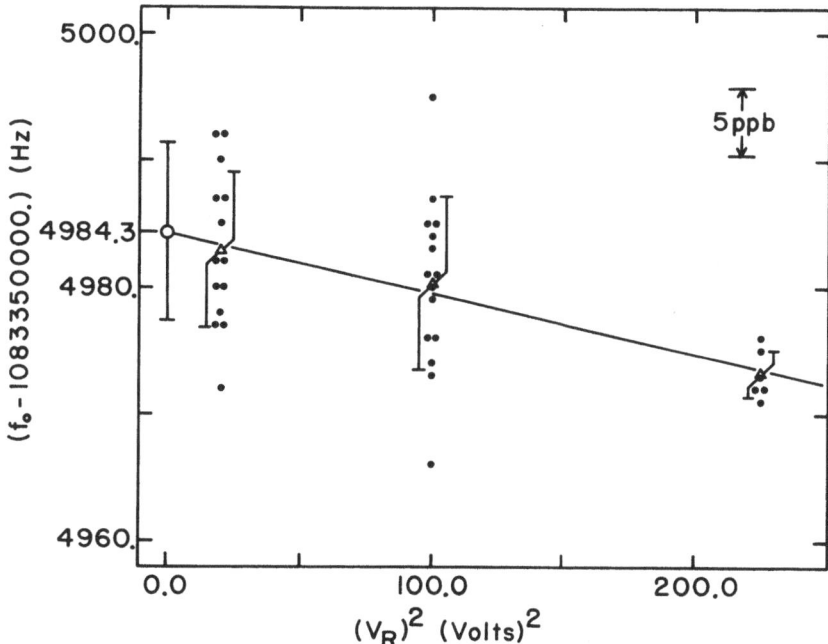

Fig. 7 Extrapolation of zero magnetic field line centers to zero rod potential. The triangles are mean values with 1σ error bars.

The theoretical expression for D_{21} is,

$$D_{21} = E_F \cdot (b_{21} + q_{21} + \delta_{21})$$

where $b_{21} = b_2 - b_1$, $q_{21} = q_2 - q_1$ and $\delta_{21} = \delta_2 - \delta_1$ are the differences in the Breit, QED and nuclear corrections for the 2s and 1s states. δ_2 and δ_1 include nuclear size, polarizability and recoil corrections. E_F is the 1s non-relativistic Fermi contact hfs for a point nucleus,

Table II Summary of Data, Corrections and Result

mean values	V_R = -15.0 V	1083 354 973.2(19) Hz
of zero field	V_R = -10.0 V	980.3(69)
line centers.	V_R = -4.5 V	982.9(61)
V_R^2 = 0 Extrapolated Value		1083 354 984.3(70)

Corrections (to be added to V_R^2 = 0 value):

a) motional averaging of inhomogeneous magnetic field	-3.2(17)
b) offset of rms averaged residual field from minimum value	-1.3(13)
c) rf Stark effect	0.2(1)
d) 2nd order Doppler shift	0.5(4)
e) pressure shift	2.0(20)

Net result: $\Delta\nu_2$ = 1083 354 982.5(76)Hz

$$E_F = \frac{8}{3}Z^3\alpha^2 R_\infty c g_I \frac{m}{M_p}(I + \tfrac{1}{2}) \left[\frac{M}{M+m}\right]^3$$

where E_F is in frequency units and m, M_p and M are the electron, proton and nuclear masses; the other quantities have their usual meanings. The Breit contribution is,

$$b_{21} = \frac{5}{8}(Z\alpha)^2 + \frac{179}{128}(Z\alpha)^4 + \mathcal{O}(Z\alpha)^6$$

The QED terms calculated to date are[19,20],

$$q_{21} = \frac{\alpha}{\pi}(Z\alpha)^2[-3.30320\ln(Z\alpha) - 5.5515].$$

The recoil terms through order $(Z\alpha)^2$ m/M and second order hfs which contribute to δ_{21} have been calculated by Sternheim[23]. δ_{21} may be partitioned as $\delta_{21} = r_{21} + s_{21}$ where r_{21} are the terms calculated by Sternheim and s_{21} is everything remaining, principally the uncalculated nuclear size contribution.

For $^3\mathrm{He}^+$ one obtains;

$$b_{21}E_F = 1.152978(1)(\mathrm{MHz}),$$

$$q_{21}E_F = 0.036026(\mathrm{MHz}),$$

$$r_{21}E_F = 0.000797(\mathrm{MHz}),$$

For a net value of

$$D_{21}\ (\mathrm{theory}) = 1.189801(1)(\mathrm{MHz}).$$

The difference D_{21} (exp) - D_{21} (theory) = 192 (62) Hz appears to be significant. It is anticipated that the next uncalculated term in q_{21} will make a contribution of order $\alpha(Z\alpha)^3 \cdot E_F = 197$Hz. It is also probable that s_{21} will begin to contribute at this level. Sessler and Foley[22] have calculated s_1 values of -180 and -143 ppm depending on the nuclear wave function used, this would mean that a $s_{21} = -1.5 \times 10^{-4} \cdot s_1$ would account for the difference in the theoretical and experimental values of D_{21}. In addition, P. Mohr[24] has indicated that for a reasonable nuclear model, $s_{21} \cdot E_F$ in $^3\mathrm{He}^+$ is of about the same absolute size as the observed difference in the D_{21} values. It probably will require an evaluation of both s_{21} and the higher order q_{21} term to establish an explanation of the difference between current theory and experiment for D_{21} in $^3\mathrm{He}^+$.

ACKNOWLEDGEMENT

Many valuable discussions with Dr. Peter J. Mohr have occurred and we acknowledge them gratefully.

REFERENCES

*Work supported by the Energy Research and Development Administration.

1. G.H. Dunn in Atomic Physics 4, G. zu Putlitz, E.W. Weber, and A. Winnacher, Editors, Plenum Press, New York and London, (1975), p. 575; see also F.L. Walls and G.H. Dunn, Physics Today, 27, 30 (1974).

2. H.G. Dehmelt in Atomic Physics, B. Bederson, V.W. Cohen and F.M.J. Pichanick, Editors, Plenum Press, New York (1969) p. 475.

3. R. Ifflander and G. Werth, Fifth International Conference on Atomic Physics (Abstracts), R. Marrus, M.H. Prior and H.A. Shugart, Editors; Berkeley, Calif. (1976), p. 404.

4. R. Van Dyck, Jr., P. Ekstrom, and H. Dehmelt, Bull. Am. Phys. Soc. 21, 818 (1976); Fifth International Conference on Atomic Physics (Abstracts), R. Marrus, M.H. Prior and H.A. Shugart, Editors; Berkeley, Calif. (1976), p. 336.

5. M.H. Prior, H.A. Shugart, Phys. Rev. Lett. 27, 902 (1971); M.H. Prior, Phys. Rev. Lett. 29, 611 (1972).

6. C. Barnett, et al., in the 1974 Review of the Research Program of the Division of Controlled Thermonuclear Research, ERDA-39, (1975) p. 143.

7. M.H. Prior and E.C. Wang, Phys. Rev. Lett. 35, 29 (1975)

8. H.G. Dehmelt Advances in Atomic and Molecular Physics Vol. 3, Academic Press, New York (1967) p. 53; Vol. 5, (1969) p. 109.

9. H.A. Schuessler, E.N. Fortson and H.G. Dehmelt, Phys. Rev. 187, 5 (1969).

10. F.G. Major and G. Werth, Phys. Rev. Lett. 30, 1155 (1973).

11. H.G. Dehmelt and F.G. Major, Phys. Rev. Lett 8, 213 (1962); H.G. Dehmelt and K.B. Jefferts, Phys. Rev. 125, 1318 (1962).

12. K.H. Kingdon, Phys. Rev. 21, 408 (1923).

13. W.G. Mourad, T. Pauly and R.G. Herb, Rev. Sci. Instr. 35, 661 (1964)

14. R. Novick and E.D. Commins, Phys. Rev. 111, 822 (1958).

15. K.B. Jefferts, Phys. Rev. Lett: 20, 39 (1968); 23, 1476 (1969).

16. H.W. Hellwig, Proc. IEEE 63, 212 (1975).

17. H. Hellwig, et al., IEEE Trans. Instr. Meas. IM-19, 200 (1970).

18. B.E. Lautrup, A. Peterman and E. de Rafael, Phys. Reports 3,
    193 (1972) .

19. D.E. Zwanziger, Phys. Rev. 121, 1128 (1961).

20. P.J. Mohr, private communication, has provided an improved
    value for a numerical integration in Ref. 19.

21. S.J. Brodsky and G.W. Erickson, Phys. Rev. 148, 26 (1966).

22. A.M. Sessler and H.M. Foley, Phys. Rev. 98, 6 (1955).

23. M.M. Sternheim, Phys. Rev. 130, 211 (1963).

24. P.J. Mohr, Private Communication.

MULTIPHOTONIC HIGH RESOLUTION SPECTROSCOPY

B. Cagnac

Laboratoire de Spectroscopie Hertzienne de l'E.N.S.,
Université Pierre et Marie Curie
4, place Jussieu, Tour 12, 1er étage, 75230 Paris- France

I wish here to review the work done during the three last years with Doppler-free multiphotonic spectroscopy. The principle of the cancelation of the Doppler effect in two-photon transitions is now well known : the laser beam which produces the two-photon transition is reflected back on itself by a mirror ; if the atom absorbs simultaneously one photon of each of the two oppositely traveling waves, the Doppler shift of one photon cancels exactly the Doppler shift of the other : all the atoms undergo the two-photon transitions for the same laser frequency whatever their velocities may be. The principle of this effect was noticed first by Vasilenko, Chebotaev and Shishaev (1). In our laboratory we arrived later, independently, to the same idea ; our calculations, performed with Grynberg and Biraben (2), showed the possibility to realize actually high resolution experiments with this method : possibility to obtain sufficient signal with the small power of dye lasers in single frequency operation - possibility to reduce the light shifts (or dynamical Stark effect) far below the natural width. Furthermore, we proposed the generalization to multiphotonic transitions (more than two photons) as we will see later.

But experimental work has been done until now nearly exclusively with two-photon transitions, and most of this talk will be devoted to Doppler-free two-photon transitions. We give first an idea of the experimental problems and then we will talk about the numerous applications already realized. I apologize to those people who could be forgotten in this talk.

I - EXPERIMENTAL PROBLEMS

The high intensity of the laser beams has permitted a great
deal of work about two-photon processes, which can be found in
review papers (3). But its application to narrow resonances in atomic spectroscopy requires tunable wavelength. The first experiment
of this kind was done by Abella with a ruby laser (4), but the generalization of such an experiment became possible only with the
advent of dye lasers (5).

In the recent experiments without Doppler broadening, we have
two rather contradictory requirements : very high light power and
extreme spectral purity. The first experiments made in Paris (6)
and in Harvard (7) used pulsed dye lasers. But the precision was
increased by the use of cw dye lasers in single frequency operation
(8) (9) (10). Due to the small power of the cw dye laser, presently
of the order of 0.1 watt, the experiments using these laser are mostly limited to the favourable cases, where there is an intermediate
level close to the half energy gap of the two-photon transition
(this fact decreases the energy defect in the denominator of the
perturbation calculation and magnifies the transition probability).It
is the reason why many interesting experiments can only be done with
nitrogen pumped dye lasers, according to Hänsch's design (11) or
other pulsed dye lasers (12).

We restrict this talk to the description of the experimental
set-up used in our laboratory and shown on Figure 1. The cw dye laser is pumped by an Argon ion laser. In order to obtain a good control of the laser frequency, we use two servo-loops, as other people
do :

1) The purpose of the first one is to maintain single frequency
oscillation of the dye laser. The Fabry-Perot etalon inside the laser cavity is modulated piezoelectrically, and the small modulation
of the laser intensity is detected by a lock-in detector with such
a phase that the etalon remains centred on the frequency of the laser cavity.

2) The second servo-loop is used to control the frequency of
the laser cavity and does not include any modulation. The laser light
passes through an external Fabry-Perot etalon. The transmitted signal is compared with the laser intensity and then amplified and
applied to the piezoceramic bearing one mirror of the laser cavity.
The working point is at half maximum of the transmission peak of
this external etalon, which has cervit spacers and is enclosed in a
vacuum box, in order to insure good stability. We can maintain the
frequency stable within a few MHz during one hour. For most experiments we need to sweep the laser frequency, and we obtain very linear scanning by pressure sweeping the external etalon. A little part

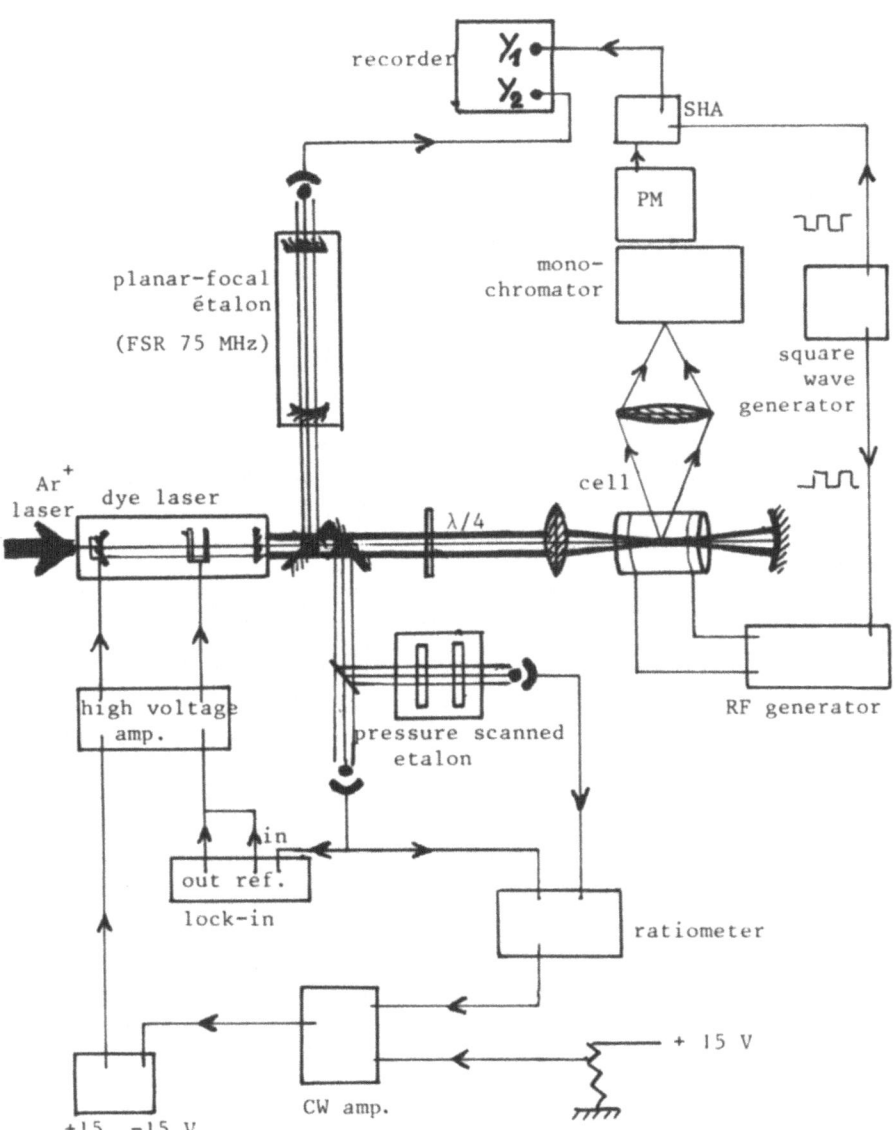

<u>Figure 1</u> - Experimental set-up for Doppler-free two-photon transitions.

of the laser light passes through a second external etalon, the
transmitted signal of which is used to scale the frequency axis of
the recorder (see bottom curve of Figure 2). This second servo-loop
has also the effect of reducing the jitter of the laser frequency
due to acoustical vibrations of the laser cavity. A residual jitter
of three MHz is still a limiting factor in our experiments. But in
other laboratories the jitter has been reduced to less than 1 MHz
(13).

The light coming from the laser is focussed into the experimental cell by a lens, in order to increase the energy density, and
then is reflected back by a concave mirror whose center coincides
with the focus of the lens. In some experiments the cell is put inside a spherical concentric Fabry-Perot interferometer (not shown
on Figure 1)   in order to obtain higher energy density. This in-
terforemeter is servo-locked to the laser frequency.

An optical isolator is needed between the laser and the experimental cell in order to prevent interference effects of the return
beam inside the laser cavity. This optical isolator can use the bi-
refringence of a quarterwave plate or the Faraday effect of a flint
glass in a magnetic field.

A photomultiplier collects the photons spontaneously reemitted
by the excited atoms. The reemitted wavelength is quite different
from the laser wavelength and the monochromator permits to eliminate
completely the stray light of the laser. The current of the photomultiplier is recorded versus the frequency of the tunable laser,
and the upper curve of the Figure 2 shows a typical record obtained
in these conditions : the two narrow peaks distant by 810 MHz correspond to the two hyperfine components of the two-photon transition
$3S \rightarrow 5S$ in the sodium atom at the wavelength of 6022.3 A. Their width
of 10 MHz (mostly jitter of the laser) can be compared to the Doppler
width of 2000 MHz.

In the experiments with neon we need another apparatus shown on
Figure 1, as the studied two-photon transition starts from a metastable level. It is necessary to induce a chopped discharge in the
experimental cell in order to produce the metastable atoms ; and a
special circuit transmits the photomultiplier current to the recorder
only during the afterglow. In these conditions were obtained the
curves of figures 3 and 7, which will be explained soon.

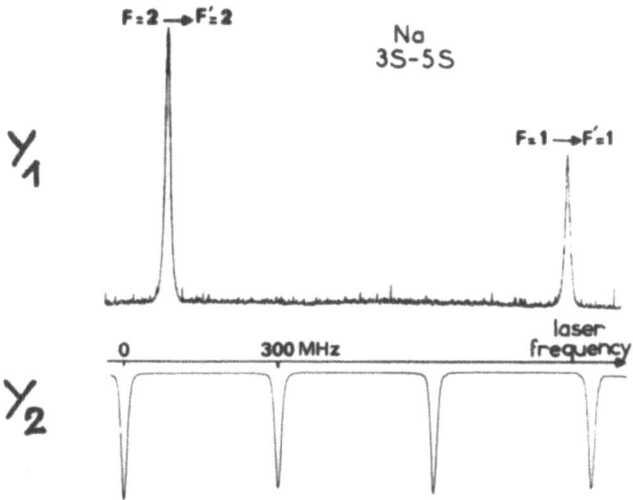

Figure 2 – Experimental recording of the 3S→5S Doppler-free two-photon transition in sodium : Y_1 photomultiplier current versus laser frequency – Y_2 transmission peaks of the calibration etalon.

II – REVIEW OF THE APPLICATION OF THE DOPPLER-FREE TWO-PHOTON METHOD

1) Measurements of Fine and Hyperfine Splittings

It is the most simple application of the method and it permits to study the highly excited states. We enumerate the levels already studied :

sodium 5S .... in references (6)(7)(10)
    4D .... (8)(9)
    6S & 5D .... (14)
    3D, 6D, 7D, 8D .... (15)
potassium nD from n=8 to 19 .... (16)
rubidium nD from n=11 to 32 .... (17)
neon 2p4d' .... (18)(19)
hydrogen 2S .... (20)(21)
thallium 7P .... (22)
mercury .... (23)

We select two experiments in that list :

a) The hyperfine structure of the odd isotope of neon (19) has been studied in the four transitions between the metastable level $2p^53s$ (J = 2) and the four levels of the $2p^54d'$ subconfiguration

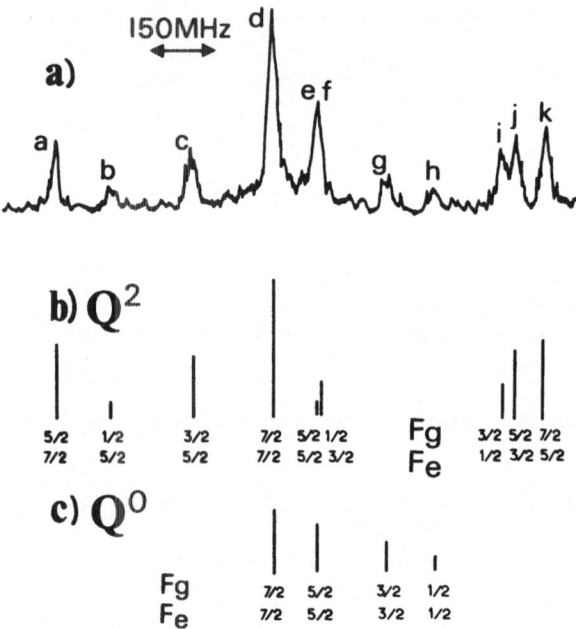

Figure 3 – Hyperfine components of the two-photon transition in the odd isotope 21 of neon (from 3s(J=2) metastable level to 4d'[5/2] (J=2) excited level)
a) experimental recording versus laser frequency
b) theoretical spectrum predicted from the quadrupolar operator Q^2
c) theoretical spectrum predicted from the isotropic operator Q^0

(one can find in Figure 6 their energy diagram and the detailed notations). Curve (a) of Figure 3 shows the experimental record in
the case of the level $2p^5 4d'$ [5/2] J = 2. It can be compared with
the theoretical positions and intensities shown below, with the distinction of the quadrupolar contribution (b) and the isotropic contribution (c).

    In these intensity calculations indeed it is useful to consid-
er the <u>symetrical two photon operator</u> introduced in reference (2)
and extensively studied in reference (24) :

$$Q = \vec{\varepsilon}_1 . \vec{D} \frac{1}{\omega - \mathcal{H}_0} \vec{\varepsilon}_2 . \vec{D} + \vec{\varepsilon}_2 . \vec{D} \frac{1}{\omega - \mathcal{H}_0} \vec{\varepsilon}_1 . \vec{D} = a \ Q^2 + b \ Q^0$$

$\vec{\varepsilon}_1$ and $\vec{\varepsilon}_2$ are the polarizations of the two oppositely traveling waves
$\vec{D}$ is the dipolar moment of the atom
ω the circular frequency of the light (the same for the two waves)
$\mathcal{H}_0$ the Hamiltonian of the atom.
This tensorial operator of rank two is the sum of an isotropic operator Q^0 and a quadrupolar operator Q^2, the coefficient a and b de-
pending on the light polarizations $\vec{\varepsilon}_1$ and $\vec{\varepsilon}_2$. In most cases, only one
of these two operators can connect the initial state g (often ground
state, but here metastable state) and the excited state e :
if $J_g \neq J_e$: $<e|Q^0|g> = 0$; the tensor of rank zero Q^0 has no ma-
trix element between the two states and the spectrum is purely quadrupolar.
if $J_g = J_e = 0$ or 1/2 : $<e|Q^2|g> = 0$; the tensor of rank 2 has
no matrix element between the two states and the spectrum is purely
isotropic (It is the case for the 3S-5S transition in sodium, and it
explains the selection rule $\Delta F = 0$ which leads to only two hyperfine components on Figure 2).
if $J_g = J_e \geqslant 1$, one obtains a mixing of the isotropic and quadrupolar spectra. It is the case of Figure 3 on which one must notice
the isotropic components g and h. They permit to measure the ratio
between isotropic and quadrupolar spectra. This ratio can be compared with the theoretical value deduced from the summation on the intermediate levels of the $2p^5 3p$ configuration (24).

    For the other level $2p^5 4d'$ [3/2] J = 2, the isotropic components
are not visible, and one calculates a destructive interference which
cancels the isotropic transition probability (24).

    b) <u>The Rydberg levels</u>, studied in potassium (16) and rubidium
(17), are of a particular interest because they permit to follow the
variation of the fine structure when the quantum number n increases.
In these Rydberg states, one single electron is rotating at large
distance from a core of charge unity. The Bohr radius increases as
n^2 and for n = 25 its value is of the order of 0.1 micron. The simplest theory predicts that the fine structure decreases as n^{-3}. It
is verified in first approximation (see Figure 4), but one observes
a small discrepancy, which can be compared with more sophisticated theories.

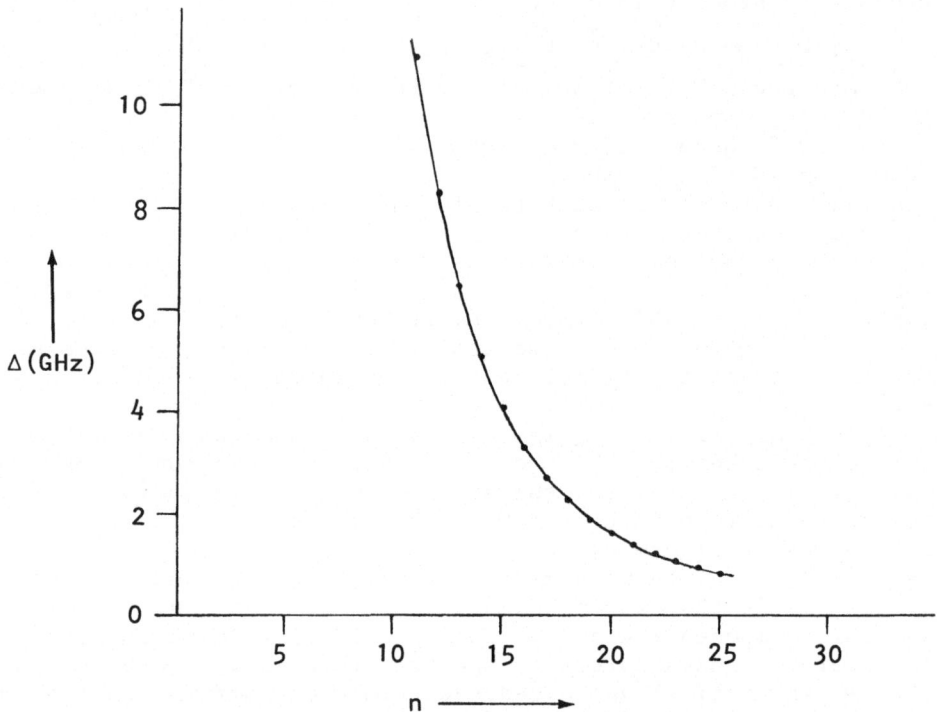

<u>Figure 4</u> – Variation of the fine structure interval Δ of Rydberg
nD states in rubidium versus the principal quantum number n, corresponding to measurements of reference (17) (courtesy from
Boris Stoicheff)

2) Separation of Zeeman or Stark Sublevels in Low Fields

This separation is possible with less than 100 Gauss for magnetic field, less than 1 kV/cm for electric field. The experiments,
in the case of sodium, are described in reference (9) and (25) for
magnetic sublevels, in reference (26) for electric sublevels. Figure 5 is reproduced from reference (26) : it shows the four hyperfine components of the two-photon transition 3S→4D in sodium
(lower curve) and their Stark decomposition in an electrical field
of 2.5 kV/cm. These experiments give not only the splitting between
the Stark, or Zeeman, components as radiofrequency experiments, but also give the absolute shift of the components. In the
case of Stark components, one can deduce the polarizabilities of
the 4D levels of sodium.

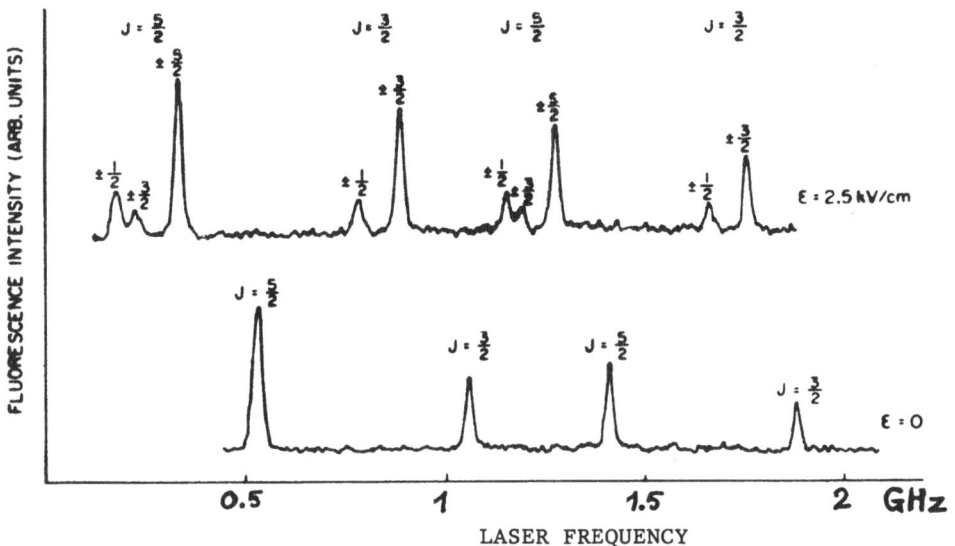

Figure 5 — Stark components of the 3S → 4D two-photon transition
in sodium in an electric field of 2.5 kV/cm (upper curve). It can
be compared with the same transition in a zero electric field (lower curve). (Reproduced from reference (26)).

3) <u>Measurements of Isotopic Shifts</u>

They are obtained in the same experiments as the measurement of hyperfine structures, but they are worthy of a particular interest because they cannot be obtained in the previous methods of Doppler-free spectroscopy such as level crossing or radiofrequency experiments. These measurements of isotopic shifts are found mostly in the references already mentioned :

^{85}Rb, ^{87}Rb .... in references (17) (46)

^{20}Ne, ^{21}Ne, ^{22}Ne .... (18) (27)

^{1}H, ^{2}D .... (21)

^{203}Tl, ^{205}Tl .... (22)

We can develop as an example the case of neon. Figure 6 gives a simplified energy diagram of the four levels $2p^{5}4d'$ which are populated by two photon transitions at wavelength 5923 Å from the metastable level $2p^{5}3s$ (j = 2) ; these levels are labelled with Racah's notation.

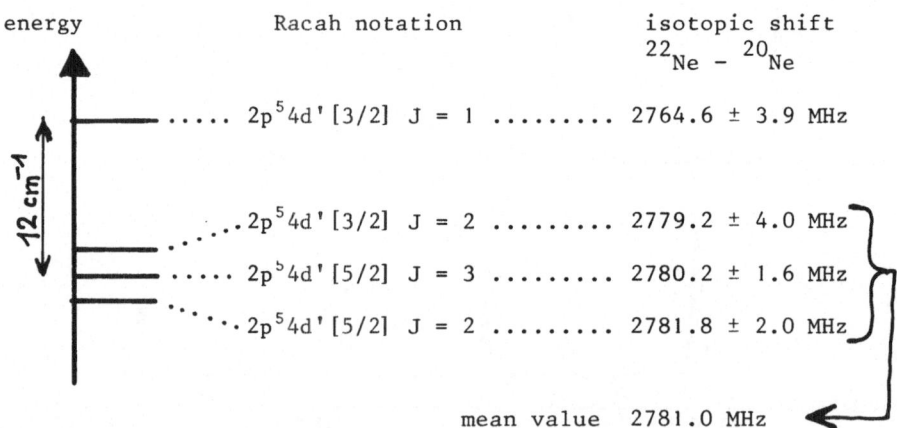

energy	Racah notation	isotopic shift ^{22}Ne – ^{20}Ne
	..... $2p^{5}4d'[3/2]$ J = 1	2764.6 ± 3.9 MHz
	...$2p^{5}4d'[3/2]$ J = 2	2779.2 ± 4.0 MHz
	..... $2p^{5}4d'[5/2]$ J = 3	2780.2 ± 1.6 MHz
	...$2p^{5}4d'[5/2]$ J = 2	2781.8 ± 2.0 MHz

mean value   2781.0 MHz

Figure 6 - Energy diagram of the $2p^{5}4d'$ subconfiguration of neon (on the left). The levels are labelled with Racah's notation. On the right, the isotopic shift ^{20}Ne – ^{22}Ne on the two quantum transition 3s (J = 2) – 4d'[K]   J is given.

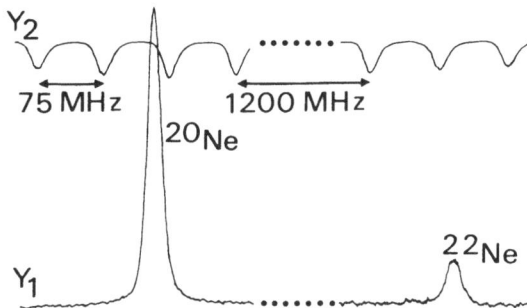

Figure 7 - Experimental recording of the metastable 3s (J = 2) -
4d' [5/2] J = 3 two-photon transition in a natural neon cell.
Y_1 : photomultiplier current versus laser frequency. Y_2 : transmission peaks of the calibration etalon.

Figure 7 gives an example of the record obtained from a natural neon
cell in the case of J = 3 level. The upper curve Y_2 is the transmitted signal of a Fabry-Perot etalon used to scale the frequency axis.
The lower curve Y_1 corresponds to the current of the photomultiplier
which collects the photons spontaneously reemitted at 5902 Å toward
the $2p^5 3p$ configuration. We see the two lines corresponding to the
even isotopes ^{20}Ne and ^{22}Ne (the small amount of the odd isotope
^{21}Ne in the natural neon does not permit to see the corresponding
signal. Figure 3 has been obtained with enriched isotope). Their
line width is of the order of 20 MHz ; half of this width is due to
collisionnal broadening, as the pressure of neon must be of the order of 1 Torr to obtain a sufficient amount of atoms in the metastable state. The isotopic shifts measured in these experiments are
reported on Figure 6. For three among these levels, the isotopic
shift is the same, but its value is definitely different for the
J = 1. This significant difference of 16 MHz cannot be explained by
the Bohr-mass contribution (+ 0.7 MHz). Its origin comes from specific mass-shift theory. The specific shift of the J = 1 level is different from those of the other levels of the 4d' subconfiguration.
Using an Hartree-Fock method, the difference can be estimated to
18 MHz (27), in good agreement with experiment.

4) Investigation of Collisionnal Broadening and Shift at Low Pressures

    It is clearly illustrated on Figure 8 reproduced from reference
(28). It shows one particular component of the two-photon transition
3S→4D in sodium : the lower curve Y_1 is recorded in a cell of pure

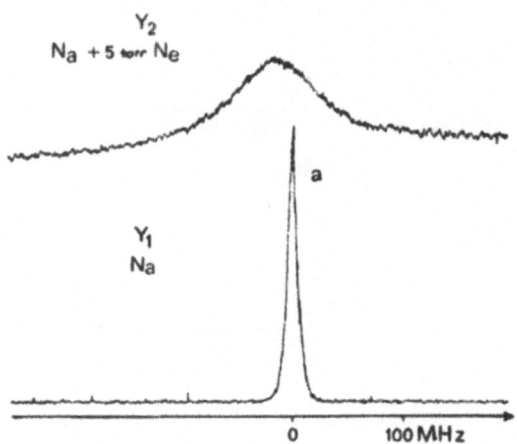

Figure 8 – Simultaneous recording of a component of the 3S – 4D two-photon transition : Y$_1$ pure sodium cell ; Y$_2$ sodium cell containing 5 Torr of neon.

sodium (it corresponds to one of the four lines of the lower curve of Figure 5). The upper curve Y$_2$ is recorded in a sodium cell containing a foreign gas (neon at the pressure of 5 Torr). You notice the broadening of the optical transition due to collisions. In this experiment, the laser beam is splitted in two parts and irradiates simultaneously two sodium cells, one with foreign gas, the other without foreign gas. The signals obtained from the two cells are simultaneously recorded and their relative positions have an actual signification : the pressure shift obviously appears.

Systematic study is under way by Biraben in our laboratory. The figure 9 shows the results he has recently obtained and which are not yet published : you see the width W and the shift S versus the atomic density n of the 3S – 4D and 3S – 5S two-photon transition for different rare gas atoms.

5) Selective Population of One Level

The selective population of one pecular level among a group of close lying levels, at distance smaller than the Doppler width, opens other possibilities. It permits to generalize all the experiments with selective pumping. We gave an example with the investigation of collisional transfer between two close lying levels of sodium $4D_{3/2}$ and $4D_{5/2}$ (29).

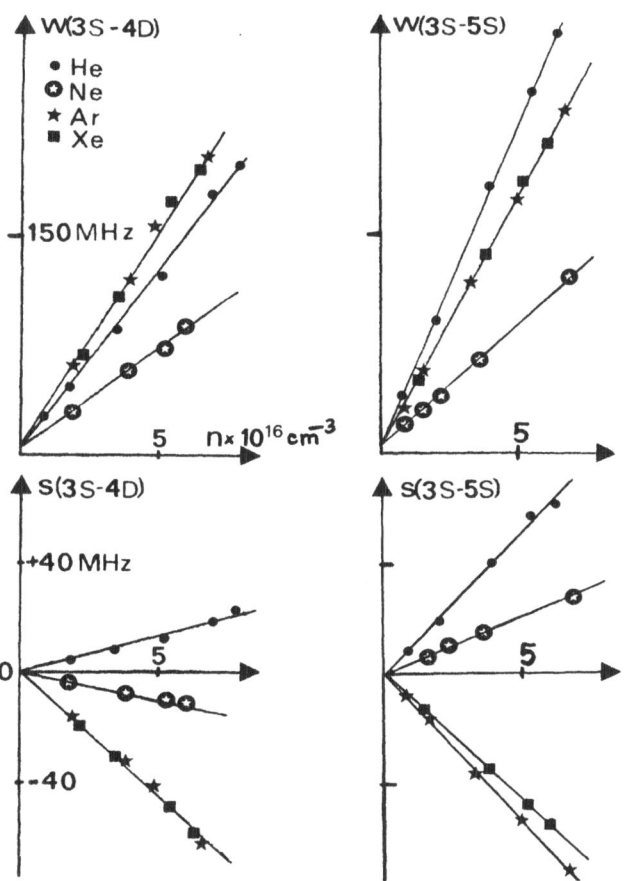

Figure 9 - Variation of width W and shift S of the Doppler-free two-photon transitions in sodium versus the atomic density n of rare gases
on the left : transition 3S→4D
on the right : transition 3S→5S

6) <u>Absolute Measurements of Energy Levels in the Optical Range</u>

It has not been worked out until now except in the experiment
by Hänsch and coworkers on hydrogen (20) (21). The interest to study
the transition from the 1S ground state to the 2S metastable state
of hydrogen has been noticed by several authors (2) (8) (30).
All the intermediate levels in the two-photon process have a higher
energy than the 2S state and the large energy defect of the intermediate process must be compensated by a high light intensity, which
is confirmed by the exact theoretical calculation (31). The desired
wavelength of 2430 Å (twice the wavelength of the Lyman α line) is
obtained by frequency doubling a 4860 Å Courmarine dye laser pumped
by a nitrogen laser. The metastable atoms are detected by collisional transfer to the 2P state, from which they emit the Lyman α line.
Figure 10 shows one experimental recording reproduced from the reference (21) : the upper curve shows the transmission peaks of the
calibration interferometer, the lower curve shows the hyperfine
components of hydrogen and the isotopic shift from deuterium to hydrogen. The new aspect of this experiment is the comparison with
the Balmer β line, which permits to measure the Lamb shift of the
1S ground state of hydrogen. With an increasing precision, this experiment could permit new tests of quantum electrodynamics. Similar
experiment could be possible in the future on the same transition
of positronium and could give a more direct test.

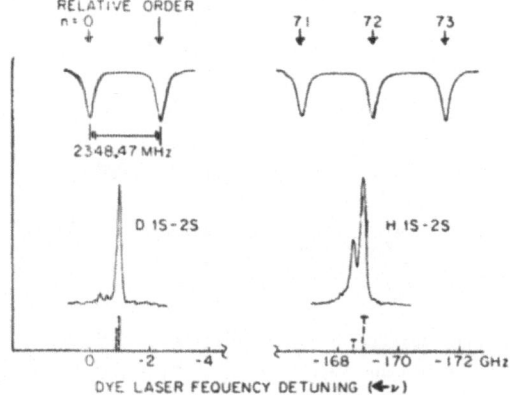

<u>Figure 10</u> – Doppler-free two-photon spectra of the 1S - 2S transi-
tion in hydrogen and deuterium (lower curves). The upper curve represents the transmission peaks of the calibration etalon (reproduced from the reference (21)).

From another point of view, Bender and coworkers in National
Bureau of Standards have the project to use two-photon transitions
toward metastable states in Ag or Hg^+ as an optical frequency standard (32).

7) Applications to Molecular Spectroscopy

Two-photon transitions are also observable in molecules in spite
of the small oscillator strength which lowers the transition
probability. The molecules already studied are :
NO .... see reference (33)
CH_3F .... (34)(35)
NH_3 .... (36)
C_6H_6 .... (23)
Na_2 .... (37)

III - DOPPLER-FREE THREE-PHOTON TRANSITION

I wish now to describe a recent experiment with three photons,
done in our laboratory by Grynberg and Biraben (38). As you know,
the Doppler shift in atomic transitions can be explained by the
conservation of momentum of the whole system atom and photons. This
explanation by momentum conservation permits an easy generalization
of Doppler-free spectroscopy to multiphotonic transitions with three
or more photons (2).

Consider the wave vectors $\vec{k}_i$ of the plane waves interacting
with the atom, and suppose that the sum of the momenta of all the
photons simultaneously absorbed by the atom is equal to zero :

$$\sum_i \hbar \vec{k}_i = 0$$

It appears that the three-photon transition is not Doppler shifted
since the first order shift is equal to the scalar product of this
vector $\Sigma \vec{k}_i$ with the velocity $\vec{v}$ of the atom. In other words, the atom
does not absorb any momentum ; its velocity does not change and there
is no modification of its kinetic energy. All the photon energy is
transferred into internal energy of tha atoms, and obeys to the relation :

$$E_e - E_g = \sum_i \hbar c k_i$$

whatever the atomic velocity may be. This three-photon absorption
is free of Doppler broadening ; as it is free also of recoil effect.

In fact, it is not necessary for all the photons to be absorbed. We can imagine for instance a three-photon transition with two photons absorbed and one emitted, as we explain on Figure 11 : k_1 and k_1' are absorbed ; k_2 is emitted. In this case, we need only to change the sign of $\vec{k}_2$ in the preceeding equations ; that is to say the vector $\vec{k}_2$ must be equal to the sum of the two vectors $\vec{k}_1$ and $\vec{k}_1'$. This is represented on the left part of Figure 11.

In our experiment on the sodium, the g and e states correspond to $3S_{1/2}$ and $3P_{1/2}$. The light beams come from two different lasers so that the two absorbed photons k_1 and k_1' have same frequency. The g and e levels are also used as intermediate relaying levels with the same energy defect $\hbar\Delta\omega$. By scanning the frequency of one laser, the frequency of the second being fixed, we can observe the Doppler-free three-photon transitions. The resonances are detected by collecting photons spontaneously reemitted from the $3P_{1/2}$ state with the wavelength of the D_1 line.

Each laser is locked on the side of the transmission peak of a pressure swept Fabry-Perot etalon. The laser powers are about 60 mW for laser 1 and 30 mW for laser 2, and the line-width is the same for both lasers : about 7 MHz. The light coming from laser 1 is focussed into the sodium cell ; then, using a set of mirrors and a lens, the transmitted light is refocussed at the same point of the sodium cell, the angle between the two beams being equal to 120°. The beam coming from laser 2 is focussed at the same point following the bissectrix of the preceeding beams. To select the fluorescent light at the atomic frequency from the Rayleigh scattering at the laser frequency, we use a good monochromator with focal length of one and half meter.

Figure 12 shows the signal recorded when the frequency of the laser 1 is scanned, while the frequency of the laser 2 is fixed 30 GHz (or 1 cm^{-1}) below the sodium resonance line. The four peaks correspond to the transitions from the two hyperfine sublevels (F = 1 and F = 2) of the ground state to the two hyperfine sublevels (F' = 1 and F' = 2) of the excited state.

The width of each resonance is about 60 MHz. That is much less than the Doppler width of the order of 2000 MHz ; but that is more than one expects from the line width of the laser and from the width of the excited state. The two main sources of broadening are :
1- a residual Doppler effect due to the fact that the angle between the laser beams is slightly different from 60° (the residual Doppler effect due to the focussing of the light beam is of the order of 2 MHz).
2- the light shifts : in the case of two-photon transitions, we showed that it is possible to find experimental conditions for which the light shift is smaller than the natural linewidth, but this state-

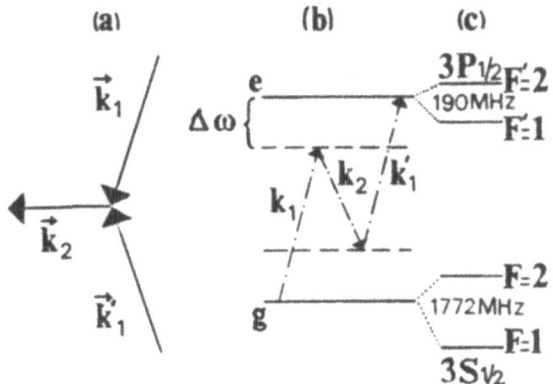

Figure 11 - Principle of a three-photon transition :
a) spatial orientation of the wave vectors in order to eliminate
   the Doppler broadening
b) scheme of the energy exchange : k_1 and k_1' are absorbed, k_2 is
   emitted
c) schematic energy levels of sodium

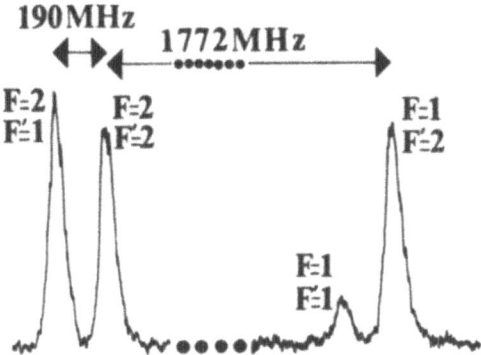

Figure 12 - Experimental recording of the Doppler-free three-photon
spectrum versus the frequency of the laser 1. The frequency of the
laser 2 is fixed 30 GHz below the sodium D_1 line.

ment is not true for the case of three or more photons (2). And the
light shifts may contribute to the line broadening for two reasons :
a) there are spatial nonuniformities of the intensity within the
region of observation ; b) the light shift varies from one atomic
velocity class to another because the Doppler shift is not very
small compared to the energy defect $\hbar\Delta\omega$ in the intermediate states
of the transition.

This last experiment demonstrates that the Doppler-free multiphotonic spectroscopy is not restricted to two-photon transitions.
By absorption of three photons, it will be possible to investigate
very high excited states whose parity is opposite to that of the
ground state ; so the three-photon absorption can be complementary
of the two-photon. Nevertheless, it must be pointed out that most
of possible experiments with three photons need high power lasers
and one has to be careful with the light shifts. It is the reason
why I think that two-photon spectroscopy will have a much wider range
of applications.

As a conclusion, it would be interesting to compare Doppler-
free two-photon absorption with other methods of Doppler-free spectroscopy with lasers :
a) Radiofrequency or level crossing after stepwise excitation (39)
b) Collimated atomic beams with one photon excitation (40) or two
   photon excitation (41)
c) Quantum beats (42)
d) Saturated absorption (43) (44)
e) Three-level optical spectroscopy (44) which is also an application of the velocity selection and has recent developments (45)

But the place is missing here for the volume of this talk. I
believe that all these methods have their own interest and they
appear in some way as complementary. I am sure that they will bring
together an important development of Atomic Physics.

REFERENCES

(1) VASILENKO L.S., CHEBOTAEV V.P. and SHISHAEV A.V. (1970) JETP
     Lett., 12, 161
(2) CAGNAC B., GRYNBERG G. and BIRABEN F. (1973) J. de Physique,
     34, 845
(3) BONCH-BRUEVICH A.M. and KHODOVOI V.A. (1965) Sov.Phys.Ups. 85, 3
     GOLD A. (1967) in the Course of the Enrico Fermi's School (Varenna)
     WORLOCK J.M. (1972) in Laser Handbook (North Holland) t.II p.1323
(4) ABELLA I.D. (1962) Phys. Rev. Lett., 9, 453

(5)   AGOSTINI P.,BENSOUSSAN P. and BOULASSIER J.C.(1972)Opt.Comm.5,293
      BONCH-BRUEVICH A.M.,KHODOVOI V.A. and KHRONOV V.V. (1971) JETP
      Lett., 14, 333
(6)   BIRABEN F., CAGNAC B. and GRYNBERG G. (1974) Phys. Rev. Lett.
      32, 643
(7)   LEVENSON M.D. and BLOEMBERGEN N. (1974) Phys. Rev. Lett., 32,
      645
(8)   HANSCH T.W., HARVEY K., MEISEL G. and SCHAWLOW A.L. (1974)
      Opt. Commun., 11, 50
(9)   BIRABEN F., CAGNAC B. and GRYNBERG G. (1974) C.R. Acad. Sc.
      Paris 279, B 51 and (1974) Phys. Lett. 48A, 469
(10) BIRABEN F., CAGNAC B. and GRYNBERG G. (1974) Phys. Lett., 49A,
      71
(11) HANSCH T.W. (1972) Applied Optics, 11, 895
      WALLENSTEIN R.W. and HANSCH T.W. (1974) Applied Optics, 13,
      1625
(12) LEVENSON M.D. and EESLEY G.L. (1976) IEEE J. Quant. Elec.,
      12, 259
(13) GROVE R.E., WU F.Y. and EZEKIEL S. (1974) Optical Engineering
      13, 531
      BARGER R.L., WEST J.B. and ENGLISH T.C. (1975) Appl. Phys. Lett.,
      27, 31
      WALTHER H. (1976) Communication to the IXth I.Q.E.C., Amsterdam
(14) LEVENSON M.D. and SALOUR M.M. (1974) Phys. Lett., 48A, 331
(15) SALOUR M.M. (1976) Optics Communications, to be published
(16) HARPER C.D. and LEVENSON M.D. (1976) Physics Lett., 56A, 361
(17) KATO Y. and STOICHEFF B.P. (1976) JOSA Lett., 66, 490
(18) BIRABEN F., GIACOBINO E. and GRYNBERG G. (1975) Phys. Rev. A12
      2444
(19) GRYNBERG G., BIRABEN F., GIACOBINO E. and CAGNAC B. (1976) Optics
      Commun., to be published
(20) HANSCH T.W., LEE S.A., WALLENSTEIN R. and WIEMAN C. (1975)
      Phys. Rev. Lett., 34, 307
(21) LEE S.A., WALLENSTEIN R. and HANSCH T.W. (1975) Phys. Rev. Lett.
      35, 1262
(22) FLUSBERG A., MOSSBERG T. and HARTMANN S.R. (1976) Phys. Lett.
      55A, 403
(23) WALLENSTEIN R. (1976) Communication to the IXth I.Q.E.C.,
      Amsterdam
(24) GRYNBERG G., Thesis Paris 1976, n° d'enregistrement CNRS 12497
(25) BLOEMBERGEN N., LEVENSON M.D. and SALOUR M.M. (1974) Phys. Rev.
      Lett., 32, 867
(26) HARVEY C.K., HAWKINS R.T., MEISEL G. and SCHAWLOW A.L. (1975)
      Phys. Rev. Lett., 34, 1073
(27) BIRABEN F., GRYNBERG G., GIACOBINO E. and BAUCHE J. (1976) Phys.
      Lett., 56A, 441
(28) BIRABEN F., CAGNAC B. and GRYNBERG G. (1975) Lettre au J. de
      Physique, 36, L41
(29) BIRABEN F., CAGNAC B. and GRYNBERG G. (1975) C.R. Acad. Sc.
      Paris, 280, B 235

(30) BAKLANOV E.V. and CHEBOTAEV V.P. (1974) Optics Commun., $\underline{12}$, 312

(31) GONTIER Y. and TRAHIN M. (1971) Phys. Lett., $\underline{36A}$, 463

(32) BENDER P.L., HALL J.L., GARSTANG R.H., PICHANICK F.M.J., SMITH W.W., BARGER R.L. and WEST J.B., Communication to the Washington Meeting of the American Physical Society (April 1976)

(33) BRAY R.G., HOCHSTRASSER R.M. and WESSEL J.E. (1974) Chem. Phys. Lett., $\underline{27}$, 167

(34) BISCHEL W.K., KELLY P.J. and RHODES C.K. (1975) Phys. Rev. Lett., $\underline{34}$, 300

(35) BISCHEL W.K., KELLY P.J. and RHODES C.K. (1976) Phys. Rev. A$\underline{13}$, 1817

(36) BISCHEL W.K., KELLY P.J. and RHODES C.K. (1976) Phys. Rev. A$\underline{13}$, 1829

(37) WOERDMAN J.P. (1976) Communication to the IXth I.Q.E.C., Amsterdam

(38) GRYNBERG G., BIRABEN F., BASSINI M. and CAGNAC B. (1976) Phys. Rev. Lett., to be published

(39) HAPPER W. in Atomic Physics 4, Proceedings of the 4th I.C.A.P. Heidelberg 1974, page 651

(40) JACQUINOT P. in Atomic Physics 4, Proceedings of the 4th I.C.A.P., Heidelberg 1974, page 615 and references herein

(41) PRITCHARD D., APT J. and DUCAS T.W. (1974) Phys. Rev. Lett., $\underline{32}$, 641

(42) ANDRÄ, H.J. in Atomic Physics 4, Proceedings of the 4th I.C.A.P. Heidelberg 1974, page 635
HAROCHE S. in High Resolution Spectroscopy, K. Shimoda Editor Springler Verlag 1976

(43) HANSCH T.W. in Atomic Physics 3, Proceedings of the 3d I.C.A.P. 1972, page 579
HALL J.L. in Atomic Physics 3, Proceedings of the 3d I.C.A.P. 1972, page 615

(44) TOSCHEK P. in Proceedings of the International Conference on "Méthodes de Spectroscopie sans Largeur Doppler", Aussois 1973 Editions du CNRS Paris, page 13, and references herein

(45) DELSART C. and KELLER J.C. (1975) Optics Commun., $\underline{15}$, 91 and (1976) Optics Commun., $\underline{16}$, 388
BJORKHOLM J.E. and LIAO P.F. (1976) Phys. Rev. Lett., $\underline{36}$, 1543

(46) ROBERTS D.E. and FORTSON E.N. (1975) Optics Commun. $\underline{14}$, 332

LASER SPECTROSCOPY AND PREDISSOCIATION OF MOLECULES

J.C. LEHMANN

Laboratoire de Spectroscopie Hertzienne de l'Ecole
Normale Supérieure
4, place Jussieu, T 12, 75230 Paris cedex 05, France

INTRODUCTION

Predissociation is a process in which a molecule in an excited state e can be transferred into a dissociative state of the same energy. Figure 1 shows a typical situation for a diatomic molecule. e is a rovibronic state labeled by its vibrational and rotational quantum numbers v and J. It belongs to an electronic excited state E. From E, the molecule can either go back to a lower electronic state G by spontaneous emission of radiation or be transferred to the dissociative state D by an internal or an externally induced

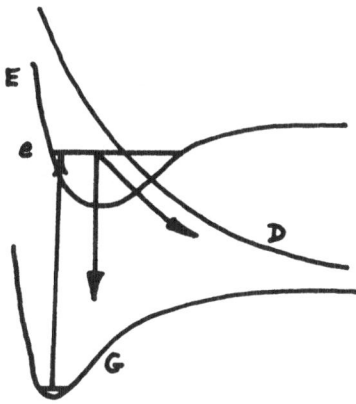

Figure 1 - Predissociation and fluorescence

coupling. To a large extent predissociation is therefore comparable to an atomic autoionization process and many conclusions that will be given in this paper can be applied to autoionization. Provided that the predissociation is weak enough to be treated as an exponential depopulation of e, one can define two time constants for the decay of e :

τ_r, the radiative decay time constant
τ_d, the predissociative decay time constant
The lifetime τ of e is therefore given by

$$\frac{1}{\tau} = \frac{1}{\tau_r} + \frac{1}{\tau_d} \tag{1}$$

Three extreme cases can then occur :

Case a: $\tau_d \ll \tau_r$. This is the case of a strong predissociation. No emission of light can then be observed from state e but the predissociation rate can be measured through the broadening of the absorption lines ending on e. Such strong predissociations have been often observed and their rates are easily measured provided that

$$\frac{1}{\tau_d} > \Delta\nu_D \qquad \text{where } \Delta\nu_D \text{ is the Doppler width of an absorption line (}[1]\text{).}$$

Case b: $\tau_d \sim \tau_r$. This is the case of a weak predissociation. An emission of light is indeed observed in this case, for example a fluorescence following a laser excitation of e. To evaluate τ_d, one can measure the lifetime τ by the exponential decay of the fluorescence. Assuming that τ_r is only slowly varying with the vibrational and rotational quantum numbers of e, any large variation of τ in a series of rovibrational levels of a given electronic state can generally be attributed to a predissociation [2][3]. We shall see an example of such a situation in the next section and how it can give informations on the origin of the predissociation process.
Another technique to observe such weak predissociations is to monitor the fragments of the predissociation (generally atoms) either by absorption or by emission spectroscopy [4].

Case c: If $\tau_d \gg \tau_r$, one has a very weak predissociation. Since lifetime measurements cannot be made with a better precision than a few percent, such effects cannot be observed by lifetime measurements. We shall see however that, if another predissociation occurs which interferes with the very weak one under study, it may become possible to enhance the very weak predissociation by interference with the larger one, and therefore to measure $1/\tau_d$, even when it is much smaller than $1/\tau_r$.

GYROSCOPIC AND HYPERFINE PREDISSOCIATION OF IODINE

   A weak predissociation effect is typically what occurs in the $B^3\Pi_{O_u^+}$ state of molecular iodine : following an optical excitation from the $X^1\Sigma^+_g$ ground state, the molecule can either emit a strong fluorescence spectrum, or be transferred to a dissociative lu state. The two potential curves cross each other for $V \sim 3$ (Figure 2).
If H_D is the perturbing Hamiltonian, the rate of predissociation is given by

$$\frac{1}{\tau_d} = \sum_{J'M'} 2\rho(E) |<B^3\Pi_{O_u^+},J,M|H_D|1u,J',M'>|^2 . |<v|v'>|^2 \qquad (2)$$

where $\rho(E)$ is the density of states in the lu state and $<v|v'>$ is an overlap integral between a vibrational wave function of e and that of the lu level of same energy (Figure 3).

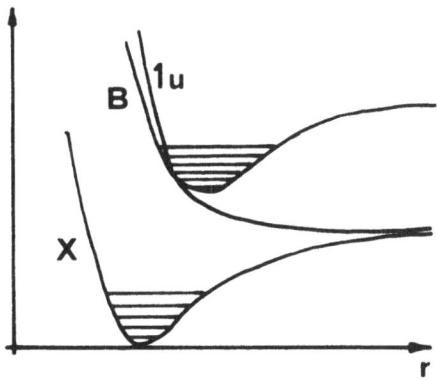

Figure 2 - Energy levels of I_2

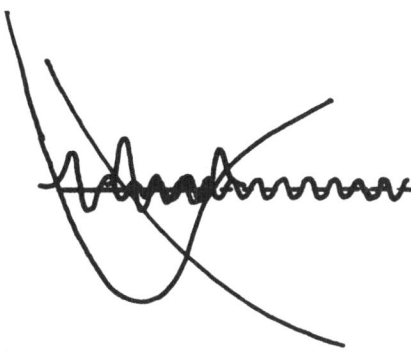

Figure 3 - $|v>$ and $|v'>$ wave functions

Depending on the nature of H_D, selection rules on $J - J'$ and $M - M'$ must be taken into account.

Obviously, $\langle v | v' \rangle$ presents a strong maximum close to the crossing point of the B and 1u potential curves.

Figure 4 shows the actual behaviour of this term with the vibrational energy E_V in the B state. The dissociation limit of B occurs for $v \sim 80$.

For symmetry considerations, H_D cannot be in the case studied here the spin orbit Hamiltonian so often involved in predissociations. Then the next term that one can think of is the rotational Hamiltonian

$$H_R = \frac{(\vec{J} - \vec{L} - \vec{S})^2}{2 \mu r^2}$$

where $\vec{J}$ is the total angular moment of the molecule, $\vec{L}$ and $\vec{S}$ being the electronic orbital and spin moments.
One can then show that ([5])

$$\langle B,J,M | H_R | 1u,J',M' \rangle = \langle B,J,M | \frac{-2\vec{J}(\vec{L}+\vec{S})}{2\mu r^2} | 1u,J',M' \rangle$$

$$= B_1 \sqrt{J(J+1)} \; \delta_{JJ'} \, \delta_{MM'} \tag{3}$$

where B_1 is somewhat similar to the rotational constant B in the $B^3\Pi_{0_u^+}$ state (some different average has to be taken of $1/r^2$).

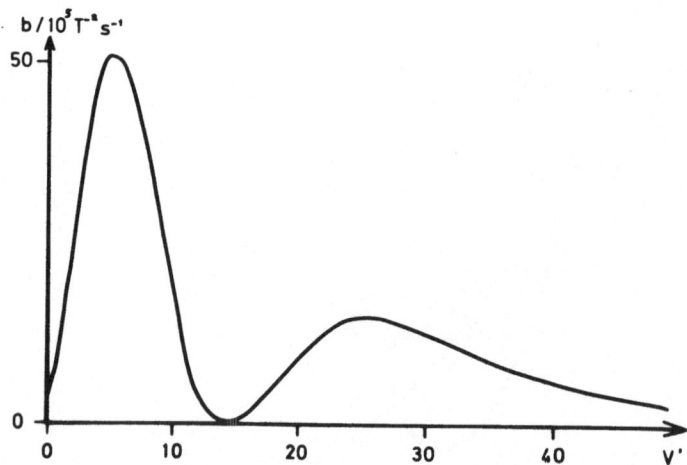

Figure 4 - $|\langle v | v' \rangle|^2$ versus the vibrational quantum number in the B state of I_2

The lifetime of a given rovibronic state is therefore given by

$$\frac{1}{\tau} = \frac{1}{\tau_r} + k_v J(J + 1) \qquad\qquad (4)$$

(collision effects are neglected in this expression and can be eliminated in the experiments by extrapolating to zero vapour pressure)
k_v is essentially proportional to $|<v|v'>|^2$. To check expression
(4), we have measured the lifetimes of series of rotational and vibrational levels of the B state of I_2. They were excited by a pulsed
dye laser with a pulse duration of about 3 nsec. The spectral
width of the laser, of the order of 1 GHz, was such that a single
rotational level could be excited, but all its hyperfine components
were excited with about the same efficiency. Simple spectroscopic
considerations on the fluorescence of the molecules gave us the
exact quantum numbers v and J of every state which was studied, and
by recording the exponential decay of the fluorescence, it was possible to measure its lifetime with an accuracy of 5 to 10%.

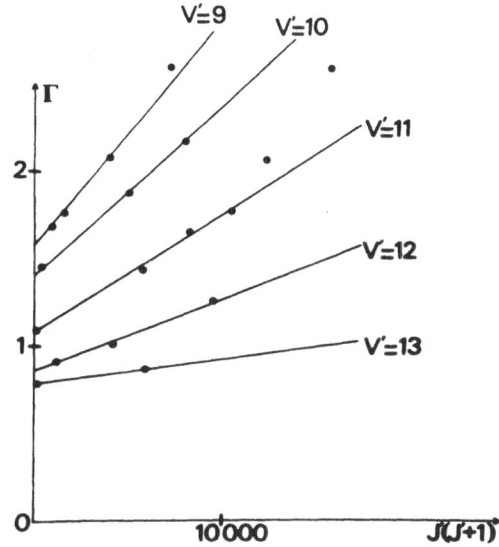

Figure 5 - $1/\tau$ versus J(J+1) for some vibrational levels of
the B state of I_2

On Figure 5 are given some values of $1/\tau$ obtained by this technique. They essentially show that it is indeed linear in J(J+1) although some deviations that we shall explain later on occur for large values of J. From the slopes of such curves, one can deduce k_v and, from an extrapolation to J=0, the value of $1/\tau_r$.

Figure 6 shows some results for k_v. They have been deduced from the slopes of the curves of Figure 5 for the largest ones, and from measurements presented in the next section for the smallest ones. They indeed reproduce the behaviour of $|\langle v|v'\rangle|^2$. Their absolute values give a good idea of the intensity of the gyroscopic predissociation.

Figure 7 shows the results obtained for $1/\tau_r$ by extrapolating $1/\tau$ to J = 0 : although it exhibits a slowly varying background which behaves as expected for $1/\tau_r$, a large variation remains which seems to reproduce the $|\langle v|v'\rangle|^2$ variations with v . This suggests that another predissociation is also present which had not yet been taken into account. The next term in the Hamiltonian that one can think of at this point is the hyperfine coupling : H_{hf}. It contains many terms which can be evaluated, but it seems that the largest one involved for this predissociation is the nuclear spin-electronic spin term. This corresponds to the fact that the dissociative 1u state is mainly a singlet state ($^1\Pi_{1_u}$), and that therefore the predissociation effects from a triplet state (the B state) require $\Delta S = 1$, which is actually possible with the spin spin hyperfine Hamiltonian.

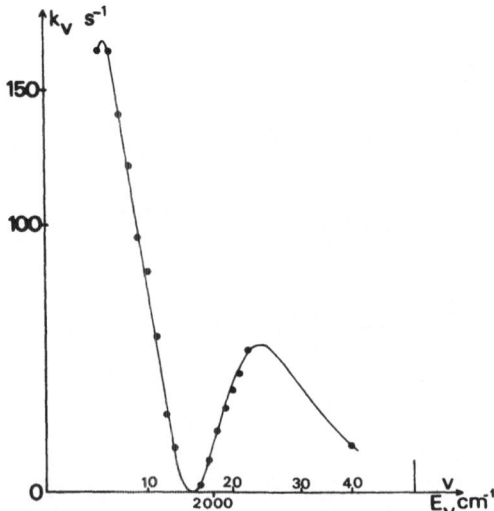

Figure 6 - k_v, gyroscopic predissociation coefficient

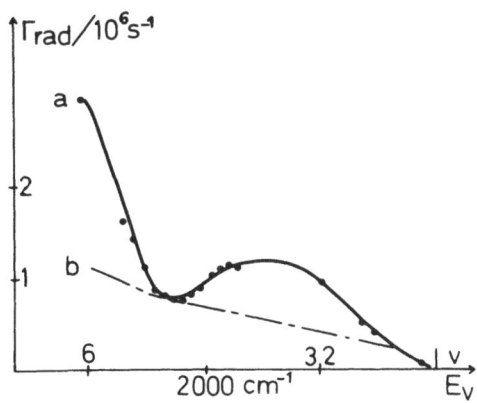

Figure 7 - Apparent radiative decay rate of the B state of I_2

A correct evaluation of the predissociation rate gives an expression proportional to ([6]) :

$$|<B^3\Pi_{O_u^+},v,J,I,F|H_R + H_{I,S}|1u>|^2$$

Three terms appear in this squared matrix element, so that

$$\Gamma_P = \Gamma_R + \Gamma_{hf} + \Gamma_{Int} \tag{5}$$

$$\begin{cases} \Gamma_R = |c_v|^2 \ J \ (J+1) & (6a) \\[2mm] \Gamma_{hf} = \dfrac{|a_v|^2}{3} \ \{\ \vec{I}^2 + \dfrac{3(\vec{I}.\vec{J})^2 + \dfrac{3}{2}\vec{I}.\vec{J} - \vec{I}^2.\vec{J}^2}{(2J-1) \ \ (2J+3)} & (6b) \\[4mm] \Gamma_{Int} = -\ \sqrt{2}\ a_v\ c_v\ \vec{I}.\vec{J} & (6c) \end{cases}$$

$|c_v|^2 = k_v$; $|a_v|^2$ is roughly proportional to k_v. Γ_{Int} is a quantum interference between the two ways of predissociation. I is the total nuclear spin of the molecule ($\vec{I} = \vec{I}_1 + \vec{I}_2$).

Since each ^{127}I nucleus has a 5/2 nuclear spin, I can take any integer value between 0 and 5. However, since the molecule is homonuclear, and taking into account the symmetry of the B state, for even values of J, one can have only I = 1, 3 or 5. They are called ortho states. Each one has 21 hyperfine sublevels labeled by the values of I and F ($\vec{F} = \vec{I} + \vec{J}$).

For odd values of J, I = 0, 2 or 4 (<u>para states</u>) and one has only 15 hyperfine sublevels.

Due to the gyroscopic and hyperfine predissociations, τ now depends not only on v and J, but also on I and F. That is <u>each hyperfine sublevel has its own lifetime.</u>

One way to observe directly the effect of the hyperfine predissociation would be to try another isotope, if possible with no nuclear spin (Notice the importance that such a hyperfine predissociation may have for isotope separation !). Unfortunately, iodine has no low nuclear spin stable isotope. However, one can show that, due to the different values of I, <u>ortho and para states behave as different isotopes and must have different average lifetimes.</u> (If each nucleus had a 1/2 nuclear spin, as in H_2, the para states for which I = 0 would not be effected by a hyperfine predissociation). Figure 8 shows the measured values of $1/\tau$ for a series of ortho and para states of the v = 7 vibrational level of the $B^3\Pi_{O_u^+}$ state of I_2. (For this level, the predissociation rate is close to its maximum). One can clearly see two series of lifetimes and this confirms the presence of a hyperfine predissociation effect. It should also be noticed that since the hyperfine structure is here of the same order as the Doppler broadening of the lines, the lifetimes measured concern always the whole hyperfine structure of a given rotational level. Since they correspond to different predissociation rates, the fluorescence decay is not exponential and the experimental observations actually fit very well with the predicted

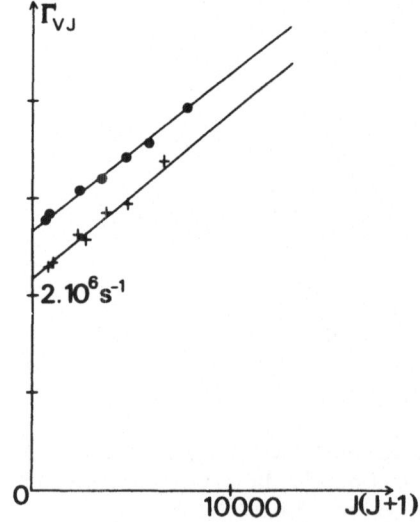

Figure 8 - Evidence of a hyperfine predissociation in the B state of I_2

behaviour. Lifetime measurements on individual hyperfine sublevels are presently on their way.

Conclusion : We have shown in this section that, by systematic measurements of lifetimes over series of rotational and vibrational states of a molecular electronic energy level, it is possible to understand the predissociation mechanism of this state. A completely new, hyperfine predissociation effect has been observed in I_2. It may have some application in isotope separation. To give some order of magnitude, for v = 7 that is close to the maximum of the predissociation rate, one has the following figures :

. radiative decay rate $\dfrac{1}{\tau_r}$ = $(1.1 \pm 0.1)~10^6~sec^{-1}$

. gyroscopic predissociation decay rate $\sim~10^6~sec^{-1}$ for J $\sim$ 60

. hyperfine predissociation decay rate $\sim$ 0.5 to $8.10^6~sec^{-1}$ depending on I, F and J. ($|a_v|^2 = (280 \pm 40)~10^3~sec^{-1}$)

MAGNETIC PREDISSOCIATION AND INTERFERENCE EFFECTS

I would now like to give an example of how very weak predissociation rates, that is when $\tau_d \gg \tau_r$, can be detected and measured.

Orientation by Predissociation

If one applies a magnetic field $\vec{B}$, it appears that the electronic Zeeman Hamiltonian

$$H_Z = \mu_o~\vec{B}.(\vec{L} + 2\vec{S})$$

also couples states B and 1u of I_2. It therefore induces a magnetic predissociation. Its rate is, close to the maximum (that is for v $\sim$ 7) of the order of $10^6~sec^{-1}$ per $(Tesla)^2$. It is then possible to tune this predissociation rate from zero up to the strong predissociation limit ($\tau_d \ll \tau_r$). This is a wellknown phenomenon and was observed as early as 1914 by R.W. Wood and G. Ribaud ([7]) as a "magnetic quenching" of the fluorescence of I_2.

However, if one comes back to the predissociation rate, it is now proportional to

$$|<B^3\Pi_{o_u+},J,M|H_R + H_{hf} + H_Z|1u,J',M'>|^2$$

Neglecting for the moment H_{hf}, one obtains three terms : the pure gyroscopic predissociation, the pure magnetic predissociation, plus an interference term. It results in an expression such as :

$$\frac{1}{\tau_d} = |c_v|^2 \ J(J+1) \ + \ |\alpha_v|^2 B^2 \ \frac{J^2+M^2}{2J^2} \ + \ \sqrt{2} \ c_v \ \alpha_v \ \vec{B}\cdot\vec{J} \qquad (7)$$

where $|c_v|^2$ is equal to k_v introduced in section I, c_v and α_v being both proportional to $<v|v'>$.

The last term of equation (7) is equal to $\sqrt{2} \ c_v \ \alpha_v$ B ℏ M and therefore introduces a rate of predissociation which depends on the <u>orientation</u> of the molecule with respect to the magnetic field.

In a steady state situation, for a molecular vapour excited by a laser and subjected to the natural and magnetic predissociations, the density matrix in the excited state has therefore non zero <u>orientation</u> terms (due to the interference term in equation (7)) and <u>alignment</u> terms (due to the pure magnetic predissociation term in equation (7)).

To observe the orientation, one can look to fluorescence circularly polarized in σ^+ or σ^- with respect to the magnetic field. Provided that a monochromator is used to observe individually the P and R lines of the fluorescence spectrum, one indeed observes a circular polarization in the presence of a magnetic field. More precisely, the fluorescence intensity polarized in σ^+ for a P line starts to increase with the magnetic field before it decreases as predicted by the magnetic quenching. However, the light polarized in σ^- decreases regularly to zero when B increases. Such a behaviour is shown on Figure 9. From the maximum rate of circular polariza-

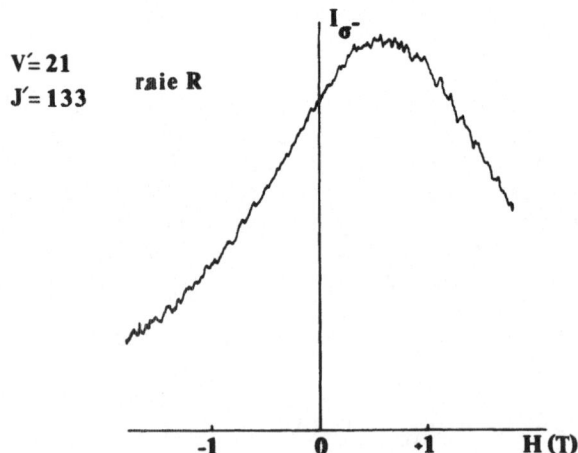

Figure 9 – Circularly polarized fluorescence light of I_2 versus the magnetic field

tion, one can obtain the value of c_v. It is easy to show that this technique, based on the interference effect between the magnetic and gyroscopic predissociations, is actually sensitive to the value of c_v / Γ_r . More precisely,

$$\frac{\tau_r}{\tau_d^G} \neq (2\rho_c)^2 \qquad \text{where } \rho_c \text{ is the rate of circular polarization}$$

close to the maximum of Figure 9, τ_r, the radiative decay rate, and τ_d^G, the gyroscopic predissociation rate.

One can therefore obtain a value of c_v even if $1/\tau_d^G = k_v J(J+1)$ is much smaller than $1/\tau_r$.

Most of the points on Figure 6 for $v > 10$ were actually measured by this technique.

With an accuracy of about 10% in the measurement of the circular polarization rate, one can detect and measure predissociation rates of the order of 1% of the radiative decay rate. It is clear that, without much difficulty, it is possible to increase the sensitivity and to measure predissociation rates of 0.1% of $1/\tau_r$. If polarization rates of 1% were measured, one would obtain a sensitivity of 10^{-4} times the radiative decay rate.

Notice that, unlike a lifetime measurement technique in which one attributes the variations of τ with v and J to predissociation effects, this interference technique gives a direct access to the weak predissociation effect provided that the large one is known.

N.B. Since the magnetic predissociation rate can be made arbitrarily high, measurements of very small values of α_v (formula (7)) can be made and this with a good accuracy. This has permitted to follow very precisely the variations of $<v| v'>$, especially around $v = 14$ for which it goes through zero. It appears actually that α_v also depends on J, and this can be explained by the deformations of the potential curves of the molecule due to centrifugal forces. These variations of α_v with J give a way to extrapolate almost continuously the values that $<v|v'>$ takes between two integer values of v. This dependance on J is also found for c_v and $k_v = |c_v|^2$, the gyroscopic predissociation coefficient. It explains partially the departure from a linear dependance in $J(J+1)$ of the curves of Figure 5. To fully explain these curves, one must also take into account the hyperfine predissociation effects, the rate of which also depends on J (formula (6)).

CONCLUSION

Selective excitation of molecular energy levels used here
appears to offer many new and exciting possibilities for molecular
physics.

It has been shown in this paper how lifetime measurements can
give informations on small predissociation effects. Most probably
many molecular energy levels which are presently believed to be
stable are actually subjected to weak or very weak predissociations.
It should be possible to study these effects and therefore to go one
step further in the understanding of simple molecules. Magnetic predissociations should also exist for many molecules and, by combining lifetime measurements with the use of high or very high magnetic fields, such effects could be studied.

Finally, one must emphasize here the very general idea that a
quantum interference effect in which a very small effect interferes
with a large one, always gives a possibility of measuring very small
quantities. This idea can be applied in many fields and, for example, was used by Bouchiat and Pottier [8] to measure very small
transition probabilities in atomic caesium.

BIBLIOGRAPHY

[1]   see for example G. Herzberg, "Spectra of diatomic molecules"
       p. 405 - D. Van Nostrand Company.

[2]   P. Erman - Proc. of the 4th Int. Beam Foil Conf., Gatlinburg,
       USA, Sept. 1975.

[3]   M. Broyer, J. Vigué and J.C. Lehmann - J. Chem. Phys. 63, 5428
       (1975).
[4]   L. Brewer and J. Tellinghuisen - J. Chem. Phys. 56, 3929
       (1972).

[5]   A. Chutjian - J. Chem. Phys. 51, 5414 (1969).
       J. Tellinghuisen - J. Chem. Phys. 57, 2397 (1972).
       J. Vigué, M. Broyer and J.C. Lehmann - J. Chem. Phys. 62, 4941
       (1975).

[6]   M. Broyer, J. Vigué and J.C. Lehmann - J. Chem. Phys. 64, 4793
       (1976).

[7]   W. Steubing - Verhandl d. Deutsch. Phys. Ges. 1181 (1913).
       R.W. Wood and G. Ribaud - Phil. Mag. 27, 1009 (1914).
[8]   M.A. Bouchiat and L. Pottier - J. de Phys. 37, 279 (1976).

TIME RESOLVED LASER SPECTROSCOPY : QUANTUM BEATS AND SUPERRADIANCE

S. HAROCHE, C. FABRE, M. GROSS and P. PILLET

Laboratoire de Spectroscopie Hertzienne de l'E.N.S.

24, rue Lhomond, 75231 PARIS CEDEX 05 - France

I. INTRODUCTION

The purpose of this paper is to describe some recently performed time resolved transient experiments in Laser Spectroscopy. Experiments in this field have developed very fast in the last few years and allowed to study a wide range of light-matter interaction effects, some of which being mere extensions to the optical range of phenomena already studied in microwave or radiofrequency spectroscopy.

A transient experiment using a laser may quite generally be described in the following way :

  i) in a first stage, one prepares with a proper laser excitation the atomic or molecular system under study in a well defined non equilibrium initial state.
 ii) The system thus prepared starts to evolve and to couple into the radiation field.
iii) One detects the temporal behavior of the emitted radiation and therefore gets useful informations about energy levels, relaxation processes or radiative properties of the system.

Such experiments may be divided in two quite different types:

a) "coherent pumping experiments" which imply, in the first stage, preparation of a macroscopic optical dipole in the sample.
b) "incoherent pumping experiments" in which only populations or radiofrequency coherences between closely spaced states are prepared and which do not require any initial optical dipole in the medium.

Photon echo ([1]), free induction decay ([2]), optical nutation ([3]), Raman beats ([4]) belong to the first category, whereas Quantum Beats ([5]) and Superradiance ([6]) are typical experiments of the second kind.

Although these two kinds of experiments have, as we will see, some common features, they may actually be distinguished according to very important theoretical and experimental criteria.

Theoretically, type a experiments may be completely understood within the frame of the semi-classical theory of light-matter inter-action, since they involve radiation from a macroscopic "classical" array of dipoles coherently prepared throughout the sample. On the other hand, type b experiments, which we will call Time-resolved fluorescence experiments, cannot be fully understood without reference to Quantum Theory of radiation, since the emission involves fluorescence from an upper state which cannot radiate according to classical theory.

Experimentally, observation of kind a effects requires the use of a narrowband single mode laser able to perform π or $\pi/2$ pulses in the system and thus to induce an optical coherence. Type b experiments may, on the contrary, be performed with rather coarse broad band lasers which provide a simple and convenient source for incoherent pumping.

It is of course not possible to review here exhaustively such a wide range of effects. I have chosen to restrict myself in the following to the description of some interesting incoherent pumping time resolved fluorescence experiments.

In a first part, I will briefly recall recent results of quantum beat experiments and describe some related modulation transient experiments. I will then show that quantum beat experiments are well suited for the observation of superradiance or collective spontaneous emission effects. This is a phenomenon which has been subject to a lot of theoretical interest in the last twenty years ([7]), although it has up to now been only scarcely studied experimentally. I will try to show in the second part of this paper that this effect may now be investigated in some very simple experimental situations in alkali atoms and may even lead to interesting applications.

2. QUANTUM BEATS AND RELATED MODULATION TRANSIENT EXPERIMENTS

2.1. Quantum Beat Spectroscopy ([5]) (some recent results)

A very interesting feature of the fluorescence transient following a pulse excitation is the possibility of observing a time modulation in the light signal. This modulation, called Quantum Beats, arises from the fact that the light pulse may prepare the atomic system in a coherent superposition of excited substates of slightly different energies (note that this is a radiofrequency type of cohe-

rence inside an atomic level and <u>not</u> an optical coherence). Provided this fluroescence is detected with a proper polarization, the
evolution of this atomic excited state coherence may reveal itself
by oscillations in the fluorescence light at the Bohr frequencies
corresponding to the excited state splittings. A typical Quantum
Beat Spectrometer is shown on Figure 1.

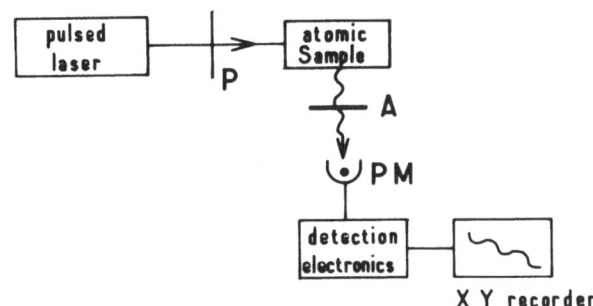

Figure 1 : *Sketch of a "Quantum Beat Laser Spectrometer"*

The laser beam produced by a pulsed laser excites the atoms in the
sample, which subsequently radiate a fluorescence signal detected
off-axis by a photomultiplier tube PM connected to a fast transient
analyzer. A polarizer P and an analyzer A are placed in front of
the laser and of the photomultipler and oriented to get optimum
modulation depth in the Beat pattern ([5]).

The analysis of the beats, which are practically not subject
to any Doppler spread , provides a very convenient way of determining with a very high resolution small energy structures in excited
or atomic molecular states. Since this effect is now well known and
is described in review articles ([5]), I will just recall here briefly some recent experimental results obtained in the study of atomic
Rydberg states of sodium.

The energy diagram of the Na atom is sketched on Figure 2.
Let us consider in this diagram the series of nD levels. In order
to study how the fine structure of these levels varies when the principal quantum number n is increased, we have performed a series of
fine structure Quantum Beat measurements in these states ([8])([9]).
The excitation of the levels is performed by a stepwise process via
the 3P state by two synchronously pulsed broadband dye lasers.
The fluorescence signal is detected on the transitions back to the
3P state. Figure 3 shows some recorded beat patterns emitted by
levels nD with n varying between 9 and 16. Note the continuous
slowing down of the beat frequency when n increases, which corresponds to the decrease of the fine structure splitting as one gets
higher in the Rydberg series. By Fourier analysis of these beats,

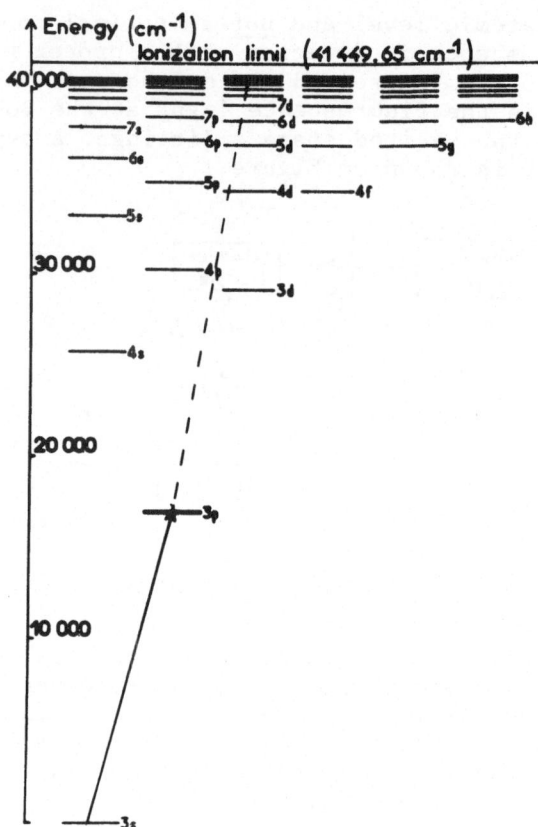

<u>Figure 2</u> : *Energy diagram of Na showing the stepwise excitation*
process which allows the preparation of nD states

one gets resonance peaks whose center frequency corresponds to the
fine structure interval of the level. Figure 4 shows such a resonance for n = 10, for different values of a small electric field
applied to the sample. From the large Stark shifts thus observed,
one can deduce the huge polarizability of the levels ([10]), and
from the sign of the shift (positive), one gets the sign of the
fine structure in the nD Na levels which happens to be negative
(inverted structure) ([9])([10]). These experiments show that Quantum
Beat spectroscopy is a very simple and elegant technique, well
adapted to the systematic study of series of very excited atomic
levels. Other closely related transient methods, which may in some
cases be more advantageous can also be developped, which we will
now try to review briefly.

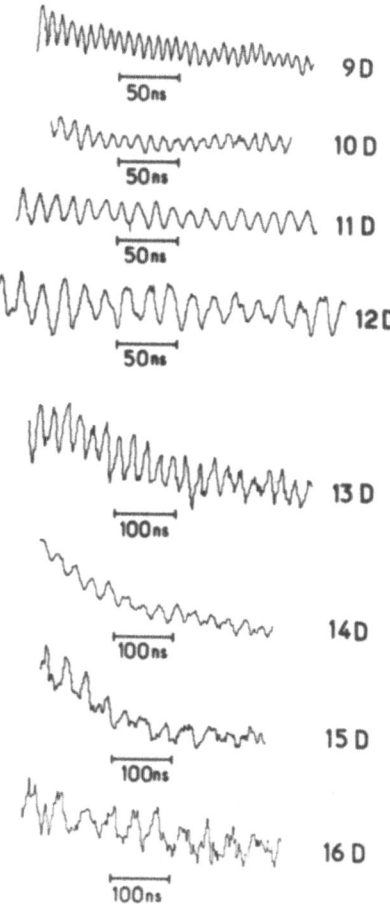

Figure 3 : *Fine structure Quantum Beats observed in levels 9D to 16D of Na (from ref. (9)).*

2.2. Related Transient Methods : Absorption and Stimulated Emission Beats

Atomic fluorescence is not the only signal which is sensitive to the radiofrequency coherence induced in an excited level by a broadband pulse. Absorption and stimulated emission processes starting from this level depend also on these coherences and the corresponding cross-sections may thus also exhibit modulation beats. The principle of these experiments is summarized on Figure 5, where we have sketched the typical level diagram of an atomic system in which all these effects could be observed : the exciting pulse

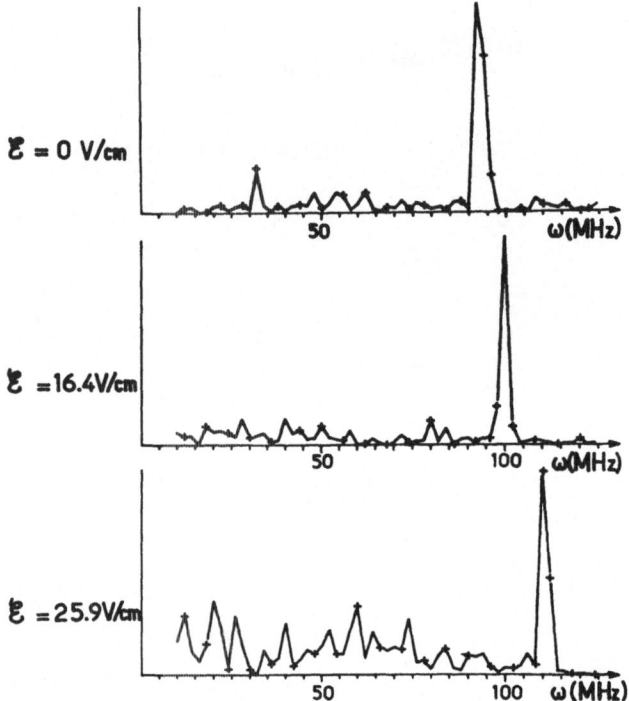

Figure 4 : *Fourier spectra of the fine structure beat signal in level 10D of Na for three values of a small applied electric field (from ref. ([10]))*

brings the atoms from ground state a to a coherent superposition of substates in b (full line arrow). This coherence may reveal itself through Quantum Beats in fluorescence down to level c (wavy line arrow n° 1) but also through modulation in the absorption cross section of a conveniently polarized probe laser beam tuned to a transition connecting level b to an upper level d (arrow n° 2). This effect has been in fact very well known for a long time in optical pumping experiments. In these experiments, precession of atomic ground state coherences (for example Larmor precession of a nuclear spin) could be detected as a modulation of the absorption of the pumping beam. The experiment sketched in Figure 5 is a mere extension of this type of optical pumping effect to excited state absorption. It has actually been recently observed by Ducas et al.([11]) in an experiment on the $3P_{1/2}$ level of Na. A first laser beam prepared the $3P_{1/2}$ state in a coherent superposition of hyperfine structure levels. Then a second laser pulse tuned to a $3P_{12}$–nS transition was sent on the system and its absorption was monitored by counting the number of ions produced by field ionization of the upper nS level. Modulation of the absorption was

displayed by delaying the probe pulse with respect to the first
one. It is worth noting that such a technique does not rely any
longer on a fast electronic detection, but that time resolution
depends only on the duration of the exciting and probe pulses.
This method should be chosen as an alternative technique for
Quantum Beats when studying long lived metastable states which do
not exhibit much fluorescence. A variant of the same experiment
could be tried in which level d would be replaced by a continuum

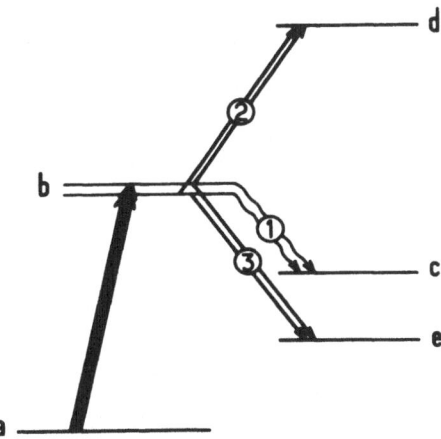

Figure 5 : *General energy level scheme of an atomic system excited
in a coherent superposition of substates in level b
which may exhibit fluorescence, absorption and stimulated emission beats (arrows n° 1, 2 and 3 respectively)*

set of states above the atomic ionization threshold : one would
then observe beats in the photoionization cross section from
level b.

     At last, one could also observe modulations in the stimulated
emission induced by a delayed probe beam on a transition to a level e below level b (arrow n° 3 in Fig.5). This experiment could
be very interesting for spectroscopy of highly excited Rydberg
states for which fluorescence is too small for ordinary Quantum
Beats to be easily observed and which have also very small cross-
sections for laser absorption into the continuum of states. For
such levels, induced emission towards lower levels could be saturated by strong dye laser irradiation and "stimulated Quantum Beats"
observed. An experiment along these lines is under way in our
laboratory.

2.3. "Anomalous" Behavior of Fluorescence in Quantum Beat
Type of Experiments at High Excitation Densities

The Quantum Beat phenomenon and the related effects discussed
above are essentially <u>single atom effects</u>. In these experiments,
it is assumed that the atoms evolve independently from each other
in the sample, the resulting signal being proportional to the number N of emitting atoms. This feature is essential in the understanding of the phenomenon. In particular, it is important to note
that the Quantum Beat modulation arises within each atom in the
sample and is in no way a beating between frequencies emitted by
different atoms at different frequencies, in which case one would
observe Doppler broadened beat patterns. As a result, when in a
Quantum Beat experiment, the density of initially excited atoms is
increased (for example by increasing the light pulse excitation),
one expects that the size of the various fluorescence signals will
increase in proportion, without any change in their time behavior.
This is true, at least until the initial excitation density reaches
a given threshold above which "anomalous" non linear effects start
to show up in the fluorescence pattern. For example, in the Na
Quantum Beat experiment, one observes a complete quenching of the
Quantum Beat pattern above an excitation density of about
10^8, 10^9 atoms/cc. At the same time, one also observes "anomalous"
branching ratios in the fluorescence emitted in cascades down from
the initially excited level. An example of such anomalous cascades
is shown in Figure 6 which represents a recording of the fluorescence versus time at 8191 Å on the 3d-3p transition after excitation of the 5s level (see scheme of levels in the inset of the
figure).. At low excitation densities (fig. 6-a ; $n_o < 5 \times 10^{10}/cm^3$)
the amount of fluorescence detected by the off-axis PM at 8191 Å
is very small as is expected from the branching ratios at
5s-4p-3d cascade (1%). Almost all the atoms cascade down either
directly from 5s to 3p or on the 5s-4p-3s and 5s-4p-4s cascades
and only a very small proportion is channelled through the 3d state.
When n is increased above $5 \times 10^{10}/cm^3$, one suddenly observes a
drastic change in the <u>fluorescence</u> from the 3d level. A burst of
fluorescent light is observed at 8191 Å, the peak of fluorescence
being delayed by a few tens of nanoseconds with respect to excitation.
This delay decreases with increasing excitation densities as it
appears on fig. 6-b, c, d.

Such a behavior of the atomic fluorescence is an evidence
that a delayed and very strong radiative transfer between levels
5s and 3d is taking place in the cell above the threshold pressure.
This transfer, which corresponds to a strong modification of the
branching ratio of ordinary transition probabilities is obviously
a collective emission effect, since it depends on the number of
initially excited atoms in the sample. Such a collective effect
in fast transient type of experiments is of course reminiscent of

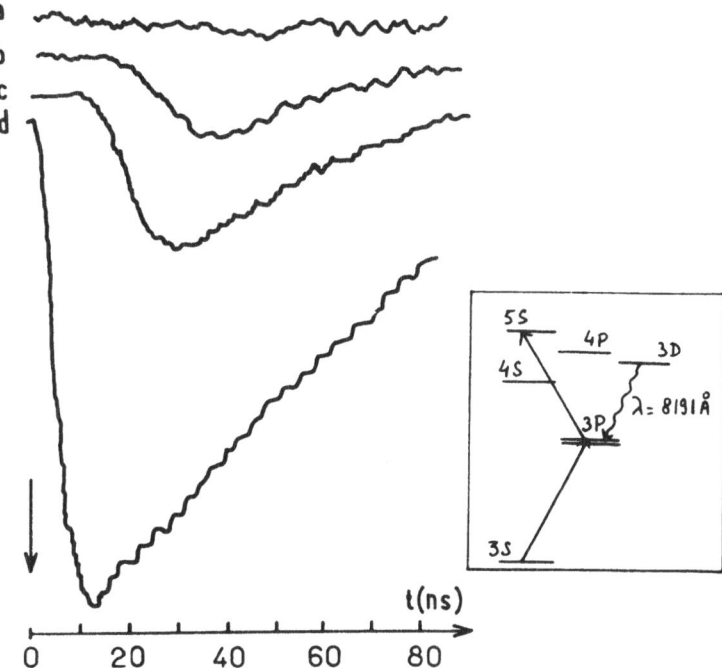

Figure 6 : *Time variation of fluorescence signal at 8191 Å (3D–3P transition) detected by off-axis photomultiplier following excitation of the 5S state at time t = 0 (arrow indicates pumping time). Traces a to d correspond to increasing pumping beam intensities at a given Na pressure.*

the superradiance phenomenon which has been long ago predicted by Dicke ([6]), but only scarcely studied in the optical domain. We show indeed in the next section that it is possible to observe an emission phenomenon closely related to Dicke superradiance in simple alkali atom systems.

3. OBSERVATION OF SUPERRADIANCE IN EMISSION FROM EXCITED ALKA-
   LI ATOMIC STATES

3.1. What is Superradiance ?

Before describing our experiments, we will recall briefly the main features of the superradiance phenomenon ([7]). It is a fast cooperative spontaneous deexcitation process in which an ensemble of atoms emit an intense burst of radiation much faster than would be expected if the atoms were emitting independently. The pheno-menon is due to the fact that the atoms develop, through their

coupling to the radiation field, strong correlations between each
others in the first stage of the emission process. These correlations induce the build up of a large macroscopic dipole proportional to the density n_o of initially inverted atoms in the sample.
This dipole radiates a light field with an intensity proportional
to n_o^2 which reacts back on the system and damps it very fast
and efficiently. As the total energy released by the system is proportional to n_o, energy conservation requires that the emission
occurs within a characteristic time proportional to $1/n_o$.

The details of the superradiant emission depend, of course, on
such parameters as the size and shape of the sample, the nature of
the homogeneous and inhomogeneous relaxation processes which de-
phase the optical dipoles and so on...

For the case of experimental interest of a pencil-shaped active
volume of length L and cross section diameter much larger than
the wavelength λ of the emitted radiation ([12]), one expects the
superradiance to occur along the axial modes associated to the
emitting volume, with two symmetric lobes of emission in opposite
directions. The evolution of the superradiant pulse depends on a
characteristic time

$$\tau_R = 8\pi \ (n_o L\lambda^2 \gamma ab)^{-1} \ , \tag{1}$$

which is equal to the ordinary spontaneous emission time $\tau_{sp}=1/\gamma_{ab}$
divided by the number of initially excited atoms in a "diffraction
volume" $L\lambda^2/8\pi$. The superradiant pulse has a characteristic duration of a few τ_R and is emitted after a delay of a few tens of τ_R.

The phenomenon can be observed only if the superradiant pulse
builds up with a time τ_R shorter than the characteristic damping
time T_2^* corresponding to the various dephasing processes which
tend to destroy the optical dipole "aligned" by superradiance. The
condition for superradiance is thus

$$\tau_R < T_2^* \ , \tag{2}$$

which sets up a threshold for the initial inversion density in the
medium :

$$n_o > 8\pi \ (\gamma_{ab} \ T_2^* \ L\lambda^2)^{-1} \tag{3}$$

Below this threshold, ordinary isotropic non cooperative fluorescence dominates; above it, anisotropic superradiant emission takes
over.

3.2. Previous Experiment

Condition (2) explains why the phenomenon has been up to now
very difficult to observe experimentally in the optical regime.
In solid state amplifying media at room temperature, T_2^* is generally very short (in the psec range). This entails very short time

scale for the superradiant emission which is hence very difficult
to study. In gaseous media, T_2^* is generally the Doppler dephasing
time, which is of the order of a nsec in the visible near infrared
range, of the order of a few tens of nsec in the far infrared
(100 μ range).

It is thus not surprising that the first direct evidence of
superradiance in the optical range has been obtained in the far
infrared ([13]) with a microsecond time scale for the superradiant
emission (experiment on rotation transition of the HF molecule).
With the nsec long pulses produced by the tunable dye laser and
the fast electronics we used for our Quantum Beats experiments,
we have been able, as we show now, to observe similar but much
faster superradiant emissions for the first time in the near
infrared.

3.3. Observation of Superradiance in Na and Cs

We have at first observed superradiance on transitions cascading down from the $5S_{1/2}$ level of Na ([14]), this level being prepared
by stepwise excitation from the ground state via the intermediate
$3P_{3/2}$ level (Fig. 7-a). The excitation is provided by two synchro-

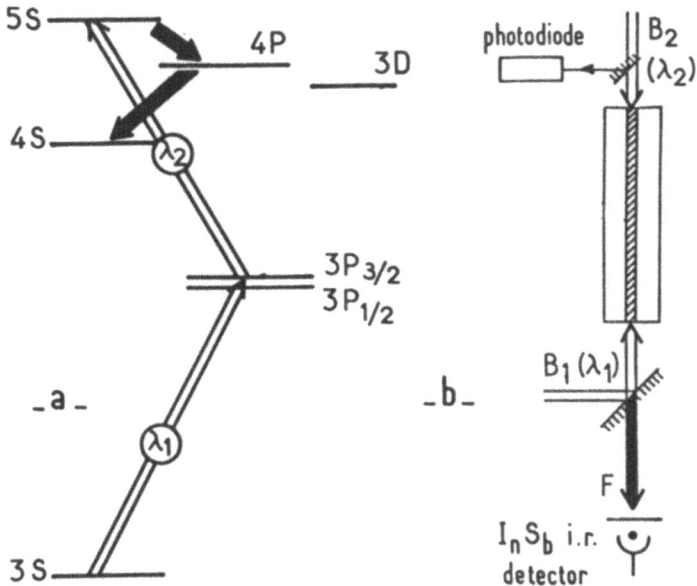

Figure 7 : (a) *Diagram of Na energy levels relevant for superradiance
experiment.   Double line arrows : pumping transition at $\lambda_1 =$
5890 Å and $\lambda_2 = 6160$ Å. Solid line arrows : superradiant infrared
transitions detected at $\lambda_3 = 34100$ Å and $\lambda_4 = 22100$Å.*
         (b) *Sketch of experimental set-up showing colinear pump-
ing beams B_1 and B_2 and on-axis InSb detector.*

nously broad band pulsed dye lasers at $\lambda_1 = 0.5890$ μ and $\lambda_2 = 0.6160$ μ. The laser pulses B_1 and B_2 are about 10 kwatt peak power, which is enough to saturate both λ_1 and λ_2 transitions so that about one quarter of the total number of atoms in the active volume can be prepared in the $5S_{1/2}$ state. The pulses propagate in opposite directions in the cell (Fig. 7-b) and overlap along a cylindrical 14 cm long active volume.

Detection of the superradiant emission is made by a fast infrared InSb detector placed on the axis of the active volume, with adequate infrared filters to select the various wavelengths expected in the emission. The detector output is connected to a dual trace fast transient recorder which records at the same time the signal from a photodiode receiving part of the exciting B_2 beam. The latter signal provides, after correction for propagation, the time origin for each sequence of pulses.

Before describing the results of the experiment, let us evaluate the order of magnitude of the threshold for superradiance on the 5s-4p transition under the above mentioned experimental conditions. We have to compare the spontaneous lifetime $\tau_{sp} = 188$ nsec and the Doppler dephasing time $T_2^\star = 1.7$ nsec of this transition. The ratio $\tau_{sp}/T_2^\star$ is about 110 and hence threshold will occur when 110 atoms will be initially inverted in a diffraction volume $L\lambda^2/8\pi = 6.10^{-8} \text{cm}^3$. This corresponds to a threshold density of 2.10^9 atom per cc. The threshold for superradiance is approximately the same for the 4P-4S transition at 2.21 μ. (In fact the actual threshold is expected to be in the range of 10^{10} cm^{-3}, somewhat higher than the above computed value, since superradiance is actually observable for T_2* larger than several τ_R [15]).

Above a Na pressure of about 10^{-6} torr($n_0 \sim 10^{10}$cm^{-3}), we have indeed detected in the output of the InSb detector two pulsed few nanosecond long directively emitted infrared signals whose wavelengths are 3.41 μ and 2.21 μ corresponding to the cascading 5S-4P and 4P-4S transitions. The 3.41 μ pulse follows laser excitation with a few nanosecond delay (longest observed delay 7 nsec). The 2.21 μ pulse is delayed by several nsec with respect to the 3.41 μ one. Figure 8 shows the exciting laser pulse B_2 (trace a) and typical recordings of the delayed pulses at 3.41 μ (trace b) and 2.21 μ (trace c). Two 3.41 μ pulses with different delays have been represented on trace b. The 2.21 μ pulse is obviously due to a cascading superradiant emission building up after the first emission at 3.41 μ has populated the intermediate 4p level.

To check the superradiant character of the observed emissions, we have systematically studied, for the 3.41 μ pulse, variations of pulses heights and delays as a function of the initial excitation density. This study is rather difficult to perform due to the

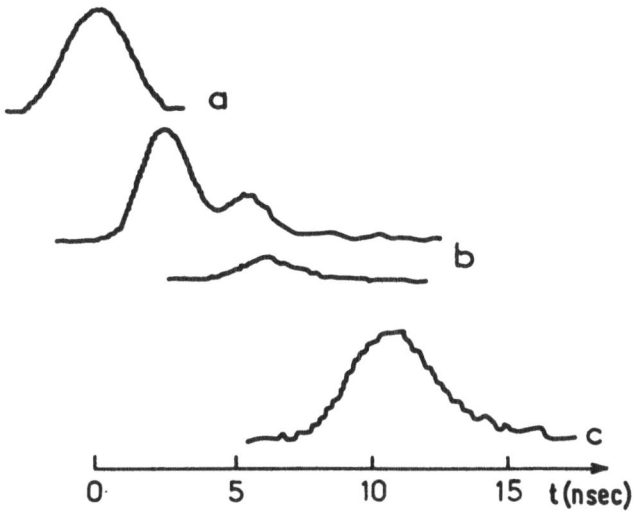

<u>Figure 8</u> : *Time variation of pulse signals emitted by the Na cell and monitored by the InSb detector. Trace a : exciting pulse B2; Trace b : 3.41 μ pulse for two different excitation intensities; Trace c : 2.21 μ pulse for same excitation as the second 3.41 μ pulse of trace b.*

fact that the superradiant phenomenon is very unstable in nature, small fluctuations in the pumping pulses intensity around threshold resulting in very large fluctuations in the intensities and delays of the emitted spikes. We have made two different and complementary checks :

a) In a first run of experiments, we have recorded a large set of pulses randomly distributed in intensities and delays and plotted the intensity h of each pulse against its delay τ_D. We have obtained the distribution of experimental points plotted in Fig. 9, which shows that h varies according to a τ_D^{-2} law (full line curve). This is the expected result for superradiance, since h should be proportional to n_o^2 and τ_D to n_o^{-1} .

b) In a second run of experiments, we have set the Na pressure at a given value around 10^{-5} torr and desaturated the pumping beam B_2 with calibrated attenuating filters.

In such a way, we prepared in the sample an initial density n_o of excited atoms proportional to the intensity I of the pumping beam B_2. To eliminate the fluctuations of the laser output, we have averaged for a few hundred runs the superradiant pulses for each value of I. We have then plotted the averaged value h versus I^2 and the averaged value τ_D versus I^{-1} (see Fig. 10).

We have checked that the signal increases as $n_o{}^2$ and is delayed as $n_o{}^{-1}$, which is a good evidence of its superradiance character.

Another interesting feature of the observed emission is the existence of ringings in the wings of the superradiant pulses which are clearly apparent in the largest 3.41 μ pulse of Fig. 8. Similar ringings have been observed by Skribanowitz et al. in the above men-

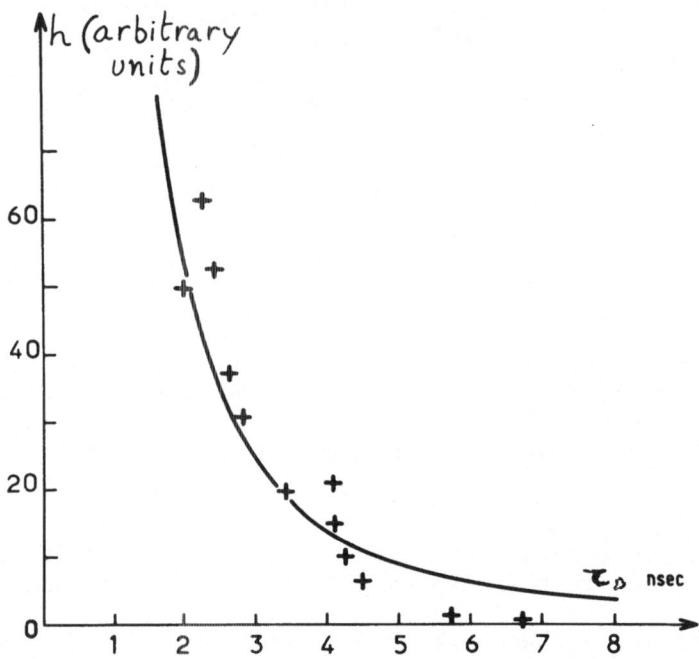

Figure 9 : *Variation of superradiant pulse heights h against their delay τ_D. Points are experimental. Solid line curve corresponds to an inverse square fit (h proportional to $\tau_D{}^{-2}$).*

tioned far-infrared superradiance experiment (^{13}) and explained by the semi-classical theory of superradiance developed by these authors (^{15}). These ringings are due to propagation effects in the superradiant sample. Computation of superradiant pulse propagation along the active medium shows that the polarization which builds up in the sample is not uniform and that at a given time the emission is produced by only a part of the cell. Other parts of the active volume can reabsorb the radiation, this phenomenon occuring at a rate corresponding to the Rabi flopping frequency of the optical dipoles in the emitted superradiant field. As a result the super-radiant pulse presents oscillations which may be understood as a kind of self-induced optical nutation. We have observed this ringing

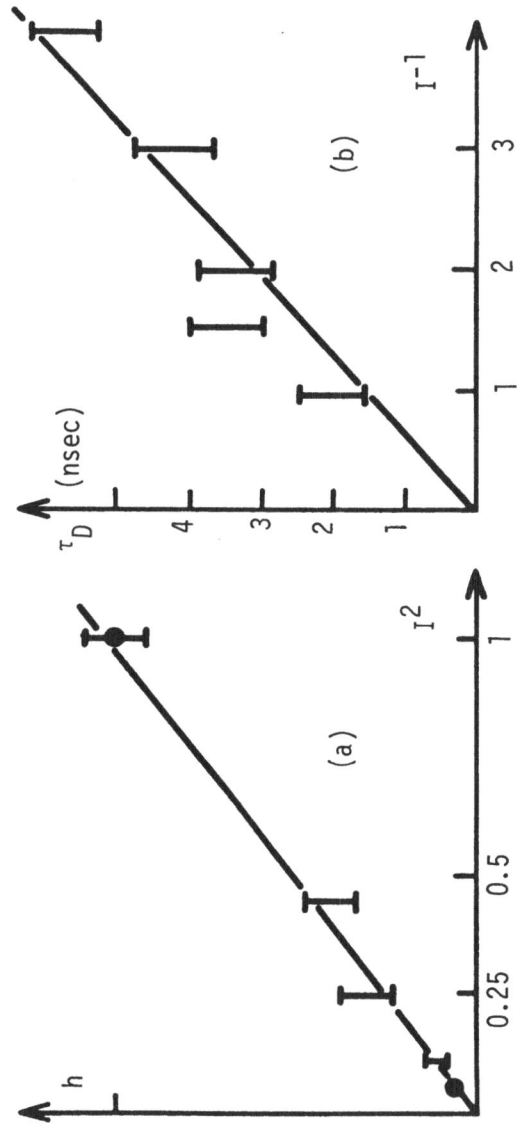

Figure 10 : (a) Height h of 3.41 μ pulse versus I², square of B₂ pumping beam intensity.
(b) Delay of 3.41 μ pulse versus I⁻¹ (I in arbitrary units). Each point results from an average of 100 runs.

effect even more clearly displayed in a similar superradiant experiment we have just set up in Cs vapor.   Figure 11 shows the energy
level scheme relevant for the experiment. A single dye laser pulse

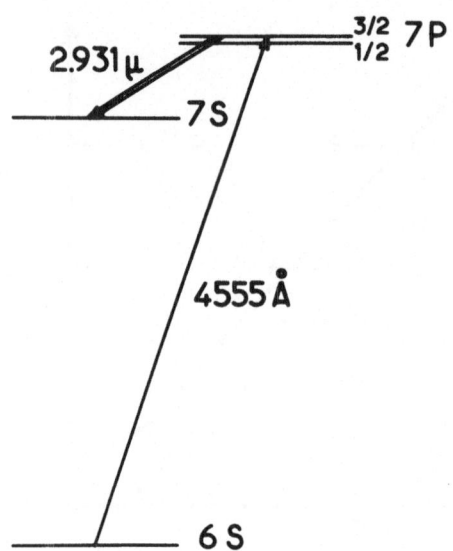

<u>Figure 11</u> : *Energy level scheme of Cs relevant for the Cs super-
radiance experiment and showing the 6S-7P pumping transition at 4555 Å and the 7P-7S superradiant transition
at 2.93 μ.*

excites in a very short time the Cs atom to the $7P_{3/2}$ level and the
superradiant emission is observed on the $7P_{3/2}-7S_{1/2}$ transition at
2.9 μ. Figure 12 shows a recording of the superradiant burst for
a large excitation density ($n_0 \sim 10^{11}/cm^3$) in a 3 cm long Cs sample
displaying very strong and fast optical nutation ringings.

All the features of the superradiant emissions we have observed
in Na and Cs are in reasonable agreement with prediction of the
semi-classical model of reference ([15]).   A detailed comparison between this theory and our experiments will be given elsewhere.

3.4. Superradiance and Anomalous Fluorescence in Na

The superradiant behavior of the Na sample described in the
previous section explains the characteristics of the anomalous
fluorescence pattern from the Na cell at high pressure (large cascade emission at 8191 Å). When superradiance threshold is reached

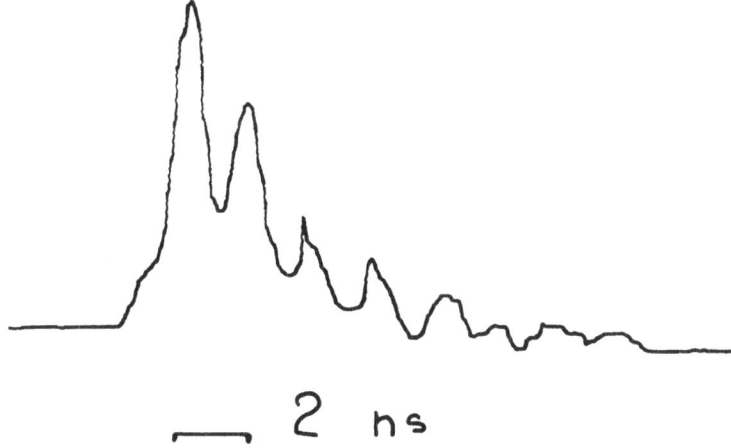

2 ns

<u>Figure 12</u> : *Time variation of superradiant signal emitted by the Cs cell. Note the fast and strong ringings due to self-induced optical nutation.*

on the 5s–4p transition ($n_o \sim 10^{10}$ cm^{-3}), there is a strong quenching on the ordinary fluorescence decay on the 5S–3P transition and a large number of atoms is channelled to the 4P state. These atoms then can "choose" between two competing transitions : the 4P–4S transition at 2.21 μ and the 4P–3D one at 9.1 μ. The former has a shorter τ_R ([14]), so that, for excitation densities in the range of 10^{10} cm^{-3}, superradiance is expected to occur rather at 2.21 μ than at 9.1 μ, which is what we have observed (see above). As the 2.21 μ emission quenches very fast the population of the 4P level, superradiance cannot at all occur at 9.1 μ and no strong population of the 3D state is expected in this case. The observed level of fluorescence at 8191 Å from the 3D state remains very low (see fig. 6–a). At higher excitation densities however ($n_o > 5.10^{10}$ cm^{-3}) the delays of the superradiant emissions at 3.41 and 2.21 μ become so short that the corresponding pulses start to occur in the wing of the exciting laser pulses. Superradiance is then counteracted by laser pumping so that atoms do not fall down any longer to the lower state of the 5S–4P–4S cascade but rather evolve towards population equilibrium between these states. As a result a large population inversion is left in the medium on the 4P–3D transition after the 3.41 μ and 2.21 μ emissions. The system can then superradiate after a longer delay at 9.1 μ. This emission cannot be directly detected by the on-axis infrared detector since the walls of the resonance cell absorb the radiation at this

wavelength. An indirect evidence for this emission is however given
by the sharp and delayed raise of off-axis fluorescence at 8191 Å
which indicates a fast and strong population of the 3D state (see
figures 6-b, c, d).

3.5. Connection with Other Collective Emission Effects

Some characteristics of the superradiant emission discussed
above are common to other well known light emission effects and
it is thus worthwhile insisting on the specificity of superradiance
as compared to these other phenomena.

At first, the n_o^2 dependence of the emission is reminiscent of
photon echo (^{1}) or free induction decay (^{2}) type of experiments
in which an array of dipoles coherently emit radiation in a given
direction. However, in these latter experiments, one has $\tau_R > T_2^\star$,
so that the decay of atomic polarization is due to dephasing relaxation processes and not to radiation reaction as it is the case
in superradiance. As a result, the damping time of the signal does
not depend, in these experiments, on the number density of excited
atoms n_o, but is characterized only by $T_2^\star$. In other words, in photon echo or free induction decay, the emission of radiation can be
considered as a probe to test the non radiative processes which
damp the system, whereas in superradiance, the radiation reacts
back on the system and damps it much more efficiently than all the
other non radiative processes.

Another important difference between superradiance and photon
echo lies in the fact that the former experiment is observable with
incoherent pumping, the dipolar polarization building up spontaneously from noise, whereas in the latter the polarization is introduced
at the beginning by a coherent optical excitation process.

It is also very instructive to compare the superradiance phenomenon described above to "amplified spontaneous emission" (^{16})
or "mirrorless laser action". At first sight, since condition (3)
for superradiance is equivalent to the condition for large gain in
the medium, one may think that the observed phenomenon is not essentially different from laser action on the infrared transitions pumped by the incoherent dye laser beam. This point of view is somewhat strengthened by the recollection that in the pioneering paper
(^{17}) on lasers by Schawlow and Townes, a pumping scheme similar to
the one of Figure 11 was proposed to get laser action in K vapor.
One thus should ask the questions : Are transient laser emission
and superradiant emission the same phenomenon ? Do all high gain
lasers behave as superradiant sources in their transient regime ?

The answer is obviously negative, since it is well known that
most transient or Q-switched lasers have an output proportional to
the number n_o of inverted atoms in the medium and not to n_o^2. To
understand where the specificity of superradiance comes from, let

us consider the most often theoretically considered case of an
instantaneously pumped medium ("side pumping" case). For such a
medium, it has been noted by several authors that superradiance
can be observed only if the amplifying medium is not too long ([18]).
The coupling between the different parts of the sample occurs indeed in a time of the order of the light propagation time L/c along
the sample. If this time is longer than the characteristic duration
time τ_D of the expected superradiant emission, the atomic dipoles
in the different parts of the cell cannot lock in phase with each
other and the cell cannot superradiate as a whole. Condition (3)
for superradiance should thus be completed, in the case of side
pumping, by condition

$$\frac{L}{c} \lesssim \tau_D \qquad\qquad (4)$$

which, since τ_D is of the order of magnitude, or shorter than $T_2^{\star}$,
sets up an absolute upper limit for the cell length

$$L \lesssim c\, T_2^{\star} \quad . \qquad\qquad (5)$$

If conditions (4) or (5) are not fulfilled, superradiance
"breaks in parts" in the cell ([18]) or even completely disappears,
although the overall gain of the sample may be very large. For a
side pumped Ruby laser rod for example, superradiance should occur
in such a short time ($\tau_R < T_2^{\star} \sim 10^{-11}$ sec) that the critical
length for superradiance is only a few mm. Hence, a ten cm long
high gain Ruby laser rod will not exhibit the superradiant behavior although it could very well lase. Our alkali atom experiments,
on the other hand, are favorable for superradiance for two reasons:

a) At first, the dephasing time $T_2^{\star}$ is rather long (a few
nsec) so that the upper limit (5) corresponds to a length of a few
tens of cm. Hence, superradiance following instantaneous excitation
can certainly be observed in a few cm long sample.

b) The actual excitation of the medium is realized by a swept
excitation, the laser light pulse sweeping the inversion along the
sample at speed of light. Such a swept excitation matches almost
exactly the propagation delay of the superradiant pulse in the
medium, so that condition (4) may actually be relaxed and superradiance should indeed be observed even in long samples ([19])
(in case of swept excitation, superradiance should however only
occur in the direction of the pumping pulse propagation and not
in both directions as it is the case for side pumping).

For the cells we were using (few cm long), the difference between swept and instantaneous excitation is negligible, since the
pulse propagation time in the medium (0.3 nsec) is much shorter
than the duration of the exciting pulse (2 nsec) and delay of the

emitted superradiant pulse. Evidence that instantaneous pumping
conditions are indeed fulfilled in our experiments is given by
the fact that we have actually observed the emission in both directions with respect to the pumping beam propagation. We thus can say
that we have indeed observed the superradiance emission as predicted
by the theories dealing with instantaneously inverted media.

To conclude this discussion, we may say that strong amplifying
media have to be short enough to be superradiant if they are instantaneously pumped, whereas all swept strong amplifiers are automatically superradiant, whatever their length is. Among the very
wide variety of instantaneous or swept amplifying laser systems
existing, there are certainly several ones operating under superradiant conditions (i.e. n_0^2 emission and strong radiation reaction
of the emitted field on the atoms). The time scale of the phenomenon is however generally so short that it is very difficult, if not
impossible to put in evidence. I hope to have shown that the simplicity of the alkali atom level scheme along with the order of
magnitude of its time evolution parameters makes the alkali an
ideal system to check superradiance phenomena in the near visible
range.

3.6. Possible Applications of Superradiance

Several possible applications of superradiance have already
been proposed, including very short pulse generation ([15]), X ray
lasers ([20]), etc... In the alkali system studied above, superradiance
could be used to produce very short infrared pulses in the pico-
second range. Figure 12 shows for example that the superradiant
emission from Cs vapor has a much shorter rise time and ringing
period than the duration of the exciting pumping pulse (2 nsec).
By further increasing the Cs pressure or the length of the swept
amplifying medium, the duration of the superradiant pulse could
certainly be reduced to much shorter values.

Another interesting application can be considered in connection with Rydberg state studies. There is a very large number of
cascading transitions falling in the microwave or far infrared
range connecting any initially excited Rydberg level to lower
lying excited states (see Fig. 2). Condition (3) shows that
threshold for superradiance is much lower at long wavelengths than
in visible or near infrared. As a result, it should be rather easy
to achieve superradiance conditions on these microwave cascading
transitions. One would thus expect the sample of atoms to emit like
a "firecracker" a whole set of cascading microwave superradiant
bursts. After these emissions, an ensemble of highly populated
Rydberg states should be prepared in the sample. One may thus use
the phenomenon either for microwave pulse generation or as a way
of populating Rydberg levels difficult to reach by other excitation mechanisms.

One may say that at the threshold excitation density for superradiance, the atomic system excited in a Rydberg level undergoes a kind of phase transition evolving from the randomly phased dipolar spontaneous emission of individual atom to a kind of "macroscopic" microwave emission emitted by strongly correlated and aligned dipoles.

A study of this phenomenon should be interesting to perform.

REFERENCES

(1)   I.D. ABELLA, N.A. KURNIT and S.R. HARTMAN, Phys. Rev. 141, 391 (1966).
(2)   R.G. BREWER and R.L. SHOEMAKER, Phys. Rev. A6, 2001 (1972).
(3)   C.L. TANG and B.D. SILVERMAN, in Physics of Quantum Electronics, edited by P.L. Kelley, B. Lax and P.E. Tannenwald (Mc Graw-Hill, New York, 1966).
(4)   R.L. SHOEMAKER and R.G. BREWER, Phys. Rev. Letters, 28, 1430 (1972).
(5)   For a complete reference survey on Quantum Beats and Quantum Beat Spectroscopy, see S. HAROCHE in High Resolution Laser Spectroscopy, K. Shimoda editor, Springer Verlag 1976.
(6)   R.H. DICKE, Phys. Rev. 93, 99 (1954).
(7)   It is impossible to give a complete reference list of all papers dealing with superradiance since Dicke first paper (ref. 6). We will only quote in the following some references which may be of interest for the reader who wishes to find more information about some specific points. For a general and critical discussion of superradiance in various experimental conditions (small or large samples, strong or small inhomogeneous broadening and so on ...), one may refer to R. FRIEDBERG and S.R. HARTMAN, Phys. Rev. A10, 1728 (1974).
(8)   S. HAROCHE, M. GROSS, M. SILVERMAN, Phys. Rev. Letters, 33, 1063 (1974).
(9)   C. FABRE, M. GROSS, S. HAROCHE, Opt. Commun. 13, 393 (1975).
(10) C. FABRE, S. HAROCHE, Opt. Commun. 15, 254 (1975).
(11) T.W. DUCAS, M.G. LITTMAN and M.L. ZIMMERMAN, Phys. Rev. Lett. 35, 1752 (1975).
(12) For sample large compared to the wavelength, the analysis of superradiance is complicated by the fact that the superradiant pulse is amplified in the medium and thus develops non-uniform polarization and population inversions in the sample. Theories of superradiance in large samples are divided in two categories : those which implicitly or explicitly ignore this complication or find it negligible and deal with uniform samples coupled to a single-mode radiation field and those which take into account the spatial variations in the sample. Among the former, one may quote : N.E. REHLER and J.M. EBERLY, Phys. Rev. A3, 1735 (1971); R. BONIFACIO, P. SCHWENDIMANN and F. HAAKE, Phys. Rev.

A4, 302 (1971; R. BONIFACIO and L.A. LUGIATO, Phys. Rev. A11, 1507 (1975).

On the other hand, spatial variation of superradiance emission along the sample have been given a great attention by N. SKRIBANOWITZ, I.P. HERMAN, J.C. Mac GILLIVRAY and M.S. FELD in Laser Spectroscopy (R.G. Brewer and A. Mooradian editors, Plenum Press, N.Y. 1975) and ref. (^{15}) below.

It is not surprising that these different theories, although they agree about the gross description of the superradiance phenomenon in an extended volume, disagree about the detailed analysis of the time behaviour of the superradiant emission. Although I don't want to enter into a detailed critical discussion of these theories, I adopt here the more realistic point of view of Skribanowitz et al. which gives obviously a better agreement with experiment (for a discussion of this point, see for example R. FRIEDBERG and B. COFFEY, Phys. Rev. A13, 1645 (1976) and R. BONIFACIO, L.A. LUGIATO and A. CRESCENTINI, Phys. Rev. A13, 1648 (1976).

(^{13}) N. SKRIBANOWITZ, I.P. HERMAN, J.C. Mac GILLIVRAY and M.S. FELD, Phys. Rev. Lett. 30, 309 (1973).

(^{14}) M. GROSS, C. FABRE, P. PILLET and S. HAROCHE, Phys. Rev. Lett. 36, 1035 (1976).

(^{15}) J.C. Mac GILLIVRAY and M.S. FELD, Phys. Rev A, to be published (1976).

(^{16}) G.I. PETERS and L. ALLAN, J. Phys. A 4, 238 (1971).

(^{17}) A.L. SCHAWLOW and C.H. TOWNES, Phys. Rev. 112, 1940 (1958).

(^{18}) See for example:
F.T. ARRECHI and E. COURTENS, Phys. Rev. A2, 1730 (1970). Different theories of superradiance in extended media give different value for the maximum length of a side pumped superradiant sample. Again, we adopt here the point of view of ref. (^{15}).

(^{19}) R. BONIFACIO, F.A. HOPF, P. MEYSTRE and M.O. SCULLY, Phys. Rev. A12, 2568 (1975).

(^{20}) F.A. HOPF, P. MEYSTRE, M.O. SCULLY and J.F. SEELEY, Phys. Rev. Lett. 35, 511 (1975).

ATOMIC-BEAM EXPERIMENTS AT THE ISOLDE-FACILITY AT CERN

Curt Ekström and Ingvar Lindgren

Experimental Physics Division, CERN, Genève, Switzerland, and Department of Physics, Chalmers University of Technology, Göteborg, Sweden

1. INTRODUCTION

The atomic-beam magnetic-resonance (ABMR) technique, invented by Rabi in 1938, has now been used for almost four decades. During the first years, diamagnetic molecules were investigated, and the main purpose was to measure magnetic dipole moments from the direct interaction with an external magnetic field. After that first period, paramagnetic atoms have been used almost exclusively - mainly for intensity reasons - for that kind of research. Such an experiment yields information about the hyperfine structure (hfs), i.e. the splitting caused by the interaction between the nuclear moments and the fields from the surrounding electrons. In order to get information about the nucleus, it is then necessary to have sufficient knowledge about the electronic structure of the atom. Alternatively, the technique can be used to study the electronic structure, if the nuclear quantities involved are known from other experiments.

So far, most hfs measurements using the ABMR method have been restricted to the region close to the beta-stability line. In conventional off-line experiments, the lower limit for the half-life is of the order of 10 minutes. This is the main limitation when investigating nuclei close to stability, where otherwise the low particle energies used for the production give highly selected reaction products. Further away from stability, the decreasing yields and simultaneous increase in the number of isotopes produced - in addition to the short half-lives - make conventional experiment very complicated. Using an atomic-beam apparatus on-line with the activity-producing source together with an intermediate isotope-separation step would not only reduce the lower limit of the half-

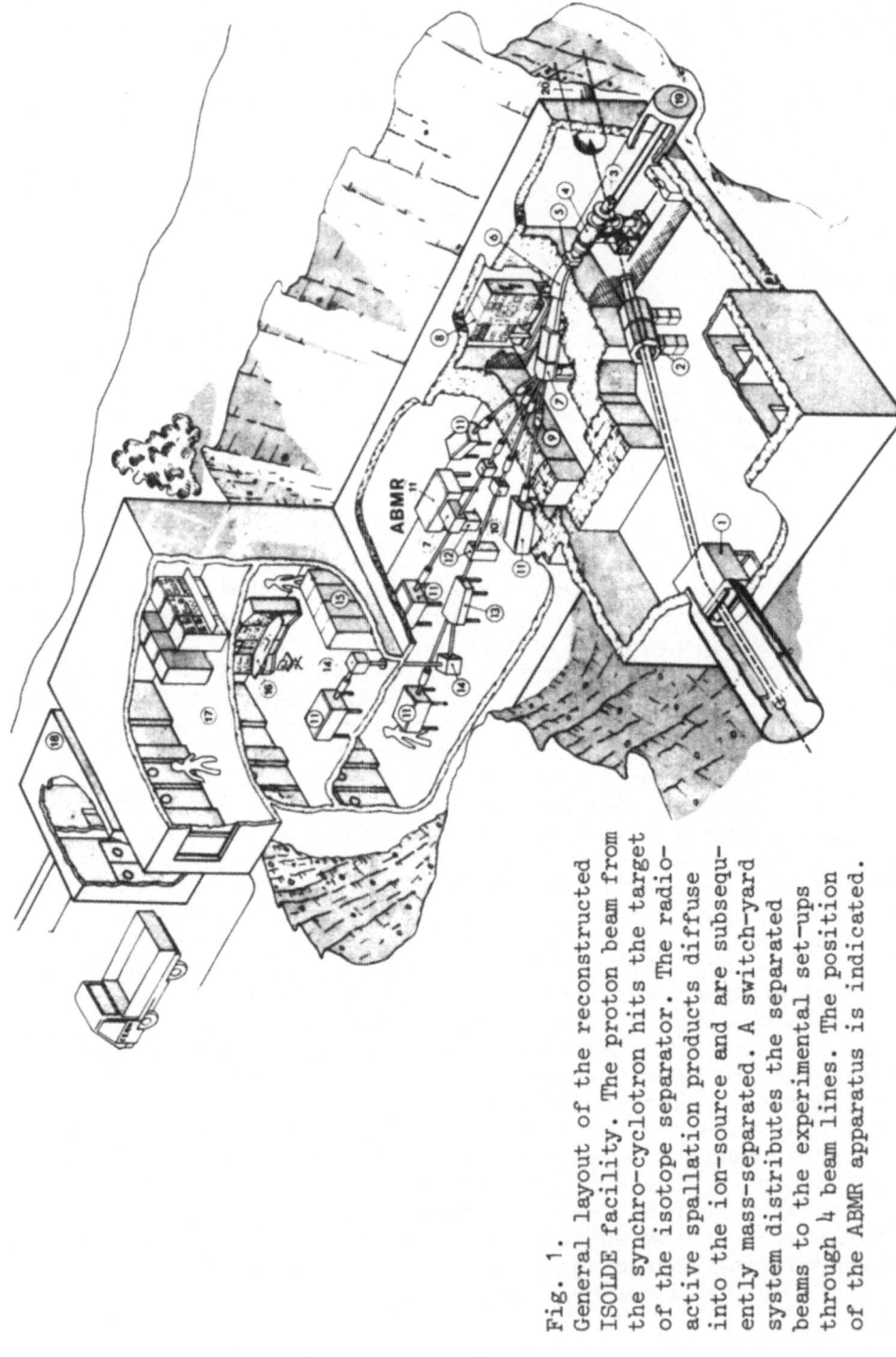

Fig. 1.
General layout of the reconstructed
ISOLDE facility. The proton beam from
the synchro-cyclotron hits the target
of the isotope separator. The radioactive spallation products diffuse
into the ion-source and are subsequently mass-separated. A switch-yard
system distributes the separated
beams to the experimental set-ups
through 4 beam lines. The position
of the ABMR apparatus is indicated.

Table 1. ISOLDE target systems

Hydroxide targets

$ZrO_2 \cdot x \ H_2O$	$Kr \rightarrow$	Br
$CeO_2 \cdot x \ H_2O$	$Xe \rightarrow$	I
$ThO_2 \cdot x \ H_2O$	$Rn \rightarrow$	At

Molten-metal targets (plasma ion source)

Ge	$Zn \rightarrow$	Cu
Sn	$Cd \rightarrow$	Ag
Pb	$Hg \rightarrow$	Au

Molten-metal targets (surface ionization source)

Y	Sr	Rb
La	Ba	Cs
Th	Ra	Fr
Gd	Eu	Sm	$\rightarrow$	Pm

Powder targets

| Nb | Sr | Rb | Kr | Br | Se | As |
| Ta | | | Yb | Tm | $\rightarrow$ | Er |

High-temperature-oxide targets

| CeO_2 | Ba | Cs | Xe | I | Te | Sb |
| ThO_2 | Ra | Fr | Rn | At | Po | Bi | Pb | Tl |

lives but also highly facilitate the interpretation of the resonance
signals observed, thus making ABMR experiments possible on nuclei
far away from the beta-stability line.

The requirements mentioned can be satisfied within the ISOLDE
(Isotope Separator On-Line) facility [1] at the 600 MeV synchro-
cyclotron at CERN. Therefore, the Swedish atomic-beam group has
reconstructed one of its machines for on-line work, and connected
it to the ISOLDE separator. The first successful experiment with
that equipment was performed in July 1975.

2. EXPERIMENTAL

2.1. General

The CERN synchro-cyclotron has recently been reconstructed in
order to yield a higher external proton beam current. During the
shut-down period also the ISOLDE facility has been thoroughly improved. A general view of the new experimental area is shown in
Fig. 1.

The different target systems, which have been in use at ISOLDE,
are collected in Table 1, together with the elements produced. The
arrows indicate daughter products, and the elements accessible to
ordinary atomic-beam studies are given within squares. During the
one and a half years of operation at the reconstructed ISOLDE facility, the target systems used in production runs have been concentrated to those giving the elements of rubidium, cesium and mercury.

A typical production-yield curve, cesium from a molten-metal
target, is shown in Fig. 2. The saturation activity is given as
function of the mass number. The peak value corresponds to about
7 nA of radioactive atoms. The horizontal line indicates the estimated lower limit of production rate required for a spin measurement, 10^6 atoms/sec. Our spin-measurements (crosses) on neutron-
deficient cesium isotopes have extended considerably the sequence
of previous measured spins (circles).

The principal design of the ABMR apparatus is shown in Fig. 3.
The apparatus has a six-pole focusing "A-magnet", which acts as a
polarizer or state selector, and a fourpole focusing "B-magnet",
acting as an analyser of the beam. The focusing magnets give a
much higher luminosity than a more conventional apparatus of dipole
type. For radioactive work this is, of course, of vital importance.
The atoms, which are focused in the first magnet and undergo a
transition in the intermediate "C-field", are defocused in the last
magnets and reach the collector. This means that the active collec-

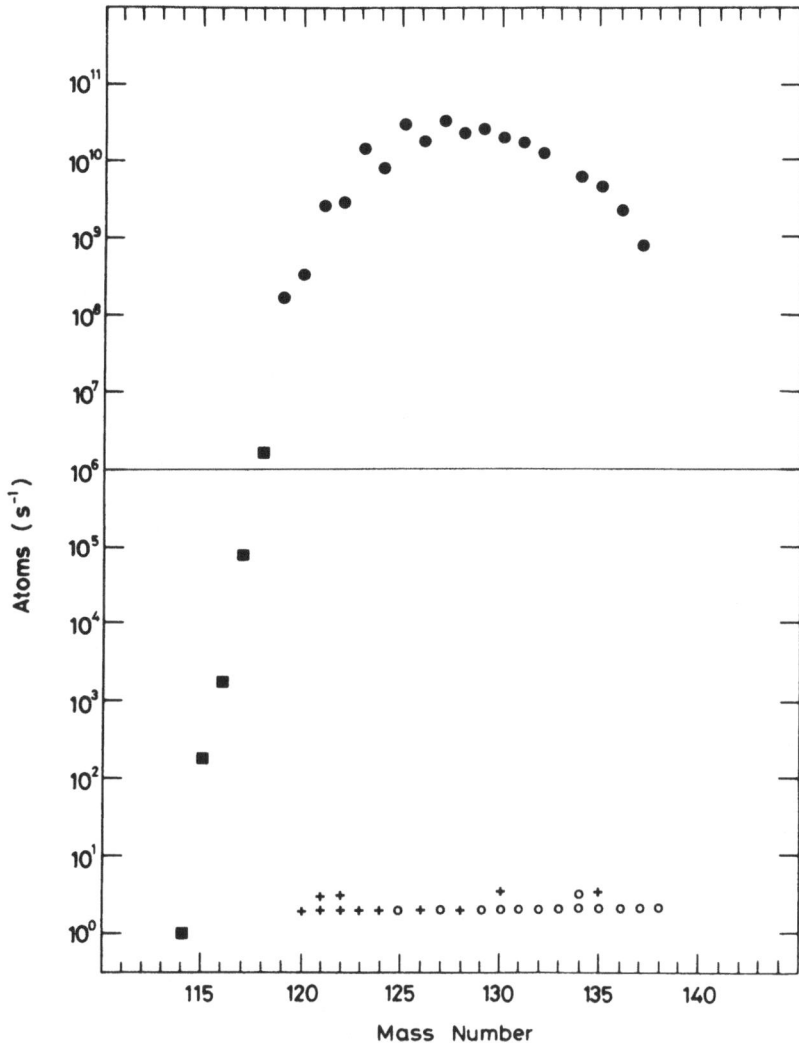

Fig. 2.
Experimental yields of cesium isotopes from a molten-lanthanum target, given as atoms per second at the collector side of the separator. Dots and squares indicate measurements using Faraday cup and
beta counting, respectively. Our spin measurements are marked by
crosses at the bottom of the figure, and previously known spins are
marked by circles.

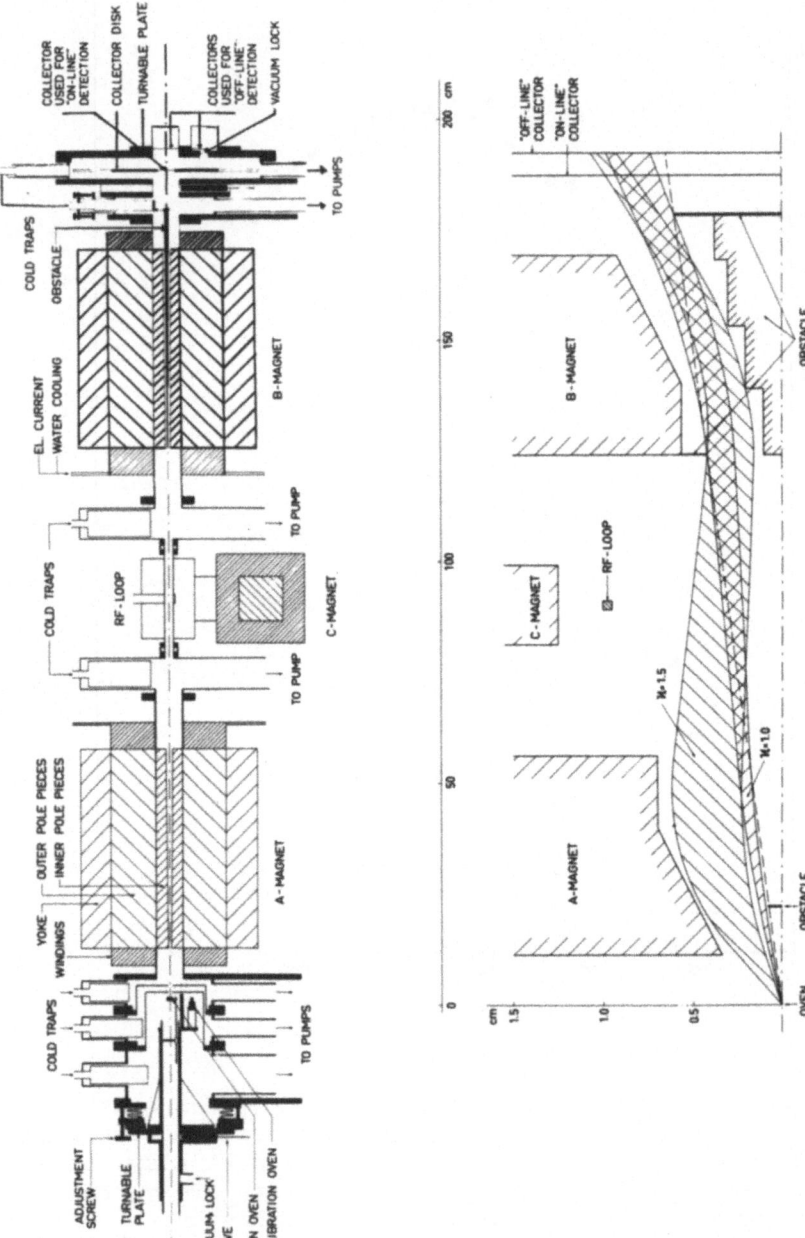

Fig. 3. Schematic cross-section of the atomic-beam magnetic-resonance apparatus at CERN. The bottom part of the figure shows calculated trajectories for atoms having undergone a transition from a focusing to a defocusing state.

tor area is ring-shaped and fairly large. The use of a four-pole
(instead of a six-pole) B-magnet has reduced the collector size
appreciably without reducing the transmission [2,3].

The ion-beam from the isotope separator is focused by an electrostatic lens to a spot of about 0.5 mm^2 at the source position of
the ABMR apparatus. Normally, the ion beam is collected on a metal
foil from which it is continuously evaporated in the form of atoms.
The problem of forming a beam of alkali atoms, having very low ionization potentials, was solved by using a tantalum foil covered by a
thin layer of yttrium.

The atoms traversing the apparatus are collected on aluminium
foils, and the activity is subsequently measured in scintillation
counters, mounted directly on the ABMR apparatus. Normally thin
plastic scintillators for β^+ and thin NaI(Tl) crystals for X-rays
following the electron-capture decay are used. Surface-barrier detectors are prepared for measurements of the α-decay in heavy elements.

A mini-computer handles the data accumulation and analysis; it
furthermore controls most of the experimental routines, such as
changing the r.f. frequencies and transporting the collector discs
within the collector-detector system.

2.2. Hyperfine Structure Measurements

In a very weak external magnetic field B (Zeeman region for
the hyperfine structure), the resonance frequency for allowed magnetic dipole transitions between magnetic sublevels of the same F
state is given by:

$$\nu = \left| \frac{\mu_B B}{h} \ g_J \ \frac{F(F+1) + J(J+1) - I(I+1)}{2F(F+1)} \right|$$

where μ_B is the Bohr magneton, g_J the atomic g-value and I, J, F the
total angular-momentum quantum numbers for the nucleus, the electrons
and the entire atom, respectively. Usually, the nuclear spin I is
here the only unknown quantity. Typical spin measurements are performed in a field of a few gauss, where the resonance frequencies
are of the order of a few MHz. Since typical line widths - which are
determined by the time the atoms spend in the r.f. field - are 20-50
kHz, the resonance frequencies corresponding to different spin values
are normally well resolved.

In stronger magnetic fields, where the deviation from the linear
Zeeman effect becomes significant, the experimental resonance frequencies yield information on the hyperfine separations. This infor-

C. EKSTRÖM AND I. LINDGREN

Table 2. Summary of results from the spin measurements
in rubidium, cesium and gold isotopes.

Isotope	Half-life	Measured spin I	
^{77}Rb	3.8 min	3/2	
^{78}Rb	17.7 min	0	a
^{78m}Rb	6.0 min	4	
^{79}Rb	23.0 min	5/2	
^{84m}Rb	21 min	6	
^{120}Cs	58 sec	2	
^{121}Cs	2 min	3/2	
^{121}Cs	2 min	9/2	
^{122}Cs	21 sec	1	
^{122m}Cs	4.2 min	8	
^{123}Cs	5.8 min	1/2	
^{124}Cs	30.8 sec	1	
^{126}Cs	1.6 min	1	
^{128}Cs	3.8 min	1	
^{130m}Cs	3.7 min	5	
^{135m}Cs	53 min	19/2	
^{185}Au	4 min	5/2	b
^{186}Au	10.7 min	3	c
^{187}Au	8.5 min	1/2	c
^{188}Au	8.8 min	1	c
^{189}Au	28.3 min	1/2	c
^{189m}Au	4.6 min	11/2	c

a) The observed resonances may also be due to a non-zero spin in
the case the magnetic moment is smaller than $5 \cdot 10^{-6}$ n.m.
b) Preliminary result. Resonances observed so far at only one
setting of the external magnetic field.
c) Result published in Phys. Lett. 60B (1976) 146.

mation can then be used to evaluate the nuclear moments in the same
way as in other types of hyperfine investigations.

2.3. Experimental Results

As mentioned, the ISOLDE runs have so far been concentrated to
a few target systems, and the ABMR experiments have for that reason
been limited to the elements of rubidium, cesium and gold. These elements have previously been studied extensively by conventional ABMR.
Our ISOLDE measurements, however, have extended the sequence of known
spins appreciably. Altogether, 22 new spins have been determined for
these elements (see Table 2).

In the elements of rubidium, cesium and gold, the neutron-deficient isotopes far away from beta-stability extend into regions of
expected nuclear deformation. The results from the systematic spin
measurements consequently give important information on the transition from essentially spherical to strongly deformed nuclear shapes.
We shall not discuss this further here, but only refer to our recent
papers [4,5], where this problem is analysed in more detail.

Our low-field measurements have recently been extended to
stronger external magnetic fields, yielding preliminary values of
the nuclear magnetic moments of a number of cesium isotopes. The
data clearly show that the moments of the doubly-odd isotopes,
having I=1, gradually decrease with decreasing mass number (A) until
A=124, while for A=122 the magnetic moment suddenly becomes much
smaller. Possible sources for this drastic change are now being
analysed.

3. FUTURE OUTLOOKS

3.1. Measurements on Metastable Atomic States

The conventional ABMR technique is only applicable to atoms
with paramagnetic ground state, i.e. with the total electronic
quantum number $J \neq 0$. This has, for instance, excluded the element
of lead (with the $^3P_0$ ground state) from this type of experiment
– an element of particular interest from the nuclear point of view.
However, the ABMR technique can be applied also to metastable atomic
states – with lifetimes of more than a few milliseconds, say – provided that the state can be sufficiently populated. Such a technique
has for a long time been employed in ABMR investigations on stable
isotopes by several groups, using thermal or electron-bombardment
excitation [6]. In Göteborg we have in recent ABMR experiments on
lead applied a plasma-discharge metastabilizer, previously used in

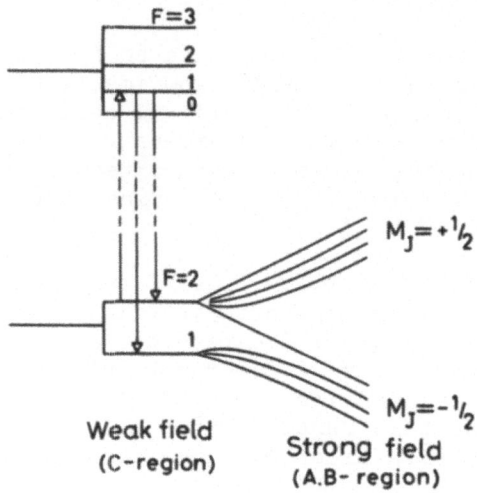

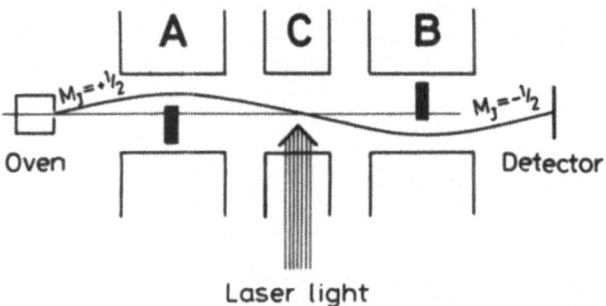

Fig. 4. Principles for optical hyperfine pumping (upper part).
       Optical transitions in weak magnetic fields change the
       population of the hyperfine levels of the ground state.
       This affects the population of the M_J states at strong
       fields, which can be detected in a conventional ABMR
       apparatus (lower part).

optical-resonance works [7]. In this way it has been possible to
measure the nuclear spin of a few neutron-deficient isotopes by
means of radioactive detection [8]. In principle, the same technique
can be applied also at ISOLDE in order to make elements with diamagnetic electronic ground state accessible to spin and hfs determinations. It would here be of particular interest to extend the
lead sequence further away from the region of stability. This would
require, though, some developments of the target technique, since
the presently available systems have insufficient yields.

3.2. ABMR in Combination with Laser Excitations

The magnetic deflection, employed in ABMR experiments, offers
a very interesting means of non-optical detection of laser-induced
transitions. This has recently been demonstrated by different groups
[9,10]. A laser beam, tuned to a specific hyperfine component of a
resonance line, can give rise to a "hyperfine pumping" of the ground
state, i.e. an accumulation in the higher or lower F states (see
Fig. 4, upper part). If such a beam is introduced into the (weak-
field) C-region of the ABMR apparatus (see Fig. 4, lower part), it
will change the population of the different M_J multiplets in the
(strong-field) B region. Such a change can be detected in the same
way as ordinary resonances, caused by radio-frequency transitions.
This technique has been demonstrated for stable as well as radioactive sodium isotopes by the Orsay group, using a surface-ionizing
detector followed by a mass-separator [9]. In this way it is possible
to measure the hyperfine splitting as well as the isotope shift of
the resonance line. In combination with the ABMR-ISOLDE facility
this technique can be extended to nuclei further away from the beta-
stability.   For elements with J=1/2 in the ground state - like the
alkalis - such measurements would be of particular importance. The
ground-state hfs, measured in conventional ABMR, gives here very
little information about the nuclear deformation. The atomic-beam-
laser technique just described, on the other hand, can be used to
determine the isotope shift as well as the quadrupole moment (via
the hfs of the excited state) - quantities that are strongly
correlated to the nuclear deformation.

Another possible use of a laser in connection with ABMR experiments is for detection of the atomic beam. This has recently been
applied at our laboratory in Göteborg in experiments on stable barium
in the metastable ^3D states [11]. Laser detection is particularly
favorable in works on metastable states, since it offers a means of
eliminating the background from unexcited atoms. It may for this
reason compete even with radioactive detection in experiments on
metastable atomic states of radioactive isotopes.

3.3. Comparison with Optical Pumping

The ABMR technique has been shown to be a useful tool for investigating ground-state properties of nuclei far from stability. It is then an alternative to the optical-pumping(OP) technique, which for some time has been employed within the ISOLDE collaboration by Otten et al.[12]. OP is essentially limited to atomic S states, while ABMR requires that the electronic state has $J \neq 0$. This implies that OP can be applied to elements with $^1S_0$ ground state, like Cd, Ba and Hg, where conventional ABMR cannot be used. On the other hand, ABMR experiments can be performed on non-S states, where OP is very difficult. Furthermore, ABMR is easier to apply than OP to elements with high evaporation temperatures. Therefore, the only overlap region of practical importance is the alkali atoms, which have $^2S_{1/2}$ ground state and low evaporation temperatures. In addition, OP is best suited for the more short-lived alkali isotopes, where ABMR suffers from intensity problems. Therefore, it is found that the two methods supplement each other in a very nice way.

SUMMARY

A sixpole-fourpole ABMR apparatus has been connected to the ISOLDE facility at CERN and used for on-line experiments on very neutron-deficient isotopes of rubidium, cesium and gold. Altogether, 22 new nuclear spins have been determined for these elements. Hyperfine-structure measurements, yielding information about the nuclear dipole moments, are in progress for these isotopes. With the target systems tested so far, it can be expected that isotopes of about 25 elements can be produced with sufficient yields for ABMR experiments.

The technique applied so far can be extended in various respects. Conventional ABMR is restricted to elements with $J \neq 0$ in the electronic ground state. Techniques for exciting the atoms into metastable states are available, which may circumvent this restriction in many cases. The combination of atomic-beam magnetic deflection and laser excitation opens very interesting possibilities to investigate the isotope shifts for long sequences of isotopes. The same technique can also be used to measure the nuclear quadrupole coupling of excited atomic states, which is of particular importance in cases where such coupling is absent in the ground state.

REFERENCES

1. H.L. Ravn, S. Sundell and L. Westgaard, Nucl. Instr. 123 (1975) 131, and further references therein.

2. M. Olsmats, B. Wannberg and I. Lindgren, Nucl. Instr. 103 (1972) 27.

3. C. Ekström, M. Olsmats and B. Wannberg, Nucl. Instr. 103 (1972) 13.

4. C. Ekström, I. Lindgren, S. Ingelman, M. Olsmats and G. Wannberg, Phys. Lett. 60B (1976) 146.

5. C. Ekström, S. Ingelman, G. Wannberg and M. Skarestad, Proc. Int. Conf. on Nuclei far from Stability, Cargèse, France, May 1976 (to appear in a CERN Yellow Report).

6. See W.J. Childs, Case Studies in Atomic Physics, 3, 215 (1972), where references to original works can be found.

7. U. Brinkman et al., Z. Physik 228, 427 (1969) and ibid. 228, 440 (1969);
    S. Garpman et al., Z. Physik 241, 217 (1971).

8. M. Gustafsson et al., to be published in Physics Letters.

9. H. Duong et al., Optic Communication 7, 371 (1973), ibid. 12, 71 (1974);
    G. Huber et al., Phys. Rev. Letters 34, 1209 (1975).

10. W. Ertmer and B. Hofer, Z. Physik A276, 9 (1976);
    W. Zeiske et al., Physics Letters 55A, 405 (1976).

11. M. Gustafsson et al., to be published in Physics Letters and Report at the Eighth EGAS Conference, Oxford, 1976.

12. E.W. Otten, presented at this conference.

HIGH RESOLUTION LASER SPECTROSCOPY OF

RADIOACTIVE SODIUM ISOTOPES

H. T. Duong, P. Jacquinot, P. Juncar, S. Liberman,
J. Pinard, J.L. Vialle
Laboratoire Aimé Cotton, C.N.R.S. II, 91405-Orsay, France

G. Huber, R. Klapisch, C. Thibault
Laboratoire René Bernas, C.N.R.S., 91406-Orsay, France

Although experimental techniques of atomic physics, and especially high resolution spectroscopic methods, have already been successfully used for a long time to get precise measurements of nuclear properties, it has appeared with the recent advent of laser spectroscopy, that these new techniques would still allow more sensitive experiments. In particular, use of an atomic beam and a tunable single mode CW dye laser, make it easy to get rid of Doppler broadening, and it seemed interesting to apply such laser methods to investigate the spectroscopic properties of unstable and rare atoms, and to study their variation along a series of isotopes. It is to be emphasized that for light nuclei such as sodium, several theoretical calculation results are available (see fig. 1), among which ab initio Hartree-Fock calculation play an important role in that sense that they predict a region of strong deformation at a neutron number of 20 (corresponding in the case of Na to a mass number of 31) [1] ; this feature has been clearly put in evidence experimentally by measuring the Na isotopes masses accurately [2], but any other confirmation appearing either with isotope shifts or with quadrupole moment measurements would give strong confidence in that sort of theoretical calculation method. On another viewpoint the sodium element is a rather good candidate for laser

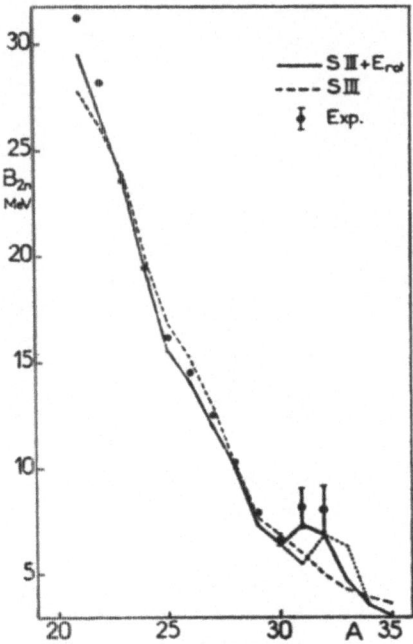

Figure 1 - Theoretical and experimental results about the
values of binding energy B_{2n} of a pair of
neutrons for the different Na isotopes.

spectroscopy because its resonance D line wavelengths match very
well the wavelength domain explorable with a rhodamine 6G tunable
dye laser. Of course for each isotope one will be interested in
getting both the hyperfine structure and the isotope shift referred
to the stable one ; since in that case the line structure is quite
smaller than the Doppler width, it is clear that one will be dealing
with very high resolution spectroscopic techniques, so the laser
has to be single mode and of course continuously tunable over a
relatively wide range of wavelength [3].

Magnetic detection of optical resonances.

Whatever the aim of the experiment is, the general principle
consists in tuning the laser frequency accross the several
hyperfine components of the D lines and then to detect the
corresponding resonances. Moreover, as it has to work with rare

atomic species, the system has to be sensitive enough to detect
signals due to small numbers of atoms. These conditions have lead
us to study and to adapt a magnetic detection method [4]. It is
based on the fact that sodium atoms have a paramagnetic ground
state and, when taking the hyperfine interaction into account, that
corresponding magnetic levels behave as shown in figure 2 in strong
magnetic field : more precisely, for instance in the case of ^{23}Na
(stable natural isotope), the ground state $3\ ^2S_{1/2}$ splits into
two hyperfine sublevels with $F=2$ and $F=1$ which themselves are
decomposed under a strong magnetic field into two groups corres-
ponding respectively to $m_J = -\frac{1}{2}$ and $m_J = +\frac{1}{2}$. It is well known
since the advent of hydrogen maser that such atoms passing through
a six-pole magnet, as long as they are not too rapid, i. e. as they
follow adiabatically the potential curves, see an inhomogeneous
strong magnetic field and undergo such forces that they deviate
from their straight line trajectory according to their m_J value :
more precisely atoms with $m_J = +\frac{1}{2}$ are focused at the output of

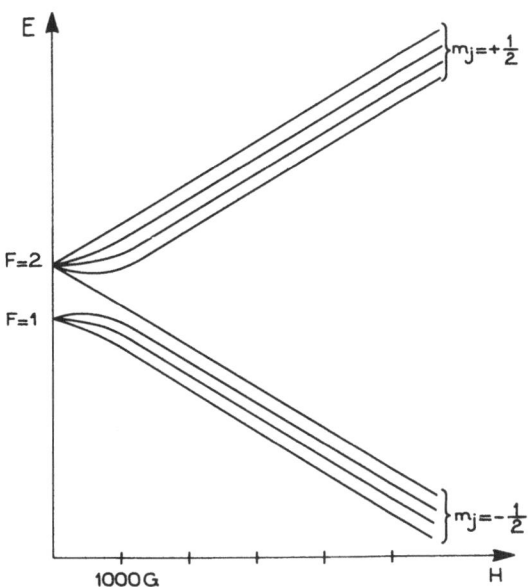

Figure 2 - Zeeman diagram of the ground state hyperfine
           sublevels of ^{23}Na $(I = 3/2)$.

the six-pole magnet whereas atoms with $m_J = -\frac{1}{2}$ are defocused.
The six-pole magnet plays the role of a magnetic state selector.

Before going accross the magnet the atoms are illuminated with
the laser light tuned at some resonance frequency. Thus the atoms
absorb the light at resonance and rise an excited state from which
they decay spontaneously and come back to their ground state. It
needs a few cycles of this optical pumping process in average for an
atom to change its effective magnetic moment, and in the case we are
dealing with, starting from $F=1$ for instance to reach $F=2$, which
comes to deplete $m_J = -\frac{1}{2}$ sublevels to the benefit of $m_J = +\frac{1}{2}$
sublevels (the reverse arising when starting from $F=2$ to reach
$F=1$, which clears out $m_J = +\frac{1}{2}$ sublevels and fills up $m_J = -\frac{1}{2}$
sublevels). Thus if one places some sort of an atom counter at the
output of the six-pole magnet, in the absence of optical interaction
the counter see a certain amount of atoms corresponding approximately to half the total number of atoms, the ones with $m_J = +\frac{1}{2}$
which are focused by the magnet. In the presence of a resonant
optical interaction, the number of atoms counted either is increased
or decreased depending upon which ground state sublevel the atoms
started from. A typical recording of the D_1 line
$(3\ ^2S_{1/2} - 3\ ^2P_{1/2})$ obtained with the natural isotope of Na is
shown in figure 3. It is to be noticed that because of the high

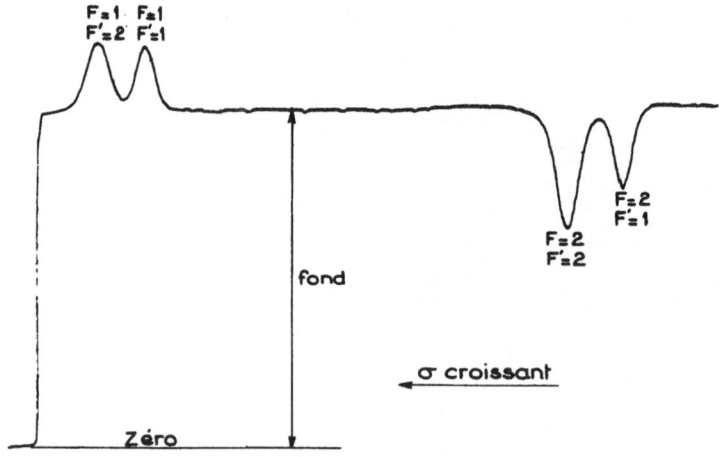

Figure 3 – Typical recording of the D_1 line hyperfine structure (^{23}Na) using the six-pole magnetic detection.

efficiency of the optical pumping a quasi total inversion can be
obtained between the two hyperfine sublevels of the ground state,
using only a few milliwatts laser beam. Having in mind that the
magnetic state selector is 100% efficient, it is easy to understand why this detection system is particularly well adapted to the
case of small numbers of atoms.

Of course there are some transition which cannot be detected
this way ; as an example with the D_2 line, no optical pumping in
the ground state can be induced through the transition :
$3\ ^2S_{1/2}$, $F=2$ – $3\ ^2P_{3/2}$, $F'=3$. However, as the $F=2$, $m_F = -2$
Zeeman sublevel corresponds to $m_J = -\frac{1}{2}$, whereas the 4 other
Zeeman sublevels correspond to $m_J = +\frac{1}{2}$, it is still possible to
induce an optical pumping in $F=2$ Zeeman sublevels using circularly polarized light and thus to measure the hyperfine interval
$F'=3$ – $F'=2$ of $3\ ^2P_{3/2}$, from which one is able to deduce an
experimental value of the nuclear quadrupole moment.

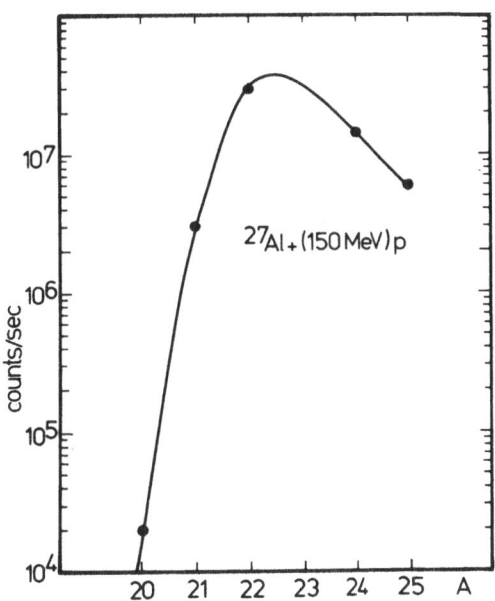

Figure 4 – Production yields of the different Na isotopes.

General arrangement of the experiment.

In the experiments that we have performed the sodium isotopes were produced by spallation of aluminium with 150 MeV protons from the Orsay synchrocyclotron [5]. A molten aluminium target heated up to 900°C was bombarded with a beam of 100 nA so that about 10^8 atoms of ^{25}Na for instance were produced per second. In fact, the spallation reaction can be labelled as : ^{27}Aℓ(p,3p+xn)$^{25-x}$Na (with $0 \leqslant x \leqslant 5$) , which means that all the sodium isotopes having their mass number between 20 and 25 were produced simultaneously but with different yields, as shown in figure 4. Consequently, in order to separate each isotope from the others, it was necessary to use a mass spectrometer (see figure 5) : in effect, all the isotopes, if not separated, contribute to the mean signal and thus to increasing the noise ; using a mass spectrometer suppressed that part of the signal coming from the other isotopes (and especially from the ^{23}Na always in high quantity in the Aℓ target) and therefore improved appreciably the signal to noise ratio. On the

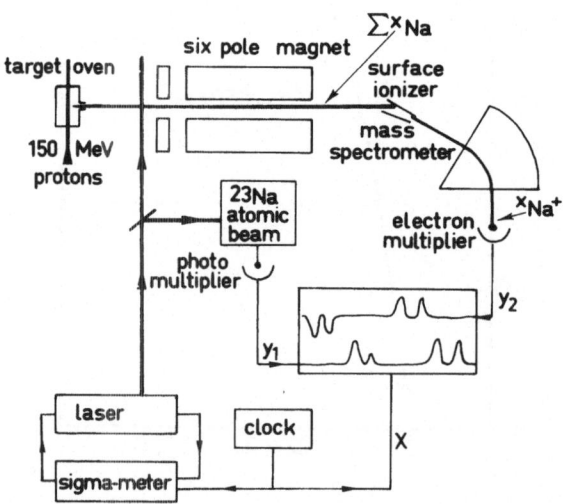

Figure 5 - General scheme of the experiment.

other hand it permits to separate the resonance signals which otherwise could be completely merged. The general scheme of the experiment was the following : the radioactive sodium atoms created in the heated target left it through a little hole and formed an atomic beam which was crossed at a right angle by the laser light beam. Then the atomic beam got through the six-pole magnet after which the atoms were ionized by impinging on a hot rhenium surface before they went in the mass spectrometer to be finally counted at the output by an electron multiplier. As we were interested in measuring, among other things, the isotope shift, each of the D_1 line corresponding to the different isotopes was recorded simultaneously with the fluorescence spectrum of an auxiliary atomic beam of ^{23}Na (stable isotope) excited by the same tunable dye laser. The laser and the so-called "sigma-meter" which served to servo-control and to tune the frequency, have been described elsewhere [3][6] ; in operating conditions the frequency of the laser mode was scanned step by step with increments of 7.5 MHz (it is to be noticed that the purely statistical uncertainties in estimating the centre of gravity of a line always came to less than 1 MHz, but little imperfections of the wavenumber control gave rise to an average uncertainty of about 6 MHz for the D_1 lines).

Recordings and results.

Figure 6 shows a display of the recordings that we obtained on the D_1 line for all the isotopes studied according to their position in the wavenumber scale with respect to the stable one. From

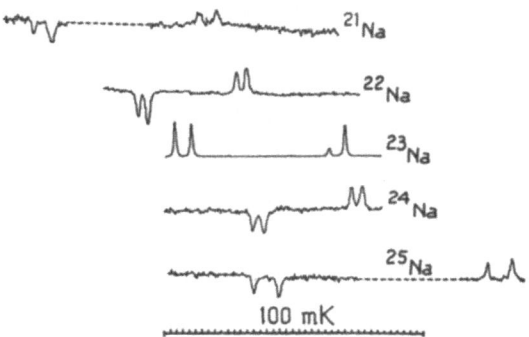

Figure 6 - Display of the recordings of the D_1 line obtained for the different Na isotopes.

these recordings one can get both isotope shifts and magnetic
hyperfine constants which give rise to the nuclear magnetic moments.
These numerical results are given in table 1.   Notice that the
measured values of isotope shifts follow the law of normal plus
specific mass effect as it is better shown on the corresponding
values of the reduced mass shift which turns out to be quite linear
within the experimental errors.   Concerning the D_2 line which,
contrary to the D_1 line, permits an experimental determination of
the nuclear quadrupole moment, examples of recordings are given in
figure 7.   The two traces correspond to σ^+ and σ^- light excitation ; one can observe in particular the expected change in sign of
the signal for the component which corresponds to the highest F
values.   The obtained Q values are also given in table 1[7]. It is
to be noticed that for ^{21}Na even the sign is not sure, whereas
for ^{25}Na our Q value does not match the most recent calculation
results [1].   Therefore, further experiments would be needed in

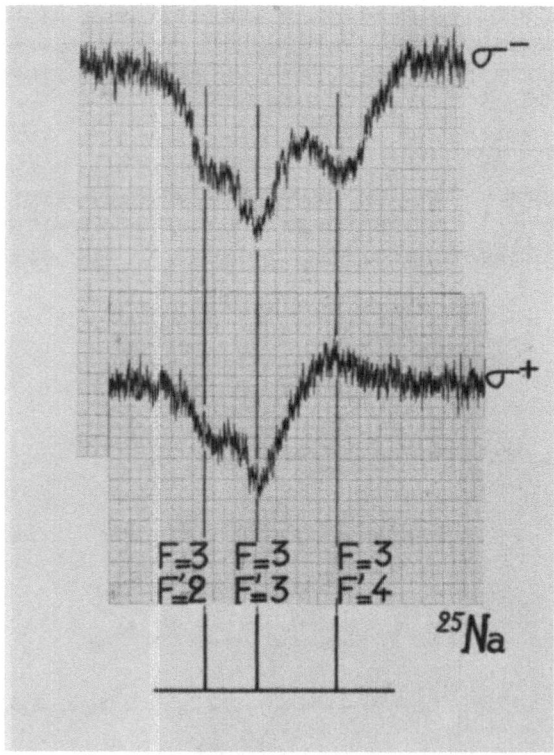

Figure 7 - Typical recordings of part of the hyperfine
          structure of the D_2 line of ^{25}Na
          using either σ^+ or σ^- excitation.

order to increase the accuracy of nuclear quadrupole moments that
we already have investigated. Moreover that kind of measurement
is going to be extended to the very neutron rich sodium isotopes,
but this will require to modify the present experiment. First of
all these heavier isotopes have to be produced by reaction of high
energy protons (about 25 GeV) from CERN proton synchrotron on an
uranium target. As these isotopes have lifetimes lying in the few
milliseconds range, the target has to be designed in such a way
that it reduces time diffusion of sodium atoms to much less than
the lifetimes. In principle this has been already achieved by using
uranium layers between thin graphite plates. In fact this diffusion
time is an essential limitation ; as an example, figure 8 gives a
recorded curve of diffusion of ^{22}Na in the molten Aℓ target
which shows that in our experiment this diffusion time, measured at
half amplitude, was about 25 sec . Secondly, the next experiment
will have to be managed taking into account the production yields
of the different isotopes which turn out to be comparable to the
ones that we obtained for ^{25}Na , but only up to ^{27}Na or ^{28}Na ,
and going decreasing rapidly for the following ones. This means
that on the one hand it would be desirable to gain a few factors in
the general efficiency of our detection system. On the other hand
it would be strictly necessary to improve the reliability of all
the apparatus and in particular of the dye laser and its servocontrol
system in order to be able to do integration of signal during very
long period of time.

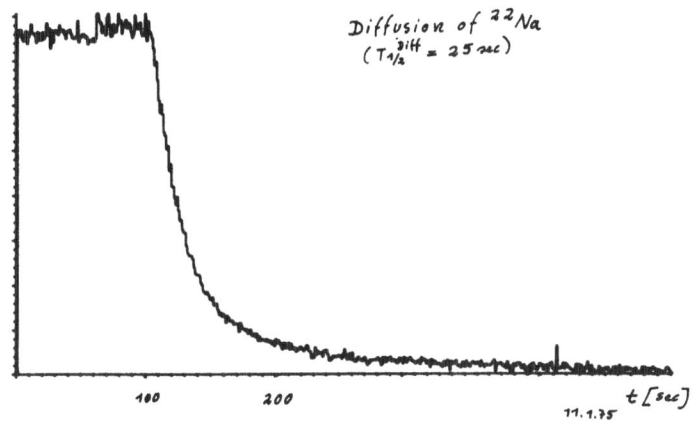

Figure 8 – Recording of the diffusion time of ^{22}Na
through the molten Aℓ target.

Despite the fact that we have at our disposal theoretical results [1], it is not possible to predict hyperfine structure since the nuclear spins are not known beyond mass number 26. In fact, in any case, the quantities that we want to measure are expected to be small. As an example, the isotope shift is well known to be the superposition of several effects : the most important being the mass shift which in the case of ^{31}Na compared to ^{23}Na can be evaluated to about 4000 MHz, the second one being the volume shift which reflects the nucleus deformation and which lies in the range of 10 MHz (see fig. 9). These numbers give an idea of high accuracy and good reproducibility that will be need to get reliable information about sodium nuclei far from stability. Nevertheless, the next experiment will first allow us to confirm what we already measured and especially the Q value for ^{25}Na ; second, it has a good probability to permit measurements of hyperfine structures as well as isotope shifts up to ^{29}Na . And third,

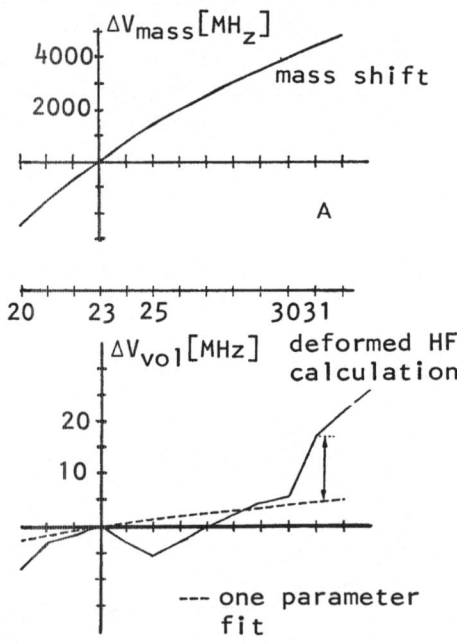

Figure 9 – Theoretical evaluation of mass effect and volume effect of isotope shifts of Na isotopes from ^{20}Na up to ^{31}Na .

in principle it would certainly be able to measure the same quantities for ^{30,31}Na which are the most interesting ones, but the production rate for these isotopes are expected to be so small that at the present time, it looks like a very difficult challenge.

Table 1

Mass number	21	22	23	24	25
Isotope shift (in 10^{-3}cm^{-1})	53.5 (2)	25.3 (2)	0	23.1 (4)	44.6 (2)
Reduced Isotope shift	2.010 (17)	0.990 (13)	0	0.990 (22)	2.00
μ/μ_N					+3.685 (022) +3.683 (004) [8]
Q (in barns)	-0.060 (0.075)				0.23 (0.08)

References

[1] X. Campi, H. Flocard, A. K. Kerman, and S. Koonin,
    Nuclear Physics, A251, 193 (1975)
[2] C. Thibault, R. Klapisch, C. Rigaud, A. M. Poskanzer, R. Prieels,
    L. Lessard, W. Reisdorf, Phys. Rev. C, 12, 644 (1975)
[3] S. Liberman, J. Pinard, Appl. Phys. Lett. 24, 142 (1974)
[4] H. T. Duong, J. L. Vialle, Opt. Commun. 12, 71 (1974)
[5] G. Huber, C. Thibault, R. Klapisch, H. T. Duong, J. L. Vialle,
    J. Pinard, P. Juncar, P. Jacquinot,
    Phys. Rev. Lett. 34, 1209 (1975)
[6] P. Juncar, J. Pinard, Opt. Commun. 14, 438 (1975)
[7] G. Huber, R. Klapisch, C. Thibault, H. T. Duong, P. Juncar,
    S. Liberman, J. Pinard, J. L. Vialle, P. Jacquinot
    C. R. Acad. Sc. Paris B282, 119 (1976)
[8] M. Deimling, R. Neugart, H. Schweickert,
    Z. Physik A 273, 15 (1975)

NUCLEAR EXCITATION BY ELECTRON TRANSITION

M. Morita

Department of Physics, Osaka University

Toyonaka, Osaka, Japan 560

ABSTRACT

When an electron hole moves from an inner orbit to an adjacent one, the energy corresponding to the difference of the binding energies for these two orbits is usually carried away by the emitted characteristic X ray or the Auger electron. In addition to these two processes, there is another possible mode of this energy release by exciting the atomic nucleus. This is called the nuclear excitation by electron transition, in short NEET. The probability of NEET is generally small compared with that of the X-ray or Auger-electron emission. It is, however, measurable in some cases. The reason why NEET has been studied theoretically is described. Preliminary data in ^{189}Os experiments are introduced. The implications of NEET are also discussed.

INTRODUCTION

In this presentation, I would like to discuss a possible new mode of nuclear excitation by deexcitation of the orbital electron system. That is, when an electron hole is created in one of the inner shells by bombarding the atom with electrons or X rays or some other reason such as the orbital electron capture, an orbital electron of the adjacent shells jumps into the vacancy immediately. The energy corresponding to the difference of the binding energies for these two shells, $E_1 - E_2$, is usually carried away by the emitted characteristic X ray or the Auger electron. On certain favorable conditions which I shall describe later, this excess energy is transferred to the nucleus so that it is excited from the ground state to one of its higher energy states. In this

process, real photons are not emitted or absorbed. Photons are
virtually exchanged between the nucleus and the relevant electron.
That is, the process is not a two-step transition, but it is a one-
step transition. We call this process nuclear excitation by
electron transition, in short NEET.

The motivation for studying NEET is to produce atomic nuclei
in their excited states by a new method other than the well-known
excitation mechanisms, such as nuclear reactions with slow neutrons
or accelerated charged hadrons, nuclear transmutations in alpha,
beta, and gamma decays. A radiochemist, Otozai in Osaka, has been
searching for a long time for a possible new method of enriching
^{235}U. He explained to me that one can separate ^{235}U from natural
uranium which is a mixture of ^{238}U, ^{235}U, and ^{234}U, if the nucleus
of ^{235}U is in its excited state.[1] The isomer state of the ^{235}U is
located at 30 eV above the ground state.[2,3] It decays back to the
ground state through the internal conversion process with a half
life of 26 min. In this case the ^{235}U atom is ionized, because
one of the orbital electrons is ejected. Consequently, it may not
be difficult to collect such atoms chemically or electrically.[1,4]
(An isotope separation using the recoil or hot-atom effect at
nuclear transformation has been known since 1934, as the Szillard-
Chalmers method.[5])

In the conventional methods for producing nuclear excited
states, we usually adopt the nuclear reactor, van de Graaff,
Cyclotron, or large-scale linear accelerator. Contrary to these,
a new method involving an electron gun is used to accelerate
electrons at relatively low energies. The accelerated electrons
interact mainly with the orbital electrons of the relevant atoms,
whose nuclei are in the ground state, and the bombarding electrons
can kick off the electrons of the inner orbits. In this way, the
energy of the bombarding electrons is stored in the orbital
electron cloud. A part of this energy is transferred to the
nucleus if there is an interaction between the nucleus and the
orbital electrons. The bombardment of electrons can be replaced
by irradiation of X-rays.

In a similar process, the nuclear fine structure in the
muonic atom was studied theoretically,[6,7] and the nuclear
excitation due to deexcitation of the muonic levels was observed
in the muonic X-ray experiment.[8]

PROBABILITY OF NEET

A simple theory of NEET is as follows.[9] See Fig. 1. We have
the ground state ψ_1 and the excited state ψ_2 with the nuclear
excitation energy E_N. The orbital electron system has an electron
hole in an inner orbit ϕ_1 with the binding energy E_1. This hole
will move immediately to the state ϕ_2 with E_2, either by a
radiative or by nonradiative process, and so on. The resultant
system Ψ of the nucleus and orbital electrons has, therefore,

a series of the states $\psi_1\phi_1$, $\psi_1\phi_2$, ..., and $\psi_2\phi_1$, $\psi_2\phi_2$, ...
In the case, which we are interested in, the nuclear excitation
energy E_N is nearly equal to the difference of binding energies
of the electron, $E_1-E_2 \sim E_N$. In this case, the energies of the
resultant system are $E(\Psi_1)=E_1$ for the state $\Psi_1=\psi_1\phi_1$ and
$E(\Psi_2)=E_2+E_N$ for the state $\Psi_2=\psi_2\phi_2$. These two states are nearly
degenerated and they may be mixed by electromagnetic perturbations,
H_I. According to the standard perturbation calculation, we have the
eigenfunctions Φ_1 and Φ_2 by diagonalizing the 2×2 energy matrix,

$$\Phi_1 = \Psi_1\cos\theta + \Psi_2\sin\theta,$$

$$\Phi_2 = -\Psi_1\sin\theta + \Psi_2\cos\theta,$$

with a mixing angle

$$\cos\theta\sin\theta = (\psi_2\phi_2|H_I|\psi_1\phi_1)/[E(\Phi_1) - E(\Phi_2)].$$

Here the denominator is nearly equal to $\Delta=(E_1-E_2)-E_N$. Now, let
us assume the case where the electron transitions, $\psi_1\phi_i \rightarrow \psi_1\phi_j$ or
$\psi_2\phi_i \rightarrow \psi_2\phi_j$, are considerably faster than the nuclear gamma decay
$\psi_2\phi_i \rightarrow \psi_1\phi_i$, or the internal conversion process $\psi_2\phi_i \rightarrow \psi_1\phi_{i'}$. The $\psi_1\phi_1$
components in Φ_1 and Φ_2 decay to the $\psi_1\phi_2$ state with the decay
constant $\lambda_{12}\cos^2\theta$ and $\lambda_{12}\sin^2\theta$, respectively, by leaving the
nucleus in its ground state, and so on. Here λ_{12} is the decay
constant for the transition $\phi_1 \rightarrow \phi_2$. The $\psi_2\phi_2$ in Φ_1 and Φ_2 also
decay to $\psi_2\phi_3$ with the decay constants $\lambda_{23}\sin^2\theta$, and $\lambda_{23}\cos^2\theta$,
respectively, by leaving the nucleus in its excited state, and so
on.

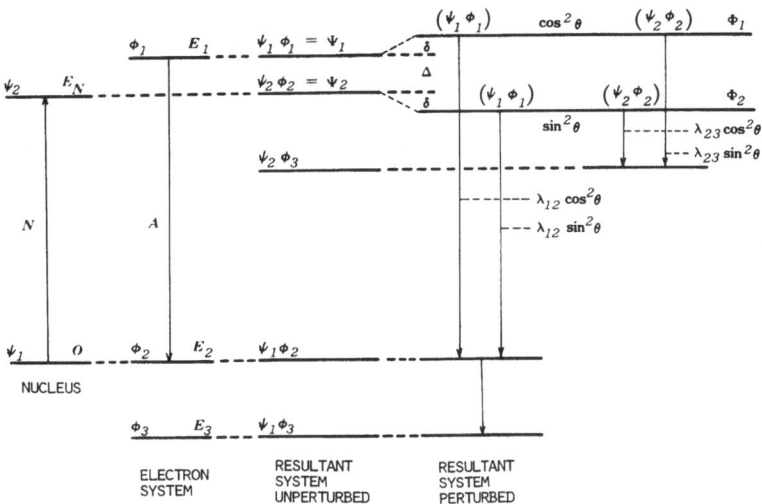

Fig. 1.   Energy levels in NEET.

At the end of these successive transitions following the formation of the electron hole, the nucleus remains in its excited state ψ_2 with the probability P,

$$P = [1 + (\Lambda_2/\Lambda_1)]\tan^2\theta \qquad \text{for } \theta \ll 1.$$

Here Λ_1 and Λ_2 are the total decay rates of the electron hole in ϕ_1 and ϕ_2, respectively.

To have a nonzero value of P for NEET, Ψ_1 and Ψ_2 should have the same spin and parity, or equivalently, the transitions A and N in Fig. 1 should have the same multipolarity and parity relation. Furthermore, P is larger if Δ is smaller. We assume the Coulomb interactions of the relevant orbital electron with the Z protons as the perturbation. The effect of the magnetic interactions is smaller.

In the case of ^{189}Os, (see Fig. 2), the interaction energy is approximately given by

$$(\psi_2\phi_2|H_I|\psi_1\phi_1) \sim -\alpha(4\pi/5)\sum_m(\psi_2|\sum_{k=1}^{Z}r_k^2Y_{2m}(\theta_k,\phi_k)|\psi_1)$$
$$\times(\phi_1|r_e^{-3}Y_{2m}^*(\theta_e,\phi_e)|\phi_2).$$

Two simple models are adopted to estimate the above expression. I. The nuclear factor is ZR^2 and the electronic factor is r_B^{-3}, R and r_B being the nuclear radius and the mean value of the Bohr radii for n=1 and 3, respectively. II. The nuclear factor is estimated with the experimental value of B(E2),[10] and the electronic factor is computed with the hydrogen-like wave functions. The interaction energy is −0.6 eV (−0.3 eV) for the first (second) case. Consequently, P is 1.4×10^{-7} (4×10^{-8}). Here the contributions come from the electronic M_{IV} and M_V states. For

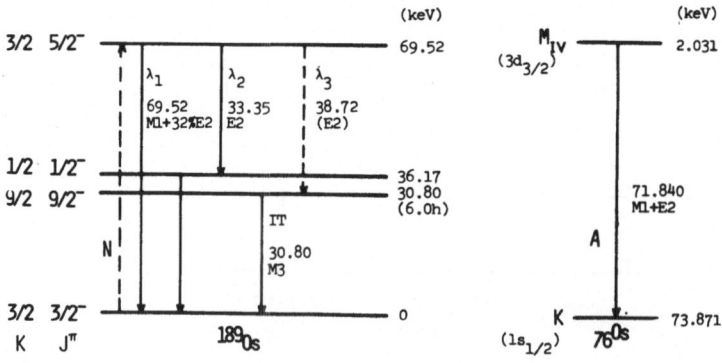

Fig. 2.   Nuclear and electronic levels of ^{189}Os.

further investigation, an effect which is similar to the penetration effect in the internal conversion process should be taken into
consideration.

EXPERIMENT ON OSMIUM 189

Experimental work on the nuclear excitation by electron
transition has been performed by Otozai and his collaborators.
As is shown in Fig. 2, the osmium nucleus has the 70-KeV state,
and this energy is nearly equal to the energy difference of the
electronic states, M_{IV} and K. Both the nuclear excitation N
and the electronic deexcitation A satisfy the spin-parity
condition for E2. Therefore, the 70-keV state may be produced by
NEET if an electron hole is formed in the K orbit by electron
bombardment.

In the first experiment,[11] they tried to find NEET in ^{189}Os
by detecting the radioactivity of the 30.8-keV isomer state. The
target was prepared by compressing 0.5-g nonradioactive osmium
metal powder in a shallow copper dish of 23 mmØ using a 26 ton
press. This target was bombarded by electrons from the electron
gun of a Hitachi HU-10 electron microscope. The electron energy
was calibrated by the standard method[12] using the diffraction ring
from the (111) plane of gold metal spattered on a plastic film.
The activity of the isomer, which consists of the L- and M-
conversion electrons of about 20 and 30 keV, respectively, was
measured by an Aloka windowless 2π gas flow GM counter of about
0.9 cpm background. The decay curve of the activity produced by
a 94-keV-100-µA-5-h bombardment is given in Fig. 3, which shows

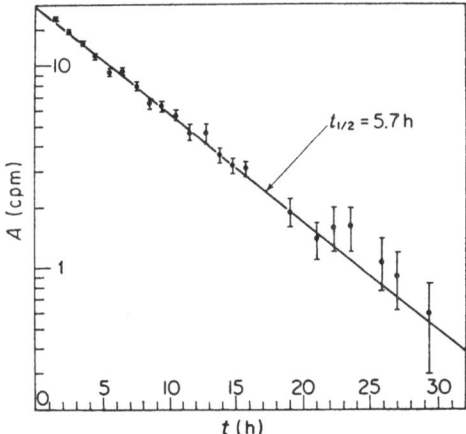

Fig. 3. Decay curve of the produced
radioactivity in ^{189}Os.[11]

the production of the isomer state clearly.[11] The relative
excitation function for the isomer production gave the threshold
energy of electrons which was not inconsistent with NEET.

In a recent experiment,[13] Otozai et al. adopt a target of a
natural osmium film on a platinum plate in a 20 mm∅ with 0.456-
mg/cm^2 thickness. After the radioactivity has been terminated,
this target is again used. The radiation damage of the metallic
osmium is negligibly small by the electron bombardment.

The electron gun of a Hitachi HU-11 electron microscope with
a beam current stabilizer is adopted in this experiment. The
electron beam is 60.0 ± 0.1 μA. The electron energy range is
72-104 keV with ± 0.3 keV uncertainty. It is calibrated with the
working curve, which has been prepared from the study of electron
diffraction ring diameters with a gold film of about 90-Å thickness.
The target is irradiated with this electron beam for four hours.
To make the temperature of the target constant, a water
circulation system is adopted.

The activity of the isomer was again measured by the same
Aloka windowless 2π gas flow GM counter as that in the first
experiment, but with about 0.8 cpm background. The isomer
production cross section $\sigma(cm^2)$ is obtained from the formula
$\sigma = A/Qnxgc[1-\exp(-\lambda t)]$. Here A(cpm) is the counting rate at the end
of bombardment, Q(min^{-1}) the incident rate of electrons, n(cm^{-3})
the atomic density of target nuclide, x(cm) the thickness of the
target, λ(min^{-1}) the decay constant of the isomer, t(min) the
bombarding time, g the probability of conversion process in the
isomeric transition, and c the detection coefficient. In this
case, g is 1.00 and c is estimated to be 0.50.

Preliminary data[13] on the counting rates A(cpm) are summarized
in Table I, where the background is subtracted. Measurements were
performed twice at each electron energy, and the mean values of two
data are listed. The errors show the standard deviations. From

Table I. Observed counting rates

Electron energy E(KeV)	Counting rate A(cpm)
72.3 ± 0.3	0.1 ± 0.1
74.1 ± 0.3	0.0 ± 0.1
77.1 ± 0.3	0.4 ± 0.1
80.2 ± 0.3	1.0 ± 0.1
85.0 ± 0.3	1.6 ± 0.1
90.1 ± 0.3	1.9 ± 0.1
94.9 ± 0.3	1.9 ± 0.1
100.0 ± 0.3	2.2 ± 0.1
103.6 ± 0.4	2.9 ± 0.1

these data, the isomer production cross sections are obtained as shown in Fig. 4.[13]

The isomer production cross section is also expressed by $\sigma = \sigma_K PB$. Here σ_K is the K-shell ionization cross section by electron bombardment, P the NEET probability and B the branching ratio of the 39-keV transition. Therefore, P can be obtained if σ_K and B are known. In fact, σ_K is estimated by the theory, e.g., ref.,[14] which explains the experiment on Au having a nearly equal atomic number to that of Os. The branching ratio B is given by $B = \lambda_3/(\lambda_1 + \lambda_2 + \lambda_3)$, with the decay rates λ_i (i=1,2,3) in Fig. 2. Here λ_2/λ_1 is known experimentally, as $118/940$.[10] The 39-keV transition is, however, not observed in experiments so that λ_3/λ_2 is around 1/50 or less. The empirical rule for the hindrance factor in the K-forbidden gamma transitions gives us also $\lambda_3/\lambda_2 \sim 1/50$. This means that B is about 2.2×10^{-3} or less. From these considerations on σ_K and B and measured values of σ, the NEET probability P is about 3×10^{-7}. This is of the right order of magnitude, which is theoretically expected, 1.4×10^{-7}. The curve in Fig. 4 refers to $PB = 6.7 \times 10^{-10}$.

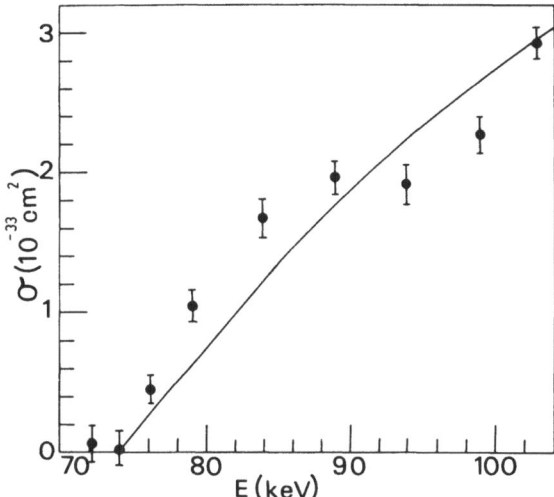

Fig. 4. Experimental isomer production cross section due to NEET in ^{189}Os.[13] The calculated curve is given for $BP = 6.7 \times 10^{-10}$.

As is seen in Fig. 4 of the excitation curve, the threshold
energy is about 74 keV in conformity with the theory of NEET. The
magnitude as well as energy dependence of the cross section is
also in favor of NEET. This experiment is still being performed.

We have to be careful of possible competitive processes for
exciting the 70-keV state. Among these, the lowest-order process
is the Coulomb excitation due to inelastic electron scattering by
the nucleus. The cross section of the Coulomb excitation with
L=2 is 6.2×10^{-32} cm^2 at 100 keV for the incoming electron energy,
based on the formula by deForest and Walecka,[15] and the experimental data on the B(E2) obtained from the 69.52-keV gamma
transition.[10] Assuming the branching ratio B=2×10^{-3}, the isomer
production cross section is 1.2×10^{-34} cm^2 which is 25 times smaller
than the observed value in Fig. 4. A similar nuclear excitation
of the M1 and E2 types due to the transversal components of the
electromagnetic interactions between the electron and nucleus can
be also calculated with the formula.[15] The effect is less than
2 % of the Coulomb excitation. The resonance absorption of the
bremsstrahlung from the scattered electron is also negligible.[16]

DISCUSSIONS

If the NEET probability is appreciable, the X-ray spectrum
becomes complex, see Fig. 5. For example, when an electron hole

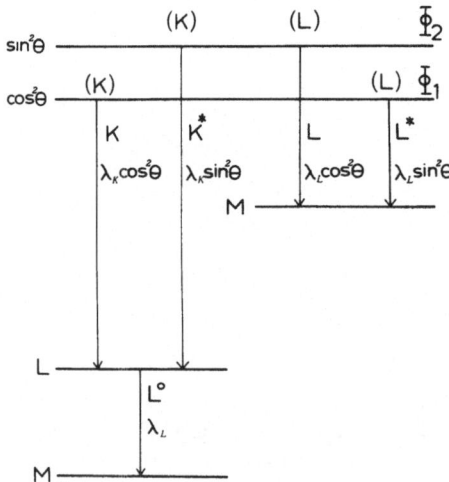

Fig. 5. X-ray spectrum in NEET. Each KX line splits
into two lines, and each LX line splits into three lines.

is created in the K orbit, the K components in Φ_1 and Φ_2 decay
into the L orbit by emitting the KX and K*X rays, respectively.
The nucleus remains in its ground state. Subsequently, the L state
emits an L°X ray and the state transforms to the M state. On the
other hand, the L components in Φ_1 and Φ_2 emit the L*X and LX
rays, respectively, and they transform to the M state. As a
result, each KX line splits into two lines and each LX line splits
into three lines, if NEET mixes the K and L states. This is the
case for ^{237}Np, which is the daughter of ^{237}Pu in its K-electron
capture. It is, however, not the case if ^{237}Np is the daughter of
the alpha decay of ^{241}Am, see Fig. 6. Otozai et al. have been
studying the X-ray spectra of ^{237}Np in these two cases.[17]

In the case of ^{235}U, the nuclear excitation energy E_N of
the first excited state coincides approximately with the energy
difference of the electronic $6d_{3/2}$ and $6p_{3/2}$ levels in Fig. 7.
If E_N is about 80 eV,[3] we may find some other favorable electronic
states. For these cases, interaction energies for NEET are
generally very small since the orbital electrons are located
at a distance far from the nucleus. The direct excitation of
the first excited state would become interesting if we can adjust
the electronic level distances by means of chemical forms.

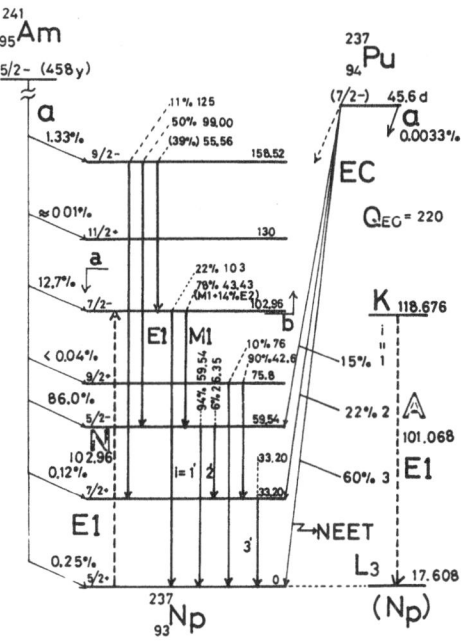

Fig. 6. Nuclear and electronic levels of ^{237}Np.

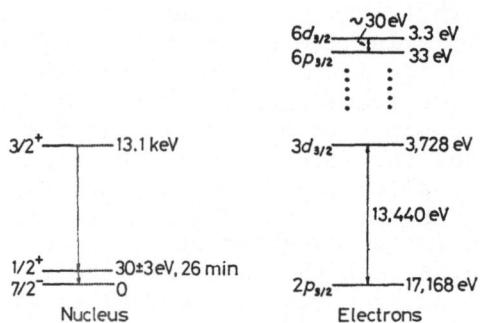

Fig. 7. Nuclear and electronic levels of ^{235}U.

The broadening of the level width and the red shift of X ray
under high pressures and temperatures may help the energy
matching condition.

Furthermore, E_N for the second excited state is approximately
equal to the energy difference of the electronic $2p_{3/2}$ and $3d_{3/2}$
states. In this case, the estimated P is about 10^{-9}. Since this
is very small, NEET is not a practical use for mass production of
the enriched ^{235}U, at the present stage of investigation. It may,
however, be useful for producing a small amount of the 100 % pure
^{235}U. It may be also useful for a calibration of the concentration
of ^{235}U if the process is established.

Finally, NEET interjoins the atomic and nuclear spectro-
scopies. NEET also provides us with a possibility of mass
separation without using mass difference. I would like to thank
Professor K. Otozai for his stimulating discussions and for showing
me his experimental results before publication.

REFERENCES

1. K. Otozai, private communication.
2. H. Mazaki and S. Shimizu, Phys. Letters 23, 137 (1966):
    Bull. Inst. Chem. Research, Kyoto Univ. 44, 394 (1966).
    A value, 23 eV or less, is given by M. A. Freedman, F. T.
    Porter, F. Wagner, Jr., and P. P. Day, Phys. Rev. 108, 836
    (1957).
3. Another value, 80 eV, is also given in Table of Isotopes by
    C. M. Lederer, J. M. Hollander, and I. Perlman, J. Wiley and
    Sons Inc., New York, 1968.
4. M. Goldhaber considered also a similar process, private
    communication.
5. L. Szillard and T. A. Chalmers, Nature 134, 462 and 494 (1934).

6. L. Wilets, Kgl. Danske Viedenskab. Selskab, Mat.-Medd. 29, (1954), No. 3.
7. B. A. Jacobsohn, Phys. Rev. 96, 1637 (1954).
8. S. Bernow, S. Devons, I. Duerdoth, D. Hitlin, J. W. Kast, W. Y. Lee, E. R. Macagno, J. Rainwater, and C. S. Wu, Phys. Rev. Letters 21, 457 (1968).
9. M. Morita, Progr. Theoret. Phys. 49, 1574 (1973).
10. S. G. Malmskog, V. Berg, and A. Bäcklin, Nucl. Phys. A153, 316 (1970).
11. K. Otozai, R. Arakawa, and M. Morita, Progr. Theoret. Phys. 50, 1771 (1973).
12. P. B. Hirsch, A. Howie, R. B. Nicholson, D. W. Pashley, and M. L. Whelan, Electron Microscopy of Thin Crystal, Butterworths, London 1965, pp. 85 and 109.
13. K. Otozai, R. Arakawa, and T. Saito, private communication.
14. A. M. Arthurs and B. L. Moiseiwitsch, Proc. Roy. Soc. (London) A247, 550 (1958).
15. T. deForest, Jr. and J. D. Walecka, Advances in Phys. 15, 1 (1966), especially Eq. (3.46).
16. W. Heitler, The Quantum Theory of Radiation, Third Edition, Clarendon Press, Oxford 1954, Sec. 25.
17. K. Otozai, R. Arakawa, H. Baba, K. Hata, and T. Suzuki, private communicaiton.

HYPERFINE AND ISOTOPE SHIFT MEASUREMENTS FAR OFF STABILITY BY OPTICAL PUMPING

E. W. Otten

Institut fur Physik, Johannes Gutenberg-Univ.

D-6500 Mainz, W-Germany

1. Introduction

In this talk I will review optical pumping (OP) experiments performed by our group at ISOLDE, CERN, in a series of Hg isotopes spanning from mass number 181 to 205 (first results in refs. [1] and [2], final in refs.[3] and [4]). I will then discuss what impact these measurements had on nuclear structure research far off stability and how the results of the latter are retroacting again on the optical work. In particular they ask for isotope shifts (IS) of light even Hg isotopes which we are prepared to measure by laser spectroscopy. Further I would like to communicate briefly first results of optical pumping experiments on "exotic" Rb and Cs isotopes. Finally I will discuss plans for laser spectroscopy on fast mass-separated beams which might favourably be applied to isotopes far off stability.

2. The application of the ß- and γ-RADOP method to mass-separated Hg-isotopes far off stability

A mass separator, set up on line with a nuclear reaction target, is ideally suited for all kinds of systematic spectroscopic work along isotopic chains extending far off stability. Especially the ISOLDE facility at CERN has been an outrider in the field attracting people from nuclear as well as from atomic spectroscopy. In the first talk of this session on atomic beam resonance far off stability, Lindgren [5] described the essentials of ISOLDE already (detailed information

on ISOLDE is given in ref.[6]). Like the atomic beam
method, OP may be adapted to the situation off stability
by tracing the signal via the nuclear ß or γ radiation
of the decaying isotope. That is enabled by the fact that
OP leads to an orientation not only of the electronic
angular momentum J but also of the nuclear spin I through
the hyperfine coupling. In case of a radioactive isotope
one has two chances, therefore, to observe the OP effect:
either by the angular distribution of optical radiation,
depending on the orientation of J, or by the angular
distribution of nuclear radiation, depending on the
orientation of I. Before discussing which of them should
be preferred we should collect few more information on
the latter method which we call ß-RADOP or γ-RADOP
(ß-(γ)- radiation detected optical pumping). The ß-RADOP
method has been developed on light, cyclotron produced
alkalies [7]. The results of this early period were
presented to this community of atomic physicists at their
second conference 1970 in Oxford [8]. γ-RADOP was tried
out first in 1971 by Cappeler and Mazurkewitz for the
example of ^{203}Hg [9].

2.1. Principles of RADOP

The angular distribution of radiation emitted by an
oriented ensemble of nuclei is given by [10]

$$W (\vartheta) = \sum_k A_k \, f_k \, P_k(\cos\vartheta) \qquad (1)$$

where A_k is a correlation factor in the decay, P_k is the
Legendre Polynomial and f_k is the k'th moment of
orientation defined by

$$f_k(I) = (^{2k}_{k})^{-1} I^{-k} \sum_m \sum_{\nu=0}^{k} (-1)^\nu \, \frac{(I-m)!\,(I+m)!}{(I-m-\nu)!\,(I-m-k+\nu)!} (^{k}_{\nu})^2 \, n_m \qquad (2)$$

n_m is the population number of the substate $m_I = m$.
Whereas f_0 measures the degree of isotropic population,
f_1 is the ordinary polarization

$$P_I = \langle I_z \rangle / I; \qquad (2a)$$

$$f_2 = \frac{1}{I^2} \sum_m m^2 n_m - \frac{1}{3} I(I+1) \qquad (2b)$$

is the alignment.

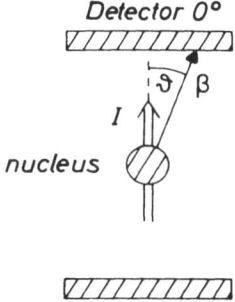

Detector 0°

β Detection - Optical Pumping tried out first for ^{37}K, ^{21}Na, by Köpf et al. (1967) [7]

Detector 180°

β - *asymmetry:*

$$W_\beta (\vartheta) = 1 + \frac{v}{c} \frac{\langle I_z \rangle}{I} A \cos \vartheta$$

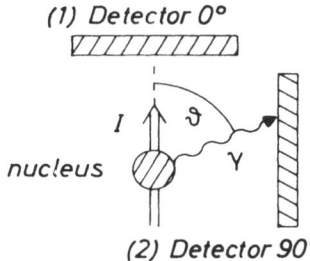

(1) Detector 0°

γ Detection - Optical Pumping tried out first for ^{203}Hg by Capeller, Mazurkewitz (1971) [9]

(2) Detector 90°

γ - *anisotropy:*

$$W_\gamma (\vartheta) = \sum_{K=even} A_K f_K P_K (\cos \vartheta)$$

Fig. 1: Geometries in ß- and γ-RADOP

In ß decay, the correlation coefficient A_1 of equ. (1) is competing with A_0 leading to the famous, parity violating $0°$ - $180°$ asymmetry (see Fig. 1 and formula therein). Higher order terms are not important. Hence the detectors are arranged under $0°$ and $180°$ with respect to the polarization axis. The sign of the polarization and hence the asymmetry can be inverted by simply turning the polarization of the pumping light from right handed (σ^+) to left handed (σ^-). Counting simultaneously under $0°$ and $180°$ one may define a relative difference signal as

$$S_\beta = (N_o^+ N_{180}^- - N_o^- N_{180}^+)/(N_o^+ N_{180}^- + N_o^- N_{180}^+) = 2a/(1+a^2) \quad (3)$$

where N_o^+ is the countrate under $0°$ and σ^+ polarization etc.

The experimental asymmetry "a" is the average

$$a = \overline{P_I A \, v/c} \qquad (4)$$

to be taken (i) over the optically pumped nuclei as well as the unpolarized background, (ii) over the velocity spectrum v/c; (iii) over different asymmetry parameters A when the ß decay runs through several decay channels.

In γ decay, only the correlation coefficients A_k with even k are different from zero, leading to a 0^0 - 90^0 anisotropy. Usually the lowest order alignment f_2 gives the dominant contribution to the anisotropy. Since $k \leq 2I$, $I \geq 1$ is required. Contrary to ß decay the investigation of the frequent I = 1/2 case by γ radiation is excluded therefore, unless one measures the circular polarization of the γ's instead of their angular distribution. But the polarimetry of γ radiation is not very efficient.

The multipolarity L of the decay imposes another restriction $k \leq 2L$ on the highest observable f_k. In case of optical radiation this would be f_2. γ-RADOP, however, may offer chances for analyzing higher orders of orientation of an ensemble of atoms undergoing spin dependent interactions like OP, relaxation, spin exchange etc. Another advantage of ß-, γ-RADOP in this respect is that they monitor the state of the system without the need of any probing light beams or rf fields. Such chances have been exploited very little so far. Some work in this direction has been reported in refs. [11],[12].

The γ RADOP signal is obtained again as the difference in W (ϑ) for two pumping modes "i" and "k" (e.g. OP with σ and π light). The signal may be defined in analogy to ß RADOP by

$$S_\gamma = (N_1^{(i)} N_2^{(k)} - N_2^{(i)} N_1^{(k)}) / (N_1^{(i)} N_2^{(k)} + N_2^{(i)} N_1^{(k)}) \qquad (5)$$

where $N_1^{(i)}$ is the countrate in detector 1 with pumping mode "i" etc. The detectors 1 and 2 form a right angle with the source. Considering only the leading f_2 term it is clear from (2b) that a concentration of population into states with high or low m leads to $f_2 > 0$ or $f_2 < 0$, respectively. This may be achieved with different OP modes. In the experiment one can switch easily and fast between different pumping modes choosing those two

modes "i" and "k" which produce a maximum change of
angular distribution and hence optimize the signal (5).

In case of pumping atoms with a diamagnetic ground
state, like Hg, only the nuclear spins are oriented and
one can apply NMR to them. NMR detected by nuclear
radiation should show some new, appealing features which
have been investigated theoretically by Matthias et al.
[13]. At high rf power for instance, the resonance curve
should split up into k bumps where k is the order of
orientation observed. Therefore ß-detected NMR will
always yield a pure Lorentzian line in ordinary NMR.
Cappeler and Mazurkewitz observed a two bumped NMR
signal in their γ RADOP experiment [9].

2.2. Comparison of RADOP and optical detection

The numbers given in table 2 serve to compare the
virtues and limits of the methods. From this table it
is clear that optical detection (col. I) has the
advantage of the highest number of detected quanta per
atom, even with an ordinary spectral lamp as assumed
here. In this case the limiting factor is usually the
high background of light scattered from walls etc.
Excitation by a tunable laser has pushed the limit of
optically detected atoms down to $100/cm^3$ [14].
Laser spectroscopy offers a wide scale of methods which
can be applied to the investigation of radioactive atoms,
the first successful example [15] being presented by
Libermann at this conference [16]. A comparison between
ß- and γ-RADOP shows that statistics is clearly in
favour of ß RADOP especially when the γ count rate is
spread over many individual lines. On the other hand it
is not guaranteed that there will be a ß asymmetry if the
decay feeds many levels of the daughter nucleus since
the asymmetry parameters of individual transitions may
differ in sign and value and thus may cancel each other
when summed up. Indeed the asymmetry observed in ß RADOP
experiments differed by a factor of 10 between isotopes
and is usually quite small. The strength of γ RADOP is
the high resolution of γ spectroscopy, avoiding this
difficulty even for complex decay schemes and high background from daughter nuclei (compare Fig. 4).

A further aspect in favour of optical spectroscopy
is that it has no limitation on nuclear spin whereas
ß RADOP needs $I \geq 1/2$ and γ RADOP $I \geq 1$.

Table 1

	optical fluorescence detected OP	ß-RADOP	γ-RADOP
number of emitted quanta[+)	30/sec	1/nucleus	$\approx$1/nucleus
detector efficiency	10 %	100 %	3-10 %
solid angle	3-5 %	10-50 %	5 %
signal[++)]	30 %	10 %	20-30 %
permissible countrate	no limit	10 MHz[+++)]	30 kHz[+++)]
selectivity against neighbouring isotopes and daughters	yes	no	yes
minimum number of accumulated atoms used in present experiments	10^{11} a)	10^4 b)	10^8 c)

a) investigating long lived Hg isotopes using a spectral
   lamp [17]

b) ref. [4], c) ref. [18]

   +) corresponding to a pumping time $T_p \approx 30$ms, as achieved
      for Hg by the spectral lamp described in ref. [3].
      For stronger resonance lines and laser light sources
      the number may go up to 10^8/s.

  ++) actually achieved in the described set-up. The
      optical signals may be higher in a set-up optimized
      for optical detection.

 +++) when background is to be rejected by coincidence or
      energy discrimination; otherwise no limit.

2.3. RADOP-experiments on Hg-isotopes

Mercury is ideally suited for the application of
RADOP for the following reasons: (i) Hg is volatile and
chemically inert. It is no problem, therefore, to keep
it quantitatively in the gas phase (which seems to be
almost impossible for alkalies, see ref. [7] and below).
(ii) Hg has a diamagnetic ground state. Hence the
relaxation time of nuclear orientation is long (of the
order of seconds) as compared to the pumping time
(≈ 30 ms). This enables a very high degree of orientation.

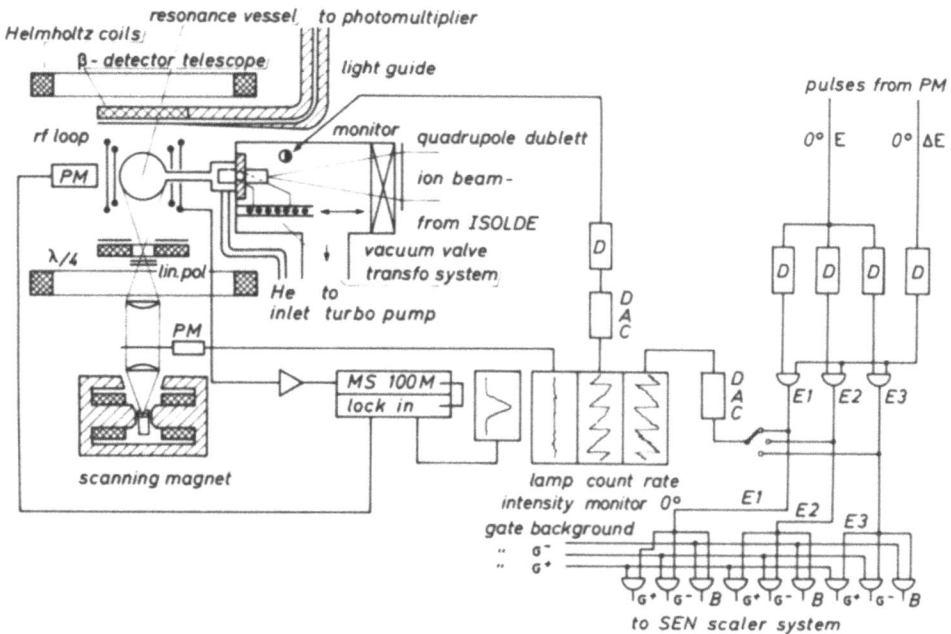

Fig. 2: Left: RADOP schematical set-up as described in
the text. The γ detectors and the 90° ß telescope
have been omitted (PM=photomultiplier).
Centre and right: Block diagram of electronics,
from left to right: OF signal detection of stable
isotopes and rf system, recorder monitoring,
energy discrimination, coincidence and gating of
pulses from one of the ß telescopes (D=discrimi-
nator, DAC=digital to analog converter).

(iii) The pumping line $\lambda = 2537\text{Å}$ to the excited $^3P_1$ state
has a hyperfine splitting (hfs) and isotope shift (IS)
much larger than the Doppler width. Hence one gets
reliable values for the nuclear moments (μ_I, Q) and the
change in nuclear charge distribution ($\langle r^2 \rangle$) even
without applying high resolution techniques.

Fig. 2 shows the set-up at ISOLDE as used for ß RADOP
experiments. A collector foil is mounted on either side
of a vacuum flange. One of them is exposed to the ion
beam and collecting activity, the other is pushed into
the bottle-neck of the resonance vessel, thereby tightening

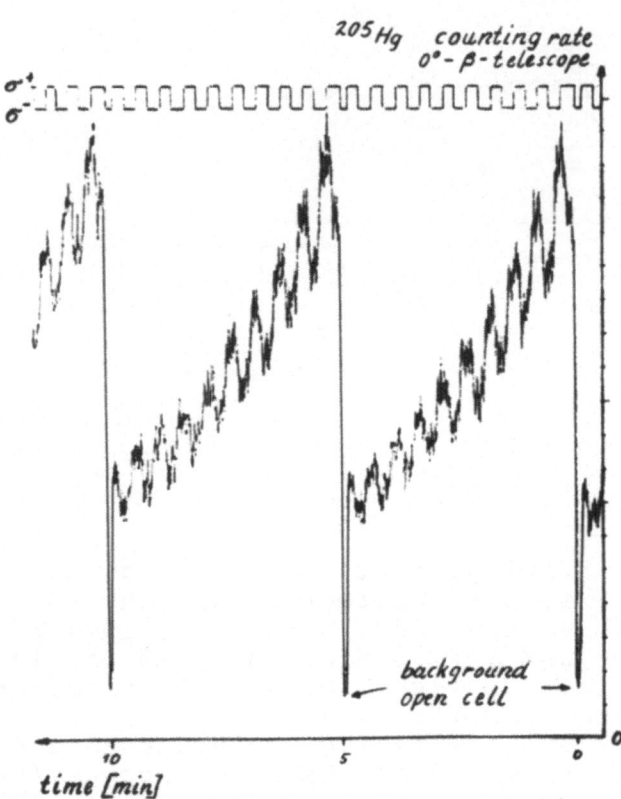

Fig. 3: Chart recording of the count rate of ß particles
emitted from polarized ^{205}Hg under 0°. Time runs
from right to left. Two full counting periods
are recorded. The increase of the count rate
when the resonance vessel is filled with a fresh
sample, the exponential decay, and the sharp drop
to the background level, when the remaining
activity is pumped off, are shown. Superimposed
to the exponential decay one sees wriggles due
to the change of sign of polarization [19].

it against the vacuum system. After a collection time of
about one half-life the two foils are interchanged by
pulling back the carriage, swiveling around the flange,
and pushing the carriage foreward. The collector foil is
directly heated for a few seconds to about 1,500°C and
a stream of He at a pressure of 100 Torr transports the
evaporated Hg activity to the cell. The pumping lamp
is subjected to a magnetic field in order to shift a
Zeeman component of the stable even isotope in the
lamp to an hfs component of the unstable odd isotope
in the vessel. The principle sequence of the measuring
cycle can be read from the recording of the 0° count
rate plotted in Fig. 3. In that favourite case the
ß asymmetry signal is a macroscopic effect to be
recognized instantly without major electronic tricks
or computer work.

An example of γ RADOP is given in Fig. 4. The signals
appear in γ lines of ^{193}Au being fed by the decay of
optically aligned ^{193}Hg^m. Obviously these signals contain
a lot of useful information for the analysis of the
decay scheme. But this field is still lying fallow since
the limited power of our group had to be concentrated
on the hfs and IS work.

Fig. 5 shows the ß decay asymmetry of ^{189}Hg as
function of the magnetic field over the lamp. The scan
is displaying the hfs and IS of the unstable isotope.
It is a peculiarity of OP in buffer gas that pumping via
the hfs state of highest angular momentum leads to a
polarization opposite to the one achieved via the other
hfs states. The final, precise value for the magnetic
moment is obtained from NMR on the oriented nuclei
destroying the RADOP signal (Fig. 5, insert).

2.4. Results

Fig. 6 contains as function of neutron number all
hfs and IS data, known for this spectral line spanning
from ^{181}Hg to ^{205}Hg. The stable isotopes are centered
around ^{200}Hg. The results from ^{190}Hg to ^{203}Hg including
isomers, were obtained by high resolution spectroscopic
methods like interferometry, level crossing and OP.
Off line techniques were applied as far as radioactives
were concerned [17], [20-22]. They established already
the widest and most complete set of hfs and IS data
collected for the isotopes of one element. This was a
very good starting position for our experiments. Except
for the case of ^{199m}Hg [18] all our hfs and IS data were
obtained by ß RADOP. First results were communicated
already at the 3. Atomic Conference in Boulder [23].
For a complete listing of data and refs. see ref. [4].

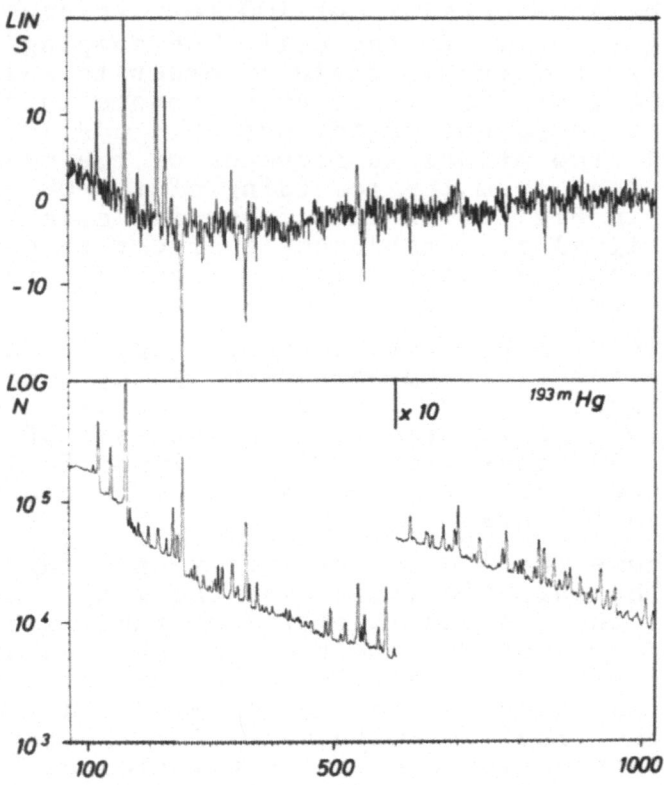

Fig. 4: Bottom: γ decay spectrum of ^{193}Hg and ^{193m}Hg
taken by a 35 cm³ Ge-Li detector. This spectrum
is obtained by σ pumping; a spectrum obtained
by π pumping is stored in another subgroup of
the memory of the analyzer.
Top: The difference spectrum divided by the
statistical error. As the ^{193m}Hg isomeric state
is selectively oriented by OP, the γ lines
showing statistically relevant anisotropy signals
are clearly fed by the decay of ^{193m}Hg.

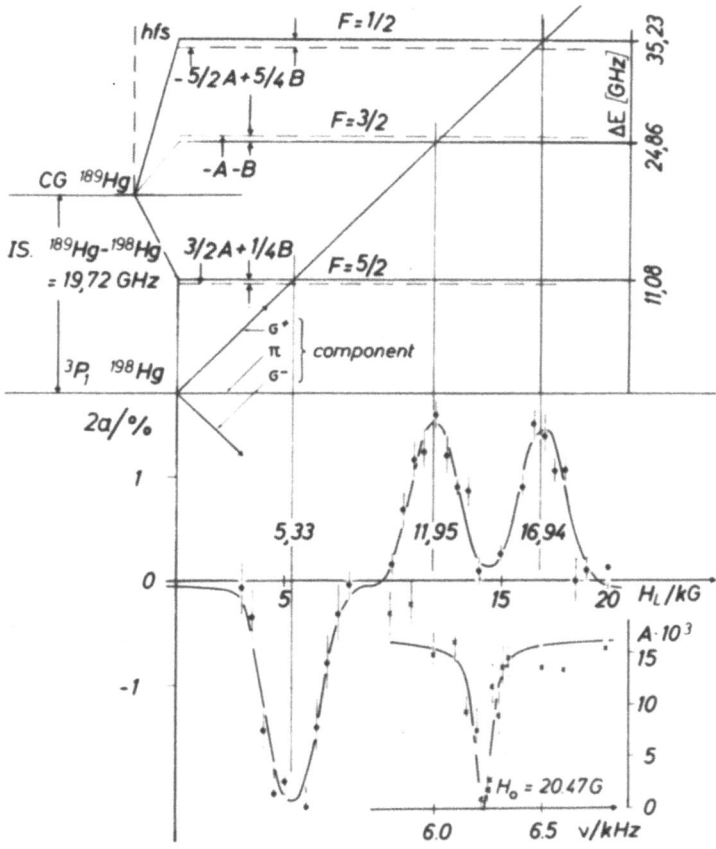

Fig. 5: Determination of hfs and g_I of ^{189}Hg by ß RADOP.
Top: Excited $^3P_1$ state of ^{198}Hg and ^{189}Hg in
their relative position to each other.
Bottom: An asymmetry in the ß decay of ^{189}Hg is
observed, whenever the Zeeman scanned σ^+
component of the ^{198}Hg spectral lamp energetically
matches one of the hfs components of ^{189}Hg.
Insert: NMR resonance of ^{189}Hg. The parameters
deduced by fitting are: I=3/2; A = 5.86(6) GHz;
B = 0.71(9) GHz; IS(189-204) = 19.72(10) GHz;
g_I = -0.3994(5). (The IS is not corrected for
pressure shift). The least squares fitted
theoretical scanning curve is obtained by
calculating the steady state solution of the
OP rate equations at each scanning point.
Fitting parameters: A, B, IS, width of (Gaussian)
emission line, ratio of pumping to relaxation
rates.

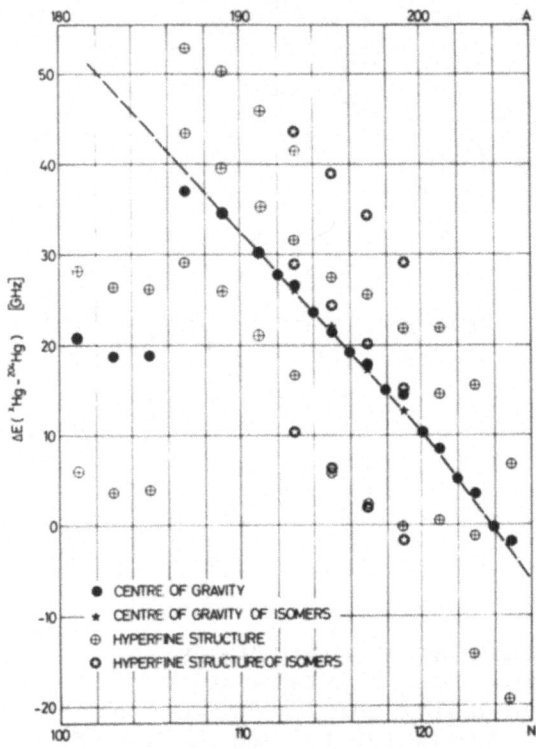

Fig. 6: Hfs splitting and IS of Hg isotopes relative to ^{204}Hg in the spectral line ($6s^2\ ^1S_0 - 6s6p\ ^3P_1$, $\lambda = 2,537$Å). The error in the position of the hfs states (and hence of IS) of the short-lived isotopes is about the diameter of the dots.

The lightest isotopes $^{181-185}$Hg have spin I = 1/2
preventing unfortunately the determination of the
quadrupole moment which would be very important for the
interpretation of the sharp jump of the IS observed
between ^{185}Hg and ^{187}Hg (see below). In spite of the
fact that nuclear structure changes drastically, the
magnetic moments of the frequent I = 1/2 and I = 3/2
isotopes are almost constant. Thus they are not very
helpful in analyzing the nuclear structure. In the
following we will confine ourselves therefore on the
interpretation of the isotope shift.

2.4.1. Isotope shift and charge radii of Hg isotopes

The isotope shift in elements as heavy as Hg is
essentially determined by the change of the nuclear
charge distribution from isotope to isotope which is
changing the Coulomb field inside the nucleus.
The second contribution, the mass effect, stemming from
the recoil motion of the nucleus contributes less than
1 % at A $\approx$ 200.

In non-relativistic theory the so called field shift
is given by

$$\delta \nu_f^{AA'} = \frac{2}{3} \pi Ze^2 \Delta |\psi(o)|^2 \delta \langle r^2 \rangle^{AA'} \tag{6}$$

where $\delta \langle r^2 \rangle^{AA'}$ is the change of nuclear rms charge
radius between the two isotopes A and A'. $\Delta |\psi(o)|^2$ is
the change of the electron density at the nucleus in the
electronic transition.

According to Bodmer [24] equ. (6) holds also for the
relativistic case when an additional factor is introduced in (6) depending only on Z and A [25]. It accounts
for the relativistic correction to $\Delta |\psi(o)|^2$ as well as
the retroaction of the extended nuclear charge on the
Dirac wave function. For very heavy elements $\langle r^2 \rangle$ should
be replaced by a power series [26]

$$\lambda = \delta \langle r^2 \rangle + \sum_{k=2}^{\infty} \frac{c_k}{c_1} \delta \langle r^{2k} \rangle \tag{7}$$

where the higher moments contribute -7 % for the case of
Hg (assuming spherical shape). The real problem is the
value of $\Delta |\psi(o)|^2$. It may be calculated from semiempiri-
cal formulas or by Hartree Fock methods both being not
very accurate. It is very helpful therefore to check
$\delta \langle r^2 \rangle$ values from optical spectra [27] with those

obtained from electronic [28] and muonic [29] X-ray
spectra as well as from electron scattering [30] which
have been compiled in a recent issue of Atomic Data
and Nuclear Data Tables. Whereas satisfying agreement
is observed in most cases, a serious discrepancy exists
in Hg between the optical value $\delta\langle r^2\rangle^{200/204}_{opt}$ = 0.236(19)
fm^2 [27] and the one from electronic X-ray work
IS $\delta\langle r^2\rangle^{200/204}_{X-ray}$ = 0.162(24) fm^2 [31]. It has been partly
removed recently by a new X-ray measurement yielding
$\delta\langle r^2\rangle^{200/204}_{X-ray}$ = 0.195(19) fm^2 [32]. A measurement of
the muonic X-ray IS in Hg is under way [32] which hope-
fully will decide this long standing controversy. For
the time being the optical value is used for the
analysis of our data. Concluding this discussion one
should stress the following points: It is very important
that either the Hartree Fock calculations of $\Delta|\Psi(o)|^2$
proceed to a satisfactory level of accuracy or that the
optical values are backed and calibrated by systematic
measurements with independent methods throughout the
chart of stable nuclei, because the optical method is
the only one having a good chance for systematic
investigation of the nuclear charge distribution off
stability.

3. Interpretation of Hg charge radii and its consequences

Fig. 7 shows the experimental λ-values in the series
of Hg isotopes. At the present state of art it is
completely sufficient to interpret them as $\delta\langle r^2\rangle$ ignoring
the contributions from higher radial moments.

3.1. Deformed liquid drop model

Using the phenomenological model of a deformed liquid
drop the change of $\delta\langle r^2\rangle$ can be separated into a volume
effect ($\sim A^{1/3}$) and a shape effect ($\sim \delta\langle \beta^2\rangle$). In first
approximation it is given by

$$\lambda \approx \delta\langle r^2\rangle = \delta\langle r^2\rangle_{vol} + \delta\langle r^2\rangle_\beta \qquad (8)$$

$$= \S\,\frac{2}{5}\,R_o^2\,\delta A/A + (3/4\pi)R_o^2\,\delta\langle\beta^2\rangle$$

with R_o = 1.2 $A^{1/3}$ fm and $\delta\langle\beta^2\rangle$ = change of the mean
squared deformation parameter. The correction factor $\S$
is the socalled IS discrepancy discussed below.

Regarding the plot of Fig. 7 the anomalous jump at the light end of the mass chain attracted most of the interest. But also the almost constant slope of the IS in the very long chain ^{187}Hg - ^{205}Hg is of considerable interest and will be discussed first. Since in Hg only two protons are lacking to close the shell Z = 82, one expects these nuclei to be spherical. Hence the first term in (8) should be responsible for the IS. However, the slope of λ_{exp} in Fig. 7 is a factor of two too flat as compared to λ_{unif} predicted from the first term in (8) with ξ = 1. A fraction of this effect can be explained by a small but steadily increasing deformation $\langle \beta^2 \rangle$ with decreasing neutron number. But at least half of it remains.

The IS discrepancy is observed as an overall effect in all mass regions but never has been checked over such a long isotopic distance.

3.2. Nuclear compressibility

Since $\xi < 1$ the charge distribution changes less within an isotopic series, than would be expected from the naive liquid drop model. A few years ago Brown pointed out that this phenomenon is a key information to nuclear compressibility [33]. The first quantitative success in this matter was recently obtained by Bainer and Lombard who calculated $\delta \langle r^2 \rangle$ between ^{187}Hg - ^{205}Hg with Hartree Fock methods using a density dependent force, this means giving up the idea of incompressibility of nuclear matter [34]. Unfortunately the perfect agreement of this calculation with experiment is questioned to some degree by the uncertainty in the evaluation of $\delta \langle r^2 \rangle$ from the optical field shift.

3.3. The phase transition in light Hg nuclei

Finally the large, very sharp jump in IS observed between ^{185}Hg and ^{187}Hg is to be discussed. An effect of this size can be explained only by a global, collective change in the proton distribution which one might call a phase transition of the nucleus. What kind of phase transition ? One possibility could be the sudden appearence of strong deformation with $\beta \approx 0.3$ as was suggested in our first paper on the subject [2]. It would increase $\langle r^2 \rangle$ via the second term in (8). One is led to this interpretation remembering the similar behaviour of IS in the rare earths between N = 88 and N = 90 which was observed in 1949 by Brix and Kopfermann and inspired the development of the concept of deformed nuclei [35].

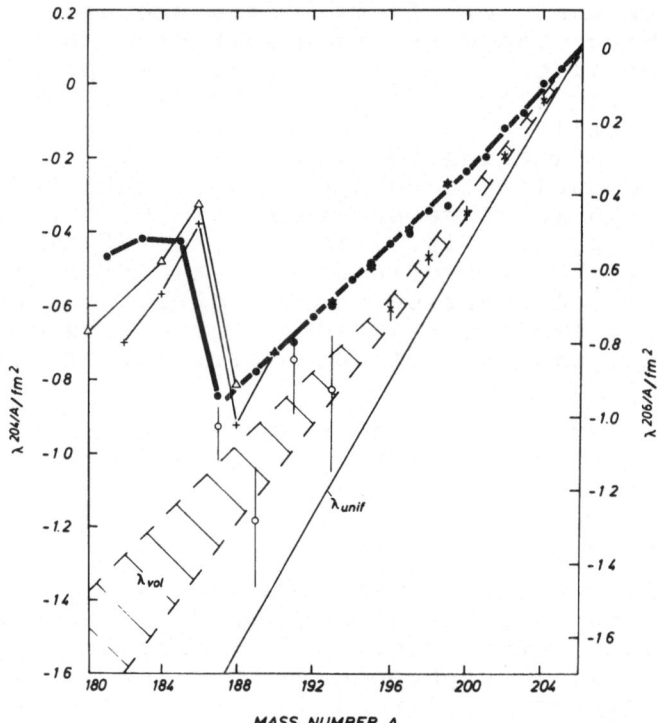

Fig. 7: Nuclear charge radius of Hg isotopes ($\lambda = \sqrt{\langle r^2 \rangle} -$ 1.1 x 10^{-3} $\sqrt{\langle r^4 \rangle} \approx \sqrt{\langle r^2 \rangle}$). λ_{exp} are the experimental values calculated from IS data. Full dots represent ground state charge radii and stars indicate those of isomers. For easy recognition they are connected by a solid line. The statistical errors of λ_{exp} caused by the experimental uncertainty of the IS measurements are about the diameter of the dots for the short-lived isotopes and much less for stable or long-lived isotopes. The dashed lines give the upper and lower limit for λ_{vol}, the change of the experimental charge radius corrected for deformation effect by $\lambda_{vol} = \lambda_{exp} - \lambda_{ß}$. To substract the deformation effect from

λ_{exp}, spectroscopic quadrupole moments (open circles) and B(E2) values (crosses) are used. Charge radii calculated by nuclear Hartree-Fock ($\triangle$) [39] and Strutinzky type calculations [38] (+) are given for comparison. The absolute values $\langle r^2 \rangle$ of these calculations are adjusted at A = 196.

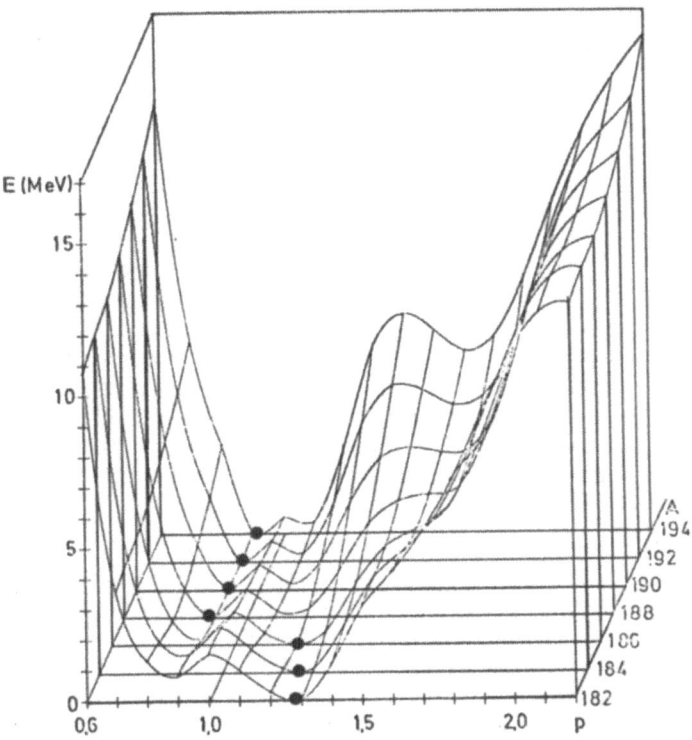

Fig. 8: Nuclear potential energy surfaces as function of deformation for even Hg isotopes in the range $182 \leq A \leq 194$ [36].

From the large number of theoretical papers having
treated the similar problem in Hg in the meantime, I
will discuss here the result of one of the earliest [36].
Using the Strutinsky method, Dickmann and Dietrich
calculated for the series of even Hg isotopes the nuclear
potential energy as function of the deformation.
The results are shown in Fig. 8. The abscissa is not ß
but the ratio of the two axis p = a/b with a being the
axis of rotational symmetry. Disregarding the secondary
minimum high up in energy on the very prolate (right)
side we observe for heavier Hg nuclei a rather steep
parabola centered around the spherical shape p = 1.
However, the bottom of the parabola is split up into
two minima, one on the oblate the other on the prolate
side with a small barrier inbetween. With decreasing
neutron number (i) the parabola gets wider, (ii) the
oblate minimum stays at $p_{min} \cong 0.9$ ($\beta_{min} \cong -0.1$) and
increases in energy, (iii) the prolate minimum runs
from small $p_{min} \cong 1.1$ ($\beta_{min} \cong 0.1$) to strong deformation
$p_{min} \cong 1.3$ ($\beta_{min} \cong 0.3$), (iiii) the central barrier gets
higher. The salient point is found between ^{188}Hg and
^{186}Hg where the absolute minimum jumps from the little
deformed oblate to the strongly deformed prolate side.
Assuming that the collective nuclear wave function is
localized at the absolute minimum (rigid, form stable
nucleus) this jump is accompanied by a large increase
of β_{min}^2, indeed.

Yet one has to consider the zero point vibration of
the nucleus in this potential. The calculation shows
that for the heavier nuclei the zero point wave function
spreads over both valleys, not recognizing the small
central barrier. But for $A \leq 188$ it is localized in the
valley of the absolute minimum. Since for a vibration
in this potential the square of the vibrational
amplitude $\langle (\beta-\beta_{min})^2 \rangle$ is different from zero, it
contributes to the total expectation value $\langle \beta^2 \rangle$ which
is entering $\sqrt{\langle r^2 \rangle}$ (see equ. 8)). Taking this contribution into account, a satisfying agreement with the
experimental findings is achieved. Other Strutinsky
calculations [37], [38], and also Hartree Fock
calculations arrived at similar results [39]. They are
compiled in Fig. 9 together with experimental results.
The plot shows the development of $\langle \beta^2 \rangle$ down the chain
of Hg isotopes. In cases the authors quoted directly
$\sqrt{\langle r^2 \rangle}$ instead of $\langle \beta^2 \rangle$ or $\sqrt{\langle \beta^2 \rangle}$, the values are plotted
in Fig. 7.

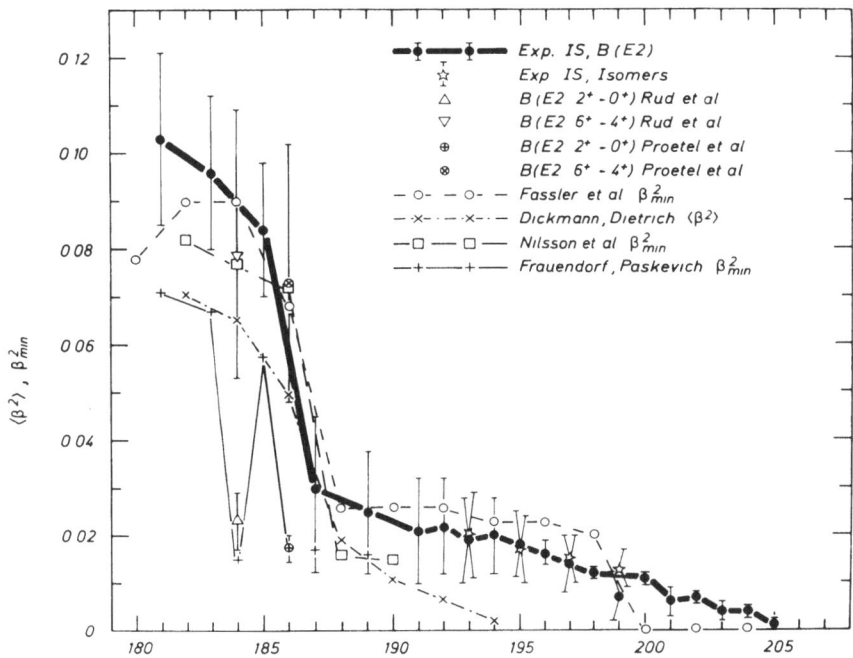

Fig. 9: Squared deformation parameters for Hg isotopes.
The experimental $\langle \beta^2 \rangle$ stem from IS and B(E2)
measurements. The theoretical values are calcu-
lated by the Strutinsky prescription. They are
obtained by Faessler et al. [37], Dickmann and
Dietrich [36], Nilsson et al. [38], Frauendorf
and Paskevich [46].

3.4. Coexistence of nuclear shapes

In the meantime experimentalists tried to confirm
this picture by searching for rotational bands in the
spectra of light even Hg isotopes in the critical region
down to ^{184}Hg. But to their great surprise they found
out that instead a vibrational band is building up on
the ground states of these nuclei being very similar to
the one known from heavier Hg nuclei which have certainly
an almost spherical shape [40-44]. One concluded therefore that the ground states of the light even Hg isotopes
are not deformed in contrary to their odd neighbours

and to the theoretical predictions. But in addition a
well developed rotational band was found indicating
clearly a deformed shape with $\beta_{min} \approx 0.25$. These phenomena
are characterized best by speaking about a coexistence
of two different nuclear shapes or phases like in the
famous fission isomers. The physical reason is found in
the two separated but almost degenerate valleys of the
potential energy surface (Fig. 8).

Theoreticians reacted on the new situation but had
difficulties in the beginning to reproduce all the
phenomena consistently with one and the same set of
interaction parameters [45]. Very recently two successful calculations were reported [46],[47]. To get a view
of the theoretical problems one must realize that a
critical phenomenon like this shape transition is
governed by differences in the binding energy of about
1 MeV compared to the total binding energy of about
1400 MeV at A = 200. The interaction parameters entering
the calculations are fitted to the experimental data in
the valley of stability because far off stability no
data were available. It is a great success therefore
that the extrapolation to nuclei of very different
composition leads not only to meaningful overall results
but comes quite close to a quantitative explanation.

4. Why and how laser spectroscopy on Hg isotopes ?

As pointed out before one concludes from isotope
shifts of odd Hg nuclei and from γ spectra of the even
ones that a huge odd-even shape staggering occurs in
light Hg isotopes. It would be very valuable for any
quantitative interpretation of the phenomenon, if it
could be observed model independently from one and the
same observable, for instance from $\sqrt{\langle r^2 \rangle}$. Since RADOP
fails in the case of even isotopes one has to perform
the measurement of the IS by purely optical means.
Excitation by a tunable Dye laser may overcome the
crucial intensity problem. We have set up an experiment
at ISOLDE, designed for laser spectroscopy of Hg isotopes
hopefully down to ^{182}Hg+). It should be suited as well
for other volatile elements with resonance lines in the
near ultraviolet like Cd, Tl, Pb.

+) This experiment has been prepared by T. Kühl,
    Ch. Duke, H.-J. Kluge, E.W. Otten and H. Rummel.

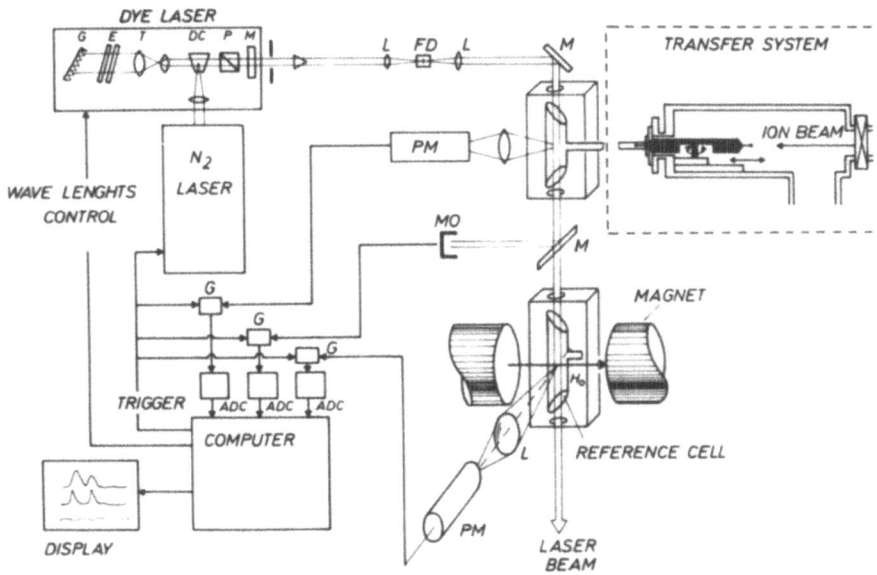

Fig. 10: Experimental set-up for laser spectroscopy on
         short-lived even Hg isotopes. G: grating,
         E: etalon, T: beam expanding telescope,
         DC: Dye cell, P: polarizer, M: mirror,
         FD: frequency doubling crystal, L: lens,
         PM: photomultiplier, MO: monitor for laser
         intensity, G: gate, ADC: analog to digital
         converter.

Fig. 10 shows a schematic view of the experiment.
The Dye laser is pumped by a pulsed nitrogen laser.
Its output is focused on a frequency doubling crystal
(ADP) for generation of the 2537 Å line. The resonance
cell contains the isotope under investigation. For on-
line work it is connected to the transfer system.
Since the lifetime of the $^3P_1$ state of Hg ($\simeq$ 1 µsec) is
long as compared to the lengths of the laser pulse
(T = 8 nsec), the fluorescence signal from the photomultiplier is gated on only for a few hundred nsec
after the light pulse. Thus, laser light scattered by
the glass walls of the cell and the dark current of the
photomultiplier are strongly suppressed. Scanning the
laser over the absorption line of the unstable isotope,
its frequency is calibrated by a reference cell

containing stable isotopes and placed in a magnetic
field in order to shift Zeeman components into the
interesting region. The system has been tested with
stable isotopes down to quite low vapor pressures
indicating sufficient sensitivity for measurements
down to ^{182}Hg. A test experiment on Cd (λ = 3261Å)
yielded also very promising results[+]).

5. ß-RADOP of exotic Rb and Cs isotopes

As mentioned before, the first on-line application of ß-RADOP concerned short lived alkalies (7).
They were produced by a cyclotron beam from noble
gas which served as target and at the same time as
buffer gas for OP. Now at the ISOLDE facility a
broad range of neutron-deficient Rb,Cs and Fr isotopes is available. In order to reach neutron-rich
alkalies as well we have set up a mass separator
on line with a fission target of the Bernas type
(48) which is placed in the core of the Triga
reactor at Mainz [++]).

All attempts to transfer the alkalies from the
evaporation to the resonance cell through a tube
(as shown in Fig. 2 for Hg) failed. In spite of
flushing with He gas and of buffering the surface,
the atoms diffused to the walls and stuck there
before reaching the resonance vessel. Therefore
the collecting foil was mounted right into the
centre of the resonance vessel and heated out
after collecting for about a halflife and filling
the vessel with 400 Torr He. Thus it was guaranteed at least that the atoms were free for the
diffussion time to the walls which is about 1s.
During that time one can observe the RADOP signal

[+] Note added in proof: In the meantime the system
   has been operated on line with the ISOLDE facility for a short run. As a first preliminary
   result it yielded the isotope shift of ^{190}Hg :
   $\delta\nu^{190/204}$ = 31.8(3) GHz.
[++] This facility has been built by W. Cürten, G.
   Huber, S. Kaufman, L. Kugler, P. Minn, E.W. Otten,
   L.v. Reisky, J. Rodriguez and J. Tegel with technical assistance by the Kernforschungszentrum,
   Karlsruhe, and by the Institut für Kernchemie,
   Mainz.

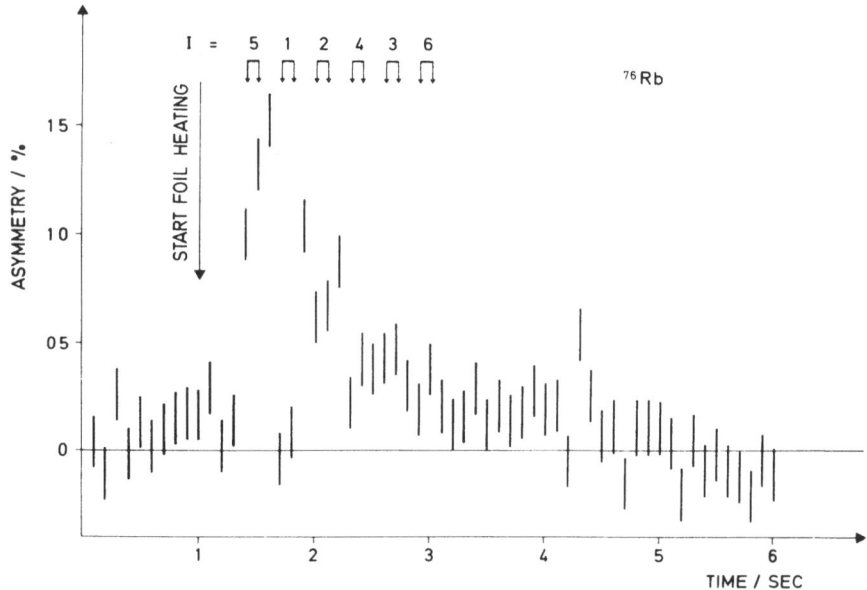

Fig. 11: Spin determination of ^{76}Rb by ß RADOP.
Applying a frequency for I = 1, the
ß asymmetry is destroyed.

(Fig. 11) and perform a spin measurement. In a weak
field the transition frequencies for each spin value
are given by

$$\nu = g_s \mu_B H_0 / h(2I+1).$$

The first results obtained on the neutron-deficient side
are[+]

$$I(^{76}\text{Rb}) = 1, \quad I(^{119}\text{Cs}) = 9/2$$

At these mass numbers the yield curve has dropped already
by a factor of 10^4 from its maximum value. Therefore it
is favourable to apply the more specialized RADOP method

[+] Ph. Dabkiewicz, H. Fischer, P. Freilinger, H.-J.Kluge,
H. Kremmling, R. Neugart and E.W. Otten

instead of the more general but less sensitive ABMR
method (compare ref. [5]). With the same method we [+)
obtained on the neutron-rich side as a first result[+)

$$I(^{91}Rb) = 3/2$$

After having determined the spins the corresponding
magnetic hfs intervals will be measured next.

6. A set-up for laser spectroscopy on fast mass-separated beams [++)

The laser spectroscopy experiment, discussed in
section 4, is motivated partly by the need to overcome
the principle limitation $I > 0$ of the RADOP method.
Other practical and much narrower limitations stem
from the concept of performing the RADOP experiment
in an atomic vapor confined in a resonance vessel:
(i) The element must be volatile; (ii) The orbital
angular momentum of the atomic ground state should be
zero, otherwise the relaxation is too fast; (iii) Chemically aggressive elements must be kept off the walls by
a buffer gas. Thereby the measuring time is automatically limited to the diffusion time to the walls.
Moreover high resolution spectroscopy of excited states
is prohibited by pressure broadening. It is proposed
therefore to perform optical experiments directly in
the fast, mass-separated beam. Such a concept would be
a way out of all the above mentioned difficulties and
would be an ideal adaptation of optical experiments to
the situation met at on-line mass separators.

The concept can be realized only by help of a cw
Dye laser (see Fig. 12). A light power of a few mW
will be sufficient in general to excite an atom at
least once along a path of about 10 cm. The fluorescent

[+) This experiment is performed by the separator group
mentioned before and J. Bonn, F. Buchinger,
W. Fellenberger, U. Kalepky, H.-J. Kluge, R. Neugart.

[++) The experiments described below are being prepared by
K.R. Anton, S.L. Kaufman, W. Klempt, R. Neugart,
E.W. Otten and B. Schinzler.

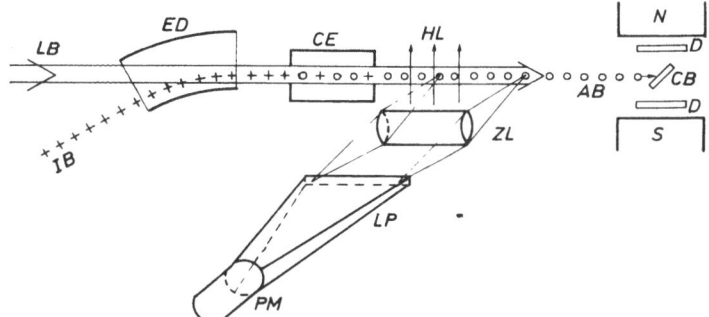

Fig. 12: Set up of fast beam laser spectroscopy with
         either optical or ß RADOP detection.
         Symbols: AB = fast atomic beam, CB = cold
         backing, CE = charge exchange cell,
         D = ß-detector, ED = electrostatic deflector,
         H_L = level crossing field, IB = mass separated
         ion beam, LB = laser beam, LP = light pipe,
         PM = photomultiplier, ZL = cylindric lens.

light may be detected with an efficiency of 1 %.
Including another diminishing factor of 10 (e.g. from
the multiplicity of ground states etc.) one may count
at least 1 photo electron per 1000 atoms passing the
laser beam. The fluorescence light may carry a level
crossing signal in order to resolve narrow hfs structures
in the excited states.

   In general the wave lengths of ionic resonance lines
are too short to be reached by present cw Dye lasers.
Thus the ion beam will be neutralized in a charge exchange
cell.

   An important consequence of aligning the two beams
is a reduction of the Doppler width by a factor of
$2\sqrt{eU/kT}$ which is about 1000 at beam energies of 40 keV.
This is easily understood remembering that the relative
spread of velocities in a certain direction is two times
the relative spread of kinetic energies in the same
direction.

   Laser excitation of a fast beam has first been
performed by Andrä et al. [49] . The concept described

above has been discussed in detail in a recent paper by one of our group (S.L. Kaufman) [50]. The velocity bunching phenomenon, used in colliding beam experiments before already (for a review see [51]) has been exploited in the meantime also in a laser spectroscopy experiment on a fast HD$^+$ beam [52].

The ultimate sensitivity of laser spectroscopy in a fast beam will critically depend on stray light problems and photomultiplier noise. At that point the RADOP principle can enter again. The fast atomic beam will be polarized by the OP action of the laser light. Then it will be stopped in a cold diamagnetic backing at a high magnetic field in order to preserve the polarization for the lifetime of the isotope. The polarization will be monitored by the ß decay asymmetry as before.

References

1   J. Bonn, G. Huber, H.-J. Kluge, U. Köpf,
    L. Kugler and E.W. Otten
    Physics Letters 36B, 41 (1971).

2   J. Bonn, G. Huber, H.-J. Kluge, L. Kugler
    and E.W. Otten
    Physics Letters 38B, 308 (1972).

3   G. Huber, J. Bonn, H.-J. Kluge and E.W. Otten
    Z.Physik A276, 187 (1976).

4   J. Bonn, G. Huber, H.-J. Kluge and E.W. Otten
    Z.Physik A276, 203 (1976).

5   E. Lindgren, this volume.

6   A. Kjelberg and G. Rudstam
    The ISOLDE collaboration, CERN Yellow Report 70-3.
    CERN,Geneva (1970).

7   U. Köpf, H.J. Besch, E.W. Otten and C. von Platen
    Z.Physik 226, 297 (1969).

8   E.W. Otten in "Atomic Physics 2", ed. G. Woodgate
    and P. Sandars, Plenum Press, London 1971, p. 113.

9   U. Cappeler and W. Mazurkewitz, Journal of Magnetic
    Resonance 10, 15 (1973).

10 S.R. de Groot, H.A. Tolhoek, W.J. Huiskamp
    In Alpha-, Beta- and Gamma-Ray Spectroscopy XIX B
    edited by K. Siegbahn, Amsterdam: North Holland
    Publishing Comp.1968.

11 Ch. von Platen, J. Bonn, U. Köpf, R. Neugart and
    E.W. Otten
    Z.Physik 244 (1971) 44.

12 H. Schweickert, H. Dietrich, R. Neugart and
    E.W. Otten
    Nucl.Phys. A246, 187 (1975).

13 E. Matthias, B. Olsen, D.A. Shirley, J.E.Templeton,
    R.M. Steffen
    Phys.Rev. A4, 1926 (1971).

14 W.M. Fairbank jr., T.W. Hänsch and A.L. Schawlow
    J.O.S.A. 65, 199 (1975).

15 G. Huber, C. Thibault, R. Klapisch, H.T. Duong,
    J.L. Vialle, J. Pinard, P. Juncar, P. Jacquinot
    Phys.Rev.Lett. 34, 1209 (1975).

16   S. Libermann, this volume.

17   P.A. Moskowitz, C.H. Liu, L. Fulop, H.H. Stroke
      Phys.Rev. C4, 620 (1971).

18   J. Bonn, G. Huber, H.-J. Kluge, E.W. Otten,
      D. Lode
      Z.Physik A272, 375 (1975).

19   J.M. Rodríguez, J. Bonn, G. Huber, H.-J. Kluge,
      E.W. Otten
      Z.Physik A272, 369 (1975).

20   W.J. Tomlinson III, H.H. Stroke
      Nucl.Phys. 60, 614 (1964).

21   S.P. Davis, T. Aung, H. Kleiman
      Phys.Rev. 147, 861 (1966).

22   R.J. Reimann, M.N. McDermott
      Phys.Rev. C7, 2065 (1973).

23   J. Bonn, G. Huber, H.-J. Kluge, U. Köpf,
      L. Kugler, E.W. Otten and J.M. Rodriguez
      Atomic Physics 3, p. 471, ed. by S.J. Smith
      and K.C. Walters
      Plenum Press, New York   (1973).

24   A.R. Bodmer
      Phys.Soc. A66, 1041 (1953); A67, 622 (1954);
      Nucl.Phys. 9, 371 (1959); 21, 347 (1960).

25   F.A. Babushkin, Sov.Phys. JETP 17, 1118 (1963).

26   E.C. Seltzer, Phys.Rev. 188, 1916 (1969).

27   K. Heilig and A. Steudel
      Atomic Data and Nuclear Data Tables 14, 613(1974).

28   F. Boehm and P.L. Lee, ibidem p. 605.

29   H. Engfer, H. Schneuwly, J.L. Vuilleumier,
      H.K. Walter and A. Zehnder, ibidem, p. 509.

30   C.W. de Jager, H. de Vries and C. de Vries,
      ibidem, p. 479.

31   R.B. Chesler and F. Boehm
      Phys.Rev. 166, 1206 (1966).

32   F. Boehm
      Californian Institute of Technology, Pasadena,
      priv.comm.

33   G.E. Brown in Facts of Physics, ed. A. Bromley
      and V.W. Hughes, Academic Press, New York,
      1970, p. 141.

34   M. Beiner, P.J. Lombard
      Phys.Lett. 47B, 399 (1973).

35   P. Brix and H. Kopfermann
      Z.Physik 126, 344 (1949).

36   F. Dickmann and K. Dietrich
      Z.Physik 263, 211 (1973).

37   A. Faessler, U. Götz, B. Slavov, T. Ledergerber
      Phys.Lett. 39B, 579 (1972).

38   S.G. Nilsson, J.R. Nix, P. Möller and I.Ragnarsson
      Nucl.Phys. A222, 221 (1974).

39   M. Cailliau, J. Letessier, H. Flocard and
      P. Quentin
      Phys.Lett. 46B, 11 (1973).

40   P. Hornshøj, P.G. Hansen, B. Jonson, A. Lindahl,
      O.B. Nielsen
      Phys.Lett. 43B, 377 (1973).

41   D. Proetel, R.M. Diamond, P. Kienle, J.R. Leigh,
      K.M. Maier, F.S. Stephens
      Phys.Rev.Lett. 31, 896 (1973).

42   N. Rud, D. Ward, H.R. Andrews, R.L. Graham,
      J.S. Geiger
      Phys.Rev.Lett. 31, 1421 (1973).

43   D. Proetel, R.M. Diamond and F.S. Stephens
      Phys.Lett. 48B, 102 (1974).

44   J.H. Hamilton et al.
      Phys.Rev. Lett. 35, 562 (1975).

45   F. Dickmann and K. Dietrich
      Z.Physik 271, 417 (1974).

46   S. Frauendorf and V.V. Pashkevich
      Phys.Lett. 55B, 365 (1975).

47   D. Kolb and C.Y. Wong
      Nucl.Phys. A245, 205 (1975).

48   I. Amarel, R. Bernas, J. Chaumont, R. Foucher,
      J. Jastrzewski, A. Johnson, R. Klapisch and
      J. Teillac
      Arkiv för Fysik 36, No 10, 77 (1966).

49   H.J. Andrä, A. Gaupp and W. Wittemann
      Phys.Rev.Lett. 31, 501 (1973).

50   S.L. Kaufman, Opt.Comm. 17, 309 (1976).

51   G. Dunn in Atomic Physics 1, ed. B. Bederson,
      V.W. Cohen, and F.M.J. Pichanick; Plenum Press,
      New York, 1969.

52   W.H. Wing, G.A. Ruff, W.E. Lamb, jr., and
      J.J. Spezeski, Phys.Rev.Lett. 36, 1488 (1976).

HIGHLY-EXCITED ATOMS

Daniel Kleppner

Dept. of Physics and Research Laboratory of Electronics

M.I.T., Cambridge, Massachusetts 02139

During the past two years the study of highly excited atoms and molecules has emerged as an area of vigorous activity in numerous laboratories. The subject is by no means new, however; its origins go back to 1884 when Balmer proposed his empirical formula for the wavelengths of atomic hydrogen. Balmer's success lay not so much in the simplicity with which he could relate the four known lines of hydrogen, but in his prediction, and the rapid discovery, of a multitude of shorter wavelength lines. By 1886 Cornu had observed Balmer lines through n = 13; in 1893 Pickering reported stellar observations of lines up to n = 31, establishing an alliance between Rydberg states and astronomy which persists to our day, and Woods, in 1906, detected Rydberg lines through n = 51 in the absorption spectrum of sodium.

Studies of the physics of atoms in high Rydberg states date from the early 1930's. Amaldi and Segré[1] measured shifts of high Rydberg lines of sodium and potassium due to rare gas at very high density — so high that the mean interatomic separation was small compared to the radius of the Rydberg electron. Fermi[2] proposed an explanation based in part on the low energy scattering length of the electron, which he treated essentially as a free particle, a point of view which has recently received considerable interest. In 1938 Jenkins and Segré[3] measured the absorption spectrum of sodium in the magnetic field of the 60" Berkeley cyclotron, then under construction about 50 meters from this conference site. The diamagnetic term, which scales as $<r^2> \sim n^4$, was clearly visible; for n = 28 it was large enough to appreciably intermix states with different ℓ.

In the early 1960's two new streams of activity emerged. The first resulted from the prediction by Kardashev[4] that recombination processes for interstellar hydrogen should give rise to a well resolved spectrum, and the observation by Hoglund and Metzger[5] of microwave lines arising from transitions among states with n ≃ 100. (States as high as n = 250 have since been identified.) The second was the discovery by Riviere and Sweetman[6] that charge neutralization of fast protons by an exchange gas yields a copious supply of hydrogen in highly excited states. Riviere and Sweetman showed that these atoms could be conveniently detected by ionization in an applied field; this technique has made possible many of the recent advances. Bailey, Hiskes and Riviere[7] published ionization rates for excited states of hydrogen which are now widely employed. A review of work on field ionization is given by Il'in in the Proceedings of the 1972 meeting of this Conference[8].

Current work with highly excited atoms and molecules is so diverse that it is impossible to provide a comprehensive review in a half hour talk. Consequently, I shall describe a few recent advances, but neglect such major areas as the fine and hyperfine structure of Rydberg states of the alkalis[9], the studies of the excited states of helium by optical-microwave and level crossing methods[10], and recent advances in the optical spectroscopy of Rydberg states of the alkaline earths[11], uranium[12], and the H_2 molecule[13].

The tunable laser has provided the opportunity for much of the current work on Rydberg states. However, W. A. Chupka[13], in a largely unpublished series of experiments, has for some years populated Rydberg states of atoms and molecules by optical excitation using a continuous source with an ultra violet monochromator. The atoms are detected by collisional ionization. Fig. 1 shows excitation spectra for Kr detected by collisional ionization with SF_6

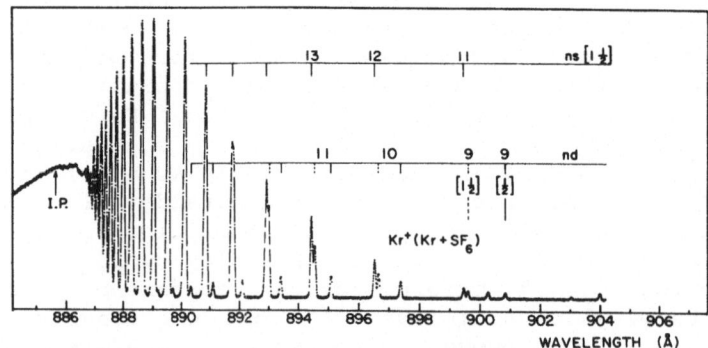

Fig. 1. Kr^+ signal vs wavelength for Rydberg states of Kr (Kr**) detected by Kr** + SF_6 → Kr^+ + SF_6^-. (From Chupka[13].)

The decreasing signal for low n states is due to the competition between the electron affinity of the SF_6 and the binding energy of Kr**; the decrease for high n is due to loss of oscillator strength.

Electron attachment rates at very low energy have been measured by West et al[14] by observing the ionization rate of Rydberg states of xenon in gases of various molecules of high electron affinity. Fig. 2 shows rate constants for electron attachment to SF_6 as a function of electron kinetic energy, including data from flowing after-glow experiments and, at lower energy, from Rydberg electrons. Matsuzama[15] has pointed out that transfer processes of this type can be understood in terms of a model in which the Rydberg electron behaves essentially as a free particle with the ionic core playing a completely passive role.

Collisions between Rydberg atoms and rare gas atoms can give rise to a variety of inelastic processes, including ionization and changes in m, ℓ, n. Gallagher, Edelstein and Hill[16] have measured the total cross section for ℓ-changing collisions in sodium by using the fact that the radiative lifetime for a given n increases rapidly with ℓ. The technique centers on measuring the fluorescense rate from sodium atoms in d Rydberg states at various buffer gas pressures. At low pressure the atom radiate freely with their spontaneous lifetime. As the pressure is increased the lifetime

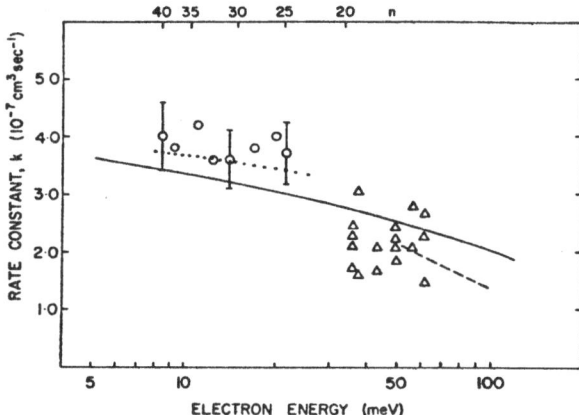

Fig. 2. Rate constants for electron attachment to SF_6. Triangles: electron swarm data. Dots: data from high Rydberg atoms. Solid line: theory, for Maxwellian velocity distribution. Dotted line: theory for velocity distribution of Rydberg atoms. (From West et al[14].)

decreases due to ℓ-changing collisions with the buffer gas. At
still higher pressure two decay rates are observed: the first is
due to the rapid depopulation of the d state as the atoms are
transfered to states of higher angular momentum (the s and p states
play little role because of their large quantum defects); the second, much smaller, is due to the repopulation of the d state as
atoms are transferred back from the reservoir of well populated
high angular momentum states. Fig. 3 shows the measured transfer
cross section for sodium with various rare gases: the solid dots
are values calculated by Olsen[17] from close coupling theory.

Charged particle collisions with Rydberg state atoms are important to astrophysics and to plasma physics. A review of theoretical cross sections for various charged particle collision processes has been published by Percival and Richards[18]. Koch and
Bayfield[19] have measured electron-loss cross sections for collisions between protons and hydrogen in the range $44 \lesssim n \lesssim 50$. The
hydrogen was formed by charge exchange from an 11 keV proton beam
which also provided the collision projectile. The center of mass
energy was varied over the range 0.4 eV to 60 eV by retarding the
protons, using the principle of merged beams. The results gave
strong evidence for validity of classical scaling laws for highly
excited states[18].

High angular momentum states of atoms have remained largely

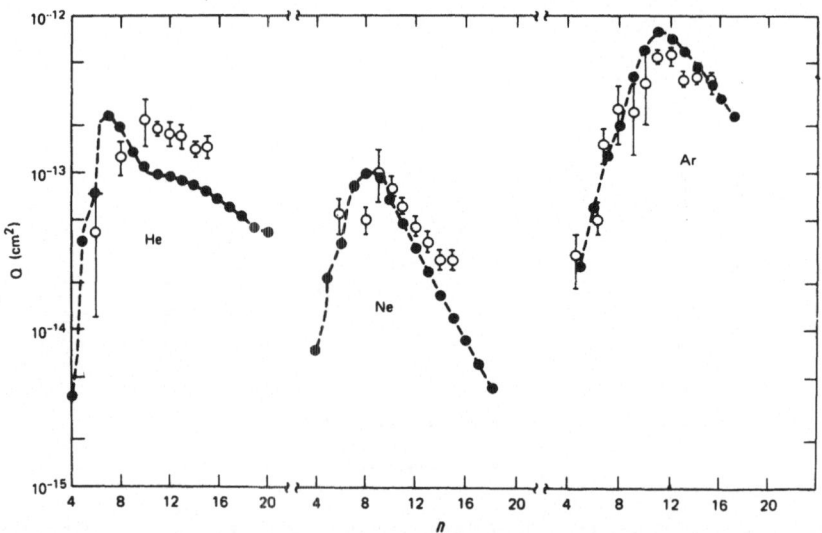

Fig. 3. Cross section for angular momentum changing collisions of
nd states of Na in various rare gases. Open circles: data.
Dots: result of close coupling calculation[17].

unstudied in the past due to the difficulty of populating or observing them by conventional spectroscopic methods. A number of
techniques have now been developed for working with high angular
momentum Rydberg states. Gallagher, Hill and Edelstein[20] have
used the scheme shown in Fig. 4 to measure the fine structures of
ℓ = 2 - 5 states of sodium using radio frequency spectroscopy. A
Rydberg d-state is populated by stepwise excitation with pulsed
lasers and the d-f dipole transition is induced by an applied rf
field. Resonance is detected by observing fluorescense on the
3d-3p transition, 8197 Å. The d-g and d-h level structures are
observed by multiphoton resonance. (The large dipole matrix elements between Rydberg states permits multiphoton processes with
only modest power.) The f, g and h fine structure splittings
were all found to be hydrogenic within experimental error. The
microwave spectra also yielded values for the quantum defects.
Freeman and Kleppner[21] have pointed out that the quantum defects
for high angular momentum states of alkalis can be related to the
polarization interaction between the valence electron and the ionic core. The perturbation is of the form

$$V_{pol} = - \frac{1}{2} \alpha'_d <\frac{1}{r^4}> - \frac{1}{2} \alpha'_Q <\frac{1}{r^6}>$$

where α'_d and α'_Q are respectively the dipole and quadrupolar polarizabilities of the alkali ion. They have shown that this interaction leads to a quantum defect

$$\delta_\ell \simeq \frac{3}{4} \alpha'_d \, \ell^{-5} + \frac{35}{16} \alpha'_Q \, \ell^{-9}$$

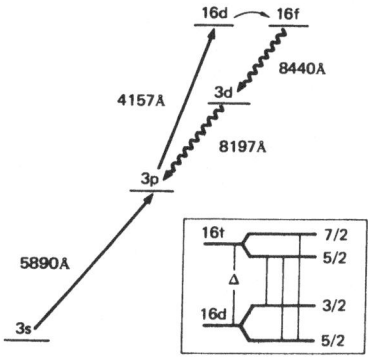

Fig. 4. Scheme used by Gallagher et al[20] to observe 16d → 16f
resonance in Na.

Using the data of Gallagher, Hill and Edelstein[20], they obtained
$\alpha_d' = 1.0015(15)$ a.u., and $\alpha_0' = 0.48(15)$ a.u. These values differ
from the static polarizabilities because of non-adiabatic effects
which are at present not well understood. The opportunity for
precision measurement of the core polarizabilities offers stimu-
lous for further theoretical work in this area.

A second technique which has recently been developed is Stark
spectroscopy which utilizes the large Stark effects possible in
Rydberg states. (The extreme shift, $\delta W_s \simeq \pm(3/2)n^2 E$, can be much
larger than the separation between Rydberg terms $\Delta W_o \simeq 1/n^3$.)
Atoms are excited from some convenient initial state to an energy
lying near a Rydberg state using a fixed, precisely known, frequency. A gradually increasing electric field is applied. An
excitation signal appears whenever a Stark subcomponent is shifted
into resonance. Ducas and Zimmerman[22] have used this method to
observe levels near the $n = 16$ state of sodium using lines from a
CO_2 laser. The energy levels are shown in Fig. 5. Atoms were
prepared in the 10s state and excited with the P(20) CO_2 line to
an energy slightly above the $n = 16$ manifold (and the nearby 17p
state). A Stark excitation spectrum for $|m| = 1$ is shown in

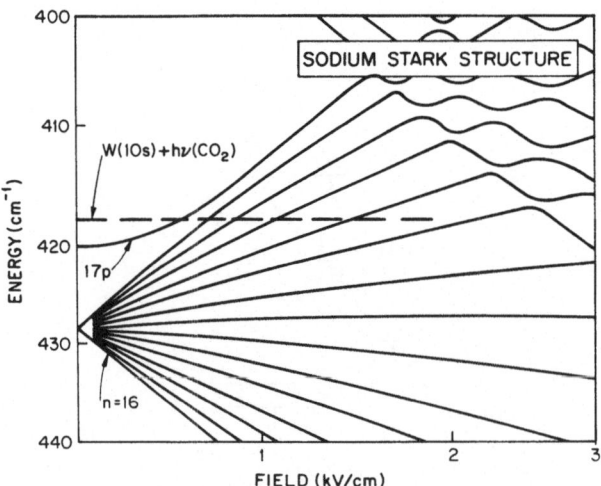

Fig. 5. Level diagram for Stark spectroscopy in Na. Solid lines:
calculated levels. Dashed line: energy of excitation from 10s
level by P(20) CO_2 line.

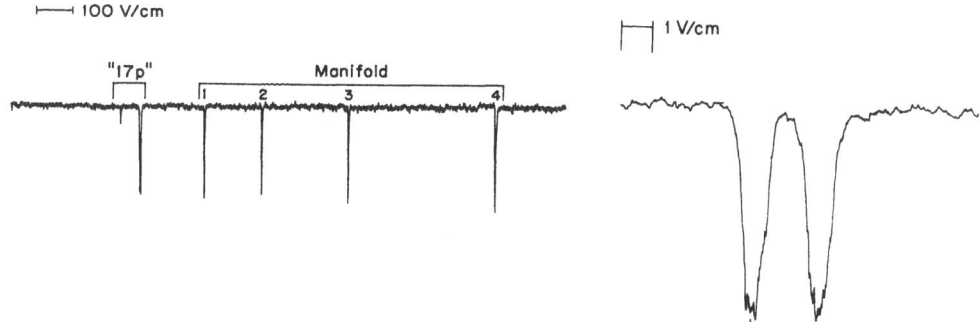

Fig. 6. Stark spectra for levels shown in Fig. 5. (a) (left)
Panoramic sweep. (b) (right) High resolution display of "17p"
fine structure doublet. Splitting is equivalent to 690 MHz.
(From Ducas and Zimmerman[22].)

Fig. 6a. (Some m = 0 features are visible due to imperfect polarization of the CO_2 radiation. The separation between the m = 0
and $|m|$ = 1 states is too small to show clearly in Fig. 5.) Each
$|m|$ = 1 state is actually a fine structure doublet; the splitting
of the "17p" state (i.e. the level which adiabatically connects
to 17p), barely discernible in Fig. 6a, is clearly displayed in
the high resolution sweep of the "17p" line in Fig. 6b. The
linewidth is due to electric field inhomogeneity, estimated at
about 1%. The Stark spectrum can be unfolded to yield values for
the fine structure splittings and quantum defects, as well as to
yield a more precise value for the term value of the initial 10s
state.

     Field ionization has proven to be a versatile and efficient
tool for the detection of Rydberg state atoms[23,24]. Early work
using fast beam has already been mentioned: what has given the
technique new importance is the ability to discriminate individual
Rydberg states, providing at once both a detector and a spectrometer. Field ionization can occur by two somewhat related processes. In the presence of an electric field, E, the potential,
V = - 1/r - Ez, has a maximum value V_{max} = - $\sqrt{2E}$. Any state with
energy W > V_{max} ionizes essentially instantaneously; this process
is sometimes called classical ionization. If W is less than, but
close to, V_{max} , the electron can escape by tunneling through the
potential barrier. The tunneling rate increases with field so
rapidly, however, that for many purposes it is sufficient to take
the rate as zero for W < V_{max} , and infinite for W > V_{max} . This
leads to the idea of a "threshold field" for ionization given by
E_o = $W^2/4$. If we take W to be the unperturbed energy of a Ryd-

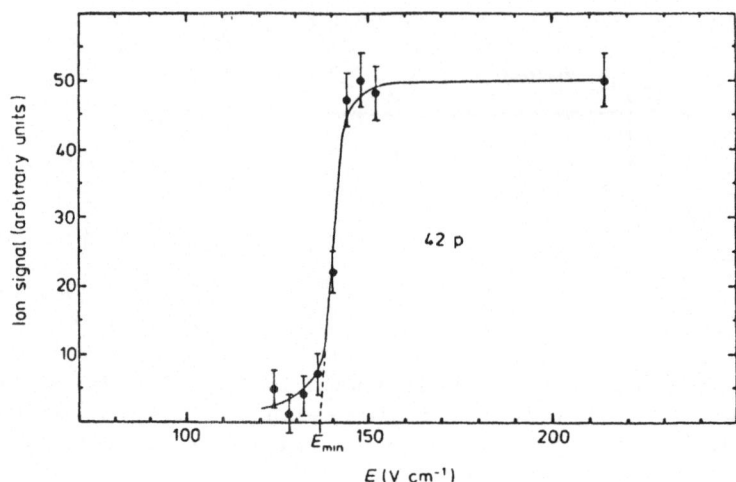

Fig. 7. Threshold ionization curve for cesium, 42p state. (From van Raan, _et al_[25].)

berg state with principle quantum number n, we have

$$E_o = \frac{1}{16n^4} \text{ a.u.} = 3.2 \times 10^8/n^4 \text{ V/cm}$$

The value $1/16n^4$ must be regarded as a characteristic unit rather than as an accurate threshold value.

In practice, a threshold for ionization is clearly exhibited for many states. Fig. 7 shows a threshold ionization curve for Rydberg states of cesium. The $1/n^4$ scaling law has so far been found to hold well[23,24], though the coefficient varies slightly from system to system.

In order to understand field ionization it is essential to understand precisely how the ionization rate varies with electric field. Fig. 8, taken from the calculations of Bailey, Hiskes and Riviere[7] (BHR), shows representative curves for hydrogen, n = 13 - 15. The edges of the shaded area are the ionization curves for the extreme Stark components of the n = 14 manifold. Curves for the intermediate Stark levels lie between in order of increasing Stark energy.

Littman _et al_[26] have developed techniques for populating and identifying individual Stark sublevels of Rydberg states of sodium, and have measured the ionization rates of selected levels.

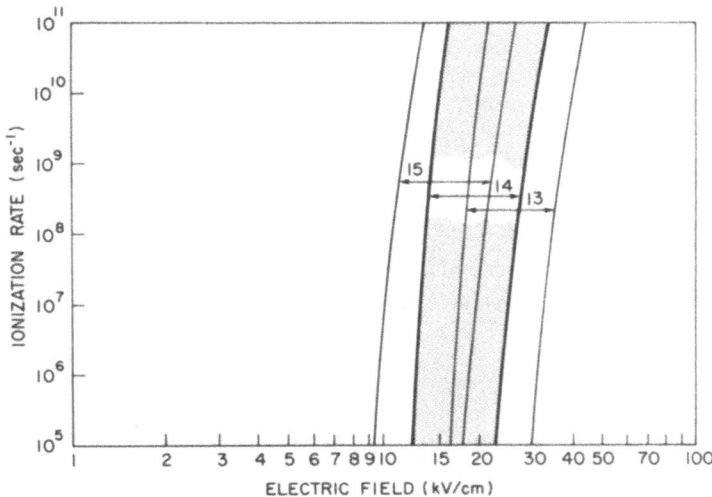

Fig. 8.   Ionization rate curves for extreme Stark components of
hydrogen, n = 13 - 15.   (From Bailey, Hiskes and Riviere[7].)

Fig. 9 shows excitation curves of Stark levels in increasing electric fields: the levels are generally well resolved and can be
identified with calculated energy levels, similar to those shown
in Fig. 5.   Using single ion timing methods, Littman, Zimmerman
and Kleppner[27] have measured the ionization rate of a number of
Stark levels in the vicinity of n = 14.   The lowest sublevels of
each term were found to be in excellent agreement with BHR.   Fig.
10a shows the ionization rate of the level (14,0,11,2) (the indi-
ces are the parabolic quantum numbers $(n, n_1, n_2, |m|)$.   The data
agrees with the calculations of BHR to within the uncertainty in
calibration of the electric field, 2%.

The results for other Stark sublevels, however, can be in
serious disagreement with BHR.   Fig. 10b shows the ionization rate
of the level (12,6,3,2).   The ionization curve is not even mona-
tonic and reaches a rate of $10^7 \mathrm{sec}^{-1}$ at less than half the field
predicted by BHR.   As Littman et al[27] discuss, the anomalous behavior arises from effects of level mixing.   Higher lying levels,
which ionize at lower fields than the level of interest, are
mixed with the latter by the dipole interaction.   The effect of
the state mixing is to decrease the radiation rate of the upper
level, and to increase the rate of the lower level.   The solid
curve of Fig. 10b is calculated from this model with no adjustable parameters.   (Two crossings were included in the calculation
with level (14,0,11,2) at 15.7 kV/cm and with level (15,0,12,2) at
17.3 kV/cm.)   These results show that the theory developed for

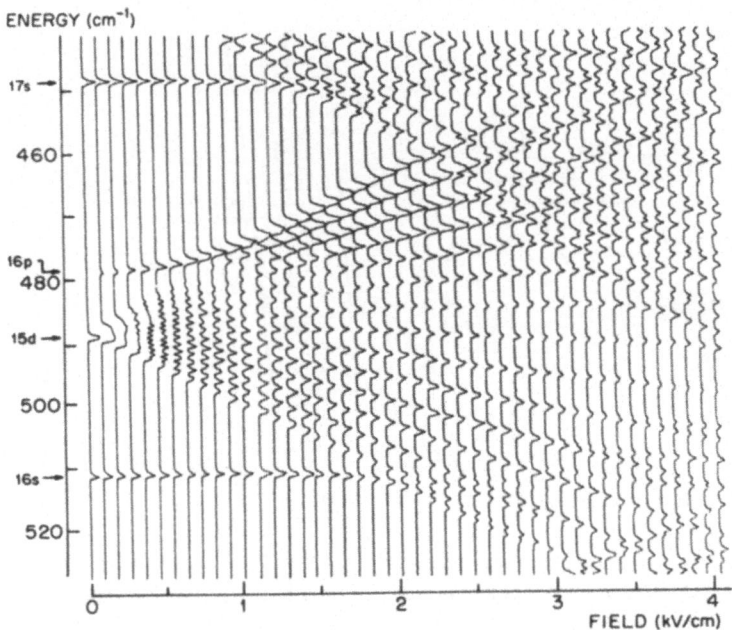

Fig. 9. Stark structure of Na. Experimental excitation curves for
Rydberg states of Na in the vicinity of n = 15. A tunable laser
was scanned across the energy range displayed (vertical axis).
The zero of energy is the ionization limit. Signals generated by
ionizing the excited atoms appear as horizontal peaks. Scans were
made at increasing field strengths and are displayed at the cor-
responding field values. Both |m| = 0 and 1 states are present.

field ionization of hydrogen cannot be applied simply to other
atoms. Also, there is some question about the effects of level
mixing in hydrogen; in the non-relativistic theory the level crossings appear to be sharp, but relativistic effects, or an external
perturbation such as a magnetic field, can produce level mixing
and seriously alter the ionization rate.

Although this brief review of some of the current lines of
research with highly excited atoms and molecules is by no means
complete, it may serve to suggest the variety of new phenomena
which have been observed, and the high level of activity in the
field.

This work was supported by the Air Force Office of Scientific
Research, and the National Science Foundation.

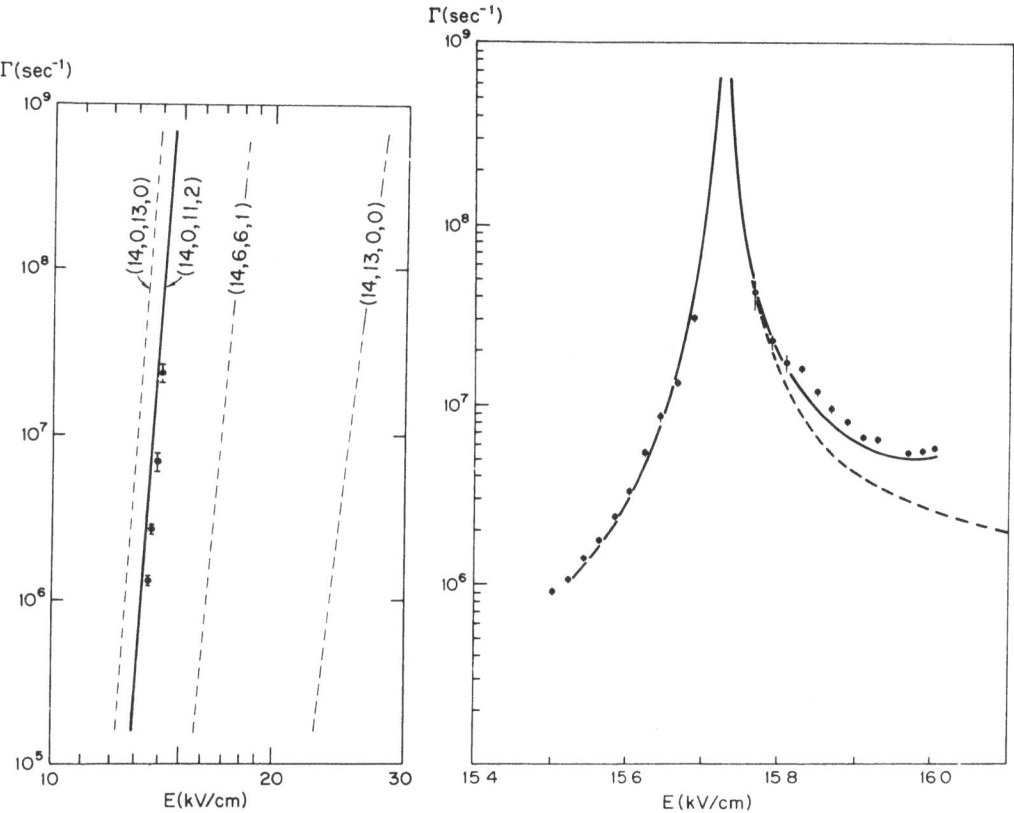

Fig. 10. Ionization rates for sublevels of the n = 14, |m| = 2 manifold of Na vs electric field. (a) (left) Rates for the lowest level (14,0,11,2). Dashed lines are calculated by BHR. Solid line is extrapolated rate for the observed state. (b) (right) Rates for the state (12,6,3,2), showing effect of level crossing at 15.7 kV/cm. Solid line is calculated rate. (From Littman et al[27].)

REFERENCES

1. E. Amaldi and E. Segré, Nuevo Cimento 11, 145 (1934).

2. E. Fermi, Nuevo Cimento 11, 157 (1934).

3. F. A. Jenkins and E. Segré, Phys. Rev. 55, 52 (1939).

4. N. S. Kardashev, Astron. Zh. 36, 839 (1959) (Soviet Astron. A J3, 813 (1959)).

5. B. Hoglund and P. G. Mezger, Science 150, 339 (1965).

6. A. C. Riviere and D. R. Sweetman, in Atomic Collision Processes, M. R. C. McDowell, ed. (North Holland, Amsterdam, 1964).

7. D. S. Bailey, J. R. Hiskes and A. C. Riviere, Nucl. Fusion 5, 41 (1965).

8. R. N. Il'in in Atomic Physics 3; S. J. Smith and G. K. Walters, ed. (Plenum Press, N.Y., 1973), p. 309.

9. S. Svanberg, P. Tseheris and W. Happer, Phys. Rev. Lett. 30, 817 (1972);
   C. Fabre, M. Gross, S. Haroche, Opt. Commun. 13, 393 (1975).

10. W. H. Lamb and W. E. Lamb, Jr., Phys. Rev. Lett. 28, 265 (1972).
    K. B. MacAdam and W. H. Wing, Phys. Rev. A12, 1464 (1975).
    T. A. Miller, R. S. Freund and B. R. Zegarski, Phys. Rev. A11, 753 (1975).

11. P. Esherick, J. A. Armstrong, R. W. Dreyfus and J. J. Wynne, Phys. Rev. Lett. 36, 1296 (1976).

12. R. W. Solarz, C. A. May, L. R. Carlson, E. F. Worden, S. A. Johnson and J. A. Paisner, Phys. Rev. A, to be published.

13. W. A. Chupka, private communication.

14. W. P. West, G. W. Foltz, F. B. Dunning, C. J. Lattimer and R. F. Stebbings, Phys. Rev. Lett. 36, 854 (1976).

15. M. Matsuzama, J. Phys. B. 8, 2114 (1975).

16. T. F. Gallagher, S. A. Edelstein and R. M. Hill, Phys. Rev. Lett. 35, 644 (1975).

17. R. E. Olsen, to be published.

18. I. C. Percival and D. Richards, _Advances in Atomic and Molecular Physics_, ed. by D. R. Bates and B. Bederson (Academic Press, N.Y., 1975), p. 1.

19. P. M. Koch and J. E. Bayfield, Phys. Rev. Lett. <u>34</u>, 448 (1975).

20. T. F. Gallagher, R. M. Hill and S. A. Edelstein, Phys. Rev. A<u>13</u>, 1448 (1976) and Phys. Rev. A, in print.

21. R. R. Freeman and D. Kleppner, Phys. Rev. A, to be published.

22. T. W. Ducas and M. L. Zimmerman, private communication.

23. T. W. Ducas, M. G. Littman, R. R. Freeman and D. Kleppner, Phys. Rev. Lett. <u>35</u>, 366 (1975).

24. R. F. Stebbings, C. J. Latimer, W. P. West, F. B. Dunning and T. B. Cook, Phys. Rev. A<u>12</u>, 1453 (1975).

25. A. F. J. van Raan, G. Baum and W. Raith, J. Phys. B <u>9</u>, L173 (1976).

26. M. G. Littman, M. L. Zimmerman, T. W. Ducas, R. R. Freeman and D. Kleppner, Phys. Rev. Lett. <u>36</u>, 788 (1976).

27. M. G. Littman, M. L. Zimmerman and D. Kleppner, Phys. Rev. Lett. <u>37</u> (August 23) (1976).

ANISOTROPY AND TIME DEPENDENCE IN ATOMIC COLLISIONS

Joseph Macek

Behlen Laboratory of Physics
The University of Nebraska
Lincoln, NE 68588

Collision excitation generally leaves atoms in anisotropic states. This anisotropy has been studied since the 1920's as a probe of the dynamics of atomic collisions.[1] Renewed interest in such anisotropy has followed the application of coincidence techniques to collision measurements,[2,3] the study of collisions between optically pumped atoms[4] and the discovery of anisotropy of foil excited atomic states.[5] This latter development was made possible by the high time resolution of the time of flight technique used in conjunction with foil excitation. In this talk, I will briefly review the connection between anisotropy and time dependence and then discuss what we have learned from recent measurements and theories of the anisotropy of collision excited states.

We say that an atomic state is aligned if the mean value of an irreducible tensor of even rank constructed from components of appropriate angular momentum operators, for example $3J_z^2 - J^2$, is nonzero. Similarly we say that an atomic state is oriented if the mean value of a tensor of odd rank, for example the vector J itself, is non-zero. The general tensor we donate simply as $T[k]q$. During an atomic collision, the nuclear spin plays no essential role; its influence is manifest later in that the hyperfine interaction perturbs the initial anisotropy of the electronic state. This perturbation is expressed by the Alder[6] formula:

$$\left\langle T^{[k]}{}_q(t) \right\rangle = \left\langle T^{[k]}{}_q(0) \right\rangle \; G^{(k)}(t) \quad , \tag{1}$$

where

$$G^{(k)}(t) = \Sigma_{F'F} \frac{(2F' + 1)(2F + 1)}{(2I + 1)} \begin{Bmatrix} F' & F & k \\ J & J & I \end{Bmatrix}^2 \cos \omega_{F'F} t \quad , \quad (2)$$

and where the bracket denotes an average over the atomic state. The average evaluated at t=o is just the initial alignment or orientation of the electronic state. The hyperfine interaction produces only a small splitting of the electronic levels and causes the anisotropy to oscillate at the hyperfine frequencies $\omega_{F'F}$. Now the intensity of the light viewed by a detector focused on the excited atoms is given by[7]

$$I = C \left\{ 1 - \frac{1}{2} h^{(2)} / j_i (j_i + 1) \left[\langle 3J_z^2 - J^2 \rangle - 3 \langle J_x^2 - J_y^2 \rangle \cos 2\beta \right] \right.$$

$$\left. + \frac{3}{2} h^{(1)} / j_i (j_i + 1) \langle J_y \rangle \sin 2\beta \right\} \quad , \quad (3)$$

in a coordinate system with z-axis along the axis of the light detector and x-axis along the major axis of the elliptical polarization selected by the detector. The parameter β specifies the degree of elliptical polarization with β = o representing linear polarization and β = π/4, circular polarization. The constants $C, h^{(2)}$ and $h^{(1)}$ depend upon the specifics of the decay and are of no importance for the discussion here. Observation of light downstream from the foil manifests the oscillation of the anisotropy as given by Eqs. (2) and (3).

Anisotropy of foil excited atomic states was first discovered by 1970 by Andra[5] through observation of a time modulation in the decay of 3^3P states in ^4He. A graph of the observed decay curve[8] Fig. (1) shows the oscillations indicative of initial anisotropy very clearly. For this system the spin orbit interaction perturbs the initial spatial alignment and is described by Eq. (2) with J replaced by L and I by S. Extensive measurements[9] on this particular system and He[3] demonstrated the applicability of the Alder formula and the assumption of spin independence of the collision excitation, just as they are applied in gas collisions.[10,11] Careful measurments of the light polarization[9,12] showed that, in a coordinate system defined by the ion beam direction, the alignment tensor had only one non-zero component $3J_z^2 - J^2$ as is approxpriate for the system with cylindrical symmetry. Additionally, D. J. Burns[12] measured the initial phase of the oscillations and found it to be 0° in agreement with Eq. (2). Owing to the fine structure interaction, anisotropy is reversibly exchanged

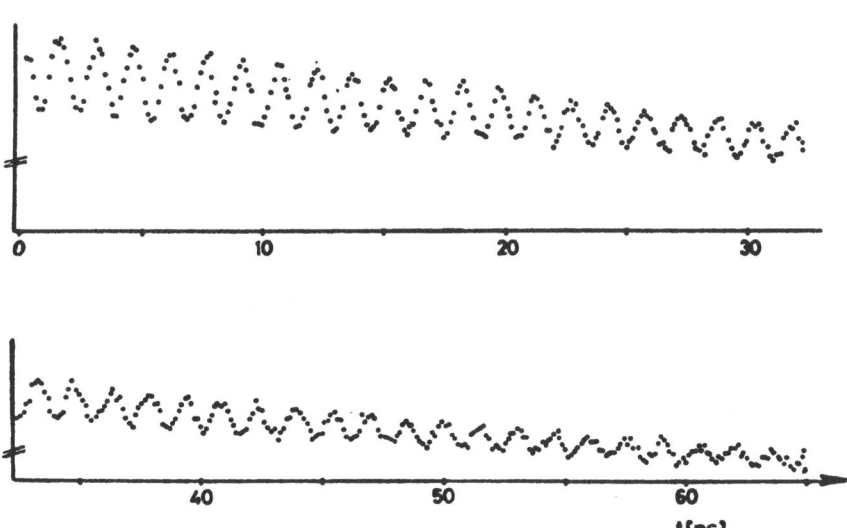

Fig. 1. ^{4}He 3 P_1 - P_2 oscillations. Ref. (8).

between spin and coordinate degrees of freedom after the collision. Since light emission leaves electronic spin unchanged, the radiation pattern reflects the spatial anisotropy only which, is largest at t=o if the electronic spin is isotropic. This 0° phase confirms rather directly that the electronic spin is not initially aligned.

The observed anisotropy, while very striking in its manifistations, demonstrated mainly the similarity of gas and foil collisions, in particular the cylindrical symmetry of the later. The collision geometry loses its symmetry if another axis is introduced for example, by detecting light in coincidence with a particle scattered through a non-zero angle,[13] or if the foil is tilted with respect to the beam so that the normal and the beam no longer coincide. Under these conditions additional tensor components appear and the orientation vector may be non-zero.[14] One of the really suprising recent results is Berry et. al.'s discovery that simply tilting the foil with respect to the beam is sufficient to excite oriented states. The observed orientation is quite sizeable, of the order of a few precent, as subsequent measurements have shown.[15] Clearly the foil surface, which is really defined only on the macroscopic scale, plays an essential role on the microscopic scale as well. One way this can come about was suggested by T. Eck,[16] namely than an electric field exists which is, on the average, normal to the foil. The field converts the initial alignment resulting from the foil collision into orientation[17] according to the relation, valid for P states

$$\langle J_y \rangle = \frac{1}{2} \langle 3J_z^2 - J^2 \rangle \sin \chi \sin 2\alpha \quad . \tag{4}$$

In Eq. (4), χ is an unknown phase related to the Stark splitting of the M = 0 and 1 eigenstates in an electric field, and α is the angle between the normal to the foil and the beam direction. According to Eq. (4) the orientation should maximize at $\alpha = 45°$ whereas later experiments[15] show that it increases at all angles between 0° and 90°. This is not in contradition with the idea of an electric field since Lombardi[18] points out that χ should be a function of α owing to the increasing time that the atoms spend in the field as α increases, and that the linear Stark effect could play a role. Both of these effects give rise to an orientation which increases for all α. Regardless of the origins of the orientation, it is apparent that its measurement will serve as a useful probe of ion-surface interactions.[19]

Eq. (4) illustrates an essential point for the orientation of excited atomic states, namely that a phase difference between the amplitudes for excitation of different magnetic substates is required. Measurements of this phase difference can only be accomplished by examining the anisotropy of atomic states. On the other hand, theories of the anisotropy must incorporate a physical mechanism for the phase difference. The Born approximation, for example, predicts no phase difference and hence no orientation. To obtain a phase difference, the wave functions of the relevant components of collision system must themselves be phase distorted. The coincidence measurements of electron excitation of He(2^1P) by H. Kleinpoppen[20] and co-workers illustrates this feature. They scattered 80eV electrons inelastically from helium and detected the electrons which had excited the 2^1P state in coincidence with the decay radiation. In this particular case, since one is dealing with P states where the magnetic substates were coherently populated, it is possible to deduce the absolute value of

$$\langle J_y \rangle = |0_1^-| \text{ from the angular distribution of the emitted light.}$$

Their measured values are shown in Fig. (2) along with the theoretical predictions of Madison and Shelton[21] based on a distorted wave Born approximation. The agreement is fairly good, particularly when one considers that the plane wave Born approximation predicts zero orientation at all angles. The origin of the phase difference of the $M_1 = 0$ and 1 amplitudes is thereby traced to a slight phase distoration of the incoming electron wave. It would appear that all distored wave approximations should predict orientation. This is apparently correct, with one notable ex-

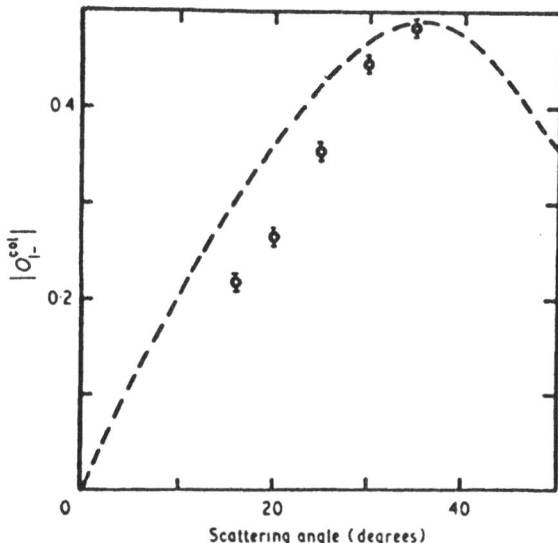

Fig. 2. Variation of orientation with electron scattering angle for He(2^1P) excited by 80 eV electrons, Ref. (20),--- Madison and Shelton, theory, Ref. (21).

ception, namely the Glauber approximation. This approximation presumes small scattering angles, where the orientation is indeed negligible since it must vanish completely at 0° for symmetry reasons. Fig. (2) provides some indication of the angular range for which the small angle approximation actually holds.

The phase distortion can be traced to specific features of the excitation process for heavy ion collisions in the velocity range where the molecular promotion model applies. Owing to the large mass of the target and projectile relative to the atomic electrons, the motion of nuclei can be treated classically. All of the dynamics, including that responsible for orientation, relates to adjustment of the atomic electrons to the changing position of the constituent nuclei. Since the orientation $\langle J_y \rangle$ measures the mean amount of angular momentum transferred from angular momentum of relative motion to electronic angular momentum we expect the so called rotational coupling to play a central role. Indeed, the rotational coupling operator $\langle J_y \rangle$ is identical to the operator that defines the orientation. The measurement of the relative phase of amplitudes for exciting the 3^3P state of He in He$^+$ + He collisions by Jaecks et. al.[22] provides an example of the interconnection of rotational coupling and phase distortion to produce orientation. Fig. (3) shows the phase difference measured for scattering angles between 1° and 2°. Despite

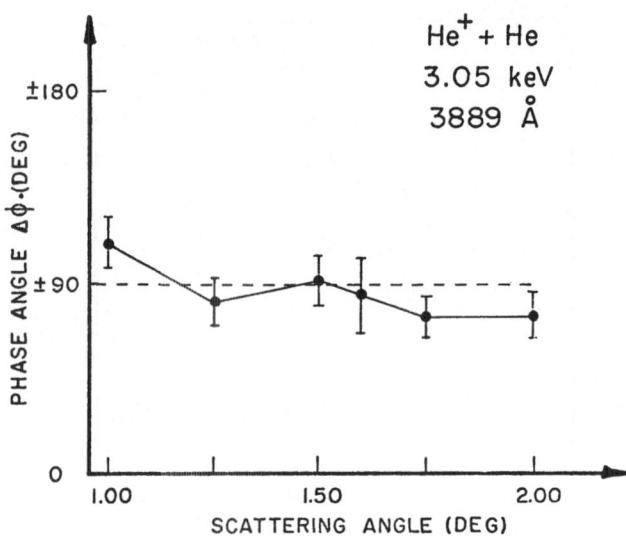

Fig. 3.   Relative Phase of the M_1 = 0 and 1 3^3P amplitudes of He
         excited by He$^+$ + He charge exchange collisions. Ref. (22).

the fact that the amplitudes varied significantly over this range
the phase difference remained approximately constant and equal
to 90°.

     Such a constant value implying a factor of i in the scatter-
ing amplitudes suggests the following model. Excitation of the
3^3P is thought to proceed through a Landau-Zener transition at the
crossing of the $1s\sigma_g(2p\sigma_u)^2$ $^2\Sigma$ and the $(1s\sigma_g)$ $4d\sigma_g$ $^2\Sigma$ molecular
levels at an internuclear distance of one atomic unit. The $4d\sigma_g$
level, which correlates with the 3P atomic levels, is populated
on the inward and outward passage of the crossing. The states
so populated differ in phase owing to the energy difference of
the two Σ levels. This phase difference normally gives rise to
Stuckelberg oscillations in the differential scatteering cross
section, but because of the near degeneracy of the $4d\sigma$, π and δ
states another effect occurs. The $4d\sigma$ level, once populated, does
not follow the rotation of the internuclear axis, rather the
electron wave function maintains the position it had at the time
of the crossing. The final state is a superposition of two 4d
wave functions of different phase oriented at angles of β and $\pi-\beta$
relative to the incoming beam, the angle β being measured from
the internuclear axis at the crossing and the incoming beam
direction. Upon superimposing the two functions and examining
the amplitude a_0 and a_1 for M_1 = 0 and 1 at large separations
one has

$$a_0 \ \alpha \ 2^{-\frac{1}{2}}(3 \ \cos^2\beta - 1) \ \cos A$$

$$a_1 \ \alpha \ i \ 3^{\frac{1}{2}} \ \sin \beta \ \cos \beta \ \sin A \hspace{2cm} (5)$$

that is, the relative phase of the two amplitudes is constant and equal to 90° independent of energy and scattering angle. Not only is this result in agreement with the data of Fig. (3), but later measurements at 1.5 keV also exhibited the constant 90° phase.[23] This simple calculation lends support to the electron promotion model of Lichten[24] and Fano and Lichten[25] in that as the nuclei spearate the system follows the diabatic $4d\sigma$ and $4d\pi$ orbitals and does not undergo appreciable changes through the crossing of a lot of other molecular levels.

Measurements of impact excited fluorescence probes only the mean values of the k = 0, 1, and 2 tensors. While these provide much information about collision dynamics, the higher order moments have not been studied. One interesting consequence of Hertel and Stoll's[4] development of collisions involving optically pumped atoms is the possibility of measuring all moments. They prepare the initial state with a known anisotropy and examine the scattering of electrons from this state. The measured electron intensity is proportional to the overlap of this initial state with the state prepared by electron scattering in the time inverse experiment. This overlap projects out of the electronically excited state that multipole moment prepared by the optical pumping,[26] thus all multipole moments are, in principle at least, measured by this technique.

Such measurements provide an opportunity and a challenge for theory. I have tried to show how anisotropy measurements probe detailed aspects of atomic collisions. Isolating the relevant physical mechanisms from the myrid of possiblities represents a most welcomed opportunity for the future.

This work is supported by NSF grant MPS 75-07805.

REFERENCES

1.   H. W. B. Skinner and E. T. S. Appleyard, Proc. R. Soc. Lond. A117, 224 (1927).

2.   D. H. Jaecks, D. H. Crandell and R. H. McKnight, Phys. Rev Letter. 25, 491 (1970).

3.   M. Eminyan, K. McAdam, J. Slevin and H. Kleinpoppen, Phys. Rev. Lett. <u>31</u> 576 (1972).

4.   I. V. Hertel and W. Stoll, J. Phys. B. Atom. Molec. Phys. <u>7</u>, 570 (1974) and 583 (1974).

5.   H. J. Andra, Phys. Rev. Lett. <u>25</u>, 325 (1970).

6.   K. Alder, Helv. Phys. Acta <u>25</u>, 235 (1952).

7.   U. Fano and J. Macek, Rev. Mod. Phys. <u>45</u>, 533 (1973).

8.   W. Wittman, K. Tillman and H. J. Andra, Nuclear Instruments and Methods <u>110</u>, 305 (1973).

9.   W. Wittmann, K. Tillman, H. J. Andra and P. Dobberstein, Z. Physik <u>257</u>, 299 (1972).

10.   J. Macek, Phys. Rev. Lett. <u>23</u>, 1 (1969).

11.   I. C. Percival and M. J. Seaton, Phil. Trans. R. Soc., London, <u>A251</u>, 113 (1958).

12.   D. J. Burns and W. H. Hancock, J. Opt. Soc. Am. <u>63</u>, 241, (1973).

13.   J. Macek and D. H. Jaecks, Phys. Rev. <u>A4</u>, 2288 (1971).

14.   H. G. Berry, L. J. Curtis, D. G. Ellis and R. M. Schectman, Phys. Rev. Lett. <u>32</u>, 751 (1974).

15.   H. G. Berry, L. J. Curtis, G. G. Ellis and R. M. Schectman, Phys. Rev. Lett. 35, 274 (1975); D. A. Church, W. Kolbe, M. C. Michel, and T. Hadeishi, Phys. Rev. Lett. <u>33</u>, 565 (1974); C. H. Liu, S. Bashkin, and D. A. Church, Phys. Rev. Lett. <u>33</u>, 993 (1974).

16.   T. G. Eck. Phys. Rev. Lett. 33, 1055 (1974).

17.   M. Lombardi, and M. Giroud, C. R. Acad. Sci. <u>B266</u>, 60 (1968).

18.   M. Lombardi, Phys. Rev. Lett. <u>35</u>, 1172 (1975).

19.   H. J. Andra, Physics Letters, <u>54A</u>, 315 (1975).

20.   M. Eminyan, K. B. MacAdam, J. Slevin and H. Kleinpoppen, J. Phys. B, <u>7</u>, 1519 (1974).

21.   D. H. Madison and W. n. Shelton, Phys. Rev. <u>A7</u>, 449 (1973).

22. D. H. Jaecks, F. J. Eriksen, W. de Rijk and J. Macek, Phys. Rev. Letter 35 (723).

23. D. H. Jaecks, F. J. Eriksen, W. de Rijk and J. Macek, Phys. Rev. to be published.

24. W. Lichten, Phys. Rev. 131, A1025 (1965).

25. U. Fano and W. Lichten, Phys. Rev. Letters, 14, 627, (1965).

26. J. Macek and I. V. Hertel, J. Phys. B7, 2173 (1974).

R-MATRIX THEORY OF ATOMIC AND MOLECULAR PROCESSES

PHILIP G. BURKE

DEPARTMENT OF APPLIED MATHEMATICS AND
THEORETICAL PHYSICS
QUEEN'S UNIVERSITY, BELFAST,
N. IRELAND

1. INTRODUCTION

In the last few years there has been considerable progress in developing new methods for electron atom and electron molecule scattering.   One approach which seems particularly promising is the R-matrix method which was first introduced by Wigner and Eisenbud (1) in fundamental papers concerned with the theory of nuclear reactions. In this review we will describe recent developments which have been made in the theory and we will consider applications not only to electron scattering but also to a broad range of other atomic and molecular processes including polarizabilities, long range atomic force constants, binding energies and photoionization cross sections.

The basic idea is that configuration space for the (N + 1)-electron-target system is divided into two regions as illustrated in figure 1.

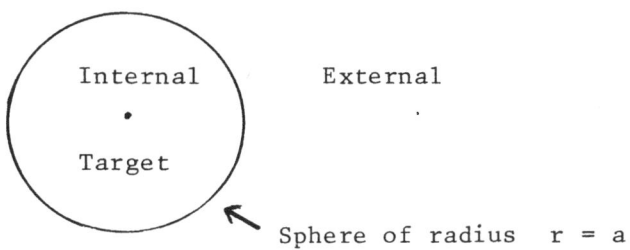

FIGURE 1.   The internal and external regions.

In the internal region the interaction is strong and difficult to calculate ab-initio.   Usually electron exchange is important in this region.   In the external region exchange can be neglected and the collision problem has an analytic solution or at least a solution which can be easily obtained by numerical methods.   The R-matrix is defined in terms of the logarithmic derivative of the radial wave function of the scattered electron on the surface of the sphere by the equation

$$u_i(a) = \sum_j R_{ij}(E) \left(a \frac{du_j}{dr} - b_j u_j \right)_{r=a} \qquad (1)$$

where the b_j are constants which can be chosen arbitrarily in each channel.   The R-matrix is easily seen to be a meromorphic function of the energy with the representation

$$R_{ij}(E) = \sum_{\lambda=1}^{\infty} \frac{\gamma_{\lambda i}\, \gamma_{\lambda j}}{E_{\lambda b} - E} \qquad (2)$$

The poles at $E = E_{\lambda b}$ all lie on the real energy axis and, most importantly the R-matrix contains no branch cuts associated with the thresholds for new channels.   The basic problem of R-matrix theory is then to calculate the reduced-width amplitudes $\gamma_{\lambda i}$ and the pole positions $E_{\lambda b}$ which together give a complete description of the collision problem in the internal region.   The reactance matrix $\underline{K}$ and the collision cross section are then simply obtained from the matrix equation

$$\underline{u}(r) = \underline{k}^{-1/2} \left(\underline{F}(\underline{k}r) + \underline{G}(\underline{k}r)\underline{K} \right) \qquad r \geqslant a \qquad (3)$$

where $\underline{F}$ and $\underline{G}$ are independent solutions of the collision problem in the external region.

The basic idea of dividing space into an internal and an external region is certainly not new in atomic and molecular physics. For example, in the case of electron ion scattering, where the long range diagonal Coulomb potential usually dominates all other potentials in the external region, $\underline{F}$ and $\underline{G}$ are just the regular and irregular Coulomb wave functions.   In this case the meromorphic property of the R-matrix coupled with the known analytic properties of the Coulomb wave functions leads immediately to the Quantum Defect Theory which has been successfully used in the analysis of electron collisions with ions and of photoionization by Seaton (2) and Fano (3).

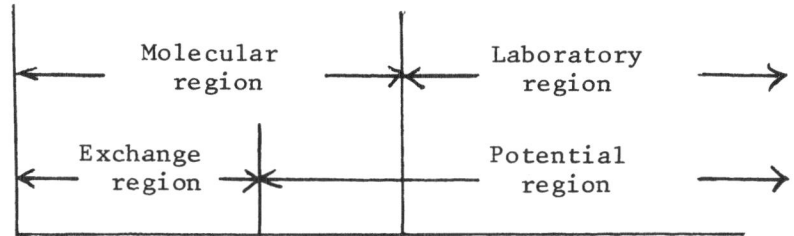

Distance of Electron from the Molecule

FIGURE 2.   Transformation between Molecular
            and Laboratory Frames

    This concept has also been of considerable value in the
development of the frame transformation theory of electron molecule
collisions by Fano and Chang (4).   We show in figure 2 the
relevant regions in this case.   The scattering of electrons by
molecules can be formulated either in the laboratory frame or in
the body-fixed or molecular frame of reference.   When the electron
is close to the nuclei it is moving fast compared with the nuclear
rotational motion and the molecular frame is most appropriate.   In
this case the wave function describing the electron molecule
interaction is similar to that encountered in molecular structure
calculations and similar computational methods can be used.   At
larger distances from the molecule, particularly for low energy
collisions, it is more convenient to use the laboratory frame making
full allowance for the rotational motion of the molecule.   It
follows that the R-matrix describing the problem in the internal
region is most appropriately calculated in the molecular frame.
If necessary it is then transformed on the boundary to the
laboratory frame using the appropriate unitary transformation and
the collision problem solved in the external region in the
laboratory frame.

    Finally we mention that the R-matrix concept of dividing
configuration space into two regions has been of value in discussing
weak interactions such as the spin-orbit interaction in light atoms
and molecules.   These interactions can often be neglected compared
with the strong Coulomb interactions in the internal region but have
to be included in the external region particularly when the collision
energy is close to threshold.

2.   GENERAL THEORY

    We will describe in this section methods which have been
developed for the ab-initio calculation of the R-matrix by solving
the (N + 1)-electron problem in the internal region.   It is

important to realize at the outset that because this region is finite the wave function can be represented by an expansion in terms of a discrete L^2 integrable basis. Thus well-known techniques developed for atomic and molecular structure problems can be used with little modification. In this respect the R-matrix method is motivated by similar considerations to other recent L^2 approaches to collision problems such as the J-matrix method of Heller et al (5) and the L^2 methods of Morrison and Lane and of Rescigno et al (6).

The operator (H - E) is not Hermitian in the space of arbitrary L^2 functions defined over the internal region. Consequently it was pointed out by Bloch (7) and re-emphasized by Lane and Robson (8) that it is convenient to write the Schrödinger equation as

$$\left(H - E + L_b\right)\Psi_E = L_b \Psi_E \tag{4}$$

where the operator L_b is expressed in terms of the channel functions $\underline{\Phi}_i$ by

$$L_b = \mathcal{A} \sum_i |\Phi_i\rangle \tfrac{1}{2} \delta(r-a)\left(\tfrac{d}{dr} - \tfrac{b_i - 1}{a}\right)\langle\Phi_i| \tag{5}$$

The operator L_b for orbitary b_i cancels the surface terms arising from H so that (H - E + L_b) is Hermitian. We can now formally solve eq. (4) giving

$$\Psi_E = \left(H - E + L_b\right)^{-1} L_b \Psi_E \tag{6}$$

and introducing the eigenfunctions $\underline{\Psi}_{\lambda b}$ and eigenenergies of (H - E + L_b) we obtain

$$\Psi_E = \sum_{\lambda=1}^{\infty} \Psi_{\lambda b} \frac{\langle\Psi_{\lambda b}|L_b|\Psi_E\rangle}{E_{\lambda b} - E} \tag{7}$$

Evaluating this equation on the surface of the sphere leads immediately to eq. (1) where the reduced-width amplitudes are defined in each channel by the projection

$$\gamma_{\lambda i} = (2a)^{-1/2} \langle\Phi_i | \Psi_{\lambda b}(r=a)\rangle \tag{8}$$

The integral in this equation is over all co-ordinates except the radial co-ordinate of the scattered electron.

In order to calculate the eigenfunctions $\overline{\Psi}_{\lambda b}$ we now assume that they can be expanded in the internal region as follows

$$\Psi_{\lambda b} = \mathcal{A} \sum_{ij} a_{ij\lambda} \, \Phi_i \, u_j^o(r) + \sum_i b_{i\lambda} \, \Phi_i \qquad (9)$$

where $u_i^o(r)$ are radial continuum basis orbitals representing the scattered electron and the ϕ_i are $(N + 1)$-electron functions constructed from the bound orbitals and possibly pseudo orbitals and representing many electron correlations. This expansion is clearly analogous to C.I. type expansions which are used in atomic and molecular structure problems. In the work of Burke et al (9) on electron atom scattering the Hamiltonian is divided into two parts

$$H = H_o + H_{int} \qquad (10)$$

where H_o is defined by the radial equations

$$\left(\frac{d^2}{dr^2} - \frac{\ell(\ell+1)}{r^2} + V(r) + k_i^2 \right) u_i^o(r) = \sum_j \lambda_{ij} \, P_j(r) \qquad (11)$$

subject to the boundary conditions

$$\left. \frac{a}{u_i^o(r)} \frac{d u_i^o}{dr} \right|_{r=a} = b \qquad (12)$$

where b is given a fixed but arbitrary value b_i in each channel ensuring that $L \overline{\Psi}_{\lambda b} = 0$. Also the λ_{ij} are Lagrange undetermined multipliers which ensure that the $u_i^o(r)$ are orthogonal to the bound orbitals $P_j(r)$. Eq.(11) is then solved numerically for a suitable potential $V(r)$ to generate the basis orbitals $u_i^o(r)$. The coefficients $a_{ij\lambda}$ and $b_{i\lambda}$ are then obtained by diagonalizing the Hamiltonian

$$\langle \Psi_{\lambda b} | H | \Psi_{\lambda' b} \rangle = E_{\lambda b} \, \delta_{\lambda\lambda'} \qquad (13)$$

The R-matrix and consequently the cross section at all energies is obtained by this one diagonalization.

The most important source of error in this approach is the slow convergence of the diagonal elements of the R-matrix as the number of terms N in eq.(9) is increased. To overcome this problem

Buttle (10) suggested that we write

$$R_{ij}(E) = \sum_{\lambda=1}^{N} \frac{\gamma_{\lambda i}\,\gamma_{\lambda j}}{E_{\lambda b} - E} + \sum_{\lambda=N+1}^{\infty} \frac{(\gamma_{\lambda i}^{\circ})^2\,\delta_{ij}}{E_{\lambda b}^{\circ} - E} \qquad (14)$$

where the terms in the first expansion are obtained from eq.(13)
while those in the second expansion are obtained using H_o. This
infinite summation can be carried out since eq.(11) can be solved
numerically at any energy. This correction has proved to be very
satisfactory in practice for electron atom collisions.

A variational correction to the R-matrix has been suggested by
Zvijac et al.(11). This corrects for that part of the potential
which is not included in the function space of eq. (9) or by the Buttle
correction. Like the Buttle correction it does not contribute when
the energy E corresponds to one of the poles $E_{\lambda b}$ in the first
expansion in eq.(14). This correction has so far not had wide
application but has proved successful in a model problem.

A basic problem with the above procedure is the need to define
an unperturbed Hamiltonian H_0. This is, as we have seen,
straightforward in the case of atoms but is not so easy in the case
of molecules where the multi-centre nature of the problem means that
the equation corresponding to (11) cannot be simply solved. In
this case it is more convenient to use analytic orbitals such as
Slater type orbitals or Gaussian orbitals for the $u_i^{\circ}(r)$. Thus
the continuum orbitals are represented in the same way as the bound
orbitals except that they do not vanish on the boundary.
Schneider (12) has recently successfully employed floating Gaussians
in a study of electron scattering by diatomic molecules by
diagonalizing the operator $H + L_b$ in a finite analytic basis. The
Block operator L_b ensures that the boundary condition of eq. (12) is
approximately satisfied for the low lying levels even though the
basis functions themselves satisfy arbitrary boundary conditions at
the R-matrix radius. Schneider also shows how an approximate Buttle
correction can be calculated, even in the absence of an unperturbed
Hamiltonian, provided that a good representation of the first N levels
is obtained.

A further development which involved replacing the single matrix
diagonalization by the solution of a set of simultaneous equations at
each energy has been considered by Shimamura et al.(13). This
approach is again designed to be appropriate for electron molecule
collisions using analytic basis orbitals. The simultaneous
equations are obtained by projecting $H - E + L_b$ on to the basis
states eq. (9) giving

$$\sum_{j} \langle \chi_i | H - E + L_b | \chi_j \rangle\, c_j = 0 \qquad (15)$$

where χ_i denote collectively $\mathcal{A} \phi_i u_j^{~o}$ and ϕ_i and c_j denote collectively $a_{ij\lambda}$ and $b_{i\lambda}$. Now it can easily be shown that if E is fixed at the energy of interest we obtain an eigenvalue problem for the b_i with n solutions, where n is the number of channels. For each solution eq.(15) reduces to a set of simultaneous equations which can be solved for the c_i. This approach corresponds to adjusting the boundary conditions b_i so that n independent R-matrix poles fall at the energy E. It has the advantage that the Buttle correction, and also of course any variational correction is zero, however since $L_b \bar{\Psi}_b \neq 0$ the wave function, as in the previous approaches is not continuous on the boundary.

We remark here that the method discussed by Oberoi and Nesbet (14) which also uses $u_j^{~o}$ satisfying arbitrary boundary conditions can be regarded as a special case of the diagonalization method with the b_i chosen to be zero.

An important feature of all the above approaches is that the wave function has a discontinuous derivative on the boundary. Although this does not appear to be important for the K-matrix, it may cause difficulties when the wave function itself is needed, as in the calculation of photodetachment cross sections where the range of integration includes the boundary. One approach which eliminates this discontinuity is the eigenchannel method first introduced in nuclear reaction theory (15) and recently applied by Fano and Lee (16) in a calculation of the Ar photoionization spectrum. In essence the method consists in choosing a set of basis orbitals satisfying given boundary conditions b_i and then varying these boundary conditions to obtain n independent eigenenergies $E_{\lambda b}$, defined by eq.(13), occurring at the energy E of interest. This procedure involves an iteration at each energy but like the previous approach has the advantage that since the energy corresponds to the R-matrix pole, the Buttle correction is zero. A "natural boundary condition" method which is numerically faster but also leads to a continuous wave function has been described by Barrett and Delsanto (17) and Ahmed et al.(18). Like the eigenchannel method this also requires an iteration at each energy and appears to give the best estimate of the cross section for the given basis set. It remains to be seen, however, whether the advantage of a continuous derivative is outweighed by the numerical inefficiency due to the iteration when the cross section is required at many energies over a broad range. In such cases the Buttle corrected R-matrix method, or the method based on the solution of the simultaneous equations (15) may well be more efficient.

3. CALCULATION OF OTHER ATOMIC AND MOLECULAR PROPERTIES

In the previous section we discussed various methods of calculating the R-matrix. As we have already emphasised this leads immediately to the K-matrix and cross section for electron collisions. In this section we will indicate very briefly how other atomic and molecular properties can be obtained.

The calculation of the dipole polarizability of an atomic or molecular system in a field of frequency ω has been discussed by Allison et al.(19).   We have

$$\alpha(\omega) = 2 \sum_s \frac{(E_s - E_o)\langle \Psi_o | M_1 | \Psi_s \rangle \langle \Psi_s | M_1 | \Psi_o \rangle}{(E_s - E_o)^2 - \omega^2} \tag{16}$$

where M_1 is the dipole length operator and the summation goes over all states including the continuum.   It is well known that this equation can be written as

$$\alpha(\omega) = \langle \Psi_o | M_1 | \theta(\omega) \rangle + \langle \Psi_o | M_1 | \theta(-\omega) \rangle \tag{17}$$

where $\theta(\pm\omega)$ are solutions of the inhomogeneous equation

$$\left(H - E_o + L_b \pm \omega \right) \theta(\pm \omega) = M_1 \Psi_o + L_b \theta(\pm \omega) \tag{18}$$

This can be solved in terms of the R-matrix basis states in the same way as eq.(4) giving

$$\theta(\pm \omega) = \sum_\lambda \frac{\Psi_{\lambda b}}{E_{\lambda b} - E_o \pm \omega} \left(\langle \Psi_{\lambda b} | M_1 | \Psi_o \rangle + \langle \Psi_{\lambda b} | L_b | \theta(\pm \omega) \rangle \right) \tag{19}$$

On substituting the first term in this equation back into eq.(17) we obtain

$$\alpha(\omega) = 2 \sum_\lambda \frac{(E_{\lambda b} - E_o)\langle \Psi_o | M_1 | \Psi_{\lambda b} \rangle \langle \Psi_{\lambda b} | M_1 | \Psi_o \rangle}{(E_{\lambda b} - E_o)^2 - \omega^2} \tag{20}$$

where the sum is now over a discrete basis and can be easily calculated once the $\overline{\Psi}_{\lambda b}$ and the $E_{\lambda b}$ are known.   The second surface term in eq.(19) is only important when ω is just below or above the ionization threshold.   It shifts the poles in eq.(20) into their correct Rydberg positions and introduces a branch cut corresponding to photoionization.   In particular the surface term

is negligible on the imaginary axis and the Van der Waals
coefficient between two atoms A and B can be obtained from the
equation

$$C_6 = \frac{2}{\pi} \int_0^\infty \alpha_A(i\omega) \, \alpha_B(i\omega) \, d\omega \tag{21}$$

using the representation of $\alpha(i\omega)$ given by eq.(20).

The imaginary part of $\alpha(\omega)$ on the branch cut is proportional to
the photoionization cross section. Burke and Taylor (20) show that
gives in the dipole length approximation

$$\frac{d\sigma_{phot}}{d\Omega} = 4\pi^2 \alpha \, a_o^2 \, \omega \, |\langle \overline{\Psi_j^-} | M_1 | \Psi_0 \rangle|^2 \tag{22}$$

where as in eq.(7)

$$\overline{\Psi_j^-} = \sum_{\lambda=1}^\infty \Psi_{\lambda b} \frac{\langle \Psi_{\lambda b} | L_b | \overline{\Psi_j^-} \rangle}{E_{\lambda b} - E} \tag{23}$$

The ingoing wave $\overline{\Psi_j}^-$ state on the boundary can be obtained using the
R-matrix calculated previously where the normalization

$$\langle \overline{\Psi_{jE}^-} | \overline{\Psi_{j'E'}^-} \rangle = \delta_{jj'} \, \delta(E - E') \tag{24}$$

has been adopted. In order to obtain reliable results by this
method the initial state $\overline{\Psi}_0$ must also be expanded in terms of the
R-matrix basis using an equation similar to eq.(23). This involves
an iterative procedure to calculate the initial state eigenenergy
consistent with the R-matrix boundary condition.

We also note that Robb et al.(21) has described an analogous
expansion for electron impact ionization in the Born approximation
using an R-matrix expansion (23) for the final state.

The extension of this method to calculate non-linear optical
coefficients can also be made in a straightforward way (22). The
infinite sums over intermediate states occurring in these expressions
can be replaced by sums over R-matrix states just as in eq.(20) for
the polarizability.

4. SOME RECENT RESULTS FOR ELECTRON SCATTERING
AND PHOTOIONIZATION OF ATOMS

The results which we describe in this section have all been
obtained using the Buttle corrected R-matrix with numerical orbitals by

solving eq.(11).   A general ASA FORTRAN program has been written
which calculates atomic polarizabilities, electron scattering cross
sections and photoionization cross sections for an arbitrary LS
coupled atom or ion.   This program has been published in Computer
Physics Communications (23) and copies are obtainable on request from
the associated program library in Belfast.   A revised double length
version of this program, compatible with the IBM 360 and 370 series
computers, will be published shortly (24).

Cross sections for electron scattering by a number of atoms and
ions including H, He, C, N, O, Ne, Mg, N^+, C^{++} and O^{4+} have been
obtained over the last two or three years.   Some of these have been
reviewed by Burke and Robb (9).   We concentrate here on some recent
results obtained by M. Le Dourneuf, Vo Ky Lan and P.G. Burke for
carbon, nitrogen and oxygen atoms.   These include polarized pseudostates in expansion (9) to allow for the polarizability of the target
ground state.   These pseudostates have been obtained by Vo Ky Lan
et al (25) by solving variationally the Schrödinger equation for the
unperturbed atom simultaneously with the inhomogeneous first-order
perturbation equation within a superposition-of-configurations
framework.   In these calculations a radius of about 10 a u is
appropriate and good convergence can be obtained with about 10 terms
in each channel.

Results obtained from a number of different approximations are
presented in the following figures.   These are

   (i)   Single configuration (SC).   This is obtained by retaining
          the channels associated with the ground state configuration
          $1s^2 2s^2 2p^q$ in eq.(9).

   (ii)   Multi-configuration (MC).   All terms of the ground and
          excited configurations $1s^2 2s^2 2p^q$, $1s^2 2s 2p^{q+1}$ and
          $1s^2 2p^{q+2}$ are included in eq.(9).

   (iii) Polarized pseudostate (PS).   The ground state
          configuration together with the polarized pseudostates
          required to give the full dipole polarizability of the
          ground state are included in eq.(9).   The target states
          are represented by CI wave functions using the pseudo
          $\overline{3s}$ $\overline{3p}$ and $\overline{3d}$ orbitals which are also used in constructing
          the pseudostates.   This approximation also includes
          between 50 to 100 terms in the second expansion in eq.(9).

   (iv)   Bethe Goldstone (BG).   Results obtained using the Bethe
          Goldstone method by Nesbet et al.(26).

We show in figure 3 the total cross section for electron carbon
atom scattering.

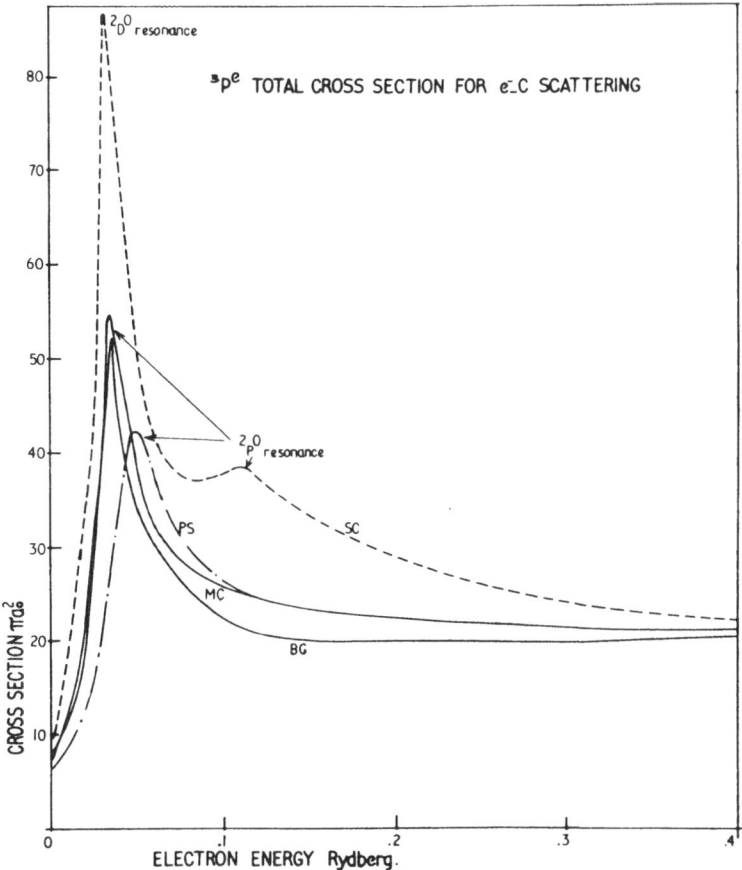

FIGURE 3.   Cross Section for Electron
Scattering from Carbon. (3pe)

The $^2D^o$ state of C^- lies in the continuum in the SC approximation whereas it is known to be bound.   The other approximations are in good accord.   The electron affinities calculated in the PS approximation are $^4S^o$ 1.406(1.268)eV   and $^2D^o$ 0.008 (0.035)eV where the numbers in brackets are the best estimates given by Hotop and Lineberger (27).

In figure 4 we show the total cross section for electron nitrogen atom scattering.

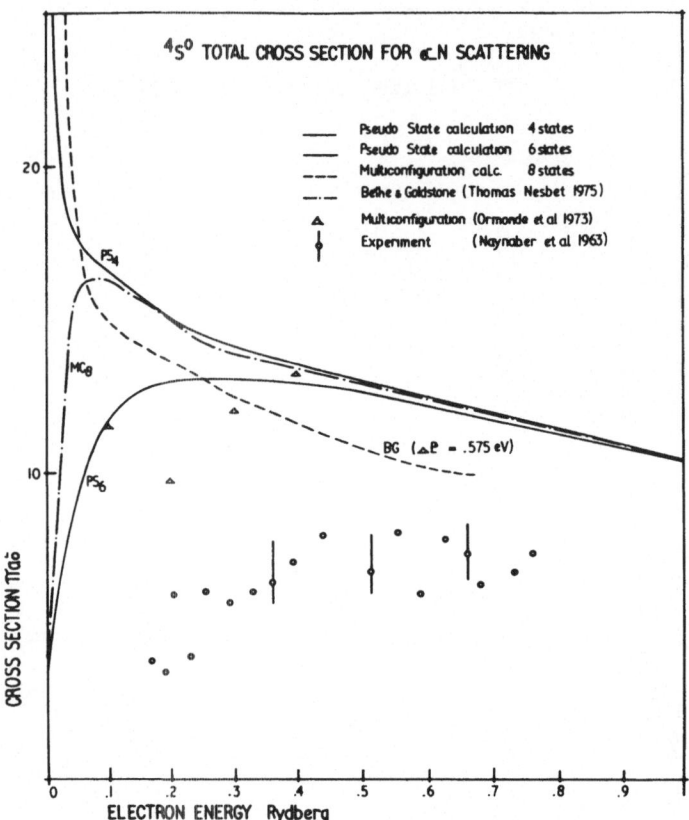

FIGURE 4.   Cross Section for Electron
Scattering from Nitrogen.   (^{4}S$^\circ$)

The most significant feature of these results is the low energy ^{3}P^e
resonance.   The position of this is very sensitive to the
approximation.   An experiment by Miller et al. (28) indicates that it
lies in the continuum.   In this case two polarized pseudostate
calculations were carried out: PS$_4$ which has already been described
and PS$_6$ in which two further excited target states were included.
The latter calculation, like the MC calculation, tends to over-
estimate the (N + 1)-electron correlation and leads to a bound ^{3}P^e
state.   The "best" calculation PS$_4$ gives a resonance with an
energy of .057eV and the BG calculation (with an adjustable parameter)
gives a similar result.   At higher energies all calculations
including MC calculations of Ormonde et al. (29) are in poor accord
with the measurements of Neyñaber et al. (30) and further experiments
would be desirable.

    In figure 5 we show the total cross section for electron oxygen
atom scattering.

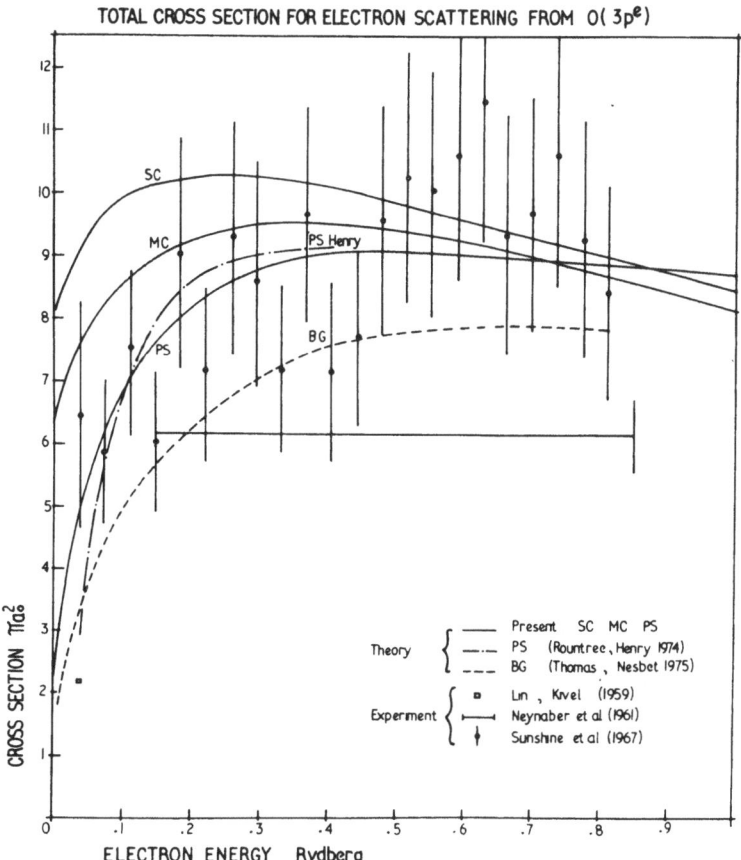

TOTAL CROSS SECTION FOR ELECTRON SCATTERING FROM O($3p^e$)

FIGURE 5.   Cross Section for Electron
Scattering by Oxygen. ($3p^e$)

The R-matrix PS calculations and similar independent calculations by
Rountree et al.(31) are in best agreement with experiment (32).   We
note that the PS approximation gives 1.480 (1.462) eV for the
electron affinity of atomic oxygen.

Turning now to photoionization, R-matrix calculations have now
been carried out for B, C, N, O, Ne, Aℓ, Ar and Cℓ as well as some
ions in the last year.   We show in the following figures a selection
of some of these results.

In figure 6 we show the inner 3s-subshell photoionization cross
section in Ar.

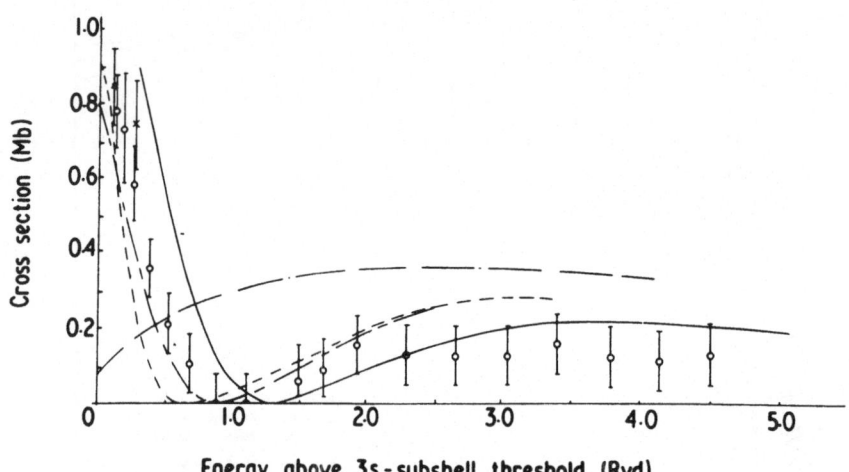

FIGURE 6.   Photoionization Cross Section for
the 3s-subshell in Argon

The R-matrix result of Burke and Taylor (20) is given by the solid
line and was obtained by coupling the ^{2}P ground state and the ^{2}S
first excited states in eq. (9).   The experiments (33) show a zero.
This zero is not obtained when this coupling is omitted as in the
Hartree Fock approximation of Kennedy and Manson (34) denoted by the
dashed dotted curve.   This zero is however correctly given by the
RPA approximation of Amusia et al. (35) denoted by the dashed curve,
and the simplified RPA of Lin (36).

In figure 7 we show the total photoionization cross section for
aluminium calculated by Le Dourneuf et al. (37).

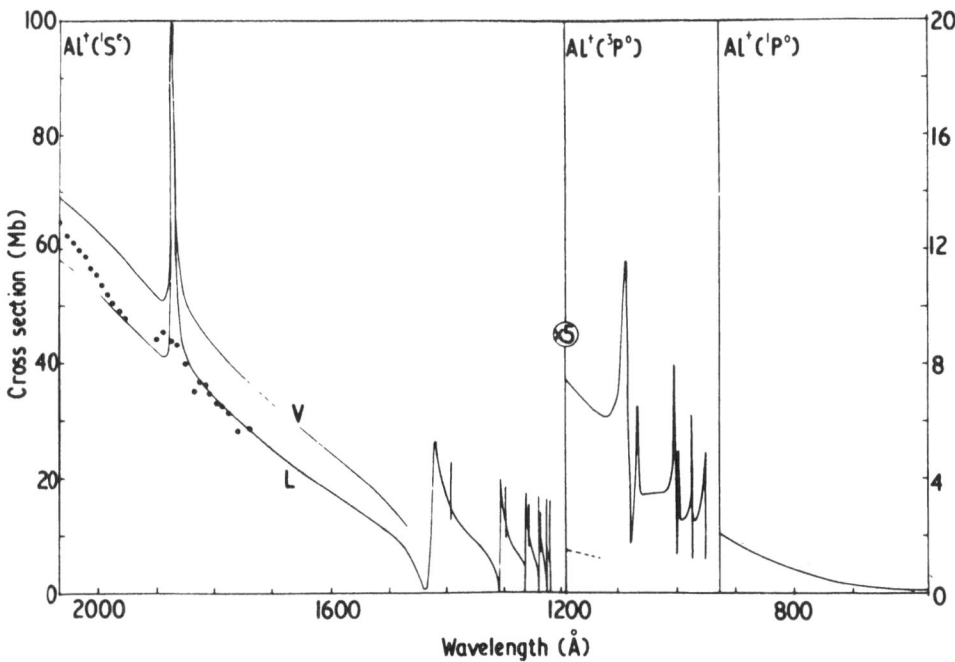

FIGURE 7. Total photoionization cross section
for Aluminium

The agreement with the experiments (38) is very satisfactory. Earlier
calculations using the quantum defect theory were about a factor of
two lower than experiment at threshold. The good agreement obtained
now can be attributed to the inclusion of channel coupling between
the three lowest ^{1}S ^{3}P and ^{1}P states of Aℓ^+. This introduces a
3s 3p^2 ^{2}D^e resonance which gives the broad rise in the cross
section at threshold and a 3s 3p^2 ^{2}S^e resonance which gives the
narrow peak near 1870 A^o.

Finally, in figure 8 we show the total photoionization cross
sections for atomic oxygen calculated by Taylor and Burke (39).

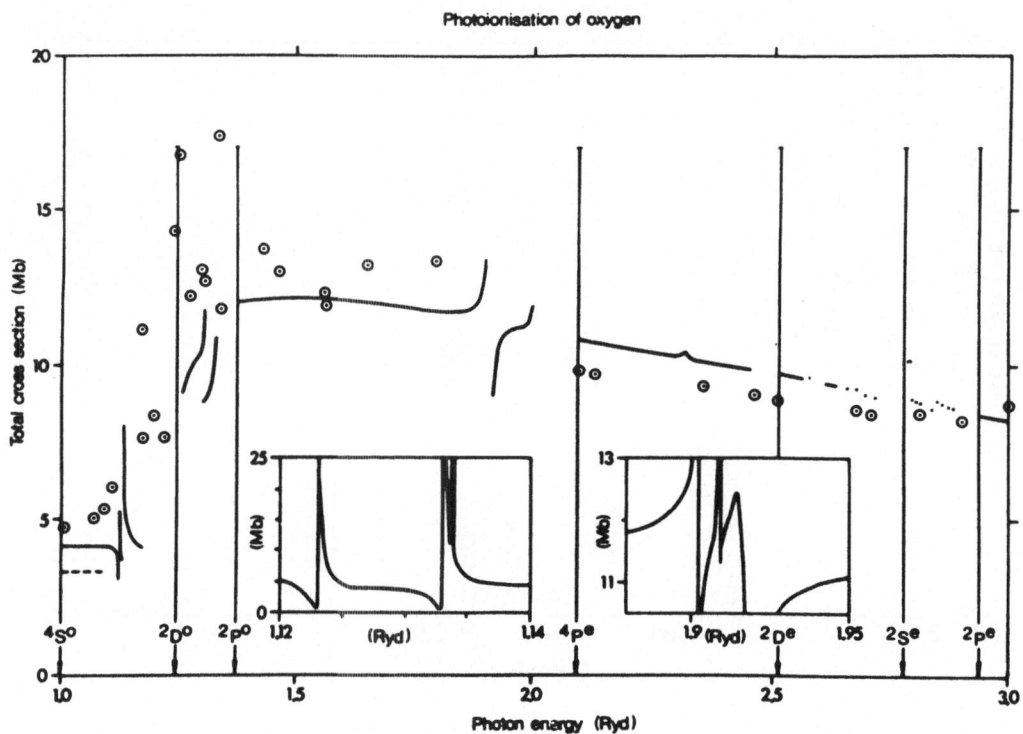

FIGURE 8.   Total Photoionization Cross Section
for Atomic Oxygen

In this calculation all terms of O^+ corresponding to the
configurations $1s^2\, 2s^2\, 2p^3$ $1s^2\, 2s\, 2p^4$ and $1s^2\, 2p^5$ were
included in eq. (9).   The agreement with the old experiments of
Cairns and Samson (40) is satisfactory.   However it would clearly
be of value for new experiments using synchrotron radiation to be
carried out.   Turning to the resonances which are a prominent
feature of the results, there is good agreement with the earlier
work of Henry (41) for the series converging to the $^2D^o$ threshold.
However there is considerable discrepancy for the series converging
to the $^2P^o$ threshold.

5. CONCLUSIONS AND FUTURE WORK

We have seen that the R-matrix method provides a general framework for the calculation of atomic and molecular processes. In the case of light atoms general computer programs have been written which will allow any of these processes to be calculated.

In the future the work will increasingly turn to heavy atoms and to molecules. In the former case the Buttle corrected numerical R-matrix method appropriate to light atoms will probably prove satisfactory but it is likely that the Dirac equation rather than the Schrödinger equation should be used. Some work by Chang (42) is already in progress in this direction. For molecules further work is necessary to decide on which R-matrix method is most economical, and much work also needs to be done to build up the appropriate computer programs.

Another area which has not been considered in detail yet is ionization by electron impact at low energies. In this case an R-matrix approach has been suggested using hyperspherical ("Fock") co-ordinates by Fano and Inokuti (43) but much further work needs to be done.

In conclusion therefore we have seen that the method has already proved successful in a number of areas and we can expect continuing attention to be given to it in the future.

6. ACKNOWLEDGMENTS

The author wishes to acknowledge with thanks the many contributions of his co-workers: K.A. Berrington, M. Le Dourneuf, I. Mackie, I. Shimamura, W.D. Robb, K.T. Taylor and Vo Ky Lan. He also wishes to thank Ugo Fano for his continued interest in the method and for his encouragement over many years.

The research was supported by the U.S. Office of Naval Research, under Contract N00014-69-0035 and by the Science Research Council.

REFERENCES

1.   E.P. Wigner, Phys.Rev. 70 15, 606 (1946).
      E.P. Wigner and L. Eisenbud, Phys.Rev. 72 29 (1947).
2.   M.J. Seaton, Proc.Phys.Soc. 88 801 (1966), Comments on Atomic
      and Molecular Physics 2 37 (1970).
3.   U. Fano, Phys.Rev. A2 353 (1970), J.Opt.Soc.Am. 65 979 (1975).
4.   U. Fano, Comm. Atom. Molec. Phys. 2 47 (1970), E.S. Chang and
      U. Fano, Phys.Rev. A6 173 (1972).
5.   E.J. Heller and H.A. Yamani, Phys.Rev. A9 1201, 1209 (1974),
      J.T. Broad and W.P. Reinhardt, J.Phys.B. (Atom.Molec.Phys.)
      9 1491 (1976) .
6.   M.A. Morrison and N.F. Lane, Phys.Rev. A12 2361 (1975),
      J.N. Resigno, C.W. McCurdy, Jr., and V. McKoy, to be published.
7.   C. Bloch, Nucl.Phys. 4 503 (1957).
8.   A.M. Lane and D. Robson, Phys.Rev. 151 974 (1966), Phys.Rev.
      178 1715 (1969).
9.   P.G. Burke, A. Hibbert and W.D. Robb, J.Phys.B. (Atom.Molec.
      Phys.) 4 153 (1971), P.G. Burke and W.D. Robb, Adv. in
      Atom.Molec.Phys. 11 143 (1975).
10.   P.J.A. Buttle, Phys.Rev. 160 719 (1967).
11.   D.J. Zvijac, E.J. Heller and J.C. Light, J.Phys.B. (Atom.Molec.
      Phys.) 8 1016 (1975).
12.   B. Schneider, Chem.Phys.Letters, 31 237(1975), Phys.Rev. A11
      1957 (1975).
13.   I. Shimamura, to be published (1976), P.G. Burke, I. Mackie and
      I. Shimamura, to be published (1976).
14.   R.S. Oberoi and R.K. Nesbet, Phys.Rev. A8 215 (1973).
15.   M. Danos and W. Greiner, Phys.Rev. 146 708 (1966).
16.   U. Fano and C.M. Lee, Phys.Rev.Letters 31 1573 (1973),
      C.M. Lee, Phys.Rev. A10 584 (1974).
17.   R.F. Barrett and P.P. Delsanto, Phys.Rev. C10 101 (1974).
18.   S.S. Ahmed, R.F. Barrett and B.A. Robson, Nuclear Phys. A257
      378 (1976), and to be published.
19.   D.C.S. Allison, P.G. Burke and W.D. Robb, J.Phys.B. (Atom.Molec.
      Phys.) 5 55, 1431 (1972), W.D. Robb, J.Phys.B. (Atom.Molec.
      Phys.) 6 945 (1973), 7 369 (1974).
20.   P.G. Burke and K.T. Taylor, J.Phys.B. (Atom.Molec.Phys.) 2620
      (1975).
21.   W.D. Robb, S.P. Rountree and T. Burnett, Phys.Rev. A11 1193 (1975
22.   T. Burnett, G. Doolen and W.D. Robb, private communication.
23.   K.A. Berrington, P.G. Burke, J.J. Chang, A.T. Chivers, W.D. Robb
      and K.T. Taylor, Comput.Phys.Commun. 8 149 (1974).
24.   K.A. Berrington, P.G. Burke, M. Le Dourneuf, W.D. Robb, K.T.
      Taylor and Vo Ky Lan, to be submitted to Comput.Phys.Commun.
      (1976).
25.   Vo Ky Lan, M. Le Dourneuf and P.G. Burke, J.Phys.B. (Atom.Molec.
      Phys.) 9 1065 (1976).

26.   L.D. Thomas, R.S. Oberoi and R.K. Nesbet, Phys.Rev. A10 1605
      (1974), L.D. Thomas and R.K. Nesbet, Phys.Rev. A11 170 (1975)
      A12 1729 (1975) A12 2369 (1975).
27.   H. Hotop and W.C. Lineberger, J.Chem.Phys. Ref.Data 4 539 (1975).
28.   T.M. Miller, B.B. Aubrey, P.N. Eisner and B. Bederson, Bull.Am.
      Phys.Soc. 15 416 (1970).
29.   S. Ormonde, K. Smith, B.W. Torres and A.R. Davies, Phys.Rev.
      A8 262 (1973).
30.   R.H. Neynaber, L.L. Marino, E.W. Rothe and S.M. Trujillo, Phys.
      Rev. 129 2069 (1963).
31.   S.P. Rountree, E.R. Smith and R.J.W. Henry, J.Phys.B. (Atom.
      Molec.Phys.) B7 L167 (1974).
32.   S.C. Lin and B. Kivel, Phys.Rev. 114 1026 (1959), R.H.
      Neynaber, L.L. Marino, E.W. Rothe and S.M. Trujillo, Phys.
      Rev. 123 148 (1961), G. Sunshine, B.B. Aubrey and
      B. Bederson, Phys.Rev. 154 1 (1967).
33.   J.A.R. Samson and J.L. Gardner, Phys.Rev.Letters 33 671
      (1974) R.G. Houlgate and J.B. West, private communication.
34.   D.J. Kennedy and S.T. Manson, Phys.Rev. A5 227 (1972).
35.   M.Ya. Amusia, V.K. Ivanov, W.A. Cherepkov and L.V. Chernyshiva,
      Phys.Letters 40A 361 (1972).
36.   C.D. Lin Phys.Rev. A9 181 (1974).
37.   M. Le Dourneuf, Vo Ky Lan, P.G. Burke and K.T. Taylor, J.Phys.B.
      (Atom.Molec.Phys.) 8 2640 (1975).
38.   J.L. Kohl and W.H. Parkinson, Astrophys.J. 184 641 (1973),
      R.A. Roig, J.Phys.B. (Atom.Molec.Phys.) 8 2939 (1975),
      J.A. Esteva, G. Mehlman-Balloffit and J. Romand, J.Quant.
      Spec.Radiat.Trans. 12 1291 (1972).
39.   K.T. Taylor and P.G. Burke, J.Phys.B. (Atom.Molec.Phys.) to be
      published.
40.   R.B. Cairns and J.A.R. Samson, Phys.Rev. 139 A1403 (1965).
41.   R.J.W. Henry, Plant.Spac.Sci. 16 1503 (1968).
42.   J.J. Chang, J.Phys.B. (Atom.Molec.Phys.) 8 2327 (1975).
43.   U. Fano and M. Inokuti, Argonne National Laboratory Report,
      ANL-76-80 (1976).

ELECTRON CORRELATION IN ATOMS FROM PHOTOELECTRON SPECTROSCOPY

D. A. Shirley, S.-T. Lee, S. Süzer, R. L. Martin,
E. Matthias, and R. A. Rosenberg
Materials & Molecular Research Division, LBL and
Dept. of Chem., U. of California, Berkeley, CA 94720

ABSTRACT

The photoelectron spectrum of an atom A yields the
eigenenergies of the ion A^+. Additional satellite peaks arise
because of electron correlation effects. Peaks observed in
spectra from Ca, Sr, Zn, Cd, and Hg have been attributed to
initial-state configuration interaction, while final-state configuration interaction gives satellite peaks in a Ba spectrum.
The neon 1s "shakeup" spectrum is a result of both effects.

I. INTRODUCTION

The photoelectric effect has been known since approximately
the turn of the century. Einstein[1] explained it in 1905 in the
work that was to be noted in his Nobel Prize citation. Nevertheless, for over a half century thereafter this effect was known
mainly as a threshold phenomenon manifested as the work function
in metals or as the ionization potential in molecules.

Only in the 1960's was photoelectron spectroscopy exploited
as an experimental method of practical use in many laboratories.
Its popularization followed the development of high-resolution
electron spectrometers together with convenient line sources.[2,3]
For its first decade, photoelectron spectroscopy was divided into
two rather distinct subdisciplines--ultraviolet photoelectron
spectroscopy (UPS) and X-ray photoelectron spectroscopy (XPS or
ESCA). With the advent of intense variable-energy photon beams
from electron storage rings, this artificial separation is now

disappearing, and the generic term photoelectron spectroscopy (PES) can be applied to the entire field.

The essential elements of a PES experiment are few and simple. A monochromatic photon beam of energy $h\nu$ impinges on a sample, ejecting electrons of kinetic energy

$$T = h\nu - E_B \quad . \tag{1}$$

Analysis of the electron intensity $N(T)$ yields peaks that can be associated with binding energies E_B of various orbitals. In this way the complete one-electron orbital spectrum of a given element can be studied, and the Aufbanprinzip can be illustrated rather directly, a feature of particular appeal in atomic physics. Thus for example PES studies on the rare gas neon $(1s^2 2s^2 2p^6; {}^1S)$ will yield the positions of the following states of Ne^+:

$$(1s^2 2s^2 2p^5; {}^2P_{3/2})$$

$$(1s^2 2s^2 2p^5; {}^2P_{1/2})$$

$$(1s^2 2s 2p^6; {}^2S)$$

$$(1s 2s^2 2p^6; {}^2S)$$

among others. Note that the ${}^2S$ states are reached, as well as ${}^2P$ states, from the ${}^1S$ ground states. We shall return to this point.

At this point one might well ask whether the PES method can yield any information not already available from X-ray physics. After all, can one not simply build up these one-electron hole levels by comparing optical and X-ray data, as has been done in various tabulations of core electron binding energies?

The answer is (surprisingly) "No". The optical data were obtained on free atoms, while X-ray spectra were taken using various condensed-phase anode materials, usually metals. It is now well-known[4] that core-electron binding energies are systematically lower in metals than in free atoms. This follows from a combination of effects, the most important of which is usually "extra-atomic relaxation" or mobile valence electrons to shield the holes formed on photoemission from core orbitals in metals. Table I illustrates the magnitude of this effect for the elements Na, Ni, and Zn, taken from a new tabulation.[5] The difference,

$$E = E_B^A - E_B^M \tag{2}$$

Table I. Binding energies of electronic
          orbitals in atoms and metals.

Element	Nl	E_B^A	E_B^M	ΔE
		— (eV) —		
Na	1s	1070	1074	5
	2s	71	66	5
	3s	5.1	3.5	1.6
Ni	$2p_{3/2}$	869	859	10
	3p	83	72	11
Zn	$2p_{3/2}$	1029	1026	3
	3p	96	93	3

between atomic and metallic core-level binding energies, runs [6] from 3 to 11 eV in these elements, a substantial effect indeed. Thus we must conclude that <u>all previous tabulations of core-level binding energies which did not distinguish between atomic and metallic samples are systematically in error</u>.

Let us now compare and contrast PES with photon spectroscopy, referring to Fig. 1. We consider an atom that has S symmetry in a ground state, and we work in the electric dipole approximation. In the presence of a radiation field represented by the vector potential $A(t)$, the electron momenta are replaced by $p - eA/c$ Standard time-dependent perturbation theory with conventional approximations yields a transition probability governed by the matrix element

$$\mathcal{M}.\ \mathcal{E}.\ \alpha\ \langle \Psi_f | \underset{\sim}{A} \cdot \underset{\sim}{p} | \Psi_i \rangle \quad . \tag{3}$$

Only states of P-symmetry are excited. For low-energy photons this means that P bound states of the N-electron (atomic) system can be found (sharp lines, Fig. 1). Above the photoionization threshold, P-symmetry N-electron states are also formed, thus satisfying the electric dipole selection rule. These give absorption edges rather than sharp lines. Further, each of these P states consists of a state of the (N - 1) electron (ionic) system, of arbitrary orbital angular momenta L', plus a continuum electron in a state of orbital angular momentum ℓ that satisfies the triangle condition

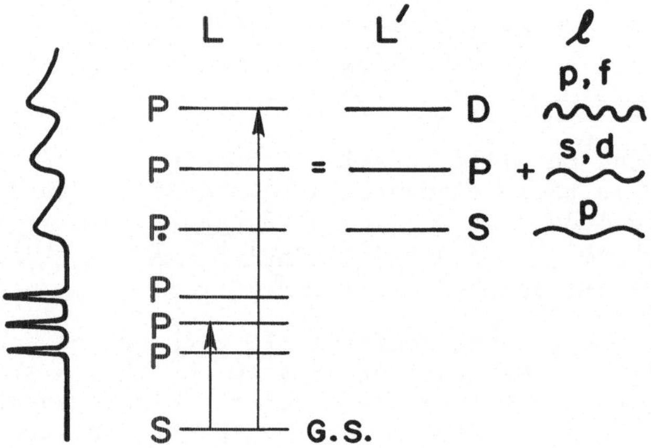

FIG. 1. Illustration of the Way in which the Electric Dipole
         Selection Rules Applied to an Atom in an S State Yields
         Only P Bound States but Ionic States of any Symmetry.

$$\underset{\sim}{L}' + \underset{\sim}{\ell} = \underset{\sim}{L} \tag{4}$$

with $|\underset{\sim}{L}| = 1$ for the P state. Thus in effect there is no absolute
selection rule for photoemission and all states are populated.

We now turn from one-electron properties (or apparent
one electron properties) to a phenomenon that is manifestly multi-
electronic in nature, namely electron-electron correlation. For
simplicity initial-state correlation effects are treated in the
next section, using the configuration-interaction formalism.
Final-state correlations are the subject of Section III. In
Section IV the neon 1s satellites, in which both initial- and
final-state correlations are important, are discussed briefly.

II. INITIAL STATE CONFIGURATION INTERACTION

A one-electron atom or ion can be treated exactly, but any
multi- (even two-) electron system must be dealt with using
approximate methods. Thus the ground state of He can be described
only roughly by the configuration $1s^2$. The e^2/r_{12} electron-
electron repulsion term in the Hamiltonian may be regarded as
leading to an explicit dependence of the electronic wavefunction
on r_{12}. It is computationally most convenient to treat correlation through configuration interaction,[7] in which eigenstates are
described by adding excited configurations with the correct
amplitude and phase relations. Thus the ^1S ground state of He is

$$(^1S) = a(1s^2) + b(2s^2) + c(2p^2) + \ . \ . \ . \tag{5}$$

In the alkaline-earth, or Group IIA, atoms, the valence configuration may be represented as (for example)

$$\Psi(Ca) = a(4s^2) + b(4p^2) + c(3d^2) + \ . \ . \ . \tag{6}$$

and similarly for Sr and Ba.

Photoemission from these ground states would yield only the
$4s^1;^2S$ state of Ca^+, etc., by one-electron processes, if there
were no electron correlation. The admixed configurations $4p^2$,
$3d^2$, etc., lead directly to observable $4p^1;^2P$ and $3d^1;^2D$ final
states on photoemission, with the line intensities being related
to admixture coefficients in the initial states. Thus photoemission spectra can give a rather direct measure of initial-state
configuration interaction (ISCI). Such spectra have now been
observed in Ca, Sr, Ba, Zn, Cd, and Hg.[8-10] The HeI resonance
line at 21.2 eV was used to induce photoemission in every case
except Ba, for which the NeI lines (16.7, 16.9 eV) were employed.

Table II lists the prominent satellites that were identified in the photoelectron spectra of these atoms.

Comparisons with theory are still in a rather crude stage. A quantitative analysis of these spectra requires both configuration interaction calculations on the initial states and estimates of the relative cross-sections for 4s, 4p, 3d, etc., photoemission. The latter are not yet available. Multiconfiguration Hartree-Fock calculations were done by Kim and Bagus[11] and by Hansen.[12] The admixture coefficients that were obtained would predict the satellite intensities to be too weak by factors between 2 and 20 before cross-section effects are taken into account. It will be interesting to learn whether the cross-section variation will resolve this discrepancy.

Because the configuration-interaction formalism is just one approach to the electron correlation problem, the reality of the configuration admixtures can be questioned. After all, one need not in principle even use the CI picture. The satellites would still be observed: what would they then mean? The answer is that operationally the one-electron states in Ca^+ (for example), together with the continuum final states are projected onto the initial state of Ca, through the dipole transition operator. This is a general statement of the process. A more specific statement, in terms of ISCI, requires care in selecting the basis functions. Once a particular set of functions is chosen, however, there exists a unique answer as to how much a given function is admixed into the initial state.

We conclude this section by pointing out a special case of ISCI involving spin-orbit coupling. The ground state of atomic Pb is normally $6s^2 6p^2$. In strong spin-orbit coupling this becomes[13] mainly $6s^2 6p_{1/2}^2$ with a small admixture of $6s^2 6p_{3/2}^2$. Süzer[14] has shown that the photoemission spectrum of atomic Pb indeed shows a very intense $(6s^2 6p_{1/2}; {}^2P_{1/2})$ peak and a much less (7%) intense $(6s^2 6p_{3/2}; {}^2P_{3/2})$ peak, as expected. Also present were all the peaks expected to arise from admixtures of $sp^2 d$ and $s^2 d^2$ into the ground state.

III. FINAL-STATE CONFIGURATION INTERACTION (FSCI)

This phenomenon is manifested as additional satellites or very unusual intensities in a photoemission spectrum. There are several known cases of FSCI. One of the most dramatic is the HeI spectrum of atomic barium, which was reported by Brehm and Höfler[15] and by Hotop and Mahr.[16] This spectrum was re-discovered in our laboratory in the course of ISCI studies in alkaline earths. The

Table II. Summary of satellites attributable to ISCI
           in photoelectron spectra of Group II elements.

Ion	Main Line	Satellites	References
$Ca(4s^2)$	4s	4p,3d,5s,4d	10
$Sr(5s^2)$	5s	5p*,4d,6s,5d,4f	10
$Ba(6s^2)$	6s	6p*,5d*,7s,6d,4f	10
$Zn(3d^{10}4s^2)$	4s	4p*	10
$Cd(4d^{10}5s^2)$	5s	5p*	·8,10
$Hg(5d^{10}6s^2)$	6s	6p*	9,10

*Doublet energies resolved.

essential unusual features of the BaI (HeI) spectrum are the
greatly enhanced 5d and 6p peaks and a number of additional nl
peaks of Ba^+ arising from resonant auto-ionization by the 21.2 eV
radiation. For example, the 6f and 8p states are clearly present.

The explanation for this phenomenon is too involved to be
presented in detail here, but basically the HeI radiation reso-
nantly excites a state of Ba that is embedded in the continuum of
Ba^+. The final N-electron state (N = 56 in this case) can be
regarded as consisting of interacting N-electron configurations,
each of which is made up of an N - 1 electron state of Ba^+ coupled
with an unbound electron in a continuum state. When the resonance
condition is satisfied, the outgoing electrons carry kinetic
energies $\{h\nu - E_i\}$, where $\{E_i\}$ are the energies of the Ba^+ states.

In Ba, autoionization occurs because of excitation of an
electron from the $5p^6$ shell to an ns or nd state. Thus the total
configuration of the state embedded in the continuum is
$5p^56s^2$(ns or nd). This may be loosely regarded as a Rydberg-like
state below the 5p photoionization threshold in Ba. Proceeding
upward in Z from Ba through the rare earths, the 5p ionization
threshold does not increase dramatically, because the additional
nuclear charge is in large measures shielded by the filling 4f
shell, and the ion core potential changes only slowly. Thus
resonant autoionization by the HeI 21.22 eV line and associated
satellite radiations persists through Sm (Fig. 2) and Eu
(Z = 62,63). It has disappeared by Yb (Z = 70).

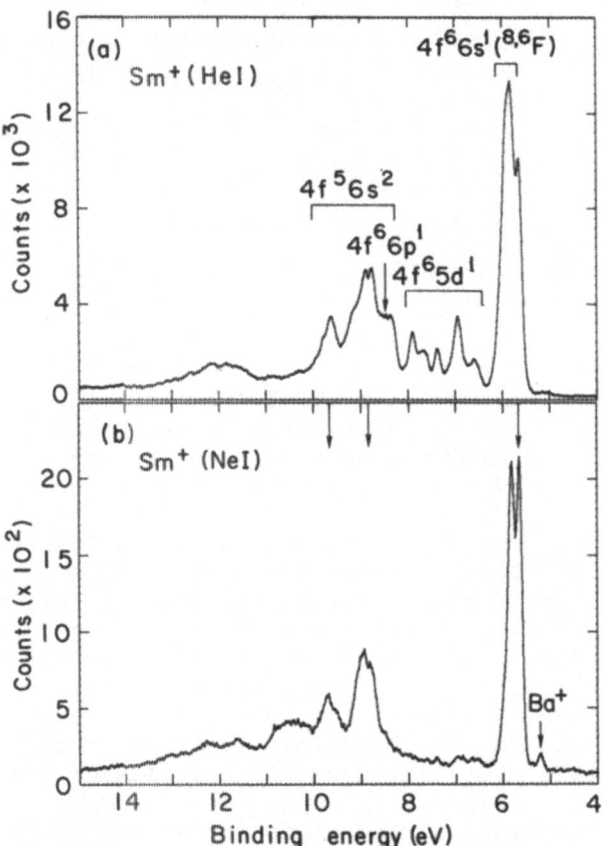

FIG. 2. The Photoelectron Spectra of Sm Vapor Excited by HeI
         (21.2 eV) and by NeI (16.7, 16.9 eV) Radiation. The
         Unusual Intensity Ratios and Extra Peaks in the HeI
         Spectrum Arise from Autoionization Following Resonant
         Excitation of the 5p Shell.

Fano and Cooper[17] have discussed configuration interaction
effects from a very general viewpoint. As tunable (synchrotron)
radiation sources become available, many instances of these
phenomena can be expected.

IV.   THE Ne 1s SATELLITES: ISCI AND FSCI

As a final example, let us consider the "shakeup" spectrum
of the Ne 1s, in which both initial- and final-state correlations
play a role. A high-resolution photoemission spectrum of the 1s
region of Ne was reported by Gelius.[18] In addition to the many
transition

$$\text{Ne}(^1\text{S};1\text{s}^2 2\text{s}^2 2\text{p}^6) \xrightarrow{-e} \text{Ne}^+(^2\text{S};1\text{s}^1 2\text{s}^2 2\text{p}^6)$$

at 870 eV, there was a large number of weak satellite lines at
about 40 eV higher energy. These arise from transitions of the
form

$$\text{Ne}(^1\text{S};1\text{s}^2 2\text{s}^2 2\text{p}^6) \xrightarrow{-e} \text{Ne}^+(^2\text{S};1\text{s}^1 2\text{s}^2 2\text{p}^5 3\text{p})$$

$$\text{Ne}^+(^2\text{S};1\text{s}^1 2\text{s}^2 2\text{p}^5 4\text{p})$$

$$\text{Ne}^+(^2\text{S};1\text{s}^1 2\text{p}^2 2\text{p}^5 5\text{p})$$

.
.
.

etc.

(It should be noted that each configuration on the right gives
<u>two</u> final states because of exchange splitting.)

There are two ways in which these transition can be understood
on the basis of FSCI alone. We note that a nonzero transition
probability requires overlap between the "passive" electron configurations. Expanding the matrix element, for photoemission,

$$\mathcal{M.\ E.}\ \alpha\ \langle x|\underset{\sim}{A}\cdot\underset{\sim}{p}|1\text{s}\rangle\langle\text{Ne}^+1\text{s}^1 2\text{s}^2 2\text{p}^5 3\text{p}|\text{Ne}1\text{s}^1 2\text{s}^2 2\text{p}^6\rangle\quad,$$

we see that the overlap product would vanish if the basis functions
in Ne^+ and Ne were the same (because $\langle 3\text{p}|2\text{p}\rangle$ would be zero). In
fact they are not: the Ne^+ $n = 2$ functions are relaxed because of
the added core charge, giving finite overlap. Alternatively we
could imagine using the basis set in Ne^+ as in Ne, but describing
the relaxation in terms of configuration mixing, again yielded
finite overlap. This effect is, however, not readily separable

from the second way in which FSCI leads to satellites; namely configuration interaction that is already present irrespective of relaxation. After all, the "final" states in Ne^+ are in fact eigenstates of a Hamiltonian with no reference to photoelectron spectroscopy.

The FSCI effects in the Ne 1s spectrum have been known for some time, but calculated spectra based on FSCI alone show much lower relative intensities in the satellites than are observed experimentally. It has been pointed out by Martin[19] that ISCI is also important in this case. In fact ISCI is responsible for about one-half the observed satellite intensities. There is a great deal of symmetry between the ISCI and FSCI contributions to the satellite intensities, because the Ne and Ne^+ states are basically quite similar, even at the level of electron correlations. Thus it even turns out that the satellite spectrum gives a fairly direct picture of configuration admixing in the ground state.

In conclusion, we have sought in this paper to summarize specific cases in which electron correlation has manifestly affected photoelectron spectra. It seems likely that photo-electron spectroscopy will prove increasingly useful in elucidating problems of this nature in atomic physics.

REFERENCES

1. A. Einstein, Ann. d. Physik 17, 132 (1905).

2. K. Siegbahn, C. Nordling, A. Fahlman, R. Nordberg, K. Hamrin, J. Hedman, G. Johansson, T. Bergmark, S.-E. Karlsson, I. Lindgren, and B. J. Lindberg, ESCA-Atomic, Molecular, and Solid State Structure by Means of Electron Spectroscopy, Nova Acta Regiae Soc. Sci. Upsaliensis Ser. IV, vol. 20, 1967.

3. D. W. Turner, C. Baker, A. D. Baker, and C. R. Brundle, Molecular Photoelectron Spectroscopy (Wiley Interscience, 1970).

4. L. Ley, S. P. Kowalczyk, F. R. McFeely, R. A. Pollak, and D. A. Shirley, Phys. Rev. B 8, 2392 (1973).

5. D. A. Shirley, Chem. Phys. Letters 16, 220 (1972).

6. This difference remains after the work function correction otherwise it would be even larger.

7. J. C. Slater, Quantum Theory of Atomic Structure (McGraw-Hill, 1960), Vol. 11, p. 40.

8. S. Süzer and D. A. Shirley, J. Chem. Phys. <u>61</u>, 2481 (1974).

9. J. Berkowitz, J. L. Dehmer, Y. K. Kim, and J. P. Desclaux, J. Chem. Phys. <u>61</u>, 2556 (1974).

10. S. Süzer, S.-T. Lee, and D. A. Shirley, Phys. Rev. A <u>13</u>, 1842 (1976).

11. Y. K. Kim and P. S. Bagus, Phys. Rev. A <u>8</u>, 1739 (1973).

12. Jørgen E. Hansen (private communication, June 1976).

13. E. U. Condon and G. H. Shortley, <u>The Theory of Atomic Spectra</u> (Cambridge University Press, 1935), p. 275.

14. S. Süzer, M. S. Banna, and D. A. Shirley, J. Chem. Phys. <u>63</u>, 3473 (1975).

15. B. Brehm and K. Höfler, Int. J. Mass Spectrom. Ion Phys. <u>17</u>, 371 (1975).

16. H. Hotop and D. Mahr, J. Phys. B: Atom. Molec. Phys. <u>8</u>, L301 (1975), and private communication.

17. U. Fano and J. W. Cooper, Rev. Mod. Phys. <u>40</u>, 441 (1968).

18. U. Gelius, J. Electr. Spectr. and Rel. Phenomena <u>5</u>, 985 (1974).

19. R. L. Martin and D. A. Shirley, Phys. Rev. A <u>13</u>, 1475 (1976).

*This work was done with support from the U. S. Energy Research and Development Administration.

POLARIZED ELECTRONS*

M.S. Lubell

Gibbs Laboratory, Yale University
New Haven, CT   06520   USA

INTRODUCTION

Since 1972, when the last survey of Polarized Electrons was
presented at an International Conference on Atomic Physics,[1] the
field has progressed to the point where it has entered a new phase,
one which, I believe, has been long awaited by everyone involved
in spin-dependent electron interactions. From the time the concept of electron spin was first introduced by Goudsmit and
Uhlenbeck[2,3] in what Sam Goudsmit has reminiscently called the
"springtime of modern atomic physics,"[4] a great many physicists
have devoted their energies to the study of phenomena which have
been related to the electron spin. These studies have spanned
virtually all of the disciplines of physics. Time and space--and
no doubt your restlessness--will not permit me to review all
these marvelous studies. With your indulgence, however, I should
like to recollect for you several of the hallmarks in the fifty
year history of the field of "polarized electrons." In using the
term "polarized electrons" I am referring now to any ensemble of
free electrons whose average spin direction is preferentially
oriented in space.

Following the introduction of electron spin in 1925-26 as a
fourth degree of freedom and the development of the quantum theory
of the electron shortly thereafter,[5] N.F. Mott applied these new
concepts in 1929 to the scattering of free electrons by an unscreened Coulomb field of a nucleus. Mott showed that when relativistic electrons are scattered by heavy nuclei, the scattering
is strongly dependent on the orientation of the electron spin
relative to its orbital momentum.[6,7] The first experimental

confirmation of "Mott scattering" was obtained by Shull, Chase and Myers in 1943 in a very difficult double scattering experiment.[8] Several years earlier, motivated by unresolved discrepancies between prior experimental results[9,10] and Mott's theory, Massey and Mohr[11] incorporated screening into the Coulomb potential. Although unable to resolve the outstanding discrepancies,[11,12] later shown by Shull, Chase and Myers to have been related to plural scattering effects,[8] Massey and Mohr found instead that spin polarization asymmetries should be expected even for non-relativistic electrons at energies as low as 100 eV. Experimental verification of Massey and Mohr's low energy predictions was provided in the 1960's by experiments carried out independently at Mainz[13,14] and Karlsruhe.[15,16] Although initially there were occasional differences between experimental results at energies below 500 eV and the detailed theoretical calculations which had been performed,[17-19] these differences were ultimately resolved by the measurements of Kessler and his co-workers at Karlsruhe.[20]

Interest in high-energy Mott scattering, on the other hand, grew following Lee and Yang's suggestion in 1956 that parity might be violated in the weak interaction.[21] The consequent discovery that space inversion symmetry is violated in beta decay from oriented nuclei[22] led to the prediction that electrons emitted in beta decay from unoriented nuclei should be longitudinally spin-polarized.[23-25] High-energy Mott scattering seemed to be an ideal vehicle for measuring the polarization of the beta-emitted electrons and indeed was successfully applied in 1960.[26] The theoretical framework for high-energy Mott scattering, at the same time, was augmented by a number of elaborate calculations, the earlier ones[27,28] using unscreened potentials and the later ones[17,29,30] using screened potentials. With the proliferation of polarized electron studies during the last fifteen years, high-energy Mott scattering has seen extensive application as a technique for polarization analysis. Detailed experimental studies[31-33] of the Mott scattering technique itself have provided the necessary technological background for reliable use of Mott scattering as a polarization detector. More complete information about the applications of Mott scattering can be found in a number of articles.[34-36]

The relationship between Mott scattering and the studies of various processes involving polarized electrons during the 1960's seems to have been almost symbiotic. In retrospect, this was very natural since no practical means of producing beams of polarized electrons had yet been developed. Thus polarized electrons could be used to study spin-dependent interactions only if the interactions resulted in the production of polarized electrons and only if a means of analyzing their polarization existed. Mott scattering satisfied the latter requirement. And

so for some 10 or 15 years Mott scattering and polarized electrons
were used to study processes which resulted in the emission of
polarized electrons.

Although the physics of many of these processes [35] held a
particular fascination for many physicists, "...colleagues from very
different fields of physics..," as J. Kessler has put it,[1] "asked,
'Would it be possible to use one or the other of these processes
to build an efficient source of polarized electrons for other
experiments?'" At the Third International Conference on Atomic
Physics in Boulder in 1972, Kessler reviewed the status of the
different sources of polarized electrons which had been studied
until that time. Implicit in his review was the assumption that
one or more of these sources would ultimately be used to study
spin-dependent interactions with polarized electron beams. Now
most appropriately on the fiftieth anniversary of the discovery
of electron spin, it is possible for me to report on the results
of some experiments which have used and the status of other experiments which are currently using beams of polarized electrons as
probes in spin-dependent processes. I would be remiss, however,
if I did not review as well the status of existing and promising
sources of polarized electrons, in order that you might gain some
appreciation for those experiments which are currently feasible and
those which are just now gleams in the eyes of optimists.

I have already given you a qualitative definition of spin-
polarization. Before embarking on a review of sources and experiments I should like to make this definition quantitative. The
polarization $\vec{P}_e$ of an ensemble of electrons is defined as the
ensemble average of the Pauli spin operator $\vec{\sigma}$:

$$\vec{P}_e \equiv \overline{\langle \vec{\sigma} \rangle} . \qquad (1)$$

Since $\vec{P}_e$ is not Lorentz invariant, the polarization of an electron
beam is defined in terms of the rest frame of the beam. If ρ is
the spin density matrix for the beam, the polarization can be
written

$$\vec{P}_e = \mathrm{Tr}\,(\rho \vec{\sigma}), \qquad (2)$$

where a normalization has been assumed such that $\mathrm{Tr}(\rho) = 1$. Alternatively ρ can be displayed in terms of the components of $\vec{P}_e$ as

$$\rho = \frac{1}{2} \begin{bmatrix} 1+P_z & P_x-iP_y \\ P_x+iP_y & 1-P_z \end{bmatrix} . \qquad (3)$$

If ρ' is the diagonalized form of ρ, ρ' can be shown to be given by

$$\rho' = P_e \begin{bmatrix} 1 & 0 \\ 0 & 0 \end{bmatrix} + (1-P_e) \begin{bmatrix} \frac{1}{2} & 0 \\ 0 & \frac{1}{2} \end{bmatrix}, \tag{4}$$

where $P_e = |\vec{P}_e|$ is properly called the degree of polarization of the beam. Often, P_e is informally called simply the polarization, although no direction is specified. Thus we see from Eq. 4 that an ensemble of electrons is characterized by an incoherent sum of a totally polarized part and a totally unpolarized part.

If a polarized electron beam interacts with an ensemble of particles through a spin-dependent interaction, the ensemble of particles can be characterized by an analyzing power, $\vec{A}$, which is defined in terms of a spin density matrix, ρ^A, for the ensemble as

$$\vec{A} = \text{Tr} (\rho^A \vec{\sigma}), \tag{5}$$

by analogy with Eq. 2. The signal detected for the interaction, I, will then be given by

$$I = I_o \text{Tr}(\rho\rho^A) = \frac{1}{2} I_o(1 + \vec{P}_e \cdot \vec{A}), \tag{6}$$

where I_o is a constant. If $\vec{P}_e$ and $\vec{A}$ are experimentally adjusted so that they are either parallel or antiparallel, Eq. 6 can be rewritten as

$$I^{\pm} = \frac{1}{2} I_o(1 \pm P_e A) \tag{7}$$

with the + sign taken for the parallel case and the − sign for the antiparallel case. An experimental asymmetry, Δ, can then be defined as

$$\Delta \equiv \frac{I^+ - I^-}{I^+ + I^-} = P_e A. \tag{8}$$

In practical terms, for the interaction under consideration the degree of electron polarization is given by

$$P_e = \frac{N^+ - N^-}{N^+ + N^-}, \tag{9}$$

where N^+ and N^- are the number of electrons with spins parallel and antiparallel to $\vec{A}$ respectively. Thus if P_e is known and Δ is measured, the quantity of interest, A, can be determined. So much for fundamentals.

SOURCES OF POLARIZED ELECTRONS

Let us now turn our attention to the current status of polarized electron sources. During the last 15 years a wide variety of
sources have been developed based upon a diversity of physical
principles. Some of these sources, while interesting because of
the techniques and physical principles involved, were not really
practicable.[37-40] Others ran into difficulty when attempts were
made to increase the polarization or the intensity to usable
levels.[41-44] In the following discussion I will restrict myself,
perforce of time, to those sources which have already been used
in experiments, those which are currently being used, or those
which have sufficient promise that they warrant serious consideration for the next generation of polarized electron experiments.

With a variety of sources available, how does one evaluate
their relative merits? A figure of merit, ζ given by

$$\zeta = P_e \sqrt{I_e} , \qquad (10)$$

where I_e is the intensity of the source, has often been used in
comparisons of sources. However this figure, based upon Poisson
statistics, simply describes the statistical accuracy which can
be achieved in an asymmetry measurement. It in no way takes into
account the special needs of an experiment. One experiment may
require a polarized beam with very small energy spread; another
may require high brightness (low phase space). In accelerator
applications pulsed sources are essential: one application, however, may require high peak intensity and low duty factor while
another may require lower peak intensity and higher duty factor.
In some cases the target may be unable to withstand a high flux
of electrons. Then a figure of merit must be calculated for the
maximum I_e allowable by the target.

Generally speaking, one must have a firm idea of the needs of
an experiment before one can evaluate the merits of a polarized
electron source. We can summarize the salient characteristics of a
source which impact on an experiment as follows:

(1) Intensity, I_e, sometimes limited by experimental require-
     ments.

(2) Degree of polarization, P_e.

(3) Figure of merit, $\zeta = P_e \sqrt{I_e}$, based upon (1) and (2).

(4) Direction of polarization and ease of reversal within
     experimental restrictions.

(5) Emittance, ε, of source at usable energy. The emittance, given by ε = ρα, where ρ is the radius of the beam at an image of the source and α is half the apex angle of the cone of the beam, is inversely proportional to the square root of the beam energy, E, and is generally expressed in mrad cm. An alternate quantity, combining intensity and emittance, is the brightness or <u>Richtstrahlwert</u>[45] given by I_e/ε^2.

(6) Pulse length.

(7) Repetition rate.

(8) Energy spread.

(9) Maximum frequency of polarization reversal.

(10) Variation of intensity and beam position under polarization reversal.

(11) Short term stability.

(12) Long term stability.

(13) Percent time available.

In Table 1 I have summarized many of these characteristics for a number of polarized electron sources. I have included one column showing the means of polarization reversal and another showing the magnetic field at the source. The latter is important because sources which employ strong magnetic fields usually suffer in characteristic (5). We can see this if we consider the generalized emittance, ε*, of a source using a magnetic field, H_o. The generalized emittance of such a source is given by[36]

$$\varepsilon^* \simeq \rho_o (E_o/E)^{1/2} + \frac{1}{2} (e/m)\, \rho_o^{\,2} H_o/v, \qquad (11)$$

where ρ_o is the radius of the effective electron production region, E_o is the mean energy of the electrons at the time of production, E is the final energy of the beam, e/m is the electron charge to mass ratio, and v is the final velocity of the electron beam. We have assumed in Eq. 11 that all quantities are in SI units. Typically E_o is ~0.1 to ~1 eV, so that the second term of Eq. 11 dominates if $\rho_o H_o > 10^{-5}$ Tesla m. Sources which employ high magnetic fields usually suffer in characteristics (9) and (10) as well, since polarization reversal generally requires magnetic field reversal, with consequent changes in electron optics. These changes may be paramount in experiments which are sensitive to small beam positional variations. Table 1 also contains entries showing the pre-

Table 1. Characteristics of Operating and Proposed Electron Sources[a]

Method	Group	Ref.	Status[b]	Mode	Pulse Length	Peak Current (e/pulse)	Rep. Rate (pps)	Avg. Current (nA)	Polarization	Polarization Reversal[c]	E(eV)	ΔE(eV)	H(G)	Emittance (mrad cm)	Brightness
Electron Scattering from Unpolarized Hg Beam	Karlsruhe-Münster	16	●	dc				$35(10^{-4})$	0.2(0.85)	θ,E	100-1700	1.0(0.6)	~0	—	High
	Münster	48	●	dc				C.1(0.01)	0.22	θ,E	80		~0	—	High
	Stanford	47,49	●	dc				10-35	0.1-0.23	θ,E	50-180	—	~0	—	High
Photoionization:															
Fano Effect Rb	Bonn	53,81	●	pulsed	11nsec	2×10^{9}	50	16	0.65	P_γ	115×10^{3}	≤2000	<500	<10	Medium
Fano Effect Cs	Yale	54	●	dc				3	0.65	P_γ	1000	~2	0.05	<20	High
Polarized Li Beam	Yale/Bielefeld	55,57	●	pulsed	1.6μsec	2.5×10^{9}	180	/2	0.95	H	65×10^{3}	1500	200	<10	Medium
	SLAC	—	●	pulsed	1.6μsec	8.15×10^{9}	180	235	0.35	H	65×10^{3}	1500	200	<10	Medium
Resonant 2-Photon Li Beam	Yale/Bielefeld/SLAC	60	O	pulsed	1.6μsec	1×10^{10}	190	288	0.85	P_γ	65×10^{3}	1500	200	<10	Medium
Resonant 2-Photon Cs Beam	FOM Amsterdam	61,62	O	dc				~100	0.3-0.6	P_γ	—	0.5	~0	—	High
Optically Pumped He Discharge	Rice	59	●	dc				$100(10^{4})$	0.5(0.3)	P_γ	500	0.5	5	5	High
Field Emission	Bielefeld	64	●	dc				20	0.9	H	2000	—	~50	—	Very High
LEED W	Rice	73	O	dc				50	0.37	θ,E	80	—	~0	—	High
Photoemission:															
NEA GaAs	ETH Zurich	68,69	●○	dc				10^{3}	0.45	P	<1	0.2	~0	2	Very High
	NBS Gaithersburg	71	●○	dc				10^{6}	0.45	P_γ		0.2	~0	2	Very High
	SLAC	72	○	pulsed	1.6μsec	1×10^{11}	180	2.3×10^{3}	<0.5	P_γ	70×10^{3}	0.2	~0	<10	Very High
EuO	ETH Zurich/SLAC	67	●	pulsed	1.2μsec	3×10^{9}	<7	3.2	0.6	H	~2	~2	21×10^{3}	1.5×10^{4}	Medium
	ETH Zurich/SLAC	67	O	pulsed	1.6μsec	3×10^{9}	180	86	0.6	H	70×10^{3}	~2	21×10^{3}	-7	Medium

[a]For sources which have two operating points corresponding characteristics are including in parentheses.

[b]Symbols used are ●: operational; ◐: prototype; and O proposed.

[c]Symbols used to indicate means of polarization reversal are θ: angle change; E: energy change; H: magnetic field reversal; P_γ: light polarization reversal.

[a]For sources which have two operating points corresponding characteristics are including in parentheses.

[b]Symbols used are ● operational, ◐ prototype, and O proposed.

[c]Symbols used to indicate means of polarization reversal are θ angle change, E energy change, H magnetic field reversal, P_γ light polarization reversal.

sent status of each source, the mode of operation, and a qualitative
description of the brightness using Kessler's classifications.[1]

Polarized electron sources fall into two general categories:
those based on atomic physics principles and those based on solid
state physics principles. Of the atomic physics techniques, those
which have been developed into operating sources are low-energy
Mott scattering,[16,46-49] photoionization of unpolarized alkali atoms
by polarized light (Fano effect),[50-54] photoionization of state-
selected alkali atoms by unpolarized light,[36,55-57] and optical
pumping of a helium gas discharge.[58,59] Resonant two-photon ionization of alkali atoms is currently receiving quite a bit of attention and appears to have promise for the future.[60-62]

Of the solid state techniques, field emission from magnetized
EuS coated tungsten,[63,64] photoemission from magnetized EuO,[65-67]
and photoemission from negative electron affinity (NEA) GaAs[68,69]
have all been implemented in prototype sources. Operating sources
based upon field emission from EuO[70] and photoemission from NEA
GaAs are in fact nearing completion at the present time.[71,72] In
addition recent studies of low-energy electron diffraction (LEED)
from a tungsten (001) crystal[73] have provided some hope for a
new type of source in the future. Let us now examine each of these
techniques a little more closely.

Electron Scattering from Unpolarized Hg Beam: Low-Energy Mott Scattering

The theory of low-energy Mott scattering was first considered
by Massey and Mohr in 1941.[11] Using a screened Coulomb potential
they were able to show that at energies as low as 100 eV the polarization of an electron beam which is scattered by a heavy nucleus
undergoes strong oscillations as a function of scattering angle.
While polarization effects at high-energies are predicted on the
basis of the interaction between the magnetic moment of the electron
and the Lorentz transformed Coulomb field of the nucleus,[34] the pronounced effects at low energies are not so easily explained. At
non-relativistic velocities one expects the interaction to be dominantly electrostatic, with the consequence that the spin of the
electron should play no role. However, at low energies, where the
DeBroglie wavelength is of the order of the extent of the atomic
potential, the field of the incident electron interferes with the
screened Coulomb field of the nucleus. At energies and angles
where this interference produces minima in the scattering cross
section, the small residual magnetic interaction becomes relatively
prominant, and polarization phenomena become pronounced. As in the
case of high-energy Mott scattering[34] the direction of polarization
is perpendicular to the scattering plane.

The observation of polarization effects in low-energy Mott scattering was first reported in 1961 at Mainz.[13] In the following years, extensive studies were carried out both at Mainz[14] and at Karlsruhe,[15] and it was recognized that low-energy electron scattering from a mercury atomic beam could provide a good source of polarized electrons. The measurements of Jost and Kessler at Karlsruhe[16] formed the basis for the later development of operating sources at Stanford[47,49] and Münster.[48] Currents produced by these sources typically range from 0.01 to 35 nA, depending upon the energy resolution desired, as shown in Table 1. The polarization obtainable at these currents is ~0.22 and can be reversed only by changing either the energy of the beam or the scattering angle. Systematic effects due to reversal are thus not easily overcome. Another low-energy Mott scattering prototype producing currents up to 20 nA with a polarization of ~0.27 was described by the Mainz group in 1969,[46] but no further development has been reported to date.

Photoionization of Unpolarized Alkali Atoms by Circularly Polarized Light: The Fano Effect

In 1951 Seaton suggested that spin-orbit coupling in continuum states could be responsible for the pronounced non-zero minimum in the photoionization cross section for heavy alkali atoms when the cross section is considered as a function of photon energy.[74] In 1969 Fano extended this reasoning to the consideration of electron polarization and suggested that photoelectrons ejected from heavy alkali atoms at certain incident photon energies might be highly polarized.[75] The argument, although complex in detail, can be understood quite simply in the following way. Since photoionization is dominantly an electric dipole process, the angular mometum of the incident photon cannot couple directly to the spin of the photo-electron. However, photoionization can be regarded as a two-step process with the first being excitation to a continuum state and the second being scattering of the unbound electron by the residual ionic core. The spin-orbit interaction which must be taken into account in the continuum state thus couples the spin of the electron to the angular momentum carried by the incident photon. Each continuum state is degenerate in total angular momentum, j, where j can take on the values 1/2 and 3/2. Pronounced polarization effects should therefore occur near photon energies where the partial photoionization cross sections, $\sigma_{1/2}$ and $\sigma_{3/2}$, pass through Cooper minima.

Experimental verification of these polarization effects was provided by studies of photoionization of K, Rb and Cs by circularly polarized light carried out at Yale,[76-78] and by independent studies of photoionization of Cs carried out at Karlsruhe.[50,79,80]

Fig. 1 shows the wavelength dependence of the photoelectron polarization for Rb and Cs for 100% circular polarization of the incident
photons, based upon the Yale work. If the electrons are extracted
along the direction of incidence of the photons, they will be
longitudinally polarized. Under these circumstances, the electron
polarization is considered positive when the electron and the
photon have the same helicity. As can be seen from Fig. 1, the
electron polarization reaches +1.0 at about 256 nm for Rb and about
297 nm for Cs. Also shown in Fig. 1 are the photoionization cross
sections for Rb and Cs as a function of photon wavelength.[81] It
is clear that wavelengths for both Rb and Cs can be chosen where
both the electron polarization and photoionization cross section
are high.

The Fano effect technique has been optimized in sources of
polarized electrons at Yale and at Bonn. The Yale source,[54] as
indicated in Table 1, is a dc source which uses a Cs atomic beam.
With a 1000W Hg-Xe lamp to provide the uv ionizing radiation, intensities as high as 3 nA have been obtained with a polarization of
0.65. The Yale source can operate continuously for 12-18 hrs.

The Bonn source,[53] shown schematically in Fig. 2, is a pulsed
source which employs 20 Rb atomic beams in a recirculating oven

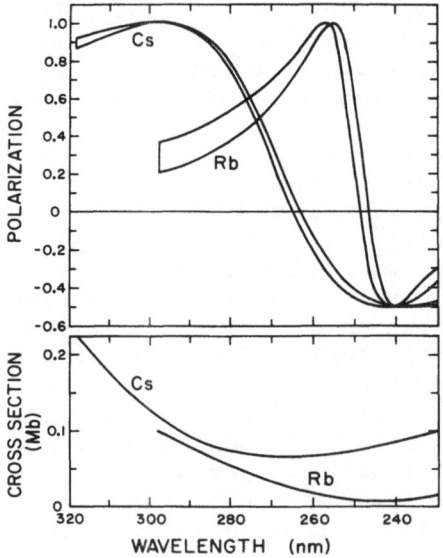

FIG. 1. Electron polarization based upon the Fano effect in Rb
and Cs, assuming 100% circular polarization of incident photons,[78]
shown together with photoionization cross sections for Rb and Cs,[81]
as a function of photon wavelength. The bands in the upper Figure
represent a 1 standard deviation uncertainty.

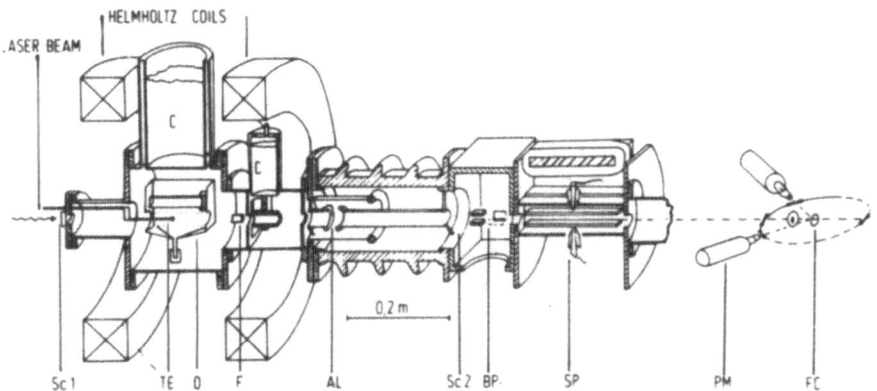

FIG. 2. Longitudinal section of Bonn polarized electron source[53] based upon the Fano effect in Rb. The symbols used are TE test electron source, O oven system, F Faraday cup, AL accelerating lens, BP beam-steering plates, Sc1 and Sc2 scintillators for laser adjustment, and C cold traps.

configuration.[52] Using a 12 g charge of Rb metal, the Bonn source can operate for about 24 hours, with 5-10 minutes required each 90 minutes to recirculate the condensed Rb. The ionizing radiation at 266 nm is provided by a quadrupled Nd-YAG laser which is Q-switched to give typically 6-8 mJ per 11 nsec pulse at 50 Hz. Under these conditions the source delivers 2×10^9 electrons/pulse with a polarization of 0.65,[82] as indicated in Table 1. A source with the same parameters is currently being built for use at the Bonn synchrotron.

A major advantage of a Fano-type source is that the electron polarization is reversed by a reversal of the helicity of the incident light. Thus the electron optical characteristics remain unchanged during polarization reversal, minimizing systematic effects associated with reversal. In addition, in pulsed sources, the polarization can be reversed on a pulse to pulse basis rather easily.

Photoionization of a Polarized Li Atomic Beam: PEGGY

In contrast to a Fano-type source, which uses an unpolarized atomic beam and polarized light, the next source we will consider uses a polarized atomic beam and unpolarized light. This technique, which was successfully applied for the first time at Yale in 1965,[83] has probably undergone more extensive development than any other

method of producing polarized electrons. Following the development
of a prototype based upon a lithium atomic beam,[36] an operational
source was designed and constructed at Yale for use at the Stanford
Linear Accelerator Center (SLAC). The source, called PEGGY (Polarized Electron Gun), was installed at SLAC during the summer of 1974
and became operational at low current in November of 1974. Since then
the current has been substantially increased, and within the last
year two high-energy particle physics experiments have been performed
using the PEGGY beam. From the time of its inception, development of
PEGGY has involved a collaboration of physicists from the University
of Bielefeld, SLAC, and Yale.

The principle of operation of PEGGY, shown in Fig. 3 has been
described in detail previously.[36] A beam of lithium atoms which
has undergone high-field state selection in a focusing six-pole
magnet is photoionized by uv light. Since the photoionization
cross section of lithium does not have a Cooper minimum, the spin-
orbit coupling in the continuum states does not act as a depolar-

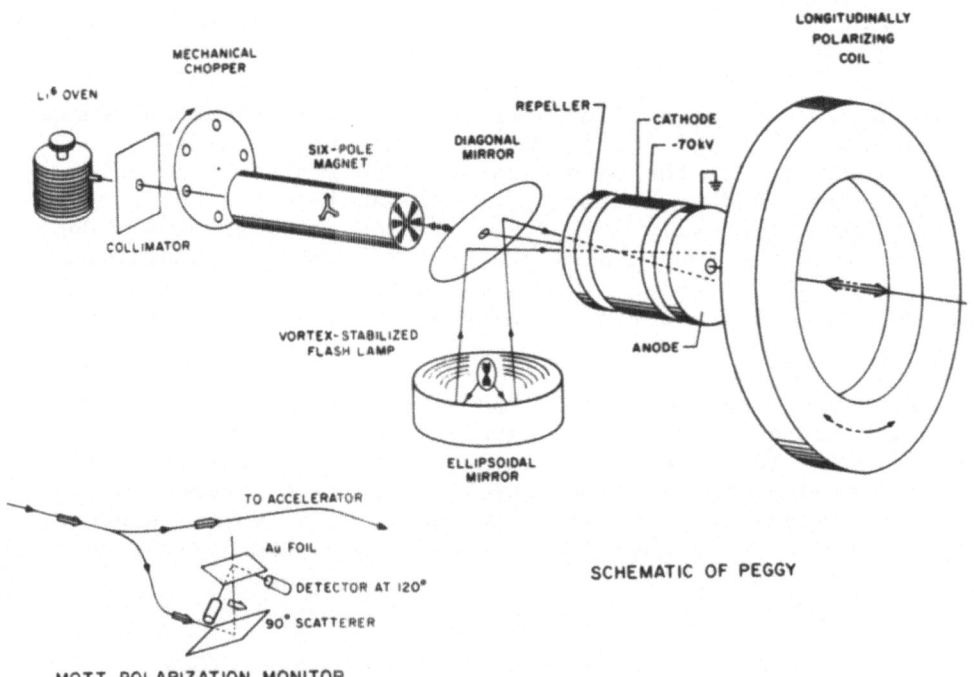

FIG. 3. Schematic diagram of the polarized electron source (PEGGY)
at the Stanford Linear Accelerator Center.[55] Inset at the lower left
shows the detail of the Mott scattering polarization monitor which
uses two scatterings, the first, a 90° scattering in transmission, to
rotate the spins from longitudinal to transverse, and the second,
120° scattering to produce the usual detected asymmetry.

izing mechanism.[78] The hyperfine interaction in the ground state, on the other hand, can indeed result in a significant depolarization. Therefore a magnetic field, chosen longitudinal for electron optical reasons, is needed in the ionization region in order to decouple the nuclear and electronic spins. In the case of the isotope Li^6-- actually an isotopic mixture of 95.6% Li^6 and 4.4% of Li^7 is used in PEGGY- -which has a relatively small magnetic moment and hence a relatively weak hyperfine interaction, a field of 200 G in the ionization region is sufficient to maintain an electronic polarization equal to 95% of the high-field state-selected value. Since the emittance of a source is linearly proportional to the magnetic field, as shown by Eq. 11, an ionizer field larger than 200 G would make the emittance of PEGGY larger than the ~10 mrad cm acceptance of the accelerator.

As Fig. 3 indicates, the uv ionizing light is provided by a vortex stabilized argon flash lamp [84] and is focused onto the atomic beam by an ellipsoidal mirror and a 45° "diagonal" mirror, both of which are aluminized and overcoated with MgF_2 for maximum reflectance in the far uv. Since the photoionization threshold wavelength for Li is 230 nm, Suprasial quartz is used for the vacuum window and the lamp envelope. The ionization region is maintained at a potential of -70 kV, and the photoelectrons are extracted to ground potential. The anode is shaped as a spherical mirror and serves to reflect the light back into the ionization region. The flash lamp, designed by E. Garwin of SLAC, provides approximately 3J of light in a 1.6 μsec pulse and operates at a rate of 180 pulses/sec. [85]

The atomic beam is modulated by a mechanical chopper wheel, in order to limit the lithium accumulation in the six-pole magnet. With a 240 g load of Li, PEGGY is capable of running for about 70 hours at an intensity of 2.5×10^9 electrons/pulse at 180 pulses/sec with a polarization of 0.85 ± 0.07, as shown in Table 1. The time necessary to reload the oven and reestablish the beam at full intensity is 24-30 hours. With the flashlamp lifetime varying from 18 to 36 hours and about one hour needed to replace the lamp, PEGGY's percent time available has been averaging about 67% during the 4 months of operation since January 1976.

A scale drawing of PEGGY is given in Fig. 4, showing the relationship of PEGGY to the SLAC injector. The fast valves (FV) and slow valves (SV) were incorporated to protect the accelerator from any catastrophic vacuum failure in PEGGY which might, under very improbable circumstances, result in Li contamination of the accelerator. As both Fig. 3 and Fig. 4 indicate, PEGGY is equipped with a Mott scattering polarization detector for on-line polarization monitoring. The Mott data acquisition electronics has recently been connected to the "counting house" (experimental control area) computer via a 2-mile data link.

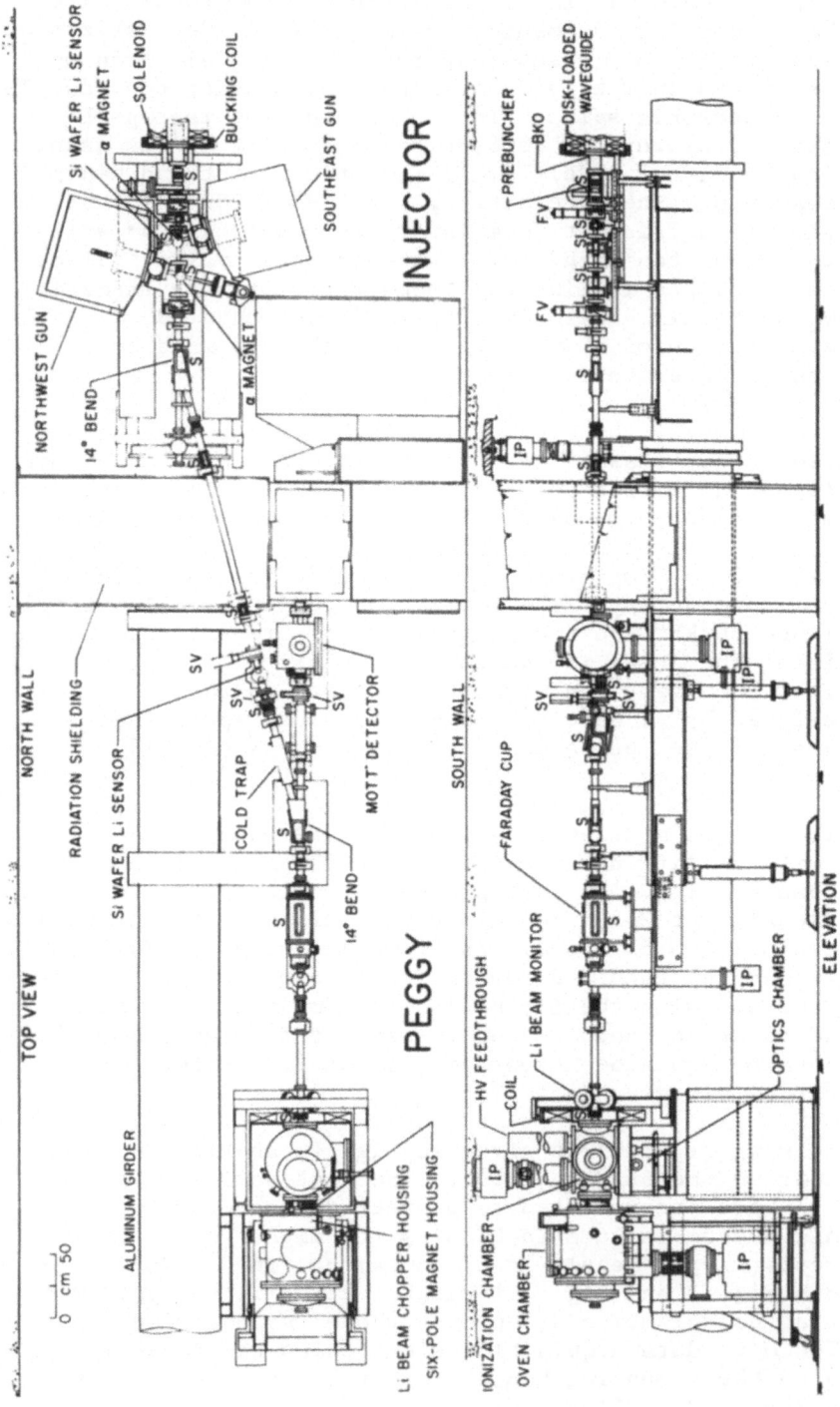

FIG. 4. Scale drawing of installation of polarized electron source (PEGGY) at the Stanford Linear Accelerator Center, showing the relative locations of PEGGY and the Injector for the accelerator. Symbols used are S steering, L magnetic lens, T toroid current monitor, SV slow valve, FV fast valve, IP ion pump, and BKO beam knock out (not used for PEGGY).

As Table 1 indicates, PEGGY provides the highest current and polarization of any pulsed source which has yet been developed. Close scrutiny of Table 1, however, will reveal that PEGGY has two modes of operation--the first providing a current I_e = 2.5 x 10^9 e/pulse and polarization P_e = 0.85 as already stated, and the second giving I_e = 8.15 x 10^9 e/pulse and P_e = 0.35. This second mode results from resonant excitation of the 2P state of Li, which occurs at 670.8 nm, as shown in Fig. 5. Photoionization can then occur directly from the 2P state with a cross section, σ_{2P}, that decreases monotonically from a measured value of ~20 Mb at threshold (349.9 nm) to ~5 Mb at 170 nm.[86] (Theoretical values of σ_{2P} [87-90] are slightly lower.) Since the photoionization cross section from the ground state, σ_{2S}, is approximately flat at ~2 Mb from threshold (230 nm) to 150 nm,[80] excited state photoionization is competitive with ground state photoionization before the 2S-2P transition is saturated, despite the 27 nsec lifetime, τ_{2P}, of the excited state.

Evidence of a resonant two photon ionization process is shown in Fig. 6. A quadratic component in the dependence of the electron intensity, I_e, on the light intensity, I_γ, is obvious. The solid line in Fig. 6 is the best fit of the data to the expression

$$I_e/I_\gamma = C_1 + C_2 I_\gamma \qquad (12)$$

with C_1 = 4.00 ± 0.13 and C_2 = 2.14 ± 0.28. From line strengths and Clebsch-Gordan algebra it can be shown that each time an atom undergoes an excitation to a bound P-state, the electronic polarization is degraded by a factor of 5/9. Here the hyperfine structure can be neglected because of the presence of the 200 G decoupling

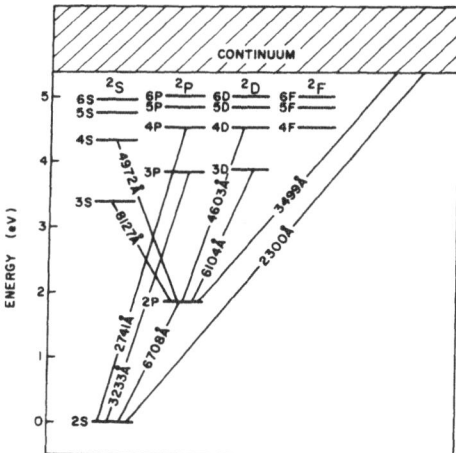

FIG. 5. Gotrian diagram of lithium showing fine-structure levels and some relevant excitation wavelengths.

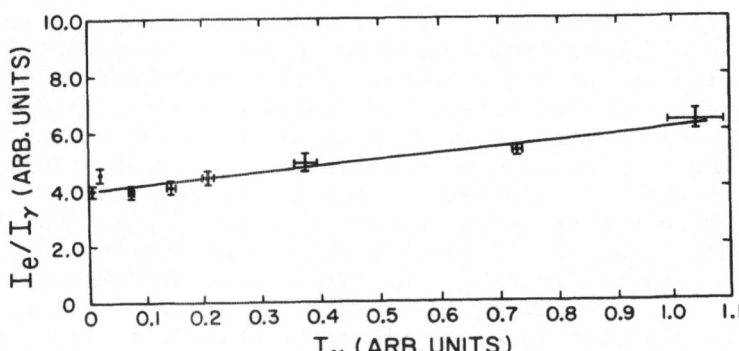

FIG. 6. PEGGY electron intensity, I_e, per unit light intensity, I_γ, as a function of I_γ, in the absence of 670.8 nm absorption filter for removing 2S-2P resonance radiation. The quadratic component in the dependence of I_e on I_γ is evident. The solid line is the best fit of the data to Eq. 12 with $C_1 = 4.00 \pm 0.13$ and $C_2 = 2.14 \pm 0.28$.

magnetic field. This depolarization effect is shown in Fig. 7, where the ratio of the measured electron polarization, P_e,[91] to the theoretically expected value, P_e^0, in the absence of 2S-2P resonance radiation is given for several different values of I_γ.

During its transit through the ionization region, an atom may be excited to the 2P state more than once before it is photoionized either from the 2P state directly or from the 2S state after decay. We can nonetheless obtain a relatively simple expression for P_e/P_e^0 in terms of the mean number of excitation-deexcitation cycles through the 2P state, α, experienced by an atom during a light pulse, if we make the following assumptions:

(1)   There is negligible depletion of the sample of atoms due to ionization.

(2)   The average time spent by an atom in the 2P state is negligible compared to the duration of the light pulse.

(3)   Only the 2P state contributes to the depolarization process.

(4)   The atoms in the ionization region are illuminated uniformly.

(5)   The relative lamp spectrum does not change with lamp intensity.

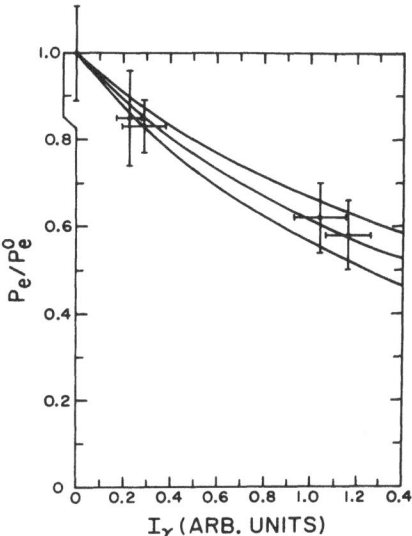

FIG. 7. PEGGY polarization as a function of light intensity. The data point at I_γ = 0 was obtained with the use of a broad band uv interence filter which absorbed the 670.8 nm 2S-2P resonance radiation. The other four data points were obtained with the filter removed. The curves are for a one parameter theoretical fit. The central curve is the mean, and the upper and lower curves represent one standard deviation errors.

These five assumptions are approximately valid for the conditions under which the measurements shown in Fig. 7 were made.[92] For a square shaped light pulse we find that the polarization of electrons produced by ground state ionization is given by[91]

$$P_{2S}(\alpha) = P_e^o \frac{1-e^{-4\alpha/9}}{4\alpha/9} .$$ (13)

Similarly we find that the polarization of electrons produced by excited state ionization is given by

$$P_{2P}(\alpha) = \frac{5P_e^o}{9} \frac{1-e^{-4\alpha/9}}{\Delta\alpha/9} .$$ (14)

With β defined by

$$\beta \equiv C_2 I_\gamma / C_1$$ (15)

we then have for the polarization of all photoelectrons

$$P_e = \frac{P_{2S} + \beta P_{2P}}{1 + \beta} = P_e{}^o \left(\frac{1+5\beta/9}{1+\beta}\right) \left(\frac{1-e^{-4\alpha/9}}{4\alpha/9}\right). \tag{16}$$

Since α is proportional to I_γ we can rewrite Eq. 16 as

$$P_e/P_e{}^o = \frac{1 + \frac{5}{9}b\, I_\gamma}{1+bI_\gamma} \frac{1-e^{-aI_\gamma}}{aI_\gamma}, \tag{17}$$

where

$$a = 4\alpha/9\ I_\gamma \tag{18}$$

and

$$b = \beta\ I_\gamma = C_2/C_1. \tag{19}$$

The value of <u>b</u> is determined by the best fit of the I_e vs I_γ data to Eq. 12. The solid lines in Fig. 7 then constitute the best one-parameter fit of the P_e vs I_γ to Eq. 17: the middle line is the mean value, and the upper and lower lines give the one standard deviation error. The value of the single parameter <u>a</u> so determined is 0.68 ± 0.19. From Eq. 18 we see that at $I_\gamma \sim 1$, ~ 1.5, thereby justifying assumption (2).

When a broad band uv interference filter was inserted to re-move the 670.8 nm resonance radiation, a value of $P_e/P_e{}^o = 1.00 \pm 0.11$ was measured. This value, shown as the data point at $I_\gamma = 0$ in Fig. 7, confirms the hypothesis that the 2S-2P resonance is respon-sible for the depolarization effect in PEGGY. With the filter re-moved, a peak intensity of 8.15×10^9 electrons/pulse was obtained at a polarization estimated (on the basis of Eq. 17) to be 0.35 ± 0.05.

Fig. 8 indicates the progress which has been made on improving the operating characteristics of PEGGY since the first successful test in March of 1974. The polarization, P_e, the peak current, I_e, and the figure of merit, $P_e\sqrt{I_e}$, are all shown. Also given are the 1969 operating conditions of the prototype, the 1970-71 design goals for PEGGY, and the new 1977 targets. In order to reach the target intensity of 1×10^{10} electrons/pulse with 0.85 polarization, PEGGY will be operated in the second mode; that is, with the reson-ant 2S-2P excitation present. However, the resonant radiation will be circularly polarized, thereby eliminating the depolarization effects. A 900 mW cw tunable dye laser at 670.8 nm containing a mixture of Rhodamine 101 and Rhodamine 6 G in ethylene glycol will

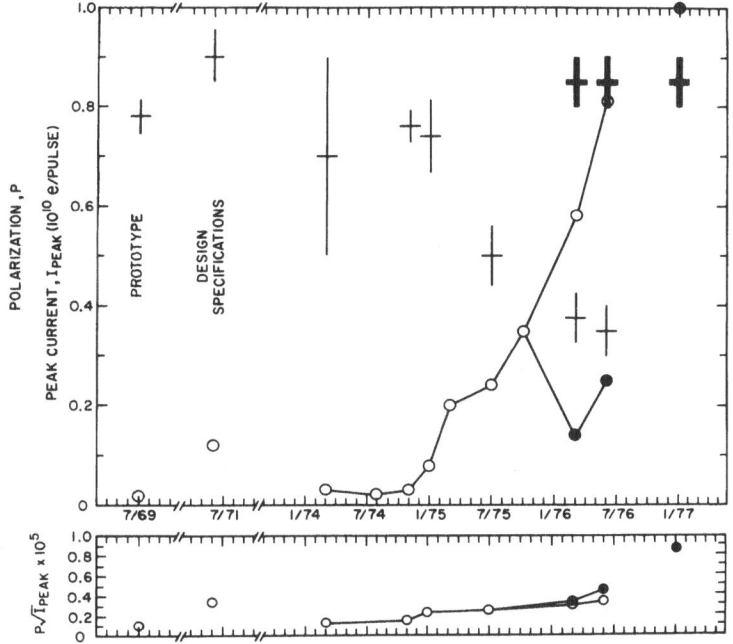

FIG. 8. PEGGY's development as a function of time and prospects
for the near future. The open circles and thin crosses in the
upper plot are respectively peak current and polarization measurements with resonant depolarization effects present. The filled
dots and heavy crosses are measurements with resonant depolarization
effects removed. The 1/77 prediction depends upon laser pumping of
the 2P state with circularly polarized light.

be used to pump the 2P state, but the existing flash lamp will
still provide the ionizing light. Since the laser will be operated
cw, pumping can take place for about 20 μsec before the ionizing
light pulse. Thus, in principle it will be possible to reverse the
electron polarization by reversal of the helicity of the laser
light, thereby eliminating the need for magnetic field reversal
which exists in the present version of PEGGY. The expected characteristics of the new version of PEGGY are listed in Table 1 in
the row labeled "Resonant 2-Photon Li Beam."

Resonant 2-Photon Ionization of Cs Atomic Beam

In 1973 the theory of multiphoton ionization of atoms was
expanded to include the effect of the spin-orbit interaction.
Lambropoulos showed that if the incident light is circularly

polarized, the photoelectrons will be highly polarized depending
upon the wavelength of the incident light.[93] The characteristics
of the wavelength dependence of the polarization are in fact somewhat similar to those of one-photon ionization, the Fano
effect,[75-78] shown for Cs and Rb in Fig. 1.

Following publication of Lambropoulos' theory, several experiments were set up to measure the electron polarization in multiphoton ionization of alkali atoms.[41,43,47,94] Off resonance, high
polarizations were observed, but the electron current was very low
since tunable high power lasers were not used. Consequently, most
of the experimental effort was confined to cases involving at least
one resonance. It was found that near resonance the electron spin
polarization was strongly dependent on laser power, suggesting
that saturation effects were involved. As the laser power was
increased and the electron current rose, the polarization dropped
dramatically. The outcome of these first studies indicated that
there were interesting effects to be examined, but the development
of a source based upon Lambropoulos' original theory was indeed
premature.

Very recently a group at the FOM in Amsterdam has found that
in resonant two photon ionization of Cs via the $6 P_{3/2}$ intermediate
state, the hyperfine interaction plays an important role whenever
the radiation for the second or ionizing step is of sufficiently
low intensity that the process can be considered a two-step
event.[61,62] (Since this condition was not met in much of the
earlier work,[44] hyperfine effects cannot account for all of the
previous discrepancies between experiment and theory.) However,
even for those cases where the hyperfine interaction plays a role,
the Amsterdam group has found that an efficient dc source of polarized
electrons can still be designed with currents of ~100 nA and polarizations of 0.3 to 0.6, as indicated in Table 1. The Amsterdam
group is currently pursuing this program.

Optically Pumped Helium Discharge

Another approach to the production of polarized electrons by
atomic physics methods is the use of an optically pumped helium
discharge. A group at Rice University has been investigating this
technique for about 10 years. In the original work, first reported
in 1969,[95] a cell containing helium gas was subjected to an electrical discharge to produce $2 {}^3S_1$ metastable atoms. Optical pumping of
either the $m_J = +1$ or $m_J = -1$ sublevel of the metastable state was
achieved using the $2 {}^3S_1 - 2 {}^3P_{0,1,2}$ transitions which were excited
by circularly polarized light from an intense helium discharge lamp.
Electrons produced during ionization in the discharge volume were
extracted by conventional electron optical means.

The Rice group found that the dominant ionization process involved collisions between two metastable atoms:

$$He^M + He^M \rightarrow He(1^1S_0) + He^+ + e^-. \qquad (20)$$

Here M denotes either the 2^3S_1 or 2^1S_0 metastable atom, the ratio of triplets to singlet being approximately 2-3 to 1 roughly in accordance with the statistical weights. Investigations of the metastable-metastable process showed that ionization persisted in the afterglow of a pulsed discharge.[58] The original Rice work resulted in a 4 μA dc electron beam with 0.08 polarization, although polarizations as high as 0.17 were observed in a pulsed mode using the discharge afterglow.

More recently,[59] the Rice group has employed a flowing afterglow with a constant microwave discharge in order to optimize the electron polarization. The microwave discharge limits the singlet to triplet ratio to 0.06 at the optimum flow-tube pressure of 0.1 Torr. As before, optical pumping of the $m_J = \pm 1$ 2^3S_1 sublevels is accomplished with circularly polarized 1.08 μ resonance radiation from helium discharge lamps. However, now two lamps instead of one are used to extend the length of the pumping region in the flowing gas, as shown in Fig. 9. With the location of the lamps

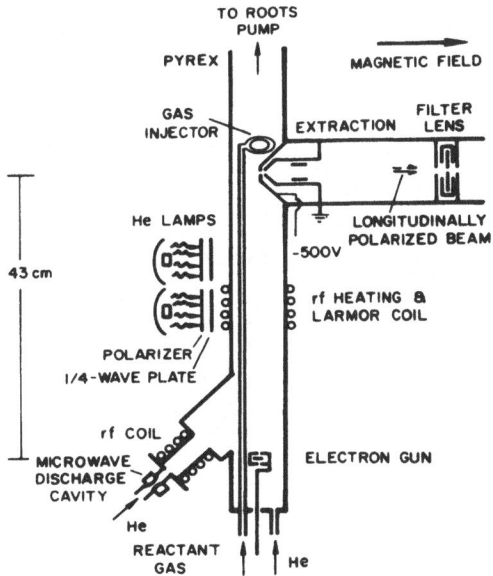

FIG. 9. Schematic diagram of the Rice optically pumped helium discharge polarized electron source.[59] With the discharge lamps oriented as shown, longitudinally polarized electrons are extracted. If the lamps are rotated by 90° to pump along an axis perpendicular to the plane of the figure, transversely polarized are extracted.

as shown, longitudinally polarized-electrons are produced. If the
lamps are rotated 90° to pump along an axis perpendicular to the
plane of the figure, transversely polarized electrons are produced.

Instead of relying on He-He metastable collisions in the after-
glow to produce electrons, the Rice group has employed chemi-
ionization of CO_2 or N_2 to enhance the ionization process. The
Rice group has further shown that spin conservation is obeyed during chemi-ionization of a variety of gases including Ar, H_2, N_2,
CO, CO_2, and N_2O, as well as during surface-electron ejection by
2^3S_1 helium.[96] Having redesigned their apparatus (shown schematically in Fig. 9) as a polarized electron source for use in other
experiments, the Rice group measured the following P_e-I_e characteristics:[97] P_e - 0.5 at I_e = 0.1 μA, P_e = 0.4 at I_e = 1 μA, and
P_e = 0.3 at 10 μA. Fig. 10 shows a number of values of P_e and I_e
obtained under a variety of operating conditions. The multiple
data points, taken under apparently identical conditions, reflect
the variations in the performance of the source. These variations
appear to constitute the only principle deficiency of the Rice
source. The remaining characteristics of the source are given in
Table 1.

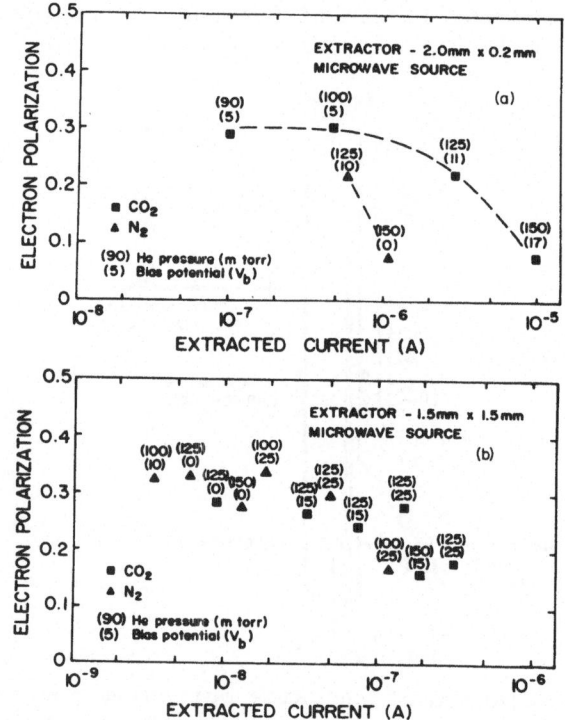

FIG. 10. Measured electron polarizations vs extracted currents for
the Rice polarized electron source.[59] Numbers in parentheses are
flow tube pressure in m Torr (upper) and bias voltage (lower).

Field Emission from Ferromagnetic Materials

We now turn to solid state techniques. The most obvious way
to produce polarized electrons using solids is to place a clean
ferromagnetic surface in a magnetic field. Band theory suggests
that for certain ferromagnets, electrons extracted from the region
near the Fermi edge should be polarized.[98] All that is needed is
some means by which the electrons can be extracted. To this end,
field emission and photoemission have been studied since the
early 1960's.

Initially, attempts to produce polarized electrons by either
field emission or photoemission were unproductive.[99,100] Finally
in 1967, a group at München reported the first successful observation of polarization of field emitted electrons.[101] Following the
suggestion of Müller, Siegmann, and Obermair,[102] the München group
used magnetized polycrystalline Gd. Unfortunately, the polarization which they measured was only about 0.08. Further studies with
Gd [103] and with Ni and Fe[104] as well, resulted in polarizations no
higher than ~0.15.

Then in 1972, Müller, Eckstein, Heiland and Zinn using EuS
coated tungsten tips cooled to below 21 K in a 20 kG field reported
observation of a polarization as high as 0.89 ± 0.07.[63] At temperatures below 16.5 K, EuS is a ferromagnetic insulator with the spin-
ordered $4f^7$ levels lying in the gap between the conduction and
valence bands. Depending upon the type of emission pattern that is
generated, high polarizations of the field emitted electrons can be
expected.

Recently a group at Bielefeld extended the field emission work
on magnetically oriented EuS-coated tungsten tip.[64] A schematic
diagram of the Bielefeld source together with polarization diagnostics is shown in Fig. 11. Oriented tungsten tips with (111) or
(110) directions parallel to the magnetic field axis were coated

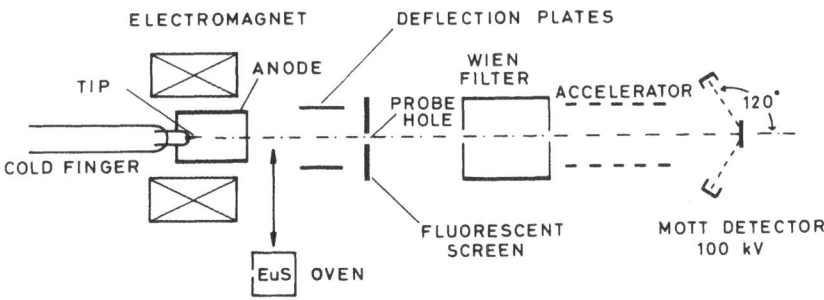

FIG. 11. Schematic diagram of Bielefeld source of polarized electrons based upon field emission from EuS coated tungsten tip.[64]

with EuS in situ by room temperature vacuum deposition. Tip radii
used ranged from 50 to 100 nm. After a 300–600°C anneal to prepare
the surface, tips were cooled by a helium flow system to temperatures down to 9 K.

Although the spin-ordered $4f^7$ levels produce a splitting of
the EuS conduction band which lowers the W–EuS barrier for one
electron spin-state and not the other, the expected unity polarization was not observed by the Bielefeld group. Instead they
measured a temperature dependent polarization which was consistent
with the spin ordering of the $4f^7$ levels. This suggests that spin-
exchange collisions between the conduction electrons and $4f^7$ electrons take place. The temperature dependence of the electron
current and polarization of the field emitted electrons is shown
in Fig. 12.

The Bielefeld source produced a maximum current of 20 nA dc
and a polarization of >0.9 at a tip temperature of 10.5 K. For all
measurements, the requisite surface cleanliness was maintained by
an ultra-high vacuum of 10^{-10} Torr at the tip. The direction of

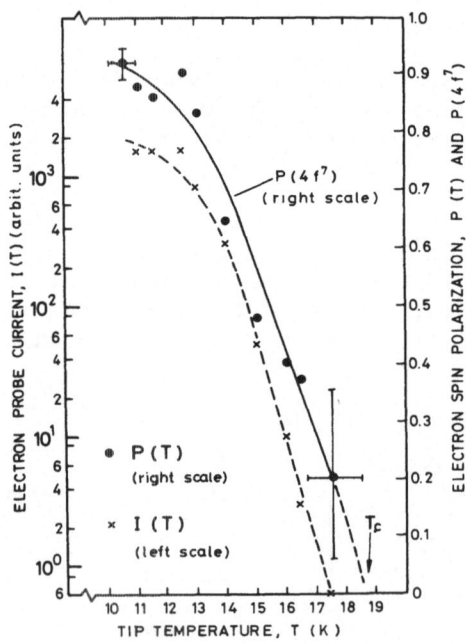

FIG. 12. Electron current, I, and polarization, P, as functions of
tip temperature, T, for the Bielefeld field emission polarized
electron source.[64]

the polarization with respect to the beam was found to vary from nearly transverse at low applied magnetic field to nearly longitudinal at high fields. In the longitudinal case, $\vec{P}_e$ was observed to be antiparallel to the external field, indicating that the magnetic moments of the emitted electrons are parallel to the direction of magnetization. The operating characteristics of the Bielefeld source, shown in Table 1, should still be considered preliminary since further studies are in progress.

Photoemission from Ferromagnetic Materials

The use of photoemission to extract polarized electrons from ferromagnetic materials has been used productively since 1969 by a group at Zurich[65] to study spin polarization in electronic energy bands. A wide variety of materials have been examined, including Gd, Ni, Fe, Co, EuO, EuS, EuSe, EuTe, GdP, and several ferrites.[65,66, 105-109] In each case, the sample studied was either vacuum deposited in situ on a helium cooled substrate or, in the case of monocrystals, cleaved in situ in an ultra high vacuum system operating at between 2×10^{-10} and 4×10^{-10} Torr. A superconducting solenoid capable of producing a 50 kG field was used to apply a strong magnetic field perpendicular to the surface, and ultra violet light was used to produce photoemission. The polarizations, measured as usual by Mott scattering, ranged from as low as 0.527 ± 0.07 in Gd[65] to as high as 0.8 in La-doped EuO.[108] A possible explanation for the high polarization value for EuO was given by Eastman in 1973.[110]

In 1974, studies were conducted jointly by the Zurich group and E. Garwin of SLAC to evaluate La-doped EuO as a possible pulsed source of polarized electrons.[67] A sealed flashlamp, originally developed for PEGGY,[55] containing 5 atm Xe was used at a repetition rate of 6.7 pulses/sec to provide $\geq$ 10 MW of peak power in a 1.2 μsec pulse. With the EuO surface maintained at 10 K, no deleterious heating effects on the electron polarization were observed. The pulsed EuO prototype produced an intensity of 3×10^9 electron/pulse and a polarization of ~0.6, as indicated in Table 1. The lower value of the polarization, compared to earlier studies, was attributed to a relatively poor vacuum of 2×10^{-9} Torr. Thus the need for extremely clean surface conditions is made abundantly clear.

Based upon the prototype results, the Zurich/SLAC collaboration concluded that a high-intensity, highly polarized, pulsed source of electrons based upon La-dope EuO was definitely feasible. The characteristics of the proposed source are shown in Table 1. A value of 0.6 for the polarization is listed there although it should be possible to reach 0.8 if the vacuum in the source is maintained at ~10^{-10} Torr.

Photoemission from Negative Electron Affinity GaAs

Both field emission and photoemission from ferromagnetic materials suffer in source characteristics (9) and (10), because they employ a relatively strong magnetic field which must be reversed in order to reverse the electron polarization. As I have already pointed out, beam stability under polarization reversal is a critical requirement of some experiments. The determination of the physically interesting quantity A, as given by Eq. 8, in some cases relies not only on the assumption that P_e is unchanged in magnitude under reversal and that I_+ and I_- can be measured accurately, but also on the assumption that the intersection region of the polarized beam and target remains unchanged under reversal. Under these circumstances the quality of the data may be severely degraded if the electron optics is altered whenever the polarization is reversed.

In the case of photoemission from GaAs, the Zurich group has shown that the magnetizing field can be removed altogether if the incident light is circularly polarized.[68,69] From band structure calculations and from experiment it is known that the wave functions at the valence band maximum and conduction band minimum have P and S symmetries respectively. The spin orbit interaction splits the valence band into $P_{1/2}$ and $P_{3/2}$ levels (just as it does in alkali atoms) with the $P_{1/2}$ level lying 0.34 eV lower than the $P_{3/2}$ level, as shown in Fig. 13. Also shown in Fig. 13 are the degenerate magnetic sublevels of the $P_{1/2}$ and $P_{3/2}$ valence bands and the $S_{1/2}$ conduction band, together with the relative transition probabilities for the transitions shown. Since GaAs is a direct band gap crystal, the mininum value of E_g occurs at Γ; that is, where the wave vector k=0. Thus if the frequency, ν, of the incident radiation satisfies the condition $h\nu \geq E_g$, only the transitions characteristic of the angular momentum states at Γ are possible.

The production mechanism for polarized electrons from GaAs should be clear by this time. If the incident radiation is σ^+ light with $h\nu \sim E_g$ the only transitions possible are

$$P_{3/2}, \ M_J = -3/2 \to S_{1/2}, \ M_J = -1/2 \qquad (21)$$

and

$$P_{3/2}, \ M_J = -1/2 \to S_{1/2}, \ M_J = +1/2. \qquad (22)$$

The relative probabilities for the transitions given by Eqs. 21 and 22 are 3 and 1 respectively. Hence the expected polarization of electrons produced by photoemission with σ^+ light is -0.5, provided emission takes place from the conduction band minimum. A coating of Cs_2O on the GaAs lowers the work function to the point

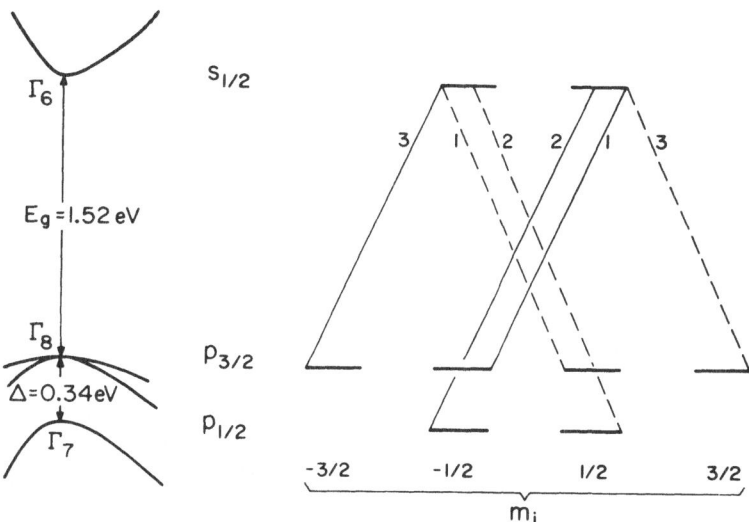

FIG. 13. On the left, a diagram showing the energy of the valence and conduction bands of GaAs as a function of wave vector, k, near Γ; that is, near k=0.[69] The energy gap, E_g, and the spin-orbit splitting, Δ, of the valence bands are shown. On the right, is a diagram of the degenerate magnetic sublevels of the $P_{3/2}$ and $P_{1/2}$ valence bands and of the $S_{1/2}$ conduction band. The numbers next to the transition lines are the relative transition probabilities which are the same as in the case of an alkali atom.

where a negative electron affinity (NEA) is achieved and emission from the conduction band minimum can in fact take place.

The apparatus used by the Zurich group in their studies of NEA GaAs shown schematically in Fig. 14, is similar to that used in the previous Zurich work with ferromagnets. The (p-type 1.3×10^{19} cm^{-3} Zn) GaAs crystals were cleaved along a (110) plane in a vacuum of 4×10^{-10} Torr. The cleaved sample was transferred through a vacuum interlock to a preparation chamber at 6×10^{-9} Torr for cesiation and oxidation. Photoemission measurements were made in the first chamber at 4×10^{-10} Torr. Using light from a Hg-Xe lamp, a conventional Zeiss M4QIII monochromator, and a circular polarizer, the Zurich group obtained polarizations and electron yields shown in Figs. 15 and 16. Note that for a slightly positive electron affinity (PEA), a polarization of 0.54 was achieved at threshold, while for the NEA case a polarization of only 0.40 was measured. The discrepancy between the theoretical value of 0.50 and the measured value of 0.40 for the NEA case is attributed by

PHOTOEMISSION

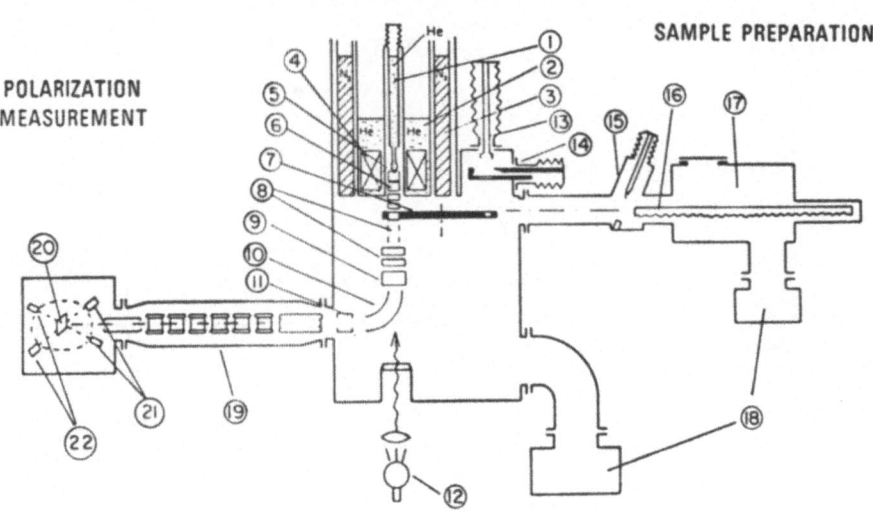

FIG. 14.   Schematic diagram of the Zurich apparatus used in polarization studies of photoemission from NEA GaAs.[69]   The numbered
items are (1) movable He cryostat with sample gripper, (2) He
cryostat, (3) liquid nitrogen, (4) superconducting coil, (5) sample
in measuring position, (6) accelerating electrodes, (7) rotatable
wheel with samples, (8) parallel beam shifters, (9) plane condenser,
(10) cylindrical condenser, (11) aperture, (12) light source, (13)
gripper for cleaving, (14) cleaving mechanism, (15) uhv valve,
(16) rack and pinion linear motion, (17) sample preparation chamber,
(18) ion-getter pumps, (19) seven-stage accelerator, (20) gold foil,
(21) detectors to measure Mott asymmetry, (22) forward detectors to
monitor beam.

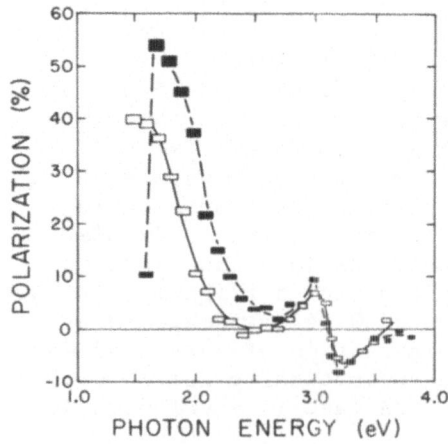

FIG. 15.   Electron polarization
as a function of photon energy for
photoemission from GaAs.[68]   The
dark and open rectangles are for
positive electron affinity (PEA)
and negative electron affinity
surfaces respectively.   The size
of the rectangles indicates the
uncertainties in the measurements.

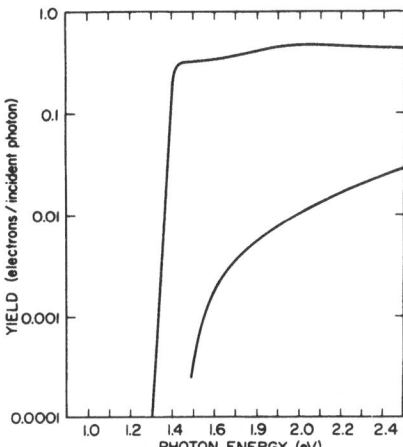

FIG. 16. Photoelectric yield of electrons per incident photon for GaAs. Higher curve is based upon the work of Bell and Spicer for an optimal photocathode.[109] Lower curve is based upon photocathode used in Zurich work.[68]

the Zurich group to recombination and spin-exchange effects. The higher value of polarization in the PEA case is attributed to restrictions in the cone angle of emission with respect to the normal of the surface and the changes in the wave functions away from the Γ point (k=0).[69]

The operating conditions of the Zurich prototype based upon photoemission from GaAs are shown in Table 1. Although the Zurich group did not attempt to optimize the source, they obtained a high current of 1 μA. They estimate their average polarization as 0.45, as shown. With a photoelectron yield as indicated by the Bell and Spicer curve[111] of Fig. 16, a current of 1 mA should be obtainable with a light power of 5 mW, assuming proper preparation of the photocathode surface. Since the threshold wavelength of 815.7 nm is within the range of tunable dye lasers, Pierce has projected the characteristics I_e=1 mA and P_e=0.45, shown in Table 1, for NEA GaAs dc source under development at the National Bureau of Standards in Gaithersburg, MD.[71] An NEA GaAs pulsed source is also under development, although this project has been undertaken at SLAC rather than NBS.[72] Using a flashlamp pumped oxazine dye laser at ~810 nm, the SLAC group projects characteristics of I_e=10^{11} electrons/pulse and P_e~0.5, with the remaining parameters as shown in Table 1.

Thus it seems fair to say at this time that photoemission from NEA GaAs represents the most promising source of polarized electrons

for the near future. Of course it must be borne in mind that the
step from prototype to operating source may have unforseen pitfalls.
We shall all eagerly await the outcome of the NBS and SLAC developments.

Low-Energy Electron Diffraction (LEED) from Tungsten (001)

Although it may come as something less than a <u>denoument</u>,
following the description of the NEA GaAs developments, I should
like to conclude this section with a few brief comments about the
prospects for producing polarized electrons by low-energy electron
diffraction (LEED) from a tungsten (001) surface. Recent experiments at Rice have shown that at incident beam energies ranging
from 45 to 190 eV and at angles of incidence ranging from 11° to
20°, polarizations of -0.35 to +0.37 were obtained as shown in
Fig. 17.[73] The intensity of the polarized beam was in excess of
50 nA dc. Although polarization reversal requires changes in
energy or angle as shown by Fig. 17, LEED from tungsten may prove
to be a valuable source of polarized electrons for experiments
where high stability of the beam under reversal is not required.
Alternatively, LEED from tungsten may find application as a simple
low-energy polarization analyzer as a replacement for Mott scattering. To this end, studies are being conducted at the National
Bureau of Standards.[112] With this hope for the future I will conclude the discussion of polarized electron sources.

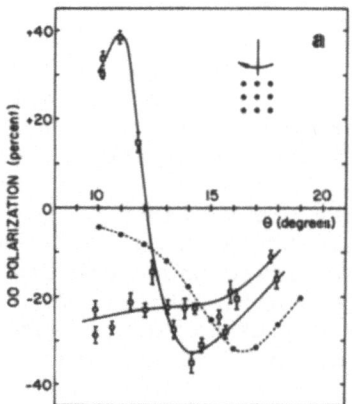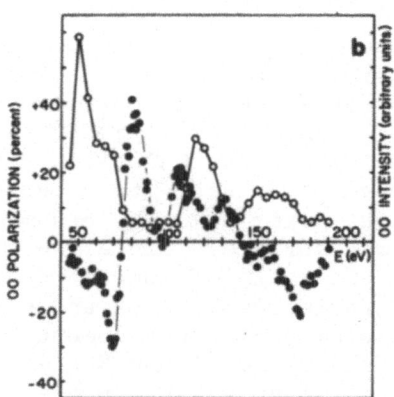

FIG. 17.(a) Electron polarization in LEED from tungsten,[73] as a
function of angle of incidence for incident energies of 69 eV
(circles) and 82 eV (squares). (b) Electron polarization (dots)
and intensity (circles) in LEED from tungsten as a function of
incident energy at an angle of incidence of 11°.

EXPERIMENTS WITH POLARIZED ELECTRON BEAMS

Having reviewed the status of polarized electron sources, I would now like to turn to a review of the status of experiments using polarized electron beams. Such experiments are, after all, the motivation for the development of sources of polarized electrons. The applications of polarized electron beams to various areas of physics has been the subject of a number of papers.[35,113-121] Nonetheless in the context of the discussion which follows it is appropriate for us to begin with a brief list of some of the possible applications, enumerated in the following table.

Table 2. Applications of Polarized Electron Beams

A. Atomic and Molecular Physics

    1. Spin exchange effects in low energy electron scattering.
    2. Spin-orbit effects in electron scattering from heavy atoms.
    3. Spin dependence in resonance scattering.
    4. Spin dependence in electron scattering by dressed atoms.
    5. Measurement of electron g-2 value.

B. Solid State Physics

    1. Spin-orbit effects in LEED to probe surface effects on electronic wave functions and barrier potentials.
    2. Spin exchange effects in electron scattering from magnetically ordered solids to probe surface magnetic order and critical phenomena.

C. Nuclear Physics

    1. Tests of parity violation in neutral weak currents using unpolarized nuclear targets.
    2. Electro-excitation of oriented nuclei.

D. High-Energy Particle Physics

    1. Quark-parton models and quark dynamics using deep inelastic scattering from polarized proton and neutron targets.
    2. Tests of scaling and sum rules using polarized targets.
    3. Tests of parity violation in neutral weak currents using deep inelastic scattering from unpolarized targets.
    4. Spin dependence in electroproduction of resonances.
    5. Photoproduction using circularly polarized bremsstrahlung.
    6. Measurement of G_E in elastic e-p scattering.

It is not my purpose here to discuss the feasibility of these
various applications, but rather to review the status of several
experiments which have been undertaken during the last few years.
As I stated in the introduction, this year is really the first
time that such a review is possible. If we exclude the Michigan
free electron g-2 experiments,[122-124] since these used very low
intensities of polarized electrons from a high-energy Mott scattering source, the first experiments to use sources of polarized
electrons were a high-energy Møller scattering (electron-electron
scattering) measurement, reported in 1975,[56] and a low energy spin-
exchange measurement in mercury, reported in 1974.[125] Thus we are
indeed considering very recent history.

Polarized Electron-Electron Scattering at GeV Energies

After PEGGY (which you may recall is the polarized electron
source based upon photoionization of state selected Li atoms) was
installed at the Stanford Linear Accelerator Center in 1974, an
experiment was designed to measure the polarization of the beam
after acceleration to high energy, in order to obtain information
about possible depolarization mechanisms in the linear accelerator.
Calculations had predicted an upper limit of 2.8% on depolarization
during acceleration,[126] but no experimental evidence existed. Spin-
dependent electron-electron scattering (Møller scattering) was
chosen for the high-energy polarization measurement,[56] because the
Møller cross section and analyzing power are large, as shown in
Fig. 18, and the process is purely quantum electrodynamic. Nonetheless it should be borne in mind that Møller scattering had never
been studied at GeV energies, although it had been used at much
lower energies to determine the helicity of electrons from β
decay[34] and muon decay.[127] Self-consistency of the results of the
experiment had to be demonstrated.

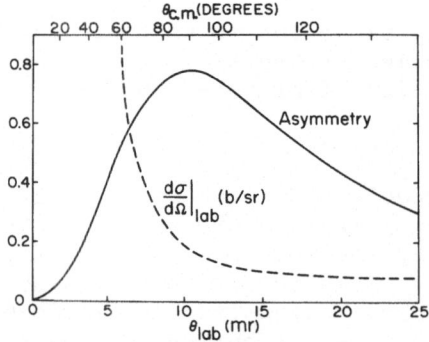

FIG. 18. The Møller asymmetry and laboratory cross section plotted
versus laboratory angle for the representative incident energy of
9.712 GeV.[56]

Since Møller scattering deals with identical particles the asymmetry and cross section shown in Fig. 18 must be calculated from the difference of two Feynman diagrams as shown in Fig. 19. [128,129] It should be noted that at the representative incident beam energy of 9.712 GeV used in Fig. 18, a center-of-mass scattering angle ($\theta_{c.m.}$) of 90°, where the asymmetry reaches a maximum of 7/9, corresponds to a laboratory angle (θ_{lab}) of only 10 mrad. Hence any Møller-scattering apparatus must be able to separate physically the scattered electrons from the primary beam.

In the 1974 measurement, the SLAC 8-GeV/c spectrometer[130] was used in an experimental configuration shown in Fig. 20. The incident beam struck a 0.025-mm-thick Supermendur target foil located 8.2 m upstream from a pivot about which the Spectrometer rotates. The foil, inclined at 20° to the beam and magnetized to saturation in a 90 G field provided an effective target polarization of 0.083 ± 0.002. As shown in Fig. 20, a C magnet, located downstream from the spectrometer pivot was used to separate the Møller scattered electrons from the primary beam. The electrons which entered the 8-GeV/c spectrometer were deflected through angles between 6° and 8°, while the primary beam was deflected by less than 2° in the fringe field. The C magnet was positioned so that particles entering the spectrometer appeared to originate from the center of the pivot at an angle θ_s from the primary beam direction. Since the spectrometer normally views a target placed at the center of the pivot, the spectrometer optics could be left unchanged from those used in a conventional high-energy experiment.

In order to reach the experimental area where the Møller measurement was made, the high-energy polarized beam had to undergo at 24.5° magnetic bend in the beam switchyard. Whenever a longitudinally polarized electron beam passes through a transverse magnetic field its spin precesses with respect to its momentum by an amount θ_a given by [131]

$$\theta_a = \gamma a \theta_c, \qquad\qquad (23)$$

where γ is the ratio of the electron energy to the electron mass, $a = (g-2)/2$ is the electron g-factor anomaly, and θ_c is the angle

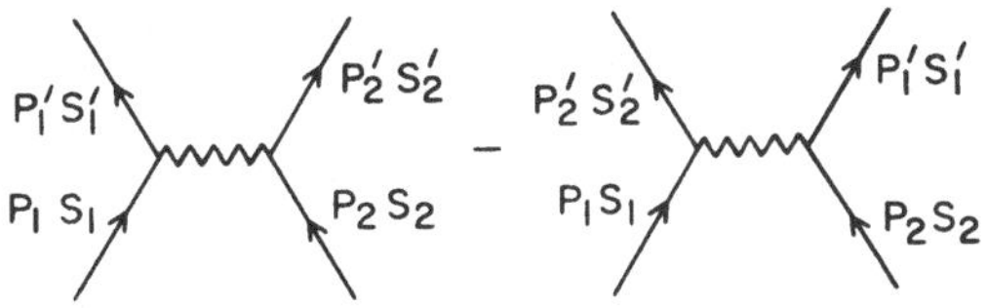

FIG. 19. Lowest order Feynman diagrams for Møller scattering. Momentum and spin are denoted by p and s respectively.

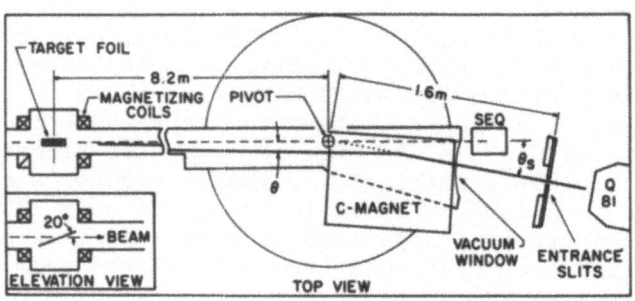

FIG. 20. Schematic outline of the experimental arrangement. The
heavy line shows the typical trajectory of a scattered electron.
Note that the trajectory after bending in the C magnet can be
extrapolated (dotted line) through the spectrometer pivot point.
The beam-line vacuum extends through the C magnet. Q81 is the
first quadrupole in the 8-GeV/c spectrometer; SEQ is a secondary-
emission quantameter used to monitor the beam.[56]

through which the particle is deflected as found from the cyclotron
frequency. Thus, if θ_a was to be restricted to multiples of π in
order to maintain longitudinal polarization, the useful beam energies had to be restricted to multiples of E_o = 3.237 GeV.

Raw asymmetries, Δ, measuring about 0.03 were converted to
beam polarizations, P_e, by use of the relationship

$$\Delta = (1-f) A_m P_T P_e, \tag{24}$$

where A_m is the Møller asymmetry for fully polarized beam and
target, P_T is the longitudinal component of the target polarization
(P_T = 0.083 cos 20°), and f is the ratio of non-Møller (background)
events to the total number events measured by the spectrometer.
The results for the longitudinal beam polarization are shown in
Fig. 21 as a function of beam energy over a range from 6.47 to 19.4
GeV. The data were found to be consistent with lowest order quantum electrodynamic predictions for Møller scattering and with a
longitudinal beam polarization of 0.76 ± 0.05 independent of energy
and sense of source polarization. The data shown in Fig. 21 are
also in excellent agreement with the accepted value of the electron
g-factor anomaly.

These Møller studies were carried out in November of 1974 at
low PEGGY beam intensities as shown in Fig. 8. More recently, in
March 1976, additional Møller measurements were made. The results
of the recent studies, which used the SLAC 20 GeV/c spectrometer
without the C magnet, are consistent with the results of the 1974

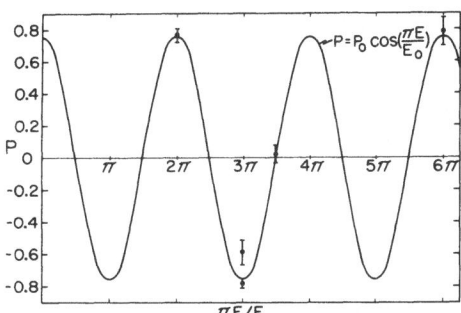

FIG. 21. The longitudinal component, P, of the beam polarization plotted versus $\pi E/E_O$ the angle through which the spin precesses relative to the momentum during the 24.5° bend into the experimental area. E is the beam energy and E_O = 3.237 GeV. The curve shown is a best fit to the data and has an amplitude P_O = 0.76 ± 0.03. P_O is the only free parameter.[56]

studies within the context of the resonant depolarization process in PEGGY which were described earlier.

High-Energy Scattering of Longitudinally Polarized Electrons by Longitudinally Polarized Protons

The Møller scattering measurements just described were a prelude to the program at SLAC for which PEGGY was really intended; namely, deep inelastic scattering of polarized electrons by polarized protons, an experiment involving a collaboration of physicists from CUNY, SLAC, and Yale in the USA; Bielefeld in Germany; and Nagoya and Tsukuba in Japan. It was recognized several years ago that the quark-parton model of the proton predicted observable spin-dependent effects in inclusive deep inelastic scattering of electrons by protons and neutrons.[117-120] Extensive studies of inclusive deep inelastic e-p scattering had previously been carried out by a SLAC-MIT collaboration,[132] but of course none of them had dealt with polarized particles. The SLAC-MIT experiments provided detailed information about two of the proton structure functions, $W_1(\nu,Q^2)$ and $W_2(\nu,Q^2)$, and led to the now famous scaling behavior:[133] as ν and Q^2 become infinite with $\omega=2M_p\nu/q^2$ held constant, W_1 and νW_2 become nontrivial functions of ω. Here $Q^2= -q^2$ is the square of the 4-momentum of the exchanged virtual photon, $\nu=E-E'$ with E and E' denoting initial and final energy of the electron, and M_p is the mass of the proton. The complete description of inclusive deep inelastic e-p scattering requires four structure functions, however, and only the use of polarized beams and targets provides access to the other two.[117]

Instead of dealing with structure functions, we can characterize the scattering process by cross sections, four independent ones
in all. Within the framework of this discussion we will restrict ourselves to the case of longitudinally polarized beams and targets and
introduce an asymmetry, A_1, for the interaction of the proton with
the exchanged circularly polarized virtual photon. We define A_1 in
terms of two spin-dependent cross sections $\sigma_{1/2}$ and $\sigma_{3/2}$ as follows:

$$A_1 = (\sigma_{1/2} - \sigma_{3/2})/(\sigma_{1/2} + \sigma_{3/2}). \qquad (25)$$

The subscript on σ denotes the sum of the z-components of the proton
and photon spins. In other words $\sigma_{1/2}$ is the cross section for photon and proton spins antiparallel and $\sigma_{3/2}$ the cross section for
spins parallel. Note that it is the photon spin and not the electron
spin that is involved in Eq. 25. I will come back to this point
shortly, but first I should point out that theoretically the asymmetry A_1, itself, is expected to obey scaling.[119]

In the most naive quark-parton model A_1 can be shown to be
equal to +5/9. That A_1 should be positive can be seen from the
fact that a proton in the naive model is made up of (ppn) quark configurations. Let us suppose that the proton has spin up. Then two
of the quarks have spin up and one has spin down. Thus in the case
of $\sigma_{3/2}$, the photon's angular momentum, which is along the spin-up
direction, can be absorbed by only the one quark with spin down. In
the case of $\sigma_{1/2}$, the photon's angular momentum, which is along the
spin down direction, can be absorbed by either of the two spin up
quarks. That A_1 should be precisely equal to 5/9 in the naive model
results from the association of a +2/3 charge with the p-type quark
and a -1/3 charge with the n-type quark, as well as from the requirement that the proton spin wave function be symmetric under quark
exchange.[117]

It would seem from the foregoing discussion that fairly large
asymmetries should have been measured at SLAC. However, the measured
asymmetry, Δ, is defined for antiparallel and parallel <u>electron</u> and
proton spins rather than <u>photon</u> and proton spins. Under these circumstances, the virtual photons are elliptically rather than circularly polarized and a kinematic factor, D, enters, which substantially
degrades the observed asymmetry. If the interference between transverse and longitudinal photon-nucleon amplitudes is neglected, Δ is
related to A_1 by

$$\Delta = A_1 D P_e P_p F, \qquad (26)$$

where P_e and P_p are respectively average electron and proton polarizations and F is the fraction of detected electrons which were scattered by free protons in the target. Under the experimental conditions, D was of the order 0.3, P_p was ~0.4, and F was ~0.11. Thus
the raw asymmetries, Δ, were of the order of a few tenths of a percent, necessitating very careful studies of false asymmetries.

The results of the first measurements, made at values of $[\omega=3, Q^2=1.680 (\text{GeV/c})^2]$, $[\omega=3, Q^2=2.735]$, and $[\omega=5, Q^2=1.418]$, are in general agreement with the quark-parton picture. In other words, with the interference between the transverse and longitudinal amplitudes neglected, A_1 was found to be large and positive, varying from ~0.3 at the $\omega=5$ point to ~0.65 at the $\omega=3$ points.[134] The consistency of the two $\omega=3$ measurements at different Q^2 is in accord with scaling.

In the course of the experimental program, elastic e-p scattering was also studied using the polarized beam and target. The theoretical asymmetry for elastic e-p scattering in the one-photon exchange approximation is given by[135]

$$A = \frac{\tau G_M}{G_E} \left\{ \frac{2M_p}{E} + \frac{G_M}{G_E} \left[\frac{2\tau M_p}{E} + 2(1+\tau)\tan^2 \frac{\theta}{2} \right] \right\}$$

$$\times \left\{ 1+\tau \left(\frac{G_M}{G_E}\right)^2 \left[1+2(1+\tau)\tan^2 \frac{\theta}{2} \right] \right\}^{-1}, \qquad (27)$$

where θ is the laboratory scattering angle, G_E and G_M are the electric and magnetic form factors of the proton for elastic scattering, and $\tau=Q^2/4M_p^2$. Thus a measurement of A determines the sign of G_E/G_M. The experimental results show that at $Q^2=0.765$ $(\text{GeV/c})^2$, G_E/G_M is positive, consistent with results from hyperfine structure interval measurements in hydrogen.[136,137]

Parity Violation in Weak Neutral Currents

Very recently a second high-energy physics experiment, using the SLAC PEGGY beam, was conducted by a collaboration involving Bielefeld, SLAC and Yale. In this experiment, which employed an unpolarized deuteron target, an interaction term of the form $\vec{\sigma}_e \cdot \vec{p}_e$ was studied to search for parity violation in weak neutral currents.[121] The data analysis has not yet proceeded sufficiently far for quotation of results.[140]

Exchange Excitation in Mercury

Although the atomic physics experiments using polarized electron beams may not be of the same fundamental nature as the high-energy physics experiments just discussed, they certainly have the same fundamental implication for theoretical calculations, in that they probe details of theory which are otherwise inaccessible. This can be seen clearly in electron-atom scattering. Two amplitudes are necessary to describe the scattering because of the indistinguishability of the target and projectile electrons. These two amplitudes, denoted by f and g, are called respectively "direct" and "exchange." In a scattering experiment involving unpolarized beams and targets no information can be obtained for f and g separately. Rather it is a linear combination of $|f|^2$ and $|g|^2$ which is measured. Only through polarization studies can $|f|$ and $|g|$ and their relative phase be determined.[114]

Although prior to 1974 several experiments had been conducted using polarized atomic beams in which the polarization of the recoiling atom or of the scattered electron was measured,[138,139] no experiment had been performed using a polarized electron beam. Then in 1974, Hanne and Kessler at Münster reported their first results of exchange excitation of the $6\ ^3P_{1,2}$ states of mercury by transversely polarized electrons.[125] Within the last few months they published a more detailed report.[48] Their experimental apparatus is shown in Fig. 22. Note that the polarized electron source uses low-energy Mott scattering from mercury. An energy of 80 eV and an angle of 80° were used to produce a polarization of -0.22 and a current of 10^{-11}A with an energy spread of 0.6 eV.

Fig. 23(a) shows the measured ratio, P'/P, of the final to initial electron polarization both $6^1S_0 \to 6^3P_1$ and $6^1S_0 \to 6^3P_2$ in the forward direction. It can be seen that at energies below 7 eV P'/P is consistent with zero for $^3P_1$ excitation as theoretically predicted.[48] Above 7 eV, however, P'/P rises rapidly reaching a value consistent with unity at energies above 9 eV. This increase is due to the 6^1P_1 admixture to the 6^3P_1 state which results from the breakdown of LS coupling in heavy atoms. In general the excitation scattering in the forward direction must be described by

$$\sigma(6^3P_1) = \sigma^o(6^3P_1) + \sigma^o(6^1P_1) \tag{28}$$

where $\sigma^o(6^3P_1)$ and $\sigma^o(6^1P_1)$ are the partial cross sections for triplet and singlet excitation scattering respectively. The ratio of $\sigma^o(6^3P_1)$ to $\sigma(6^3P_1)$ can be expressed in terms of P'/P by

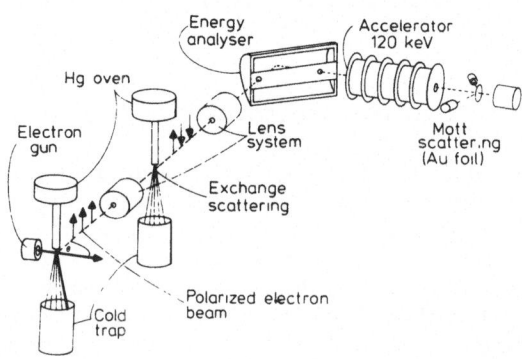

FIG. 22. Schematic diagram of Münster apparatus for studying exchange excitation of the $6^3P_{1,2}$ states of mercury by transversely polarized electrons.[48]

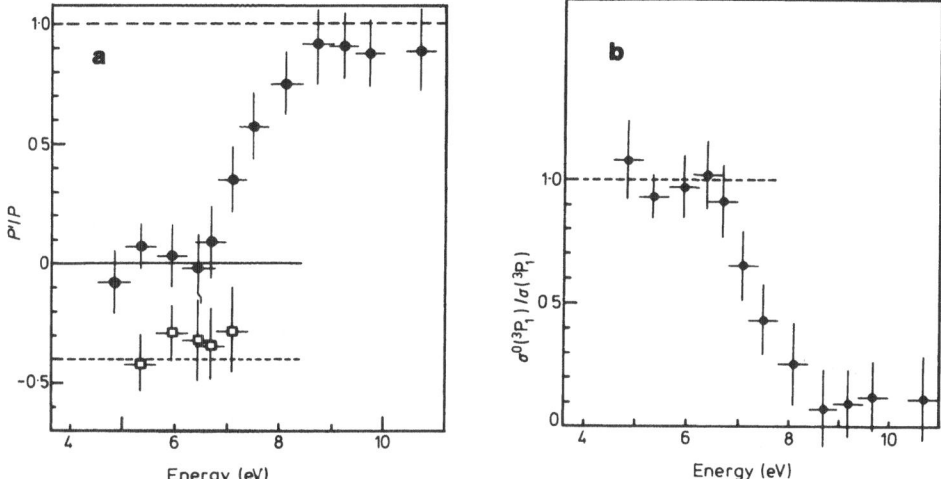

FIG. 23(a). Ratio P/P ' in the forward direction as a function of incident energy in the Münster experiment.[48] The dark circles are for the $6^1S_0 \to 6^3P_1$ process and the open squares are for the $6^1S_0 \to 6^3P_2$ process. (b) Ratio of $\sigma^0(^3P_1)$ to $\sigma(^3P_1)$ based upon Eq. 29 and the data of (a).

$$\sigma^0(6^3P_1)/\sigma(6^3P_1) = 1-P\,'/P. \qquad (29)$$

This ratio, calculated from Eq. 29 is shown in Fig. 23(b). As can be seen, triplet scattering dominates at energies below 7 eV, while at energies above 9 eV, singlet scattering dominates.

Spin-Exchange in Low-Energy Elastic Scattering of Polarized Electrons by Polarized Hydrogen Atoms

An experiment is currently in progress at Yale University to measure spin-exchange in low-energy elastic scattering of polarized electrons by polarized hydrogen atoms.[54] A Cs Fano source provides a dc beam of polarized electrons with a polarization of 0.65 and an intensity up to 5 nA. A schematic diagram of the crossed-beams apparatus is shown in Fig. 24. The longitudinally polarized electron beam is seen intersecting the state selected, transversely polarized hydrogen atomic beam at right angles. Since low energy electrons are involved, the magnetic field in the scattering region must be kept small, typically ≲ 50 mG. Thus, in the scattering region the polarization of the high-field state selected hydrogen beam is only 50% of its high field value. At an incident energy of ~10 eV, the raw asymmetry formed for spins antiparallel and spins

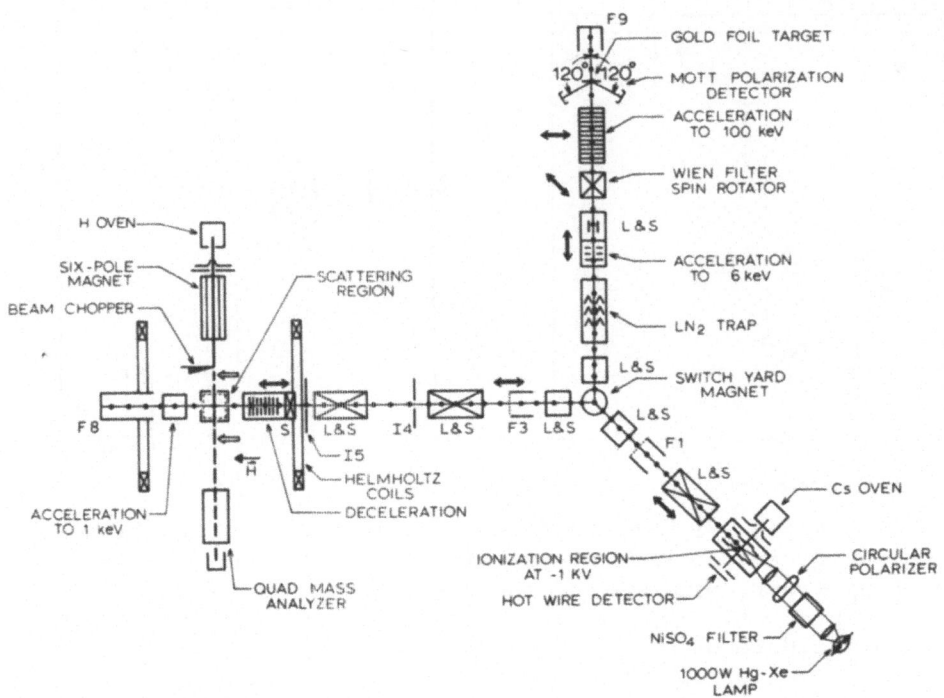

FIG. 24.   Schematic diagram of Yale crossed-beams apparatus for
measuring spin-exchange in low-energy elastic scattering of polarized electrons by polarized hydrogen atoms.   Lenses and steering
elements are denoted by L and S respectively; current sensors and
Faraday cups are labeled I and F respectively.

parallel,

$$\Delta = \frac{N\uparrow\downarrow - N\uparrow\uparrow}{N\uparrow\downarrow + N\uparrow\uparrow} , \tag{30}$$

is thus expected to be only 0.01-0.02 for 90° differential scattering.[35]   The Yale group has reported observation of asymmetries,
but the experiment is still in its preliminary stages and no quantitative results have been published.

THE FUTURE

Although I usually reframe from prognostication, I should like to
observe that the future of experiments using polarized electron beams
appears very promising. Groups at NBS, NYU and Bielefeld are planning

work on low-energy electron-alkali scattering, and the NBS group is
also planning surface physics studies. Groups at SLAC are planning to
extend the deep inelastic studies over larger kinematic regions
and to probe more deeply for parity violation in weak neutral
currents. The group at Bonn is planning to use their Fano source
at the Bonn synchrotron, although they haven't told us what experiments
are planned, and last but not least, King Walters has
informed me -- I am beginning to feel like a gossip columnist --
that the Rice group is planning to extend recent work in biophysics
carried out at Stanford.[49] They will be looking for evidence to
prove the hypothesis that parity violation in the weak interaction
is responsible for the origin of optically active organic molecules.
I kid you not! And on that merry note, I will conclude.

ACKNOWLEDGMENTS

    I would like to thank the various groups working with polarized
electrons for sending me results of their recent work prior to
publication. I would also like to thank P.W. Wainwright for his
careful reading of the manuscript and Laurie Liptak for her
invaluable help in its preparation.

REFERENCES

*The preparation of this report was supported in part by the NSF
(MPS75-02376), the USONR (N00014-76-C-0077), and the USERDA (Yale
Report No. 148) (E(11-1)3075).

1. J. Kessler in Atomic Physics 3, edited by Stephen J. Smith and
    G. King Walters (Plenum Press, New York, 1973), pp. 523-541.

2. S.A. Goudsmit and G.E. Uhlenbeck, Naturwiss. 13, 953 (1925).

3. S.A. Goudsmit and G. Uhlenbeck, Nature 117, 264 (1926).

4. Samuel A. Goudsmit and George E. Uhlenbeck, Physics Today,
    June 1976 29, 40 (1976).

5. P.A.M. Dirac, Proc. Roy. Soc. (London) A117, 610 (1928).

6. N.F. Mott, Proc. Roy. Soc. (London) A124, 425 (1929).

7. N.F. Mott, Proc. Roy. Soc. (London), A135, 429 (1932).

8. C.G. Shull, C.J. Chase and F.E. Meyers, Phys. Rev. 63, 29 (1943).

9. E.G. Dymond, Proc. Roy. Soc. A145, 657 (1934).

10. H. Richter, Ann. Physik 28, 533 (1937).

11. H.S.W. Massey and C.B.O. Mohr, Proc. Roy. Soc. (London) A177, 341 (1941).

12. H.S.W. Massey and C.B.O. Mohr, Proc. Roy. Soc. (London) A182, 189 (1943).

13. H. Deichsel, Z. Physik 164, 156 (1961).

14. H. Steidl, E. Reichert and H. Deichsel, Phys. Lett. 17, 31 (1965); H. Deichsel and E. Reichert, Z. Physik 185, 169 (1965); E. Reichert, Z. Physik 173, 392 (1963).

15. K. Jost and J. Kessler, Phys. Rev. Lett. 15, 575 (1965); J. Kessler and H. Lindner, Z. Physik 183, 1 (1965).

16. Klaus Jost and Joachim Kessler, Z. Physik 195, 1 (1966).

17. G. Holzwarth and H.J. Meister, Nucl. Phys. 59, 56 (1964); Tables of Asymmetry Cross Section and Related Functions for Mott Scattering of Electrons by Screened Au and Hg Nuclei (Univ. Munich, Munich, Germany, 1964).

18. P.J. Bunyan and J.L. Schonfelder, Proc. Phys. Soc. (London) 85, 455 (1965).

19. J.L. Schonfelder, Proc. Phys. Soc. (London) 87, 163 (1966).

20. W. Eitel, K. Jost and J. Kessler, Phys. Rev. 159, 47 (1967).

21. T.D. Lee and C.N. Yang, Phys. Rev. 104, 254 (1956).

22. C.S. Wu, E. Ambler, R.W. Hayward, D.D. Hoppes and R.P. Hudson, Phys. Rev. 105, 1413 (1957).

23. J.D. Jackson, S.B. Trieman and H.W. Wyld, Phys. Rev. 106, 517 (1957).

24. R.B. Curtis and R.R. Lewis, Phys. Rev. 107, 543 (1957).

25. K. Alder, B. Stech and A. Winther, Phys. Rev. 107, 728 (1957).

26. J.S. Greenberg, D.P. Malone, R.L. Gluckstern and V.W. Hughes, Phys. Rev. 120, 1393 (1960).

27. N. Sherman, Phys. Rev. 103, 1601 (1956).

28. N. Sherman and D.F. Nelson, Phys. Rev. 114, 1541 (1959).

29. Shin-R Lin, Phys. Rev. 133, A965 (1964).

30. W. Bühring, Z. Physik 212, 61 (1968).

31. L. Mikaelyan, A. Borovoi and E. Denisov, Nucl. Phys. 47, 328
     (1963).

32. J. van Klinken, Nucl. Phys. 75, 161 (1966).

33. W. Eckstein, Institut für Plasmaphysik, München Report No.
     IPP 7/1, 1970 (unpublished).

34. H. Frauenfelder and A. Rossi, in Methods of Experimental
     Physics, edited by L.C.L. Yuan and C.S. Wu (Academic Press,
     New York, 1963), Vol. 5, Pt. B, pp. 214-274.

35. W. Raith, in Atomic Physics 1, edited by V.W. Hughes, B.
     Bederson, V.W. Cohen and F.M.J. Pichanick (Plenum Press, New
     York, 1969), pp. 389-415.

36. V.W. Hughes, R.L. Long, Jr., M.S. Lubell, M. Posner, and W.
     Raith, Phys. Rev. A 5, 195 (1972).

37. B. Donnally, W. Raith, and R. Becker, Phys. Rev. Lett. 20,
     575 (1968).

38. R. Krisciokaitis and W.K. Peterson, DESY Report 73/63
     (Deutsches Elektronen-Synchrotron, Hamburg, Germany, December,
     1973).

39. D.M. Campbell, H.M. Brash and P.S. Farago, Phys. Lett. 36A,
     449 (1971).

40. U. Heinzmann, H. Heuer, and J. Kessler, Phys. Rev. Lett. 34,
     441 (1975).

41. H.D. Zeman, U. Heinzmann and D. Schinkowski, Fourth Inter-
     national Conference on Atomic Physics, Abstracts of Contributed
     Papers, Heidelberg, 1974 (unpublished), p. 394.

42. Herbert Zeman, in Electron and Photon Interactions with Atoms
     (Proceedings of the International Symposium on Electron and
     Photon Interactions with Atoms, Stirling, Scotland, 1974),
     edited by H. Kleinpoppen and M.R.C. McDowell (Plenum Press,
     New York, 1976), pp. 581-594.

43. P. Lambropoulos and M. Lambropoulos, in <u>Electron and Photon Interactions with Atoms</u> (Proceedings of the International Symposium on Electron and Photon Interactions with Atoms, Stirling, Scotland, 1974), edited by H. Kleinpoppen and M.R.C. McDowell (Plenum Press, New York, 1976), pp. 525-552.

44. P. Lambropoulos, private communication.

45. J.A. Simpson, in <u>Methods in Experimental Physics</u>, edited by L. Marton, V.W. Hughes and H. Schultz (Academic Press, New York, 1967), Vol. 4, Pt. A, p. 84.

46. M. Wilmers, R. Haug, and H. Deichsel, Z. Angew. Phys. <u>27</u>, 204 (1969).

47. H.D. Zeman, K. Jost and S. Gilad, in <u>Abstracts of the VIIth International Conference on the Physics of Electronic and Atomic Collisions, Amsterdam, 1971</u> (North Holland, Amsterdam, 1971), p. 1005.

48. G.F. Hanne and J. Kessler, J. Phys. B <u>9</u>, 791 (1976), and 805 (1976).

49. W.A. Bonner, M.A. VanDort, M.R. Yearian, Nature <u>258</u>, 419 (1975).

50. U. Heinzmann, J. Kessler and J. Lorenz, Z. Physik <u>240</u>, 42 (1970).

51. G. Baum, M.S. Lubell and W. Raith, Bull. Am. Phys. Soc. <u>16</u>, 586 (1971).

52. W.V. Drachenfels, U.T. Koch, R.D. Lepper, T.M. Müller and W. Paul, Z. Physik <u>269</u>, 387 (1974).

53. W.V. Drachenfels, U.T. Koch, Th.M. Müller and H.R. Schaefer, Phys. Lett. <u>51A</u>, 445 (1975).

54. P.F. Wainwright, M.J. Alguard, G. Baum, V.W. Hughes, J.S. Ladish, M.S. Lubell and W. Raith, Bull. Am. Phys. Soc. <u>21</u>, 573 (1976).

55. M.J. Alguard et al., in <u>Proceedings of the Ninth International Conference on High Energy Accelerators, Stanford, California, 1974,</u> CONF740522 (Stanford Linear Accelerator Center, Stanford, CA, 1974), p. 313.

56. P.S. Cooper et al., Phys. Rev. Lett. <u>34</u>, 1589 (1975).

57. M.J. Alguard et al., Bull. Am. Phys. Soc. <u>21</u>, 35 (1976).

58. M.V. McCusker, L.L. Hatfield and G.K. Walters, Phys. Rev. A 5, 177 (1972).

59. P.J. Kehiler, R.E. Gleason and G.K. Walters, Phys. Rev. A 4, 1279 (1975).

60. M.J. Alguard et al., Bull. Am. Phys. Soc. 21, 98 (1976).

61. E.H.A. Granneman, M. Klewer, K. Nygaard and M.J. Van der Wiel, J. Phys. B 9, L1 (1976).

62. E.H.A. Granneman, M. Klewer, and M.J. Van der Wiel, submitted to J. Phys. B.

63. N. Müller, W. Eckstein and W. Heiland, Phys. Rev. Lett. 29, 1651 (1972).

64. E. Kisker, G. Baum, A.H. Mahan and W. Raith, Phys. Rev. Lett. 36, 982 (1976).

65. G. Busch, M. Campagna, P. Cotti and H. Ch. Siegmann, Phys. Rev. Lett. 22, 597 (1969).

66. G. Busch, M. Campagna and H.C. Siegmann, J. Appl. Phys. 41, 1044 (1970).

67. E. Garwin, F. Meier, D.T. Pierce, K. Sattler, and H.-C. Siegmann, Nucl. Instr. Meth. 120, 483 (1974).

68. D.T. Pierce, F. Meier and P. Zürcher, Appl. Phys. Lett. 26, 670 (1975).

69. Daniel T. Pierce and Felix Meier, submitted to Phys. Rev.

70. G. Baum, private communication.

71. Daniel T. Pierce and R. Celotta, private communication.

72. C.K. Sinclair, private communication.

73. M.R. O'Neill, M. Kalisvaart, F.B. Dunning and G.K. Walters, Phys. Rev. Lett. 34, 1167 (1975).

74. M.J. Seaton, Proc. Roy. Soc. (London) A208, 408 (1951).

75. U. Fano, Phys. Rev. 178, 131 (1969).

76. M.S. Lubell and W. Raith, Phys. Rev. Lett. 23, 211 (1969).

77. G. Baum, M.S. Lubell, and W. Raith, Phys. Rev. Lett. <u>25</u>, 267
     (1970).

78. G. Baum, M.S. Lubell, and W. Raith, Phys. Rev. A <u>5</u>, 1073 (1972).

79. J. Kessler and J. Lorenz, Phys. Rev. Lett. <u>24</u>, 87 (1970).

80. U. Heinzmann, J. Kessler, and J. Lorenz, Z. Physik <u>240</u>, 42
     (1970).

81. G.V. Marr and D.M. Creek, Proc. Roy. Soc. (London) <u>A304</u>, 233
     (1968).

82. W. Paul, priviate communication.

83. R.L. Long, Jr., W. Raith, and V.W. Hughes, Phys. Rev. Lett.
     <u>15</u>, 1 (1965).

84. M.E. Mack, Appl. Opt. <u>13</u>, 46 (1974).

85. E. Garwin, to be published.

86. D.E. Rothe, J. Quant. Spectrosc. Radiat. Transfer <u>11</u>, 355 (1971).

87. Yu. V. Moskvin, Opt. Spectrosc. <u>15</u>, 316 (1963).

88. B. Ya'akobi, Proc. Phys. Soc. <u>92</u>, 100 (1967).

89. T.C. Caves and A. Dalgarno, J. Quant. Spectrosc. Radiat.
     Transfer <u>12</u>, 1539 (1972).

90. D. Norcross, private communication.

91. The data point at $I_\gamma = 0.288$ was obtained by the E80 collabora-
     tion at SLAC using Møller scattering, Ref. 56. The other four
     data points were obtained by the E95 collaboration at SLAC and
     have not yet been published.

92. J.E. Clendenin, et al., in preparation.

93. P. Lambropoulos, Phys. Rev. Lett. <u>30</u>, 413 (1973).

94. M. Lambropoulos, S.E. Moody, W.C. Lineberger, and S.J. Smith,
     Bull. Am. Phys. Soc. <u>18</u>, 1514 (1973).

95. M.V. McCusker, L.L. Hatfield and G.K. Walters, Phys. Rev.
     Lett. <u>22</u>, 817 (1969).

96. P.J. Kehiler, F.B. Duhning, M.R. O'Neill, R.D. Rundel and G.K.
     Walters, Phys. Rev. A <u>11</u>, 1271 (1975).

97. G.K. Walters, private communication.

98. G. Obermair, Z. Physik 217, 91 (1968).

99. H. von Issendorf and R. Fleischmann, Z. Physik 167, 11 (1962).

100. R.L. Long, Jr., V.W. Hughes, J.S. Greenberg, I. Ames, and
     R.L. Christensen, Phys. Rev. 138, A1630 (1965).

101. M. Holmann, G. Regenfus and O. Schärpf, Phys. Lett. 25A,
     270 (1967).

102. N. Müller, H. Siegmann and G. Obermair, Phys. Lett. 24A, 733
     (1967).

103. G. Chrobok, M. Hofmann, and G. Regenfus, Phys. Lett. 26A,
     551 (1968).

104. W. Gleich, G. Regenfus, and R. Sizmann, Phys. Rev. Lett. 27,
     1066 (1971).

105. U. Bänninger, G. Busch, M. Campagna, and H.C. Siegmann, Phys.
     Rev. Lett. 25, 585 (1970).

106. G. Busch, M. Campagna and H.C. Siegmann, J. Appl. Phys. 42,
     1779 (1971).

107. G. Busch, M. Campagna and H.C. Siegmann, J. Appl. Phys. 42,
     1781 (1971).

108. K. Sattler and H.C. Siegmann, Phys. Rev. Lett. 29, 1565 (1972).

109. S.F. Alvarado, W. Eib, H.C. Siegmann and J.P. Remeika, Phys.
     Rev. Lett. 35, 860 (1975).

110. D.E. Eastman, Phys. Rev. B 8, 6027 (1973).

111. R.L. Bell and W.E. Spicer, Proc. IEEE 58, 1788 (1970).

112. R. Celotta, private communication.

113. J. Kessler, Rev. Mod. Phys. 41, 3 (1969).

114. H. Kleinpoppen, Phys. Rev. A 3, 2015 (1971).

115. H. Kleinpoppen, "Analysis of Electron Atom Collisions,"
     (Center of Theoretical Studies, University of Miami, Coral
     Gables, Florida, 1969), Center for Theoretical Studies
     Report No. CTS-AP-75-L.

116. L.J. Weigert and M.E. Rose, Nucl. Phys. $\underline{51}$, 529 (1964).

117. J. Kuti and V.W. Weisskopf, Phys. Rev. D $\underline{4}$, 3418 (1971).

118. J.D. Bjorken, Phys. Rev. D $\underline{1}$, 1376 (1970).

119. F. Gilman, in Proceedings of Summer Institute on Particle Physics, Stanford Linear Accelerator Center, July 9-28, 1973 SLAC Report No. 167, 1973 (unpublished), Vol. 1, p. 71.

120. F. Close, Nucl. Phys. $\underline{B80}$, 269 (1974).

121. S.M. Berman and J.R. Primack, Phys. Rev. D $\underline{9}$, 2171 (1974).

122. W.H. Louisell, R.W. Pidd and H.R. Crane, Phys. Rev. $\underline{94}$, 7 (1954).

123. D.T. Wilkinson and H.R. Crane, Phys. Rev. $\underline{130}$, 852 (1963).

124. J.C. Wesley and A. Rich, Phys. Rev. A $\underline{4}$, 1341 (1971).

125. G.F. Hanne and J. Kessler, Phys. Rev. Lett. $\underline{33}$, 341 (1974).

126. R.H. Helm and W.P. Lysenko, SLAC Report No. SLAC-TN-72-1, 1972 (unpublished).

127. D.M. Schwartz, Phys. Rev. $\underline{162}$, 1306 (1967).

128. A.M. Bincer, Phys. Rev. $\underline{107}$, 1434 (1957).

129. See, for example, J.D. Bjorken and S.D. Drell, Relativistic Quantum Mechanics, (McGraw Hill, New York, 1964), p. 140.

130. SLAC Users Handbook.

131. V. Bargmann, L. Michel, and V.L. Telegdi, Phys. Rev. Lett. $\underline{2}$, 435 (1959).

132. R. Taylor, in Proceedings of the 1975 International Symposium on Lepton and Photon Interactions at High Energies, Stanford University, August 21-27, 1975 (Stanford Linear Accelerator Center, Stanford University, Stanford, CA, 1975), edited by W.T. Kirk, p. 679.

133. J.D. Bjorken, Phys. Rev. $\underline{179}$, 1547 (1969).

134. M.J. Alguard et al., submitted to Phys. Rev. Letters.

135. N. Dombey, Rev. Mod. Phys. $\underline{41}$, 236 (1961).

136. S.J. Brodsky and S.D. Drell, Ann. Rev. Nucl. Sci. $\underline{20}$, 147 (1970).

137. H. Grotch and D.R. Yennie, Rev. Mod. Phys. $\underline{41}$, 350 (1969).

138. D. Hils, M.V. McCusker, H. Kleinpoppen, and S.J. Smith, Phys. Rev. Lett. $\underline{29}$, 398 (1972).

139. B. Bederson, in **Atomic Physics 3**, edited by S.J. Smith and G.K. Walters (Plenum, New York, 1973), p. 401.

140. Although the deep-inelastic scattering experiment with the polarized proton target did not have a parity violation search as its primary objective, a value of the asymmetry, r, defined by $r = (d\sigma^- - d\sigma^+)/(d\sigma^- - d\sigma^+)$, was observed, where $d\sigma^-$ and $d\sigma^+$ are the inelastic differential cross sections for negative and positive electron helicities respectively. An experimental upper limit of $r < 5 \times 10^{-3}$ was determined at a 95% confidence level for values of Q^2 between 1.42 $(GeV/c)^2$ and 2.74 $(GeV/c)^2$. However, gauge theories (Ref. 121) predict values of $r \sim (10^{-5}$ to $10^{-4}) Q^2/M_p^2$, or in other words about two orders of magnitude lower than the sensitivity of the experiment. The unpolarized target experiment should improve this experimental sensitivity by about one order of magnitude, and future experiments are expected to increase the sensitivity to the point where values of $r \sim (10^{-6}$ to $10^{-5}) Q^2/M_p^2$ will be observable, well below the gauge theory predictions.

ROLE OF IMPURITIES IN MAGNETICALLY CONFINED

HIGH TEMPERATURE PLASMAS*

C. F. Barnett

Oak Ridge National Laboratory
Oak Ridge, Tennessee 37830

In the late 1950's there was overwhelming optimism in controlled fusion research that high temperature plasmas could be
achieved and fusion energy would be a reality. This optimism gave
way to a period of pessimism in the 1960's when all plasmas were
beset by instabilities, including a universal instability. Beginning in 1970 theoretical studies indicated that most of these instabilities did not exist or ways could be found to reduce the
instabilities to an acceptable level. These studies, along with
the success of the tokamak plasma experiments, have initiated a
period in which extrapolations have been made to thermonuclear type
reactors. In 1976 the cycle has been completed and a high degree
of optimism exists. Since the tokamak approach has been chosen as
the most likely method to achieve thermonuclear type plasmas using
magnetic confinement, the role that impurity atoms and ions play
in limiting the plasma temperature has been studied intensely
during the past few years. This paper summarizes the atomic
physics concerned with plasma cooling by impurities and the limiting effect that impurities may have on heating of plasmas by
neutral injection.

A general description is given of the tokamak concept and the
present and next generation experiments are described. The time
and spatial behaviour of O and Mo multicharged ions in present
hydrogen plasmas is presented. This is followed by a discussion
of the power loss from a plasma containing one per cent Fe.
Finally, the limitation of plasma heating by energetic H or D
injection is summarized.

*This work was supported by the U. S. Energy Research and Development
Administration under contract with Union Carbide Corporation.

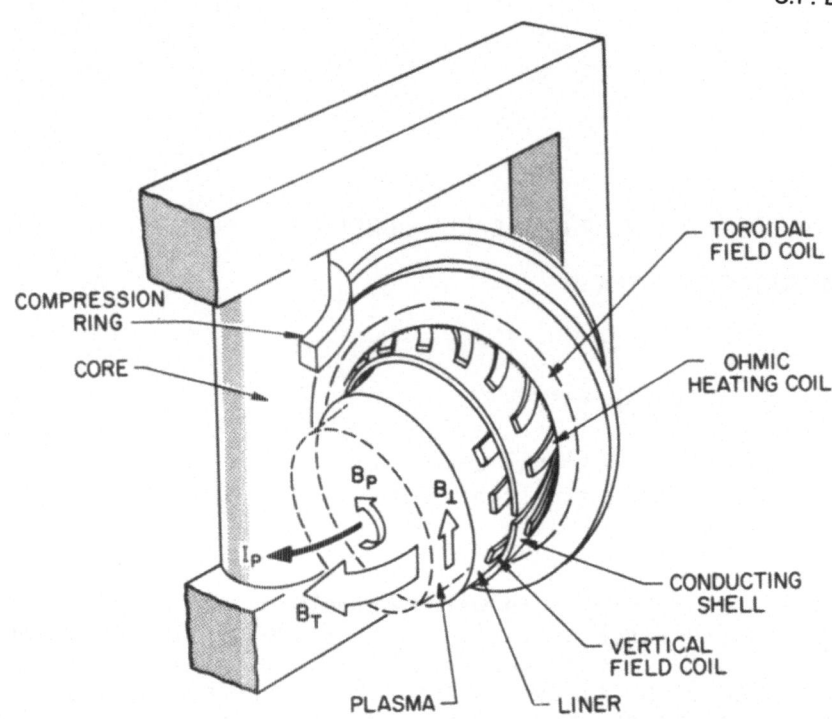

COMPRESSION RING

CORE

B_P $B_\perp$

I_P

B_T

PLASMA

LINER

TOROIDAL FIELD COIL

OHMIC HEATING COIL

CONDUCTING SHELL

VERTICAL FIELD COIL

Fig. 1: Schematic diagram of the tokamak fusion plasma experiment.

Shown in Fig. 1 is a schematic diagram of a tokamak plasma. The toroidal field coil generates a magnetic field B_T of 25-50 kg around the torus. H_2 gas at a pressure of approximately 10^{-4} torr is admitted to the torus and the application of a rf voltage produces a glow discharge. Applying a pulsed voltage to the ohmic heating coil induces a voltage around the torus of 400-500 V which drives the plasma current I_p. A large iron core yoke increases the coupling between the ohmic heating winding and the plasma. As the plasma conductivity increases due to the heating by I_p, the induced torus voltage decreases to a few volts. To prevent the shorting out of the induced voltage, a ceramic break must be inserted in the liner or the liner must be made of very thin SS with a high resistivity. Since the magnetic field decreases with the major torus radius, a vertical field coil creates a B field which centers the plasma in the liner. Also, since a cylindrical

plasma tends to expand radially, a conducting shell is placed
around the plasma to force the plasma inward. The large current
flow I_p creates a magnetic field B_p when combined with B_T results
in a helical magnetic field that gives the plasma its dynamic
stability. Not shown is a 1-2 cm annular W or Mo ring, known as
the limiter, that prevents the plasma from coming into direct
contact with the walls.

Table I

TOROIDAL PLASMA PARAMETERS

T_e (Electron Temperature)	1-2 keV
T_i (Ion Temperature)	0.5-2.0 keV
N_e (Electron Density)	10^{14} cm^{-3}
a (Minor Torus Radius)	10-45 cm
R (Major Torus Radius)	0.5-1.5 m
I_p (Ohmic Heating Current)	100-1000 kA
τ_e (Energy Containment Time)	5-60 msec
T (Pulse Time)	50-400 msec
P_H (Power of Injected Neutrals)	100-400 kW

    In Table I the parameters of the present generation tokamak
plasmas are tabulated. Without neutral particle injection heating
the maximum ion temperature has been less than 1 keV. The higher
temperatures have been obtained with this additional heating.
Table II tabulates the anticipated parameters of the next generation
of machines which are designed to use D-T as a fuel and produce
power but not economically. The JET experiment is the joint
European tokamak, TFTR is to be constructed and installed at Princeton, T-20 is the Soviet facility and the JT-60 is the Japanese entry
into the field.

Table II

NEXT GENERATION TOROIDAL MACHINES

	JET	TFTR	T-20	JT-60
R (m)	2.96	2.48	5.0	3.0
a (m)	1.25	0.85	2.0	1.0
I_p (MA)	4.8	2.5	6.0	3.3
T_i (keV)	5.0	6.0	7-10	5-10
T_e (keV)	5.0	6.0	7-10	5-10
N_e (cm^{-3})	5×10^{13}	4×10^{13}	5×10^{13}	$2\text{-}10 \times 10^{13}$
$N \tau_e$ (cm^{-3}sec)	5×10^{13}	1.5×10^{13}	1×10^{14}	$2\text{-}6 \times 10^{13}$
T (sec)	15	1.0	5-20	10
Neutral Injection Power (MW)	3-25	12-40	60	10-20
D^o Energy (keV)	160	120	160	100

Plasma impurities arise from electromagnetic radiation and energetic neutral particles impinging on the confining walls which are usually fabricated of SS or Au. Adsorbed gases and the wall materials are either sputtered or desorbed and flow into the plasma where they are immediately ionized. In addition to this source of impurity, the plasma comes into direct contact with the limiter and sputters or evaporates W or Mo into the plasma. Table III lists the impurities that have been found in the Oak Ridge ORMAK. The most abundant impurities are C and O. These impurities have been identified by measurements of line radiation. At the present time the energy levels of Au and W multicharged ions are not known with sufficient accuracy to be used to identify these elements in the complicated spectra usually observed in these high temperature plasmas.

Table III

IMPURITIES FOUND IN ORMAK PLASMA

Element	% Abundance
He	trace
Be	< 1.0
C	1.0 - 5.0
N	trace
O	1.0 - 2.0
Na, Cl, Ca	< 0.1
Si	< 1.0
S	$\sim$ 0.1
Cr, Mn, Ni	unknown
Fe	< 1.0
Cu, Ag, W, Pt, Au	unknown

In the interpretation of optical radiation and the modeling of plasma dynamics, the assumption is usually made that corona equilibrium exists in the plasma. In steady state corona equilibrium the assumptions usually made are:

1. Instead of each collisional process being balanced by its inverse process, as in local thermodynamic equilibrium, the balance is between collisional ionization and excitation and recombination, including both radiative and dielectronic recombination, and spontaneous decay.

2. Plasma is optically thin.

3. Free electrons have Maxwellian distribution.

4. Only a negligible number of plasma ions in excited states.

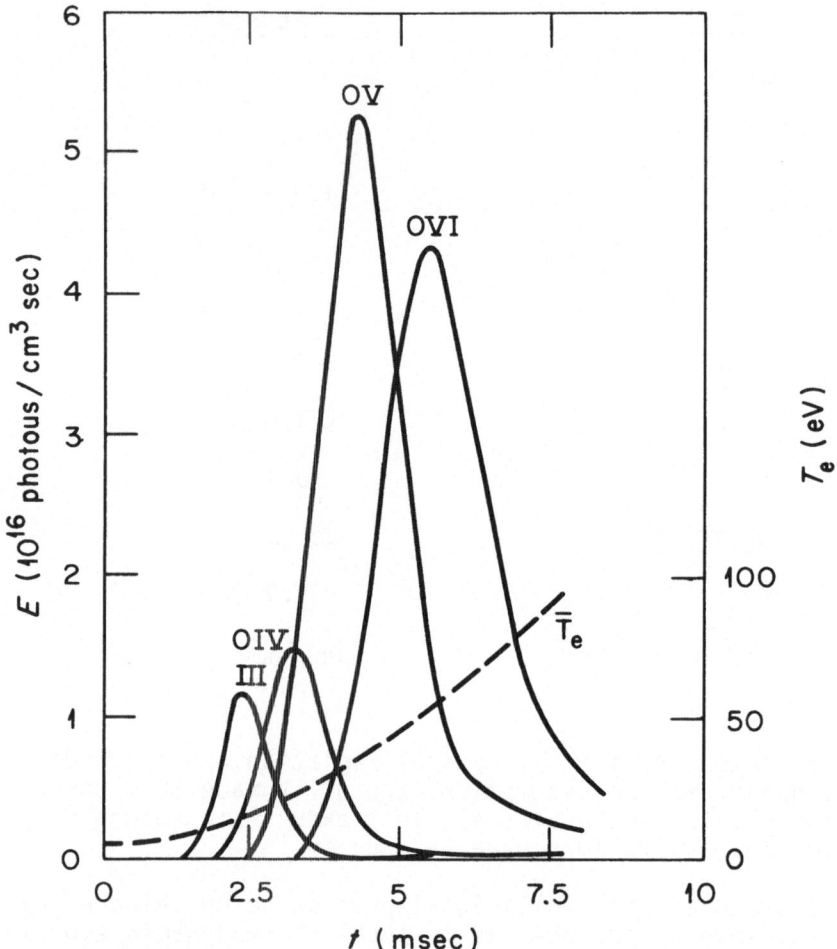

Fig. 2: The evolution of charge states of O with time for the FTR plasma experiment. Also shown is the plasma electron temperature during the formative stage of the discharge.

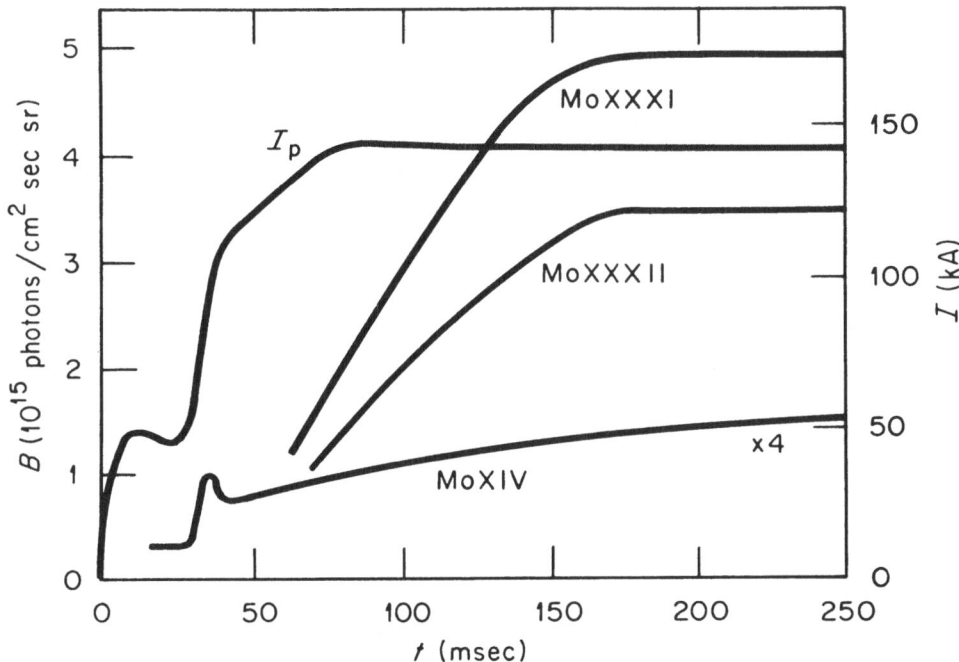

Fig. 3:   The appearance of highly ionized states of Mo during the
FTR discharge. Shown as I_p is the time dependence of the
plasma current used to ohmically heat the ions.

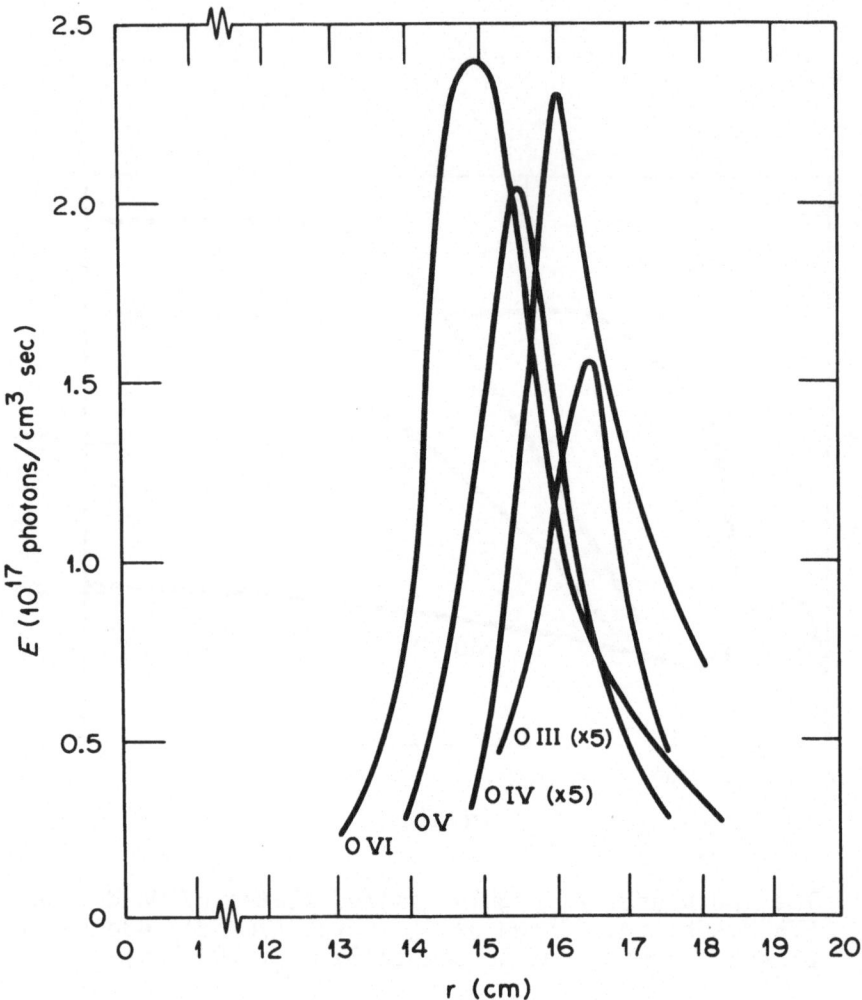

<u>Fig. 4</u>:   Radial profile of O ions in the FTR plasma. These profiles
         were taken at 200 msec after the initiation of the
         discharge.

Using this corona model, optical radiation has been used to study the time and radial distributions of impurities. Figs. 2-4 are measurements of oxygen and molybdenum impurities found in the plasma of the French tokamak, FTR.[1] In Fig. 2 are shown the OIII to OVI profiles as a function of time at the beginning of the discharge. During the first few msec of the discharge, higher charge states of oxygen appear. At an electron temperature of 100 eV most of the 0 is in charge states greater than 0^{5+}. Lines of OVII - OVIII are difficult to measure due to background interference. The time dependence of some highly ionized Mo lines are shown in Fig. 3 along with the time dependence of the plasma current. Lines of Mo^{30+} and Mo^{31+} were observed only after 50 msec indicating the time necessary for step-by-step ionization. The radial profiles of 0 ions are shown in Fig. 4 at a discharge time of 200 msec which was sufficient time for plasma equilibrium to be established. Each charge state exists in a shell within a few cm of the periphery of the plasma where the electron temperature was approximately 100 eV. Fig. 5 indicates the radial distribution of iron ions as calculated by Hogan[2] for coronal equilibrium. The ratio of the ionic charge species to the total Fe present is plotted as a function of the plasma radius for a central electron temperature of 750 eV; at this temperature approximately 30% of the ions in the central core are in charge state 13.

In understanding the operating characteristics of present plasmas and in modeling future machines, a detailed knowledge of atomic physics of the plasma ions is required. Merts, Cowan, and Magee[3] have made computations of the power loss from a hydrogen plasma seeded with 1% iron for relatively low electron temperatures up to temperatures of 10 keV. Atomic processes they included in the calculation were collisional excitation and ionization, both radiative and dielectric recombination, and free-free or bremsstrahlung radiation. Since experimental cross sections are not known for the multicharged Fe ions, they relied on theoretical predictions. For ionization the ionization rates given by Lotz[4] were used; electron excitation cross sections were those predicted by Bely[5] and Van Regemorter[6]; dielectric recombination rates from Burgess[7] and Jordan[8]; and radiative recombination rates from Kramer's formula[9]. The results of these calculations are tabulated in Table IV. The total power radiated from the plasma varies from 4.1 w/cm^3 for a low temperature plasma down to 0.42 w/cm^3 for a reactor-type plasma with an electron temperature of 10 keV. These power losses can be compared to 1 w/cm^3 power production rate in a D-T reactor. In the table collisional excitation refers to the radiative de-excitation following collisional excitation. For plasma electron temperatures below 2 keV the dielectronic recombination loss is greater than the radiative recombination loss. The cross sections and reaction rates used have large uncertainties, such that the radiated power is known no better than a factor of two. Raymond et al[10] have performed similar calculations with the predicted power loss

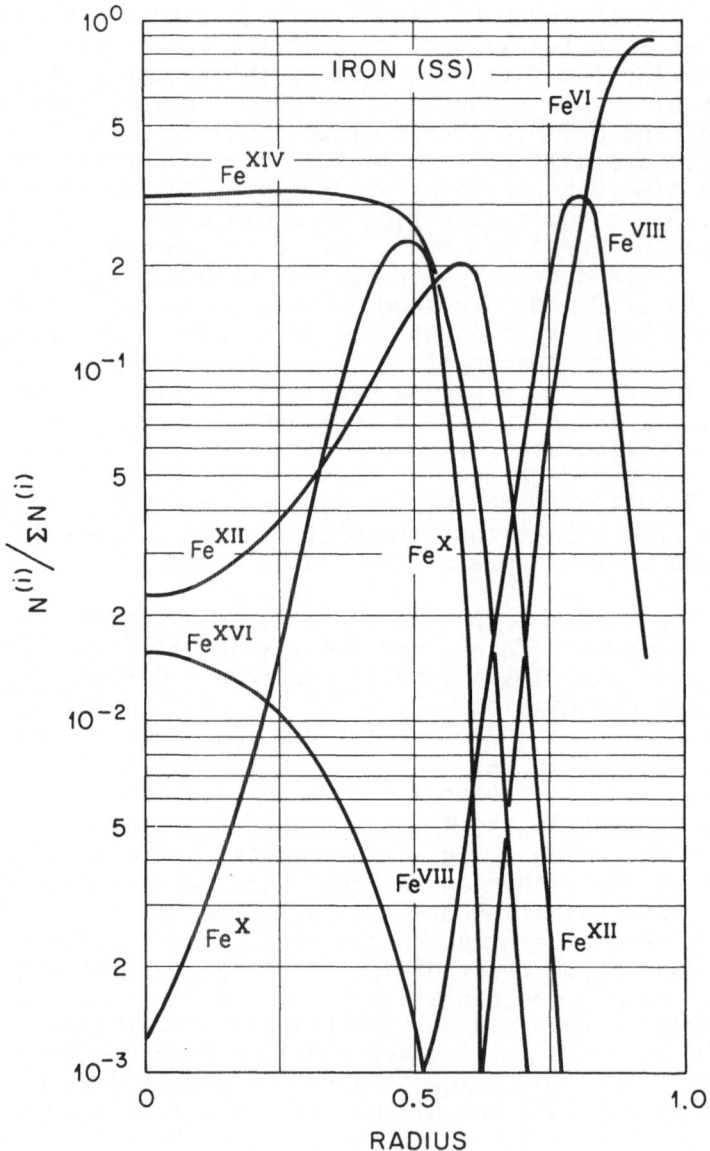

<u>Fig. 5:</u> Calculated radial profile of Fe ions in a 750 eV plasma
assuming corona equilibrium.

being a factor of two higher than the results of Merts et al. Computations such as these illustrate the serious need for atomic data to accurately predict the feasibility of economic power production from a thermonuclear reactor with impurities present.

Table IV

POWER LOSS FROM FE-SEEDED PLASMA

$$N_e = 10^{14} \text{ cm}^{-3} \qquad N_{Fe} = 10^{12} \text{ cm}^{-3}$$

T_e (keV)	Total Power Loss w/cm³	Coll. Excit $\Delta n = 0$	$\Delta n \neq 0$	Recombination Diel.	Rad.	Free-Free
0.8	4.1	2.2	1.1	0.78	0.04	0.02
1.5	1.5	1.1	0.20	0.12	0.09	0.04
2.0	0.76	0.46	0.12	0.05	0.08	0.04
5.0	0.42	0.05	0.17	0.02	0.11	0.07
10.0	0.42	0.004	0.14	0.01	0.16	0.11

In a tokamak plasma the plasma ions are heated by coulomb collisions between electrons and ions. As the plasma electron temperature approaches 1 keV, the collision cross section becomes small, resulting in inefficient heating of the ions. To overcome this difficulty, plasma heating to higher temperatures is accomplished by injecting large currents of energetic H or D neutrals which are trapped by ionizing collisions. These trapped energetic ions transfer their energy to the cooler plasma ions again through coulomb collisions. During the past few months theoretical studies have indicated that if the impurity ions in the plasma have a large H or D ionization cross section, a beam-deposition instability will develop. The H or D atoms will be trapped at the plasma edge where the ion will be readily lost to the plasma confining walls by charge exchange, since the neutral pressure is high in the plasma periphery. These particles will desorb and sputter more impurities which flow into the plasma boundary. This process of impurity buildup near the edge may cascade and lead to no plasma penetration or reflection of the injected neutral beam.

The mean free path of a neutral particle injected into the plasma is given by

$$1/\lambda_0 = \frac{n_e \langle \sigma v \rangle_{ion}}{v_0} + n_i \sigma_{ex} + n_i \sigma_i + \sum_k n_k \sigma_k$$

where the trapping or ionization is due to electron impact, charge exchange, and impact ionization on plasma protons, respectively. The last term is the impact ionization by impurity ions. The assumption was made that σ_k is given in terms of the impurity charge Z_k such that $\sigma_k = Z_k^2 \sigma_i$ where σ_i is the proton impact ionization cross section. Ionization of the neutral by charge exchange with impurity ions has been neglected. In plasma physics the impurity concentration is denoted by Z_{eff} which is equal to $(n_i + \Sigma_k n_k Z_k^2)/n_e$ such that

$$1/\lambda_0 = \frac{n_e \langle \sigma v \rangle_{ion}}{v_0} + \frac{n_i}{n_e} \sigma_{ex}(v_0) + Z_{eff} \sigma_i(v_0)$$

where v_0 is the injected neutral velocity. Using this expression we may calculate the fast ion density profile $H(r)$ in the plasma. The results of the calculation by Hogan[11] are shown in Fig. 6 where the fast beam profile and plasma ion temperature profile are plotted as a function of the radius for the proposed TFTR facility. The different curves are for the effective Z which in present plasmas varies between 2 and 10. As Z_{eff} becomes greater the fast ion profile peaks near the plasma edge, the peak ion temperature moves outward. Application of these results to present and future experiments indicate that the present small machines have no penetration problems; the PLT machine being placed into operation at Princeton may have problems; and the future machines will have serious problems even for low effective Z.

The curves in Fig. 6 were obtained under the assumptions that charge exchange cross sections were negligible and the impurity ionization trapping cross sections were equal to the ion charge squared times the proton impact ionization cross section of H atoms. Cross section data pertinent to neutral beam trapping are shown in Fig. 7. Cross sections are plotted as a function of the H ion or atom energy. Shown are the cross sections for trapping by resonant charge exchange; electron ionization of H° in plasmas whose electron temperatures are 1 and 10 keV; ionization of H° by H⁺; and ionization or stripping of H° by oxygen atoms. These cross sections have been known for several years. Also, plotted in Fig. 7 are some recent unpublished preliminary charge exchange cross sections of Kim, Bayfield and Stelson[12] for the reaction $H + Fe^{10+} \rightarrow H^+ + Fe^{9+}$ at equivalent H energies of 70 and 250 keV. The large cross section of 6.6×10^{-15} cm² for Fe^{10+}

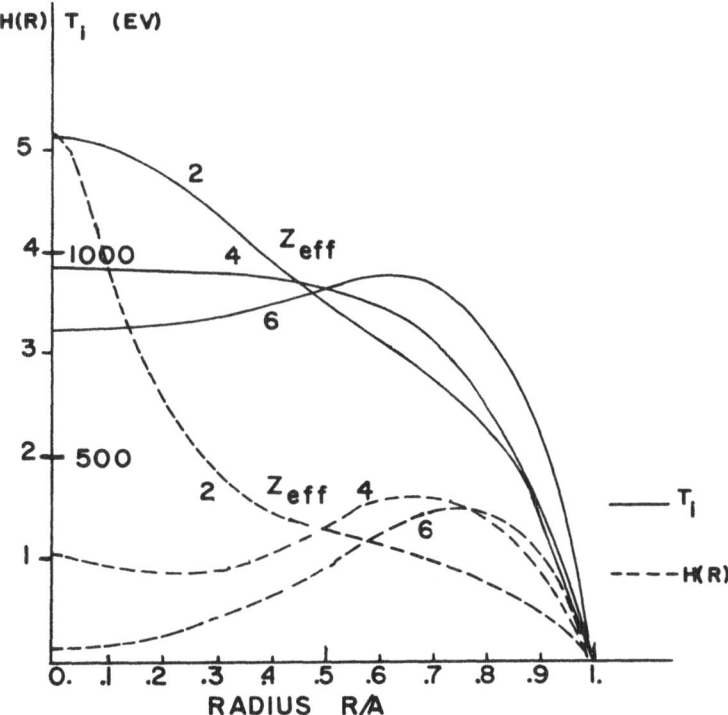

Fig. 6: The radial profile of the fast ion density and ion
         temperatures in the proposed TFTR as the impurity
         effective charge changes.

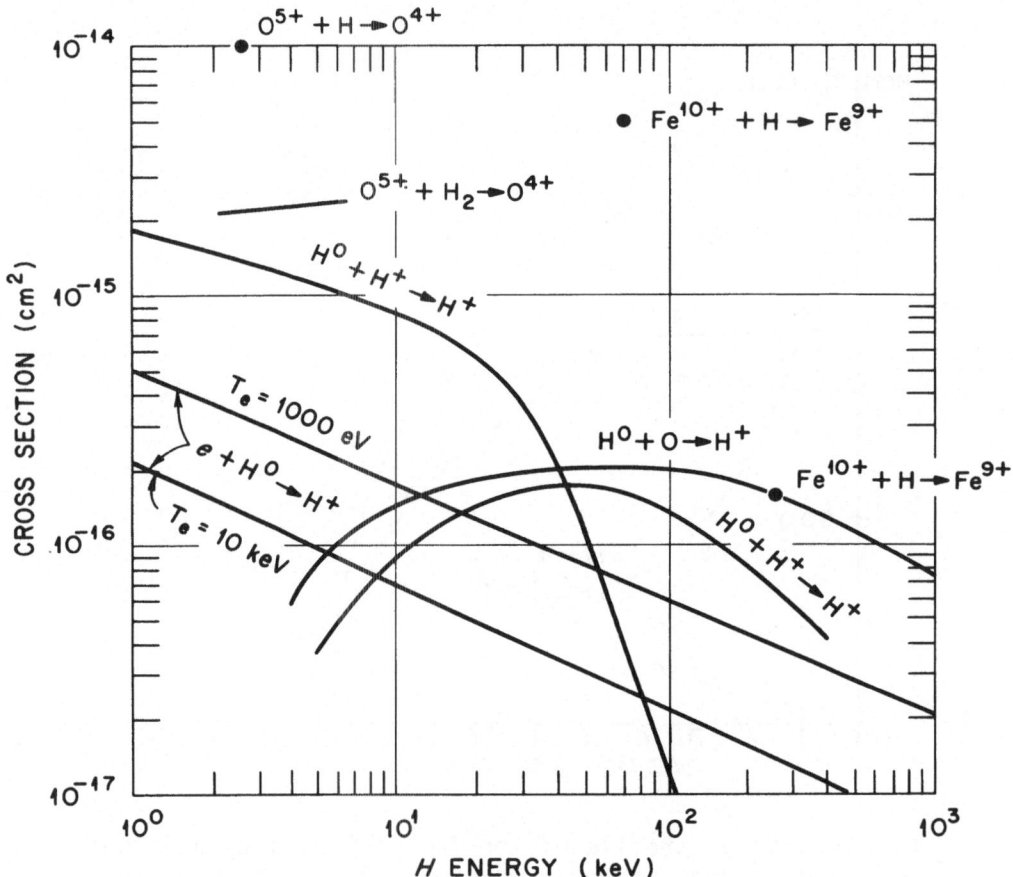

Fig. 7: Atomic cross sections pertinent to the trapping of a neutral H beam in high temperature plasmas.

indicates that charge exchange trapping of H cannot be ignored in considering beam penetration into plasmas. On this graph the planned injection D° energy of 120 keV for the Princeton PLT plasma experiment would be at 60 keV. To indicate the magnitude and importance of the cross sections involving multi-charged ions, some recent data are also shown at the lower energies. Bayfield at this conference has reported on measurements of light ions in atomic H. At an equivalent energy of 2.5 keV the charge exchange cross section of O^{5+} in H was $10^{-14} cm^2$. The data for O^{5+} in H_2 are from some recent unpublished work of Crandall et al[13]. The large sketchy cross section data of multi-charged ions clearly indicate the future difficulties of heating plasmas by neutral injection and the increased plasma cooling by charge exchange which has been ignored in the past.

With the available experimental data and theoretical predictions, the future of obtaining high temperature plasmas is doubtful unless some way is found to eliminate impurities or to maintain a low impurity level.

REFERENCES

1. FTR Group, Nucl. Fusion 15, 1053 (1975).

2. J. T. Hogan, ORNL Thermonuclear Division Progress Report, ORNL-5035 (1974), p. 43.

3. A. L. Merts, R. D. Cowan and N. H. Magee, Jr., Los Alamos Scientific Report LA-6220-MS (1976).

4. W. Lotz, Z. Physik 206, 205 (1971); Z. Physik 216, 241 (1968).

5. O. Bely, Proc. Phys. Soc. (London) 88, 587 (1966).

6. H. Van Regemorter, Astrophys. J. 136, 906 (1962).

7. A. Burgess, Astrophys. J. 139, 776 (1964); Ann. D'Astrophys. 28, 774 (1965); Astrophys. J. 141, 1588 (1965).

8. C. Jordan, Mon. Not. R. Astr. Soc. 142, 501 (1969); Mon. Not. R. Astr. Soc. 148, 17 (1970).

9. H. A. Kramers, Phil. Mag. <u>46</u>, 836 (1923).

10. J. C. Raymond, D. P. Cox, and B. W. Smith, Wisconsin Astrophysics, No. 12 (1975).

11. J. T. Hogan, private communication.

12. H. Kim, J. Bayfield and P. H. Stelson, unpublished data.

13. D. H. Crandall, D. Kocur and M. Mallory, submitted for publication in Phys. Review A.

DIAGNOSTIC PROBLEMS OF LARGE TOKAMAKS

Kenneth M. Young

Plasma Physics Laboratory, Princeton University

Princeton, New Jersey 08540

I. INTRODUCTION

The experimental study of high temperature plasmas in tokamaks has recently gained momentum with the occurrence of the first discharges in the two rather similar American PLT (Princeton Large Torus) and Russian T-10 Tokamaks towards the end of last year. Furthermore, the final design stage of the TFTR (Toroidal Fusion Test Reactor) was begun in March this year; the TFTR is roughly twice as large as the other two. Because of its size, the expected high electron temperatures, the very large neutral beam system to provide auxiliary heating and the intention to use tritium as a filling gas as target for deuterium beams with the resultant fusion product particles, TFTR poses some clear problems for diagnostics of the plasma behavior.

The aims behind the construction of TFTR can be summarized:[1]

i) The attainment of ion and electron temperatures of 5-10 keV ($\sim 10^8$ °K) in substantially pure hydrogenic plasmas.

ii) The maximization of plasma density within the constraint of retaining gross stable plasma confinement.

iii) The attaining of the product of density, n, and energy containment time, τ_E, $n\tau_E \sim 10^{14}$ cm^{-3} sec under favorable circumstances.

iv) The generation of reactor-like fusion power densities under conditions of approximate break-even in plasma energy in operation with deuterium and tritium (D-T).

The high temperatures are well within the levels computed for plasmas where auxiliary heating has been applied and where the

impurity level remains low ($n_{Fe} \lesssim 0.1\%$ n_e). A central density as
high as 1×10^{14} cm^{-3} is an achievable goal but it is intended to
raise the density by injection of gas during the discharge (and
lower the relative impurity level) following the successful experiments in Alcator[2] and Pulsator[3] Tokamaks. High density operation
ab initio for the discharges has always led to bad disruptive
behavior. The energy containment requirement $n\tau_E \sim 10^{14}$ cm^{-3} sec
is of interest as this represents the break-even value for D-T
fusion in a 10 keV thermal plasma (known as the Lawson criterion[4]).
It is a somewhat uncertain prediction in that the extrapolation
involves theoretically predicted diffusion mechanisms which have
not been met experimentally in the more collisional plasmas so far
measured in tokamaks. However, the criterion for break-even is
relaxed for neutral beam injection, or two-component plasmas, so
that the value of the experiment is maintained.

The final intention of reactor-like fusion power densities
means that the gas filling for the Ohmically heated plasma in the
toroid will be tritium and the injected beams will be deuterium
atoms, after successful operation of the device has been achieved
and the plasma reasonable understood. The major purpose of the
device will be to study the fusion aspects, particularly to look at
the heating. The maximum flux of virgin 14.1 MeV neutrons is expected to be 6×10^{12} n cm^{-2} sec^{-1}. This flux places extreme
requirements on radiation shielding but although it leads to acti-
vation of the vacuum vessel it is insufficient to produce significant
sputtering of impurities off the walls. However, diagnostic systems
will have to operate in a hostile environment and will have to be
shielded and located appropriately.

Section II provides a brief description of the operation of a
Tokamak with the inference on the large TFTR device and discusses
the role impurities play in the shaping and heating of the plasma.
Section III describes some recent results with the inference on
information obtained about impurity behavior. Section IV will discuss the problems associated with the extrapolation in scale and
energy of the large future tokamaks. These problems are not related
to the radiation and damage effects of the high neutron flux mentioned above though this is a severe problem.

II. OPERATION OF A TOKAMAK

II.1 Description of the Tokamak and Its Auxiliary Heating

There have been two recent comprehensive review articles about
Tokamaks.[5,6] For this paper, a much simplified description will be
given to remind readers of the tokamak principles.

The simplest tokamak confinement system consists of nested toroidal surfaces of helical magnetic field lines. These are created by an Ohmic current driven in a plasma in a strong toroidal field. Figure 1 is a trimetric drawing of the TFTR which serves to show the necessary field coil arrangement. The toroidal primary windings of the, in this instance air-cored, transformer to drive the Ohmic current are placed outside the toroidal field coils.

Since the toroid of plasma tends to expand in major radius because of the plasma thermal pressure and the poloidal field pressure, an additional transverse field is necessary to hold the plasma in equilibrium. In early Tokamaks where the duration of the current pulse was short, this equilibrium could be provided by image currents in copper shells outside the vacuum vessel. The plasma is produced inside a stainless steel vacuum vessel by the Ohmic current after a preliminary production of electrons by some means such as an rf oscillator. The plasma is normally kept clear of the vacuum vessel wall by an aperture limiter made of a refractory metal occupying most of the poloidal circumference but only a minute fraction of the toroidal circumference. Charged plasma particles diffusing out from the plasma, both thermal particles and high energy runaway electrons, collide with this limiter. Radiation from the plasma and fast neutral atoms, created by charge-exchange between atoms moving into the plasma and high energy plasma ions, carry energy onto the vacuum vessel walls. Between one quarter and half the energy lost goes to the limiter under standard operating conditions in the ST[7] and T-3[8] Tokamaks.

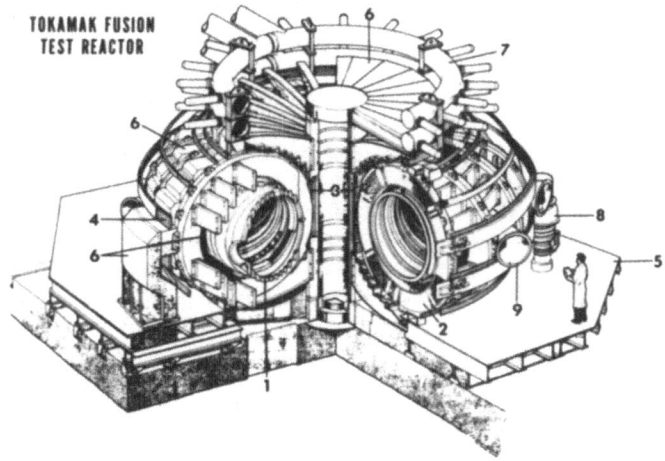

Fig. 1. Trimetric drawing of the TFTR Tokamak: 1. Stainless steel vacuum vessel containing the plasma; 2. toroidal field coil; 3. Ohmic heating field coil; 4. equilibrium field coil; 5. device substructure; 6. radiation shielding; 7. water header for cooling of field coils; 8. vacuum pump; 9. neutral injection duct.

The tokamak is called a low-β device because β, the ratio of the plasma pressure to the magnetic field pressure, is limited by loss of stability. Roughly

$$\beta \lesssim 0.1(a/R)$$

where a is the minor and R is the major radius of the plasma toroid. Thus to obtain high plasma temperatures, high toroidal magnetic fields and as tight a toroid as possible are desirable.

Table 1 provides a comparison of three well-documented tokamaks with PLT whose early results are given and with TFTR for which predictions are made. The ST Tokamak[9] was a thoroughly diagnosed device while the relatively high density achieved in Alcator[2], and relatively low impurity level have encouraged a trend to high current densities and high toroidal fields. The TFR Tokamak[10] had produced the best quality plasma parameters before the end of last year. The PLT results[11] are preliminary in that no detailed optimizing of conditions has yet been made. The TFTR predictions[1] are produced from sophisticated computer code studies[12] involving transport models based on diffusion and instability theories, influx of neutral gas and impurities whose reliability can only be checked by more experimental data from PLT and T-10. The dominant feature of the TFTR will be the neutral beam auxiliary heating which will provide about an order of magnitude more power than the Ohmic heating.

A high energy neutral beam entering a plasma in a magnetic field rapidly becomes ionized by electron bombardment and by charge-exchange with plasma ions. Provided the mean free path is short enough, the energy is effectively absorbed into the plasma. Experiments on the ORMAK[13] and ATC[14] Tokamaks have shown effective trapping and ion heating at low power levels. In addition, one can make use of the beams to benefit the fusion power output of a Tokamak since the peak of the fusion cross section for Deuterium-Tritium reactions is at 120 keV whereas the thermal energy of the plasma is only a few keV. Hence by injecting beams of higher energy one can enhance the fusion output considerably. Figure 2 shows a calculation for the slowing down of a 180 keV deuterium beam in a 5 keV tritium plasma.[1] Initially the energy is mostly transferred to the electrons but as it slows down the energy is transferred to the tritons by elastic collisions. The figure also shows the fusion probability and the energy multiplication factor Q.

$$Q \equiv \text{Integrated fusion probability} \times 17.6 \text{ MeV}/W_o$$

where 17.6 MeV is the fusion energy yield and W_o is the initial beam energy.

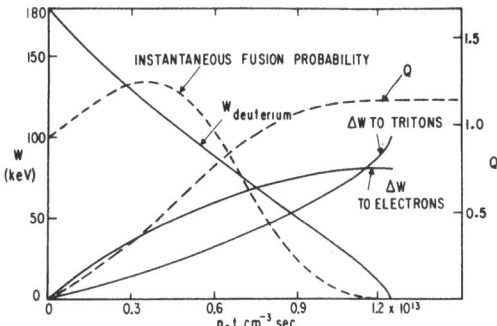

Fig. 2. The classical slowing down of an energetic deuteron beam in a 5 keV tritium plasma.[1] The curves show the transfer of energy to the electrons and tritons, the instantaneous fusion probability and the energy multiplication factor Q.

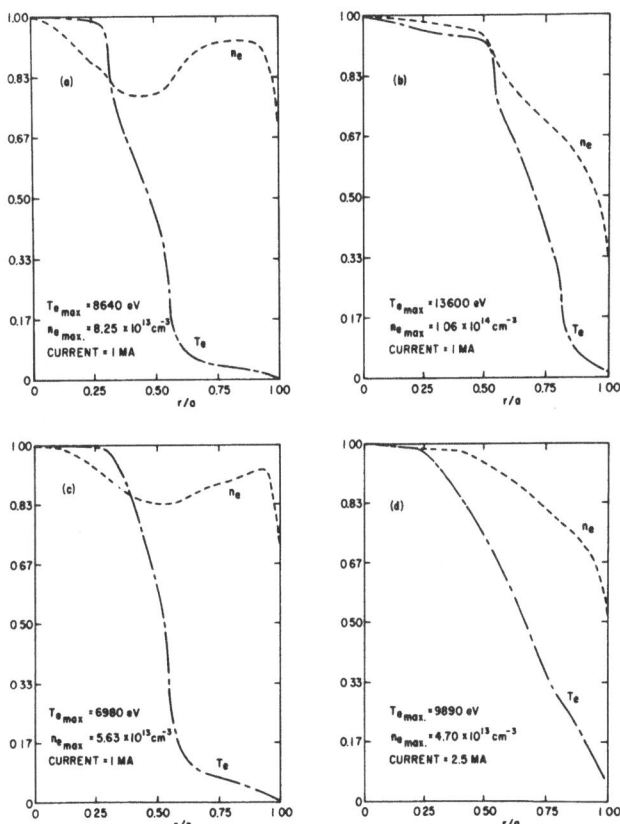

Fig. 3. Computed radial profiles of electron density for four possible operational modes of TFTR.[1] a) Weakly compressed mode. b) Strongly compressed mode. c) Low-current quasi-static mode. d) High-current quasi-static mode. These are optimal results after neutral beam injection and compression (in cases a and b).

TABLE I   Plasma Parameters of Some Tokamaks

	ST	TFR	ALCATOR	PLT (preliminary data)	TFTR (predicted)
Major Radius R(m)	1.09	0.98	0.54	1.3	2.5
Plasma Minor Radius a(m)	0.14	0.20	0.12	0.45	0.85
Toroidal Field Strength B_T (W/m²)	5.0	0.60	7.0	3.5	5.2
Plasma Current I(MA)	0.12	0.20	0.15	0.6	2.5
Electron Temperature T_e (keV)	~2.0	~2.0	~1.0	~2.0	5–10
Ion Temperature T_i (keV)	~0.5	1.1	~0.5	1.5 estimate	5–10
Electron Density n_e (cm⁻³)	5×10^{13}	6×10^{13}	5×10^{14}	3×10^{13}	$>10^{14}$
Neutral Beam Power (MW)	–	~0.6	–	(1.5) not yet installed	20.0
Length OH Pulse (sec)	0.05	0.4	0.1-0.6	1.0	>1.0

Other forms of auxiliary heating have been used on Tokamaks. Heating by rf power and by adiabatic compression have shown promise and the capability for the latter is being incorporated for the TFTR device. The simplest compressional heating makes use of the fact that the toroidal field is inversely proportional to major radius, $B_T \propto 1/R$. The plasma is first formed at large R. It is then forced inwards to smaller R by increasing the vertical field. The maximum compression achievable in TFTR is about 1.5 raising the density by over 100% and the bulk plasma temperature by over 70%. If the neutral beam heating is applied before compression, there is an enhanced heating effect. Figure 3 shows computer derived radial temperature and density profiles for different modes of

operation. The plasma parameters from these figures establish the
parameters for the provision of diagnostic apparatus although it
is probable that the recent good high density results from
Alcator[2] will lead to operation at densities up to 5×10^{14} cm^{-3}
by injection of additional neutral gas. The impurity concentration
is also raised by compression.

II.2 Effect of Impurities on the Plasma

The importance of atomic physics in the fusion program lies
in enabling understanding of the effects impurities have on the
plasma by determining the ionization states and spatial distribution of the ions. For the very high temperature discharges,
$T_e \gtrsim 5$ keV, most of the impurities at the center of the plasma will
be fully stripped except for relatively heavy atoms such as molybdenum or tungsten. These materials will probably be used in
limiters or in protection plates against accidental neutral beam
bombardment of the wall.

The dominant effects of impurity ions coming from evaporation
or sputtering of the vacuum vessel walls or from the limiter can
be summarized as follows. For relatively low power tokamaks, the
impurity effect has probably been beneficial in that the plasma
resistance is higher, and hence the Ohmic power input is higher,
than for pure hydrogenic plasmas. ($\bar{Z}_{eff} \equiv \Sigma n_i Z_i^2 / \Sigma n_i Z_i$ has been
typically 3-5 and is the relative enhancement factor.) However,
there are significant detrimental effects which become more
serious for larger high temperature plasmas. By cooling the outer
edges of the plasma they cause the current density to shrink
toward the center, which can lead to a disruptive instability.
They make the plasma transport properties predominantly collisional
when the ions of a pure hydrogenic plasma should be essentially
collision-free. Classical theory, and more sophisticated theories
including toroidal effects, particle-mirroring along field surfaces
and instability driven transport, predict peaking of the impurity
ions at the axis.[6] These impurities will cool the center by the
enhanced bremsstrahlung and, if they are not fully stripped, by
line radiation. The peaking of the impurity concentration was
observed on the T-4 Tokamak[15] but not on the ST Tokamak.[16] This
difference may be caused by a strong mixing effect inside the
unstable $q = 1$ surface in the ST Tokamak for which the diffusion
theories are not applicable. An experiment by Cohen et al.[17],
injecting small laser-produced bursts of aluminum into the ATC
Tokamak discharge agreed well with the predicted theoretical
penetration to the center.

An example of the effects can be seen in Fig. 4 which shows a
set of computer derived radial profiles of density and temperature
for plasmas with different peak densities at the same plasma

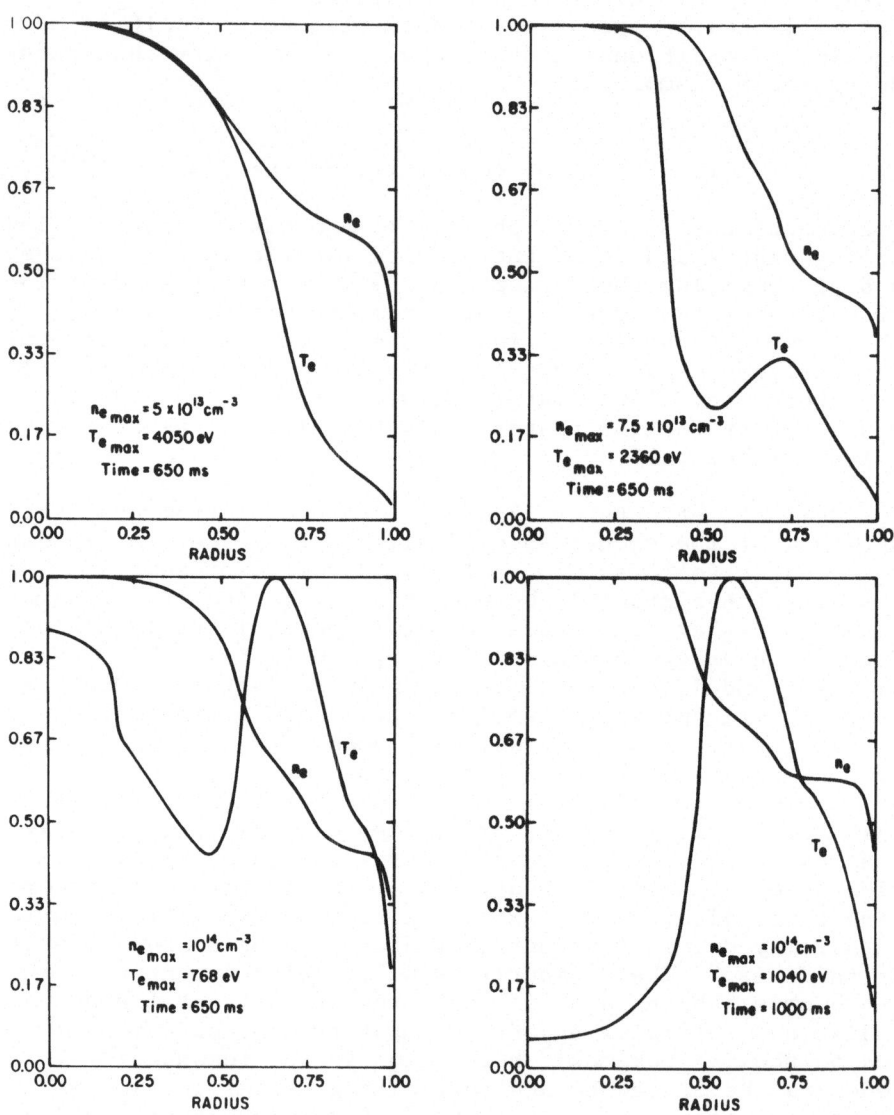

Fig. 4.   Computed radial profiles of electron density and
temperature for high current operation in TFTR.[1]   The central density has been varied.   The lower pair of pictures are at different
times in the same discharge.

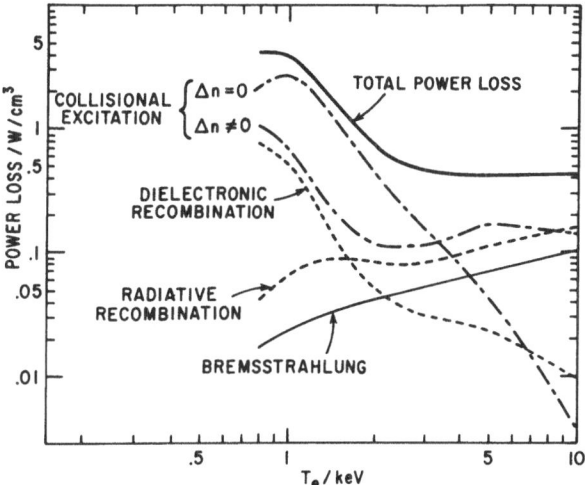

Fig. 5. Radiation power loss of iron impurities calculated by Merts et al.[21] ($n_e = 10^{14}$ cm^{-3}, $n_{Fe} = 10^{12}$ cm^{-3}). (From Ref. 22.)

current ($I_p = 2.5$ MA) in TFTR.[1] An initial oxygen concentration of 1.5% and an iron concentration of 0.1% was assumed. At the lowest central electron density of 5×10^{13} cm^{-3}, the electron temperature profile is roughly parabolic but as the density increases and hence the radiation loss increases, relative to the power input, the profile changes to a hollow profile with a rather cool peak temperature. The lower pair of curves show two times in the same discharge and reveals the development of this radiation-caused thermal instability. It is this type of result which causes the interest in experiments with magnetic divertors which essentially retract the limiter into a separate pumping chamber and can also decrease the probability of heavy sputtered atoms from the walls getting into the hot plasma.[18,19,20]

The radiative power loss due to iron at high levels of ionization has only recently been calculated by Merts et al.[21] The results can be summarized in Fig. 5[22] which is based on a coronal equilibrium model for which the ion transport times are assumed to be long. (Coronal equilibrium can be applied if $n\tau \gtrsim 10^{11}$ cm^{-3} sec where τ is the ion containment time.) At low temperature the loss is dominated by line radiation and dielectric recombination while at the higher temperatures predicted for future Tokamaks, radiative recombination and bremsstrahlung dominate. The presence of impurities can also lead to defects in heating by high energy neutral beams. The beams themselves will tend to sputter impurities off

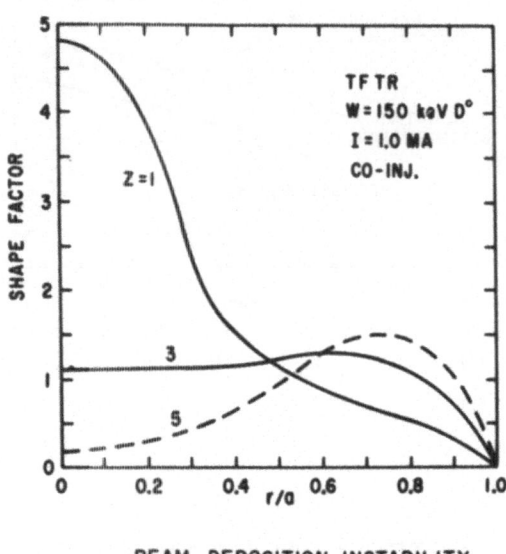

BEAM-DEPOSITION INSTABILITY

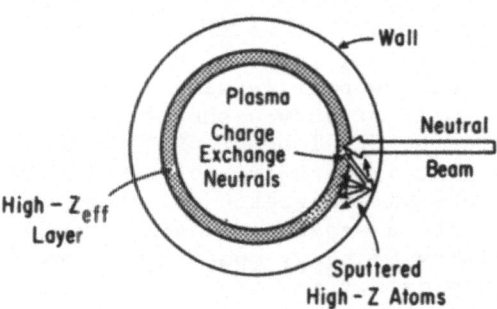

Fig. 6.   Schematic representation of effect of plasma impurities on neutral beam deposition.   The upper figure shows profiles of beam-deposited energetic ion density for various values of $\overline{Z}$. The lower figure shows the growth of the outer impurity layer.

the walls of the vacuum vessel to some extent. With injection
tangential to the major circumference, only a small fraction, of
order 10^{-4}, of the injected neutrals will hit the wall. However,
another effect is possible which is shown schematically in Fig. 6.[1]
As the beam enters the plasma, there is local ion heating at the
edge which enhances the charge-exchange in the edge region. The
product neutrals can reach the walls locally to cause sputtering
with the release of iron or other impurity into the edge region.
This enhanced impurity level at the edge increases the impact
ionization of the beam so that there is a growing beam deposition
instability curtailing the usefulness of the injection. The upper
figure shows that with $\bar{Z} = 5$, the deposition is strongest in the
plasma edge region.

The impurities will have a significant effect on the fusion
yield since the number of fusing particles is necessarily depleted
by their presence. For the TFTR, the fusion yield may be reduced
to make the attainment of $Q = 1$ impossible.

From the point of view of diagnostics, the impurities are
interesting in themselves and offer the possibility of measuring
ion temperatures fairly deep into the plasma by Doppler broadening
of spectral lines of highly ionized ions (e.g., Fe^{23+}, Mo^{30+}).
They provide an enhancement of the continuum radiation in the x-ray
region over the pure hydrogenic bremsstrahlung which should enable
electron temperature measurements by x-rays to be made in the presence of the high prompt flux of γ-photons caused by the fusion
neutrons colliding with the nuclei in the vacuum vessel wall and
the radiation shielding surrounding the device. The impurities
also strongly affect the scattering of far infrared laser light
which may be used in techniques for measuring the ion temperature
and the current density distribution. Evans and Yeoman[23] have
shown that there is a large enhancement at the center of the ion
feature due to the impurities, this distortion being as much as a
factor 2 for a fractional impurity level as low as 10^{-4} of fully
ionized molybdenum.

III.   SOME RECENT RESULTS

Extensive studies have been made of the radiation from the
plasma in the ST Tokamak in both the vacuum ultraviolet[24] and the
x-ray[22] regions of the spectrum. Amongst the most important work
have been studies of the iron spectra and the interpretation of the
enhancement of the continuum x-ray radiation over the pure hydrogenic levels. The x-ray signals could also be used for looking at
plasma fluctuations near the center of the plasma and for measuring
the radial electron temperature distribution. Recent results show
that fluctuations correlated with the x-ray fluctuations can also

be seen on highly ionized molybdenum lines in TFR (Mo XXXl) at an
electron temperature of about 2 keV.[25]

As an example of the ultraviolet spectroscopy, iron has been
selected although the larger radiation loss is due to oxygen.
Figure 7 shows the time behavior of various iron impurity lines in
the ST Tokamak integrated along a line of sight.[24] The average
density and three electron temperature points at about 1.8 keV are
also shown. The limiter radius was opened to a = 13 cm to let the
plasma be closer to the stainless steel wall; all the iron impurity
comes from that source. The Fe XV line behavior is an indication of
iron influx during the pulse. Both highly ionized lines have been
identified in solar flare operation so that the identification is
fairly certain. From the approximate rate coefficients for excitation at 1.8 keV for these lines an increase of about a factor 3 in
iron concentration is found to occur between 8 msec and 70 msec.
However, spatial distributions of the lines are essential for conclusions on the ion accumulation rate in future tokamaks.

The group working on TFR have also been working on the impurity
problem, and in particular on oxygen, iron, and molybdenum.[26]
Figure 8 shows calculated curves for the fractional abundance of
highly ionized states of iron as a function of temperature. The
upper curve includes dielectronic recombination relative to the lower
curve which only assumes radiative recombination. Dielectronic
recombination has the effect of raising the temperature at which any

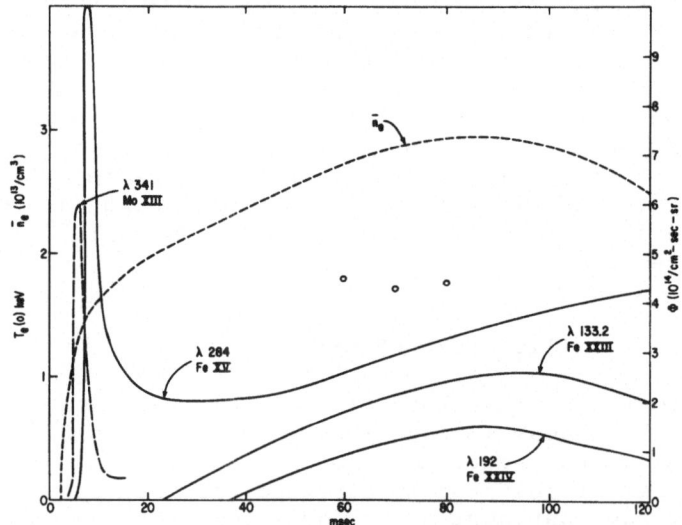

Fig. 7.   Time behavior of iron resonance lines in an ST
Tokamak discharge.[24] The average density and the MoXIII resonance
line are also shown. The circles indicate measurement of the peak
electron temperature.

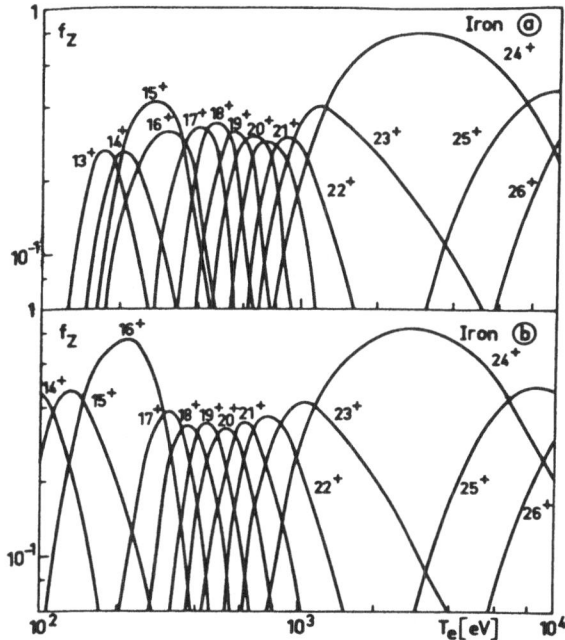

Fig. 8. Calculated curves of the fractional abundance of ionization states of iron.[26] The upper curve includes dielectronic recombination and radiative recombination, the lower curve only includes radiative recombination.

ionization state is most abundant to higher values. These curves show that, while Fe 23^+ is the dominant impurity level at the center of TFR, the helium-like and hydrogen-like states will dominate in TFTR. By producing a radial scan of a number of lines of different iron ionization states, a quantitative picture of the iron influx and its effect could be built up.

The presence of iron should also be identifiable from a spectrum taken in the x-ray region. In the ST Tokamak, a spectrum was taken with a low resolution pulse-height analysis system and is shown in Fig. 9.[22] The observed spectrum (a) is compared with a calculated spectrum by Merts et al. (b), for electron temperatures $T_e = 1.0$ and 1.2 keV and which accounts for the instrumental resolution.[21] The assumption of coronal equilibrium may not apply and this may explain the rather better agreement of the theory at $T_e = 1.0$ keV with the experiment at $T_e = 1.2$ keV, this temperature having been obtained by Thomson scattering.

The impurities also affect the quantitative emission of x-rays from the plasma. Since the bremsstrahlung and recombination

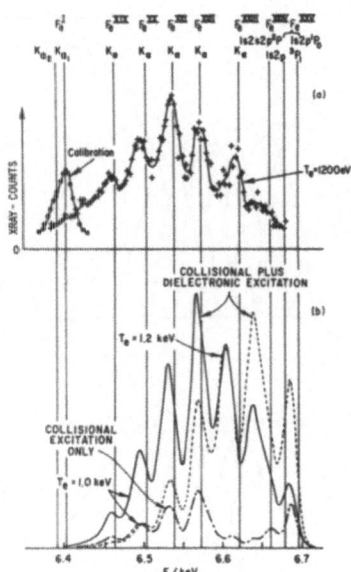

Fig. 9. Comparison of the K_{α}-line structure of iron
a) measured in the ST Tokamak for central electron temperature
1.2 keV[22] with b) computations of the K_{α}-line by Merts et al.[21]
for $n_e = 10^{14}$ cm^{-3}, $n_{Fe} = 10^{12}$ cm^{-3} for two values of temperature
1.0 and 1.2 keV. The contribution of direct-collisional excita-
tion at 1.0 keV only is also shown. (From Ref. 22).

radiation have almost the same functional dependence on the electron
temperature[16] and are proportional to $\Sigma n_i Z_i^2$ the x-ray continuum
shows a considerable enhancement over a pure hydrogenic plasma. For
conditions in the ST Tokamak, where the Thomson scattering system
indicates a temperature of 1.45 keV, the x-ray continuum enhancement over hydrogenic bremsstrahlung was 18 with iron the dominant
impurity.   The pulse height analysis curve is shown in Fig. 10.[16]
K_{α} lines of some other impurities in the plasma can also be seen.
Full analysis using the computed iron and oxygen enhancement factor
produced good agreement for the average impurity $\overline{Z}$ with that
obtained from the resistance measurement.

     Detailed studies of x-ray fluctuations which produced con-
siderable insight into the stability behavior of the plasma near the
hot center have been made in the ST,[22] ACT,[27] and TFR[10] Tokamaks.
These results will not be discussed here, but because of their
importance in understanding the magnetohydrodynamic stability of
the central core of the plasma and the inferences that can be
drawn from them about the current density distribution, their
measurement is considered essential for large tokamaks.

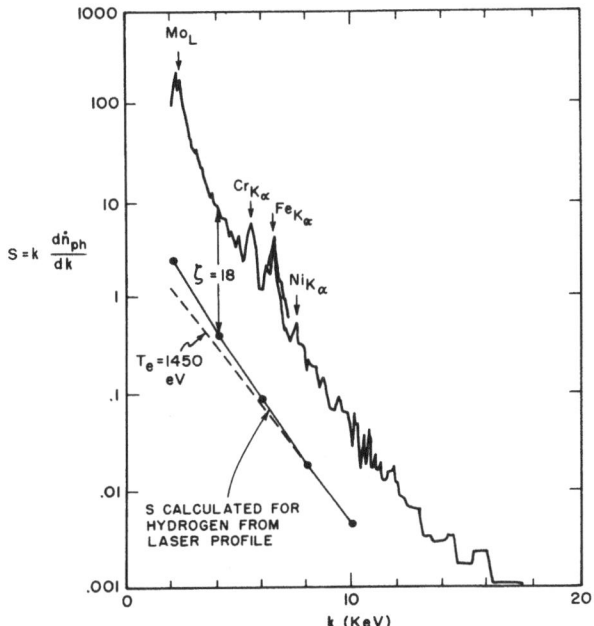

Fig. 10. The observed pulse height analysis x-ray spectrum for an ST Tokamak discharge. It is compared to the computed hydrogenic bremsstrahlung for the measured electron temperature and density profiles.

IV. PROBLEMS ASSOCIATED WITH EXTRAPOLATION TO LARGE TOKAMAKS

In considering larger tokamaks, and specifically TFTR, many technical problems appear. Diagnostics have to be capable of producing useful information from plasmas with large minor radius ($a \lesssim 85$ cm), high peak density ($n_e(o) > 10^{14}$ cm^{-3}) and high peak temperature ($T_e(o) \approx T_i(o) \lesssim 10$ keV). They must be capable of following a compression of the plasma and must share space around the tokamak with neutral beam lines, vacuum pumps and radiation shielding. Because of pulsing rate limitations due to the high magnetic field energies involved and total pulse limitations set by the permissible radiation levels, the diagnostics must collect large amounts of data in each pulse. It will be desirable to obtain spatial and temporal behavior of most plasma parameters concurrently during a single discharge. Rough estimates made to evaluate a computer data handling system suggest that about ten million words of data will be collected for each discharge at full power.

Another requirement for the TFTR is that some measure of feed-back control of the plasma will be necessary. The high energy densities of the plasma and the poloidal flux require that the energy be released in as gentle a fashion as possible for protection of the vacuum vessel and the coil systems. In particular, the rapid disruption of the current common to all tokamaks at certain plasma conditions should be prevented. The diagnostic techniques for measurement of plasma current, electron density, plasma position, and large scale magnetic fluctuations for providing data for the feedback system will be quite conventional. The development problem lies in determining the precursor identifying symptoms of the disruption, determining how to integrate them into a control system, and determining the proper response.

Some of the plasma parameters for which measurement is necessary to ensure the usefulness of the large TFTR as an experimental device are summarized in Table II. Specifically the problems of measuring the power input directly from the measured voltage and current, the electron temperature by forward scattering of ruby laser light rather than 90° scattering, the plasma density distribution by interferometer and analysis of wall surfaces have been excluded as these are largely technological problems.   The most used techni-que for measuring the ion temperature has been energy analysis of charge-exchanged neutral atoms emerging from the plasma. Figure 11 shows the penetration length (e-folding length) as a function of the emergent neutral hydrogen atom's energy using the cross

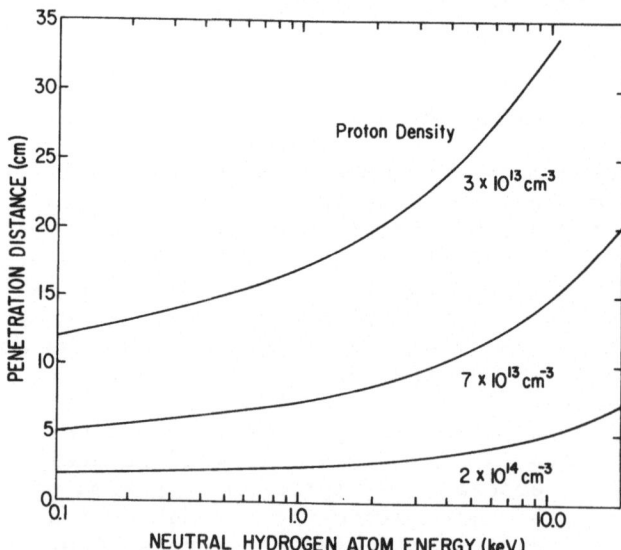

Fig. 11. The penetration length for neutral hydrogen atoms emerging from inside the plasma as a function of energy. Electron ionization collisions have been ignored.

section measurements of Stebbing et al.,[28] the three lines being for different proton densities. For the projected discharges in TFTR the range of a neutral atom emerging from the axis is about 10 cm so that there may be as many as eight mean free paths in its trajectory outwards. Since the neutral atom density will be less than 10^8 atoms cm^{-3} (or charge-exchange will dominate the ion energy loss at the axis) measurement becomes impossible without an artificial enhancement such as that proposed by Eubank et al.[28] for PLT where a modulated atom beam will be injected into the plasma. The measurement of neutral hydrogen density will have to be done independently and a resonance fluorescence scheme has been proposed by Koopman et al.[29] This group is currently developing a laser system involving a dye laser followed by a frequency doubler and a tripling stage to produce radiation at 1216 A°.

Other techniques for measuring the ion temperature have been proposed. If sufficient suitable forbidden lines in the ultraviolet spectrum can be found, then Doppler broadening measurements will be made. Doppler shift measurements can also identify gross plasma motions. Lines of Fe XX, XXI, and XXII have so far been identified and Hinnov[30] plans to attempt measurement of the ion temperature distribution by this technique on PLT. It may be necessary to seed the plasma with test impurities to find lines relevant to the higher temperatures of TFTR. Von Goeler et al.,[31] are studying the possibility of a curved crystal spectrometer for looking at Doppler broadening of iron impurity lines in the x-ray region in PLT. To extend this measurement, it is essential for measurements of cross sections and spectra of other relevant highly ionized atoms to be determined.

It is also possible to obtain the temperature from collective scattering of laser radiation with $\lambda \gtrsim \lambda_D$. λ is the incoming far infrared wavelength and λ_D is the plasma Debye length, typically of order 50 μm. Methyl fluoride (CH_3F)[32] and carbon dioxide (CO_2)[33] laser systems have been proposed but neither has been tried on a tokamak plasma. Trials on Alcator, Dite, and PLT are expected within the next two years.

Studies of the neutrons will obviously play a major role in determining the success of the heating of TFTR. The large flux ($\sim 6 \times 10^{12}$ neutrons cm^{-2} sec^{-1}) of virgin neutrons predicted for the operation with tritium and smaller flux ($\sim 6 \times 10^{10}$ neutrons cm^{-2} sec^{-1}) predicted for operation with deuterium[1] must be used for detailed measurements of the ion temperature and slowing down of the beam particles. The developments required are predominantly adaptations of current nuclear physics methods with, however, an extended intense source, long times of operation and very high background fluxes of scattered neutrons slightly degraded in energy.

TABLE II   Diagnostic Development Required

Parameter To Be Measured	Problem with Present Techniques	Proposed New Techniques	Research and Development Required
Ion Temperature	Mean free path limits charge-exchange neutrals.	Enhance neutral density by modulated 20 keV beam.	Check feasibility on PLT.
		Doppler broadening of impurity lines in UV.	Identify suitable forbidden lines. Develop scanning UV spectrograph.
		Doppler broadening of impurity lines in x-ray region.	Develop curved crystal spectrometer. Identify highly ionized Mo, W spectra.
		CH_3F laser scattering (496 µm)	Develop laser, detector, optics system.
		CO_2 laser scattering (10 µm)	Develop detector, optics system.
		Neutron yield and line width.	Develop collimator system, adapt spectrometer systems. Measure TT cross section.
Quantitative Impurity Measurement	Low mass impurities fully ionized for most of plasma cross section.	Bremsstrahlung enhancement (x-ray).	Ionization rates, cross section measurement. Develop curved crystal spectrometer.
		Intensities of high Z lines in UV and x-ray.	See ion temperature.

TABLE II Diagnostic Development Required (cont.)

Parameter To Be Measured	Problem with Present Techniques	Proposed New Techniques	Research and Development Required
Neutral Hydrogen Density	Mean free path limits charge-exchange neutrals.	Lyman α fluorescence	Develop light source at 1216 A°. Prove feasibility in presence of plasma light.
Current Density Distribution	No satisfactory measurement.	Heavy neutral particle probe.	Develop 2 MeV source of neutral Thallium.
			Establish detector system. Prove sensitivity.
		Modulation of CO_2 laser at ω_{ce}.	Set up laser-detector system. Prove feasibility on PLT.
		Polarization of Zeeman split Li impurity line.	Develop Li beam source, optics. Prove feasibility on Pulsator.
		Faraday rotation of HCN laser light.	Develop complete measurement system. Prove feasibility.

The ultraviolet spectral and x-ray spectral measurements for identifying impurity ions have very largely the same requirements as for the ion temperature measurement. To obtain a measurement of the average impurity level, or $\bar{Z}$, it is desirable to measure the bremsstrahlung enhancement γ factor discussed earlier. However, the interpretation is clearly very difficult and a much greater knowledge of the radiation from heavy non-ferrous impurities is necessary. The actual measurement of the continuum will be difficult during operation with tritium because of the high photon fluxes from the shielding materials and from the vacuum vessel wall facing the detector. Instead of using a pulse height analysis system with direct line-of-sight viewing of the plasma by the detectors, a low resolution crystal spectrometer might be considered. Figure 12 shows a possible arrangement of the connection of a multi-position spectrometer to the vacuum vessel where the spectrometer would be underneath six feet of shielding under the tokamak and the top tubes are placed to provide a considerably reduced promptly flux from the vacuum vessel in the line of sight. The scattered γ photon flux at the spectrometer is estimated to be of order 10^8 photons cm^{-2} sec^{-1}; the bremsstrahlung signal is of order 10^9 photons cm^{-2} sec^{-1}. To allow reasonable counting rates and dynamic range additional shielding will have to be incorporated in the spectrometer design.

For laser scattering from an impurity contaminated plasma, the ion feature can be dominated by an impurity peak at the center of the scattered line. The line shape is dependent on the average impurity concentration, $\bar{Z}$, and α, the ratio of the scattering scale length to the Debye length. By using two scattering angles to produce different values of α, $\bar{Z}$ can be determined.[23] The most probable choice of laser wavelength is 496 µm. A trial is expected shortly.

The measurement of current density distribution gains importance for TFTR because of the high power neutral beam auxiliary heating which will modify the poloidal flux and because of the difficulty of measuring the power input using a measurement of voltage around the toroid. No universally satisfactory technique has been achieved though a heavy ion beam was used on the ST Tokamak[34] and its development will be extended on PLT. In this technique, the singly ionized thallium ions are ionized by collision to Tl^{2+}. They then move parallel to the toroidal minor axis because of the poloidal field. In practice, the measurement was better at determining the radial electric field in the plasma by the change in energy of the Tl^{2+} ions.[35] For TFTR, the energy of the thallium beam must be increased and the penetration can only be achieved by means of a neutral thallium beam. The technique of looking at the Thomson scattering of 10 µm radiation from a CO_2 laser is being developed but has not been proven feasible in tokamaks.[36,37] In a magnetic field there is a qualitative change

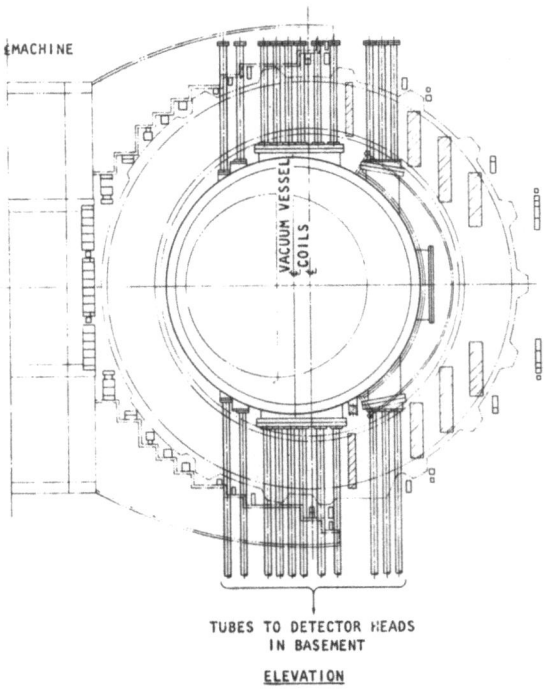

Fig. 12.   A view of a section of the TFTR Tokamak showing a
speculative arrangement of a multichannel x-ray spectrometer at the
TFTR vacuum vessel.   The lower set of tubes passes through the
coils and six feet of shielding material to the spectrometer in a
basement.   The upper set of tubes passes beyond the radiation
shielding to cut γ-photon flux from the line-of-sight to a minimum.

in the spectrum of Thomson scattered light when the difference
between the incident and scattered wave vectors is perpendicular to
that field.   In this case the Gaussian spectrum becomes a series
of sharp lines separeted by the electron gyrofrequency.   Sufficient
accuracy in determination of the line separation allows measurement
of the poloidal field.   McCormick[38] is developing a system for use
on Pulsator using an injected lithium beam and looking at the
polarization of the components of the Zeeman triplet.   The polarization is dictated by the local magnetic field direction, from
which the strength of the poloidal field can be found.   The
feasibility of this technique awaits trial and it is doubtful
whether the high density expected in TFTR will allow its use.
Faraday rotation of submillimeter laser light in the poloidal field
(rotation $\alpha \int n_e B_z dl$ where B_z is the magnetic field in the direction
of propagation and the integration is over the path length in the

plasma) has been studied theoretically[39] but has not been used in a tokamak.

It is unlikely that more than one of the above techniques for measuring any one of the plasma parameters listed in Table II will be installed on the TFTR device but is is a measure of the difficulty of the diagnostic problem that so many varied techniques are considered worthy of study. It is certainly probable that other techniques will be proposed and tested by 1981, the projected start-up date for TFTR.

V.   CONCLUSIONS AND ACKNOWLEDGMENTS

In a paper addressing the problems of plasma diagnostics in large tokamaks it is inevitable that many topics receive short shift. In this paper the problems most relevant to atomic physicists have been given most weight. Some results from recently operating tokamaks have been given to show the current state of the art in ultraviolet spectroscopy and x-ray spectroscopy.

The TFTR Tokamak is expected to start operation by mid-1981. By that time, many of the diagnostic problems should have been solved. This will require some research in the area of atomic physics. Specifically for the heavier metals in highly ionized states there is very little relevant atomic data.

The list of references in this paper is far from comprehensive and much comparable work is being done by other people. I apologize for not referring to all the available literature. My special thanks go to S. von Goeler and E. Hinnov for discussions relevant to this paper. This work was supported by the U. S. Energy Research and Development Administration Contract E(11-1)-3073.

REFERENCES

[1]Tokamak Fusion Test Reactor: Final Conceptual Design Report, Vol. 1 (Princeton Plasma Physics Laboratory, 1975).

[2]R. R. Parker, Bull. Am. Phys. Soc. 20, 1372 (1975).

[3]B. Cannici, W. Engelhardt, J. Gernhardt, E. Glock, F. Karger, O. Klüber, G. Lisitano, D. Meisel, P. Morandi, and S. Sesonic, Proc. VII European Conf. on Controlled Fusion and Plasma Physics (Lausanne) 1, 136 (1975).

[4]J. D. Lawson, Proc. Phys. Soc. (London) B70, 6 (1957).

[5]L. A. Artsimovich, Nucl. Fusion 12, 215 (1972).

[6]H. P. Furth, Nucl. Fusion 15, 487 (1975).

[7]E. Hinnov, J. Nucl. Mat. 53, 9 (1974).

[8]L. L. Gorelik, S. V. Mirnov, V. G. Nikolevsky, V. V. Sinitsyn, Nucl. Fusion 12, 185 (1972).

[9]D. L. Dimock, H. P. Eubank, E. Hinnov, L. C. Johnson, and E. B. Meservey, Nucl. Fusion 13, 271 (1973).

[10]Equipe TFR, Proc. VII European Conf. on Controlled Fusion and Plasma Physics (Lausanne) II, 1 (1975).

[11]W. Stodiek, private communication.

[12]D. E. Post, P. H. Rutherford, H. P. Furth, and R. R. Smith, Princeton Plasma Physics Laboratory MATT-1262 (1976).

[13]L. A. Berry, C. E. Bush, J. L. Dunlap, P. H. Edmonds, T. C. Jernigan, J. F. Lyon, M. Murakami, and W. R. Wing, Plasma Physics and Controlled Nuclear Fusion (IAEA, Vienna) I, 113 (1974).

[14]K. Bol, J. L. Cecchi, C. C. Daughney, R. A. Ellis, Jr., H. P. Eubank, H. P. Furth, R. J. Goldston, H. Hsuan, E. Mazzucato, R. R. Smith, and P. E. Stott, Plasma Physics and Controlled Nuclear Fusion (IAEA, Vienna) I, 77 (1974).

[15]V. A. Vershkov and S. V. Mirnov, Nucl. Fusion 14, 383 (1974).

[16]S. von Goeler, W. Stodiek, H. Eubank, H. Fishman, S. Grebenshchikov, and E. Hinnov, Nucl. Fusion 15, 301 (1975).

[17]S. A. Cohen, J. L. Cecchi, and E. S. Marmar, Phys. Rev. Lett. 35, 1507 (1975).

[18]Princeton Plasma Physics Laboratory Annual Report MATT-Q-32 (1974).

[19]C. Colven, A. Gibson, and P. E. Stott, Proc. V European Conf. on Controlled Fusion and Plasma Physics (Grenoble) 1, 6 (1972).

[20]Y. Shimomura, Proc. VII European Conf. on Controlled Fusion and Plasma Physics (Lausanne) II, 81 (1975).

[21]A. L. Merts, R. D. Cowan, and M. H. Magee, Jr., Los Alamos Scientific Laboratory Report LA-6220-MS (1976).

[22]S. von Goeler, Proc. VII European Conf. on Controlled Fusion and Plasma Physics (Lausanne) II, 71 (1975).

[23]D. E. Evans and M. L. Yeoman, Phys. Rev. Lett. 33, 76 (1974).

[24]E. Hinnov, Princeton Plasma Physics Laboratory MATT-1240 (1976).

[25]J-P Girard, private communication.

[26]Equipe TFR, Nucl. Fusion 15, 1053 (1975).

[27]R. R. Smith, Princeton Plasma Physics Laboratory MATT-1150 (1975).

[28]H. P. Eubank, private communication.

[29]D. W. Koopman, T. J. McIlraith and V. P. Myerscough, Bull. Am. Phys. Soc. 20, 1321 (1975).

[30]E. Hinnov, private communication.

[31]S. von Goeler, private communication.

[32]D. L. Jassby, D. R. Cohn, B. Lax, and W. Halverson, Nucl. Fusion 14, 745 (1974).

[33]N. Bretz, private communication.

[34]R. L. Hickok and F. C. Jobes, Bull. Am. Phys. Soc. 16, 1231 (1971).

[35]F. C. Jobes and J. C. Hosea, Proc. VI European Conf. on Controlled Fusion and Plasma Physics (Moscow) I, 199 (1973).

[36]N. Bretz, Applied Optics 13, 1134 (1974).

[37]M. Murakami and J. F. Clarke, Oak Ridge National Laboratory Report ORNL-TM-3128 (1970).

[38]J. Fujita and K. McCormick, Proc. VI European Conf. on Controlled Fusion and Plasma Physics (Moscow) I, 191 (1973).

[39]F. deMarco and S. E. Segre, Plasma Physics 14, 245 (1972).

RELATIVISTIC MAGNETIC-DIPOLE TRANSITIONS IN ATOMS, IONS AND PSIONS

Joseph Sucher

Department of Physics of the University of Maryland

College Park, Maryland 20742

I. INTRODUCTION

Since my talk will illustrate, to some extent, the enormous scope and power of quantum electrodynamics, it may be appropriate to begin by noting that the receipt date of Dirac's paper[1] founding the subject is February 2, 1927. If we assume that it took more than 32 days of creation, the year 1926 marks at least the conception if not the birth of Q.E.D. So we are at the golden anniversary year (See Fig. 1).

Now let's consider the title of the talk. What is a relativistic magnetic-dipole (M1) transition? Why isn't it something you've always wanted to know about, but were afraid to ask? Why should you care now? What are psions?

Briefly, a relativistic M1 transition is a radiative M1 transition of an S-state to another S-state with an orthogonal radial wavefunction -- a highly forbidden transition in atomic physics. Interest in forbidden transitions was for a long time restricted to astrophysics, where collisional de-excitation can be sufficiently slow to allow metastable states to live out their natural life, to then die a natural death with the emission of one (or more) photons. Subsequently, they became of interest for solar physics, in connection with the study of the soft X-ray spectrum of the solar corona. In addition, some indication arose of the presence of such transitions in laboratory plasmas. However, the subject really came into its own with the development of atomic beam-foil spectroscopy, associated with the construction of heavy-ion linear accelerators. In 1970 this led to the first direct detection in the laboratory of the decay of an atomic state via a forbidden transition, namely $2^3S_1 \rightarrow 1^1S_0 + \gamma$ in Ar^{+16} (helium-like argon) and, subsequently, to the measurement of the lifetime of the state. Since then, there has been a host of further measurements of this kind, in both H-like and He-like ions, ranging from sulphur (Z = 16) all the way to krypton (Z = 36). This has been accompanied by a corresponding increase in theoretical activity, some of which I shall try to describe.

Finally, by one of those twists which illustrate the unity of nature and of science, the spectacular discovery of the new very narrow resonances J/ψ and ψ', first produced in p-Be collisions at BNL and first formed in electron-positron collisions at SLAC, has led to an interest in both ordinary and relativistic M1 transitions in elementary particle physics. This is because in one of the most popular views of these new particles they are pictured as nonrelativistic bound-states of a heavy quark "c" and an antiquark "$\bar{c}$". For reasons connected with the theory of weak interactions, these quarks are thought to carry a new quantum number, ordinarily called "charm." The "psions" ψ and ψ', both J = 1 states, are then regarded respectively as ground and excited $^3S_1$ states of a bound $(c\bar{c})$ system, called "charmonium", in analogy with positronium, the bound (e^+e^-) system. The charmonium model predicts, among other things, the existence of a $^1S_0$ state, the "hyperfine" partner of ψ, into which ψ and ψ' should decay via ordinary and relativistic M1 transitions respectively. Thus we may be on the threshold of an entirely new "atomic" spectroscopy.

We see that an understanding of these transitions is in fact now of interest for a considerable class of observed physical phenomena, as symbolized by Fig. 2.

Let me now outline the remainder of this talk. I will first give a quick historical survey of events, through 1969, leading up to the present interest in relativistic M1 transitions in atomic physics. I will then describe the results of atomic beam-foil experiments during the past five years (1970-1975), sketch the theory of M1 transitions for He-like ions and review the present state of the comparison of theory and experiment.[2] In the second part of my talk I will give you a bird's eye view of the charmonium model and all that, sketch the extension of the theory of M1 transitions to this case, and describe the current status of the psion radiative decays.

HAPPY GOLDEN ANNIVERSARY!

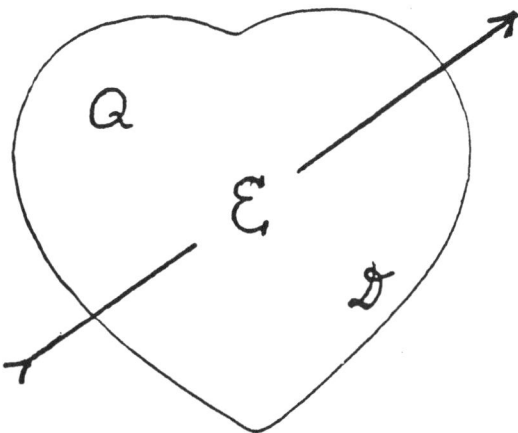

Fig. 1.   50th Anniversary of Q.E.D.   (Heart, courtesy of Anne Sucher).

1930's: Astrophysics

1960's: Solar physics, plasma physics

1970's: Beam-foil spectroscopy
         Particle physics:
            New charms in spectroscopy?
            Parity violations in radiative decay?

Fig. 2.   Growth of interest in M1 transitions.

II. HISTORICAL SURVEY: HOW DO METASTABLE S-STATES DECAY?

A. Preliminaries

The selection rules for transitions of the type $|\psi_i\rangle \rightarrow |\psi_f\rangle + \gamma$ are not only well-known and found in most text books dealing with such matters, but better yet, easily derivable by expanding the wavefunction $\vec{\varepsilon}\, e^{-i\vec{k}\cdot\vec{r}}$ of the emitted photon in spherical harmonics:

$$\vec{\varepsilon}\, e^{-i\vec{k}\cdot\vec{r}} = \sum_{\ell=0}^{\infty} (2\ell+1)\,(-i)^{\ell}\, j_{\ell}(kr)\, P_{\ell}(\hat{k}\cdot\hat{r})\, \vec{\varepsilon}\; .$$

They are: E1: $\Delta J = 0, \pm 1$, "YES"; M1: $\Delta J = 0, \pm 1$, "NO", etc. Let us apply these to the simplest excited state in atomic physics, the first excited S-state of hydrogen.

How does the $2\,S_{1/2}$ state decay? Parity conservation implies that the transition $2\,^2S_{1/2} \rightarrow 1\,^2S_{1/2} + \gamma$ must be of the magnetic multipole type. Let us treat the electron in the Schroedinger-Pauli (S.P.) approximation. The interaction with the quantized transverse radiation field $\vec{A}_T(\vec{r})$ is then given by:

$$H' = -e\,\frac{\vec{P}}{m}\cdot\vec{A}_T - \vec{\mu}\cdot\vec{H}_T$$

where $\vec{\mu} = g\left(\frac{e}{2m}\right)\vec{s}$ is the electron magnetic moment and $\vec{H}_T = \vec{\nabla} \times \vec{A}_T$. Apart from a factor $-e/\sqrt{2\omega}$, the matrix element for a transition with $\Delta J = 0$, no parity change, is given by $M = M_{orb.} + M_{spin}$, where $M_{orb.}$ arises from the first term in H' and

$$M_{spin} = \Phi_f \left| \frac{g}{2m}\, \vec{s}\cdot(-i\vec{k}) \times \vec{\varepsilon}\, j_o(kr) \right| \Phi_i \rangle\; ,$$

with the Φ's eigenfunctions of $H_{n.r.} = (p^2/2m) - Z\alpha/r + H_{f.s.}$, and $H_{f.s.}$ the fine structure operator. The general form of Φ for an S-state is $\Phi = \chi R(r)$, with χ a Pauli spinor. Since $\vec{\varepsilon}\cdot\vec{k} = 0$, $M_{orb.} = 0$ for the transition of interest and, with $g \approx 2$, we get

$$M_{spin} = -i(\Sigma/2)\, \langle R_f | j_o(kr) | R_i \rangle, \quad \Sigma \equiv \langle \chi_f | \vec{\sigma}\cdot\frac{\vec{k}}{m} \times \vec{\varepsilon} | \chi_i \rangle\; .$$

But $\langle R_f | R_i \rangle = 0$ and $j_o(\rho) = 1 - \frac{\rho^2}{6} + \dots$, so that the leading term is

$$M^{S.P.}(2\,^2S_{1/2} \rightarrow 1\,^2S_{1/2} + \gamma) \equiv \frac{i\Sigma}{12}\, \langle R_f | k^2 r^2 | R_i \rangle\; . \tag{2.1}$$

Since $kr \sim \alpha Z$, $k/m \sim (\alpha Z)^2$, we have

$$M \sim (\alpha Z)^4$$

and the transition is highly suppressed relative to an El transition for which M ~ αZ, or an ordinary Ml transition for which M ~ $(\alpha Z)^2$. Note that an extra factor of $(\alpha Z)^2$ arose from the orthogonality of R_f and R_i. This is our first example of a "relativistic" Ml transition.

The adjective is somewhat misleading, because as the above calculation shows, the amplitude, although small, is non-zero even when the electron is treated nonrelativistically. However, as we shall see, the above treatment, although correct as to order of magnitude yields the wrong answer (it is too small by a factor of 2). To get the right answer, the electron must be treated relativistically -- more precisely, relativistic effects which contribute to the leading term in M are missed if one treats the electron in the Schrödinger-Pauli approximation.

B. History: t = $-\infty$ to 1969

1930: Maria Goeppert-Mayer[3] suggests that the dominant decay mode for the $2^2S_{1/2}$ state is in fact two-photon emission (each photon being emitted via a virtual El transition), i.e.

$$R_{2El} \ (2^2S_{1/2} \to 1^2S_{1/2} + 2\gamma) \gg R_{Ml} \ (2^2S_{1/2} \to 1^2S_{1/2} + \gamma).$$

-- 10 YEARS PASS --

1940: G. Breit and E. Teller,[4] motivated by astrophysical considerations, re-examine the problem of the decay modes of low-lying S-states in both H and He.

(a) For H, they estimated

$$\tau_H(2^2S_{1/2} \to 1^2S_{1/2} + 2\gamma) \approx \tfrac{1}{7} \text{ sec.}$$

To discuss the Ml transition they began with the amplitude given by Dirac theory:

$$M^D = \ < \psi_f |\vec{\alpha} \cdot \vec{\epsilon} \ e^{-i\vec{k} \cdot \vec{r}} |\psi_i > \ , \tag{2.2}$$

where the ψ's are Dirac wavefunctions, and evaluated M^D approximately to get

$$\tau_H(2^2S_{1/2} \to 1^2S_{1/2} + \gamma) \approx 2 \text{ days} ,$$

confirming the conclusions of Maria Goeppert-Mayer.

(b) For He, recall first the familiar division of the states
into parahelium (spin-singlet states, S=0) and orthohelium (spin-triplet states, S = 1). The lowest lying S-states are shown in
Fig. 3. First how does 2^1S_0 decay? The one-photon decay $2^1S_0 \rightarrow 1^1S_0 + \gamma$ is now strictly forbidden (0 $\neq$ 0); Breit and Teller
estimated, as one would expect,

$$\tau_{He}(2^1S_0 \rightarrow 1^1S_0 + 2\gamma) \sim \tau_H(2^2S_{1/2} \rightarrow 1^2S_{1/2} + 2\gamma).$$

Next, how does 2^3S_1 decay? This is the question of greatest
interest for this talk. Breit and Teller concluded that the 2E1
decay dominates the M1 decay (by a huge factor):

$$R_{He}(2^3S_1 \rightarrow 1^1S_0 + 2\gamma) \sim 10^{20} R_{He}(2^3S_1 \rightarrow 1^1S_0 + \gamma),$$

and that the lifetime for the 2γ decay was $\sim 10^{-5}$ sec.

-- 29 YEARS PASS - ENTER THE SUN --

<u>1969</u>: (α) A. H. Gabriel and Carole Jordan[5] note that certain lines
seen in the soft X-ray spectrum of the solar corona correspond in
wavelength to the transition

$$2^3S_1 \rightarrow 1^1S_0 + \gamma \tag{2.3}$$

in certain <u>He-like ions</u>: C V, O VII, Ne IX, Mg XI, Si XIII.

(β) H. Griem[6] shows that the Gabriel-Jordan interpretation of
these lines is plausible by pointing out that Breit and Teller
vastly underestimated the rate for the transition in question (for
helium). An estimate of the M1 decay rate (patterned after the
original Breit-Teller calculation for hydrogen) indicates that the
rate is sufficiently large, compared to collisional de-excitation
rates, to explain the strong emission observed, and to make 2γ
decay negligible, for Z $\leq$ 20.

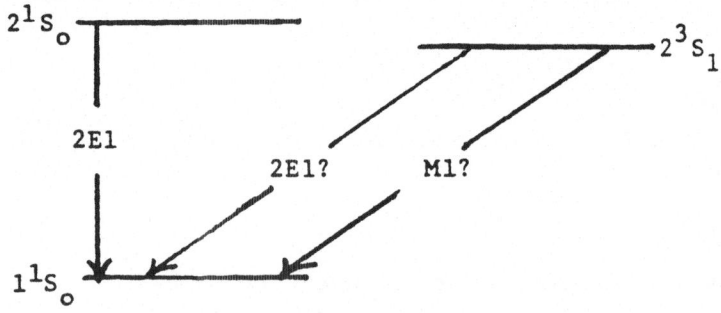

Fig. 3. Lowest-lying S-states of He-like ions.

This completes our survey of the early theoretical work. We should add that Gabriel and Jordan also developed a scheme to determine electron densities in the solar corona, based on the observed intensities of the lines in question and on theoretical decay rates.[5] It thus became of interest to carry out more refined calculations of these rates. Such calculations soon became of even greater interest with the advent of accurate experimental results from beam-foil spectroscopy for decays with $Z \geq 16$, beginning in 1970 with the observation of (2.3) in Ar^{+16} and with measurement of the rate by Marrus and Schmieder.[7] One might say that with these measurements the subject had finally come down to earth.

III. $2^3S_1 \rightarrow 1^1S_o + \gamma$ IN He-LIKE IONS: EXPERIMENT AND THEORY (1970-1975).

A. Experiment

"In spite of the fact that the hydrogen and helium spectra can probably qualify among the most studied problems in the history of physics, there is relatively little information concerning the forbidden decays, i.e. those decays which proceed by other than an allowed electric dipole transition."

This statement, the opening sentence in a 1972 paper of Marrus and Schmieder,[8] is happily no longer descriptive of the situation, at least as far as the M1 decays for large Z are concerned. The lifetime for the 2^3S_1 state has now been measured for many of the He-like ions, ranging from Z = 16 (sulfur) to Z = 36 (krypton). All these measurements have been carried out by the atomic beam-foil method in which ions, emerging from a HILAC (heavy-ion linear accelerator) are sent through thin foils, thereby stripping them of still more electrons, and creating a beam of He-like ions, some of which are in the 2^3S_1 state. Measurement of the intensity of the photons from the M1 decay of this state as a function of the distance from the foil then determines the lifetime. The most recent results of such measurements for helium-like sulfur,[9] chlorine,[9] argon,[10] titanium,[11] vanadium,[12] iron,[12] and krypton[13] (how exotic can you get?) will be displayed later, after a discussion of the theory of these decays.

It should be noted that the lifetimes range from $\sim 10^{-6}$ sec. for Z = 16 to $\sim 10^{-10}$ sec. for Z = 36, over more than three orders of magnitude. This corresponds to the fact that for large Z, the rate goes as Z^{10} and $(36/16)^{10} \sim 3 \times 10^3$. These values of Z are near the limits of what can be achieved by these methods: for much larger Z one gets too close to the foil, and for Z < 16 one

begins to run out of real estate.   [For v/c ~ .1, a decay length
is 30 m when $\tau = 10^{-6}$ sec.]

It would be a pity if one could not add the first member of
the helium iso-electronic sequence, namely ordinary helium, to the
list of atoms for which the lifetime is measured. The long lifetime
($\sim 10^4$ sec) makes the detection of this decay exceptionally
difficult. The 654 Å X-ray expected from the transition $2^3S_1 \rightarrow$
$1^1S_0 + \gamma$ in HeI was first seen in the after-glow from a helium
radiofrequency discharge and the decay rate was roughly determined
by Woodworth and Moos; quite recently the error in this deter-
mination was greatly reduced, giving $\tau^{-1} = 1.10 \times 10^{-4}$ sec^{-1}, with
a 2σ uncertainty of $\pm$ 30%.[14]

Because of the role played by the solar corona in our story,
and the fact that helium itself was first discovered in the sun, it
seems only poetic justice that we can indeed add HeI to our list.
This extends the range of the lifetimes in question to over
fourteen orders of magnitude and greatly increases the severity of
a comparison of experiment and the theory, to which we now turn.

B.   Theory

The computation of the matrix element M for the transition
$2^3S_1 \rightarrow 1^1S_0 + \gamma$ in He and He-like ions is, in principle, a
straightforward exercise in quantum electrodynamics. In practice,
it is rather delicate for HeI and low-Z ions, since one cannot then
treat the electron-electron interaction as a small perturbation.
Because of the highly relativistic nature of the decay, one needs
a formalism in which this interaction is treated relativistically.
I will sketch the approach used by G. Feinberg and myself, based
on quantum electrodynamics, to evaluate $M = M(\alpha, \alpha Z)$ to lowest order
in α and αZ (but to all orders in $1/Z$).[15] The same result for the
leading term was obtained independently by G. W. F. Drake using
semiclassical radiation theory.[16]

1. <u>H-like ions</u>. As a warm-up we shall first consider the M1
decay $n^2S_{1/2} \rightarrow n'^2S_{1/2}$ in H-like ions. Although in this case M^D
(Eq. (2.2)) may be evaluated exactly,[17] it is more instructive, in
preparation for He-like ions, to find the leading term in an
expansion in powers of αZ. The general form of the Dirac wavefunction ψ for an $n^2S_{1/2}$ state is

$$\psi = \frac{1}{\sqrt{4\pi}} \begin{pmatrix} g(r) \chi \\ -if(r) \vec{\sigma} \cdot \vec{r} \chi \end{pmatrix}, \quad f = \frac{g'(r)}{E+m-V}$$

where χ is a Pauli spinor. Then M^D reduces to

$$M^D = -i \ \Sigma \ T_H \tag{3.1a}$$

where

$$\Sigma = m^{-1} \ \langle \chi_F | \vec{\sigma} \cdot \vec{k} x \vec{\epsilon} | \chi_I \rangle \tag{3.1b}$$

and

$$T_H = -(m/k) [\ \langle g_F | j_1(kr) | f_I \rangle + \langle f_F | j_1(kr) | g_I \rangle] \ . \tag{3.1c}$$

If one replaces $g(r)$, $f(r)$, and $j_1(kr)/k$ by their leading approximations, $g^{(o)} = R(r)$, $f^{(o)} = R'(r)/2m$, $(j_1/k)^{(o)} = r/3$ where R is the non-relativistic radial wavefunction, then

$$T_H \rightarrow const. \quad \langle R_F | R_I \rangle = 0 \ ,$$

which is the origin of the characterization of the transition as "relativistic". Writing $g = g^{(o)} + \delta g$, $f = f^{(o)} + \delta f$, $j_1(k) = (r/3) + \delta(j_1/k)$, we find, correspondingly, three contributions to T_H.

Neglecting δf and $\delta(j_1/k)$ one gets a term $\langle g_F | g_I \rangle / 2m$, which is non-vanishing, because only $\langle \psi_F | \psi_I \rangle = \langle g_F | g_I \rangle + \langle f_F | f_I \rangle = 0$. Since $\langle f_F | f_I \rangle \simeq \langle g_F' | g_I' \rangle / 4m^2 = \langle g_F | p^2 / 4m^2 | g_I \rangle$ one finds a "non-orthogonality correction" T_g, shown below. Similarly, neglecting δg and $\delta(j_1/k)$, one gets a small-component wavefunction correction term, T_f, and neglecting δg and δf one gets a retardation term $T_{ret.}$:

$$T_g = \langle R_F | -\vec{p}^2 / 8m^2 | R_I \rangle \ , \quad T_f = \langle R_F | -5p^2 / 24m^2 + rV'(r)/6m | R_F \rangle \ ,$$

$$T_{ret.} = \langle R_F | -k^2 r^2 / 12 | R_I \rangle \ .$$

Note that $-i\Sigma \ T_{ret.}$ coincides with $M^{S.P.}$, Eq. (2.1). Using a double commutator to rewrite $T_{ret} = \langle R_F | (\vec{p}^2/6m^2) - rV'(r)/6m | R_I \rangle$, one finds

$$T_H^{(o)} = T_g + T_f + T_{ret.} = \langle R_F | -\vec{p}^2 / 6m^2 | R_I \rangle \ , \tag{3.2}$$

which is __independent__ of the form of V. For a Coulomb potential, $rV' = -V$ and, using $\langle V \rangle = \langle -\vec{p}^2/2m \rangle$ one sees that $T_f + T_g \rightarrow 2T_g = 3/2 \ T_H^{(o)}$, while $T_{ret.} = -1/2 \ T_H^{(o)}$. Thus we note, on the one hand, that as advertised the Schrödinger-Pauli approximation, $T_{ret.}$, is off by a factor of 2. On the other hand, the amplitude found in Ref. 4, in which the effects of retardation were not included, is essentially $T_f + T_g$, which is off by almost as large a factor.

The decay rate for the transition in question is given by

$$R_{H-like} = \frac{4\alpha k^3}{m^2} |T_H|^2 . \tag{3.3}$$

To lowest order in αZ, $k \to k^{(o)} = \frac{3}{8} (\alpha Z)^2 m$ and $T_H \to T_H^{(o)} = (-4\sqrt{2}/81)(\alpha Z)^2$ so that

$$R_{H-like}^{(o)} (2^2S_{1/2} \to 1^2S_{1/2} + \gamma) = \frac{1}{972} \alpha (\alpha Z)^{10} m . \tag{3.4}$$

2. <u>He-like ions</u>. Let us first neglect <u>all</u> electron-electron interaction. Then, using antisymmetrized products of Dirac wave-functions for both initial and final states, one finds, e.g. for the one-photon transition from the initial 2^3S_1 state ψ_I with $J_z = +1$ to the 1^1S_0 ground state ψ_F, that the matrix element

$$M_D = \langle \psi_F | \vec{\alpha}_1 \cdot \vec{\varepsilon} \, e^{-i\vec{k}\cdot\vec{r}_1} + \vec{\alpha}_2 \cdot \vec{\varepsilon} \, e^{-i\vec{k}\cdot\vec{r}_2} | \psi_I \rangle \tag{3.5}$$

reduces to

$$M_D^{(o)} = -i \langle \chi_{-1/2} | \vec{\sigma} \cdot \frac{\vec{k}}{m} \times \vec{\varepsilon} | \chi_{1/2} \rangle T_H ,$$

where T_H is just the hydrogenic radial matrix element given by Eq. (3.1c). Thus, to lowest order in αZ, one finds

$$R_{He-like}^{(o)} (2^3S_1 \to 1^1S_0 + \gamma) = \frac{2}{3} R_{H-like}^{(o)} (2^2S_{1/2} \to 1^2S_{1/2} + \gamma). \tag{3.6}$$

The factor 2/3, sometimes associated with the ratio of the statistical weight of the $^2S_{1/2}$ state in H to that of the $^3S_1$ state in He, may be simply understood as follows. Its origin is that the rate for the spin-flip transitions ($s_z = 1/2$ to $s_z = -1/2$) is <u>twice</u> that of the rate for non-spin-flip transition ($s_z = 1/2 \to s_z = 1/2$) in H-like ions. In He-like ions only the spin-flip process is available, because of the Pauli exclusion principle -- the outer electron with spin up can only join the spin-up electron in the inner shell by flipping its spin. Thus the ratio of the rates is 2:(2+1) = 2/3.

To include the effects of electron-electron interaction one must choose an appropriate starting point. The approach of Ref. 15 is based on the recognition that both virtual-pair effects induced by Coulomb interactions and effects arising from the exchange of transverse photons can be treated by perturbation theory. Thus we take as the zero-order Hamiltonian

$$H_{++} = \vec{\alpha}_1 \cdot \vec{p}_1 + \beta_1 m + \vec{\alpha}_2 \cdot \vec{p}_2 + \beta_2 m + \Lambda_{++}(V_1 + V_2 + V_{12})\Lambda_{++} , \tag{3.7}$$

where $\Lambda_{++} = \Lambda_+(1) \Lambda_+(2)$ and $\Lambda_+ = (E(\vec{p}) + \vec{\alpha}\cdot\vec{p} + \beta m)/2E(\vec{p})$ is the positive-energy projection operator. The zeroth-order eigenfunctions

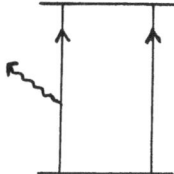

Fig. 4.   Time-ordered diagram representing M_{++}, the no-pair, no-
transverse photon part of the amplitude M for radiative
decay of a He-like ion.

ψ then satisfy $H_{++} \psi = E\psi$ and $\Lambda_{++} \psi = \psi$, and M_{++}, the no-pair
no-transverse photon part of M, corresponding to Fig. 4, is
given by

$$M_{++} = 2 \langle \psi_F | \vec{\alpha}_1 \cdot \vec{\epsilon} \ e^{-i\vec{k}\cdot\vec{r}_1} | \psi_I \rangle \ .$$

$M_{++}^{(o)}$, the value of M_{++} to lowest order in αZ but to all orders in
Z^{-1}, is then found to be

$$M_{++}^{(o)} = -2i \langle \chi_F | \vec{\sigma}_1 \cdot \frac{\vec{k}}{m} \times \vec{\epsilon} | \chi_I \rangle \ T_{++}^{(o)} \ ,$$

where

$$T_{++}^{(o)} = -\frac{1}{3} \langle \phi_F | \frac{\vec{p}_1^2}{m^2} + \frac{k^2 r_1^2}{4} | \phi_I \rangle \ ,$$

with χ denoting a singlet or triplet spin wavefunction and the ϕ's
denoting the completely non-relativistic spatial wavefunctions.
Thus, remarkably, the electron-electron Coulomb interaction appears
only implicitly -- its effect being fully contained in the non-
relativistic wavefunction, to this order. The Breit interaction
also makes no contribution in this order.

The additional contributions to $M^{(o)}$ all come from time-
ordered diagrams involving the creation of virtual electron-
positron pairs, shown in Fig. (5a,b,c). These give, respectively,

$$T_a^{(o)} = \frac{1}{6m} \langle \phi_F | r_1 V_1' | \phi_I \rangle \ ,$$

$$T_b^{(o)} = \frac{1}{6m} \langle \phi_F | \vec{r}_1 \cdot \vec{r}_{12} V_{12}' | \phi_I \rangle \ ,$$

and

$$T_c^{(o)} = \frac{1}{6m} \langle \phi_F | \vec{r}_1 \cdot \vec{r}_{12} (\alpha/r_{12}^2) | \phi_I \rangle \ ,$$

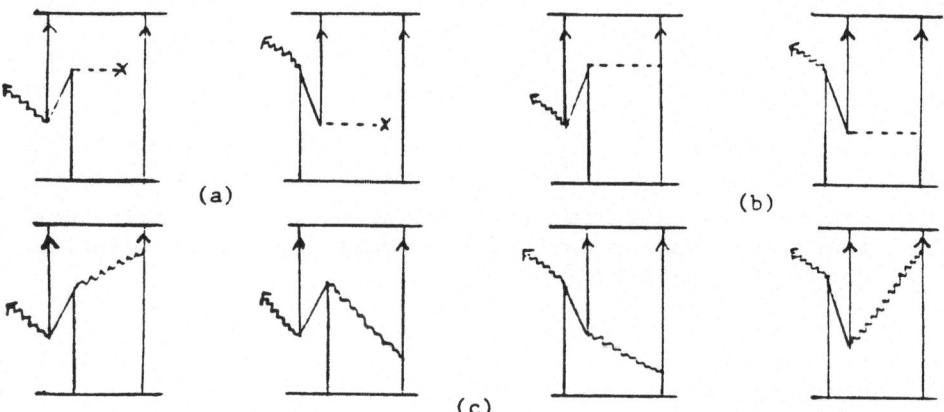

Fig. 5.   Time-ordered diagrams representing (a) one-pair, external
          Coulomb-field corrections; (b) one-pair, electron Coulomb-
          field corrections; (c) leading one-pair transverse-photon
          corrections.

where V_1 and V_{12} are the static interaction of electron "1" with
the external field and the second electron, respectively. The
total $T^{(o)}$ is then

$$T^{(o)} = T^{(o)}_{++} + T^{(o)}_a + T^{(o)}_b + T^{(o)}_c .$$

With $V_{12} = \alpha/r_{12}$, $T^{(o)}_b$ and $T^{(o)}_c$ are seen to cancel exactly, so that
for $V_1 = Z\alpha/r_1$, one gets

$$T^{(o)}_{He} = -\frac{1}{3} <\phi_F| \frac{\vec{p}_1^2}{m^2} + \frac{k^2 r_1^2}{4} + \frac{V_1}{2m} |\phi_I> . \qquad (3.8)$$

The corresponding decay rate is given by

$$R_{He\text{-like}} = \frac{2}{3} \cdot \frac{4\alpha k^3}{m^2} |\sqrt{2}\, T^{(o)}_{He}|^2 . \qquad (3.9)$$

If the electron-electron interaction is neglected, $|\sqrt{2}\, T^{(o)}_{He}| \to$
$|T^{(o)}_H|$ and we recover (3.6).

     The equations (3.8) and (3.9) were derived independently by
G. F. W. Drake, using semi-classical radiation theory.[18] Drake
also evaluated the rate, using many-parameter wavefunctions obtained
by him via an expansion in powers of Z^{-1}. For Z = 18, corresponding
to Ar^{+16}, he found τ = 212.7 n sec to be compared with τ_{exp} =
172 ± 30 n sec found by Marrus and Schmieder in 1970. To quote
from Drake's review of the subject at the III International Atomic
Physics Conference in 1972, "The discrepancy, if real, is one of
the few remaining examples of a disagreement in one- and two-
electron systems...".[2]

As is well known, discrepancies are a Good Thing, providing a stimulus to further work by both experimentalists and theorists. This discrepancy was no different. On the experimental side, a number of new experiments on other species: S, Ti, Va, Fe gave results which were largely consistent, within the errors, with the predictions of Eq. (3.9). However, a new discrepancy emerged for the case of Cl^{+15}, while the error bar for Ar^{+16} grew smaller.

3. <u>Higher-order corrections</u>. On the theoretical side, a number of efforts were therefore undertaken:

(a) Radiative corrections were briefly considered in Ref. 15, where it was noted that one expects these to be of order $\alpha \log (Z\alpha)^{-1}$, relative to the leading matrix element $M^{(o)}$. If the coefficient were positive and, e.g., of the order of 2 to 3, then for $Z \sim 20$ one would get a 10% or so decrease in the lifetime. However, a detailed analysis by D. G. Lin and G. Feinberg[19] for the case of H-like ions shows that although individual diagrams give $\alpha \log Z\alpha$ corrections, these all cancel on addition so that one gets zero. The correction due to the anomalous electron magnetic moment, of relative order α, is also too small to count -- in fact, for the case of the $2^2S_{1/2} \rightarrow 1^2S_{1/2} + \gamma$ transition in H-like ions it is cancelled by another $O(\alpha)$ effect[19] so that there are no corrections of relative order α at all![20] In He-like ions there is a surviving radiative correction $\delta M^{(o)} = .97\alpha/\pi Z$, which is however much too small to be of interest, at present.[19]

(b) Recoil and retardation corrections to the approximations made in Ref. 15, especially in diagrams involving transverse-photon exchange, were examined by E. Kelsey in his 1974 Maryland Ph.D. thesis.[21] Although there are effects of relative order αZ arising from individual diagrams, these cancel on addition. Thus these corrections are at best of relative order $\alpha^2 Z$, and therefore again unimportant.

(c) There are relativistic corrections of order $(\alpha Z)^2$ relative to the leading term in M which we have not yet considered. For the hydrogenic $2^1S_{1/2} \rightarrow 1^1S_{1/2} + \gamma$ transition these may be found by keeping the next terms in an expansion in powers of αZ of the energy difference k and of the radial wavefunctions g and f, entering Eqs. (3.1 a,b,c) for M^D. Such a computation was carried out by D. G. Lin in his Ph.D. thesis.[19] The result is that the r.h.s. of Eq. (3.4) should be multiplied by a correction factor

$$1 + c_2(\alpha Z)^2 \tag{3.10}$$

where

$$c_2 = 1.07 . \tag{3.11}$$

Even for $Z = 20$ this represents only a 2% correction to Eq. (3.4) and the corresponding correction for He-like ions will be equally small.

For the hydrogenic case, as mentioned earlier M^D may in fact be evaluated exactly in terms of hypergeometric functions, as shown by W. R. Johnson.[17] Comparison of the exact values with those obtained via (3.4) and (3.10) shows that the terms of order $(\alpha Z)^4$ and higher are negligible even for Z as large as 40; this fact will be relevant for us shortly. Expansion of the exact formula for M^D yields[22]

$$c_2 = (19/20) + \ln(9/8) \tag{3.11'}$$

which implies $c_2 = 1.0678$, in agreement with (3.11).

Work to include such higher-order αZ effects for the He-like decays has been based on finding self-consistent relativistic Hartree-Fock wavefunctions, starting with the relativistic two-body Hamiltonian

$$H_{Dirac} = \sum_{i=1}^{2} (\vec{\alpha}_i \cdot \vec{p}_i + \beta_i m - Ze^2/r_i) + e^2/r_{12} \tag{3.12}$$

and then directly evaluating the matrix element (3.5) with these wavefunctions. The results found this way[23] agree substantially with those found by the evaluation of the reduced form of the matrix element with the nonrelativistic variational wavefunctions. From the above remarks concerning the smallness of such effects in the H-like case, this is not surprising.

We should mention, in passing, that the Hamiltonian (3.12), unlike (3.6), is unlikely to have normalizable eigenstates, so that results obtained from it must be treated with caution.[24]

4. Wavefunction accuracy. At this stage, the numerical accuracy of the variational wavefunctions used by Drake in his evaluation of the decay rate via Eqs. (3.8) and (3.9) remained the only aspect of the calculation which had not been fully nailed down. Experience with another problem -- the exact evaluation[25] of the London-van der Waals constant C_{HH}, entering the H-H potential C_{HH}/r^6 -- suggested that the first-order contribution of the electron-electron Coulomb interaction ought to be exactly claculable. Such a calculation was undertaken by Kelsey and myself with the result that

$$R_{He-like} = R^{(o)}_{He-like} [1 + b_1/Z + b_2/Z^2 + \ldots] , \tag{3.13}$$

where b_1 is exactly expressible in terms of (many) hypergeometric functions.[26] Numerical evaluation of the unbelievably lengthy formula for b_1 yields

$$b_1 = -4.099 . \tag{3.14}$$

This is in extremely close agreement with the value of b_1 which can be inferred from Drake's variational calculation and confirms the accuracy of the wavefunctions used by him. The variationally determined value of b_2 is 6.7; we may use this to write

$$R_{He-like} \approx R^{(o)}_{He-like} \ (1 - 4.10/Z + 6.7/Z^2) \ . \tag{3.15}$$

The right-hand side of (3.15) reproduces the numerical values given in Ref. 16 to better than 1% for $Z > 16$; this is simply a consequence of the fact that the terms of order Z^{-3} or higher, as determined by the variational wavefunction, make only a very small contribution to R.

A completely different confirmation of the fact that the evaluation of Eq. (3.8) had been carried out to sufficient accuracy was obtained by Anderson and Weinhold.[27] Using theorems which enable one to put bounds on the error made in evaluating an off-diagonal element of an operator when inexact eigenfunctions are employed, these authors were able to show that, in the case of Ar^{+16}, the value obtained by Drake was good to better than .3%.

C.   Comparison of Theory and Experiment

This seems an opportune time to attempt a more systematic comparison of theory with experiment than has been presented heretofore. Rather than use a mixture of tabulated numerical results from variational and/or Hartree-Fock type calculations, we note that the theoretical results described in Sec. B. may be used to write an explicit formula for the M1 decay rate, in terms of α and Z, which can be expected to be of high accuracy for $Z \gg 1$.

Let us write, for the hydrogenic decays,

$$R^{th.}_{H-like} = R^{(o)}_{H-like} \ [1 + c_2(\alpha Z)^2 + d_1 \ \alpha \ \log \ (Z\alpha)^{-1} + d_2\alpha$$
$$+ \ \Delta] \ . \tag{3.16}$$

The c-term is the leading correction to the lowest order result $R^{(o)}_{H-like}$ (Eq. (3.4)), involving in part relativistic corrections to the initial and final nonrelativistic wavefunctions, relativistic corrections to the energy difference, etc. The d-terms represent the "leading" radiative corrections. The Δ-term consists of terms at most of order $(\alpha Z)^4$ or of order $\alpha(\alpha Z)\log \ (Z\alpha)$ and so can be expected to be less than 1% even for, say, Z as large as 40.

Let us further write, for the helium-like decays,

$$R^{th.}_{He-like} = R^{th.}_{H-like} \cdot \eta \tag{3.17}$$

where η is a correction factor. To lowest order in α, the correction factor η can be read off from Eq. (3.13)

$$\eta = \frac{2}{3} [1 + b_1/Z + b_2/Z^2 + \ldots] . \tag{3.18}$$

Using (3.4), (3.10), (3.17) and (3.18) we see that

$$R^{th.}_{He\text{-like}} = \frac{2}{3} R^{(o)}_{H\text{-like}} [1 + b_1/Z + b_2/Z^2 + c_2(\alpha Z)^2 +$$
$$d_1 \alpha\log Z\alpha + d_2\alpha + O(Z^{-3}) + O(\alpha^2 Z) +$$
$$O(\alpha Z^{-1})] . \tag{3.19}$$

For the case at hand, we have, as discussed above, the exact results $d_2 = d_1 = 0$, $c_2 = 1.07$, and $b_1 = -4.10$. Further, $b_2 \approx 6.7$, according to the variational calculation. Thus we write

$$R^{th.}_{He\text{-like}} \approx \frac{2}{3} \frac{\alpha(\alpha Z)^{10} m}{972} [1 - 4.10/Z + 6.7/Z^2 + 1.07(\alpha Z)^2] \tag{3.20a}$$

and

$$\tau^{th.}_{He\text{-like}} = (R^{th.}_{He\text{-like}})^{-1} . \tag{3.20b}$$

We expect, conservatively, that (3.20) is accurate to better than 2% in the entire range $15 \leq Z \leq 40$.

Table I. Experimental and theoretical lifetimes for the decay $2^3S_1 \to 1^1S_0 + \gamma$ in He-like ions. For Z = 16 to 36, $\tau^{th.}$ was computed from Eq. (3.20) of the text; for Z = 2, $\tau^{th.}$ is taken from Ref. 15.

Element	Z	$\tau^{exp.}$ (n sec)	$\tau^{th.}$ (n sec)	$\tau^{exp.}/\tau^{th.}$
S	16	706 ± 86 [a]	697	1.01 ± .12
Cl	17	354 ± 24 [a]	373	.95 ± .06
A	18	202 ± 20 [b]	207	.98 ± .10
Ti	22	25.8 ± 1.3 [c]	26.4	.98 ± .05
V	23	16.9 ± 0.7 [d]	16.8	1.01 ± .04
Fe	26	4.8 ± 0.6 [d]	4.78	1.00 ± .13
Kr	36	.20 ± .06 [e]	.170	1.18 ± .35
He	2	$.91^{+.40}_{-.20} \times 10^4$ sec [f]	$.841 \times 10^4$ sec	$1.08^{+.48}_{-.24}$

a) Ref. 9   b) Ref. 10   c) Ref. 11   d) Ref. 12   e) Ref. 13   f) Ref. 14

Table I shows the comparison between $\tau^{exp.}$ and $\tau^{th.}$ as computed from Eq. (3.20).

As can be seen, the agreement between theory and experiment is now very satisfactory, largely as a result of the fact that re-measurements in the cases of Cl^{+15} and Ar^{+16}, further downstream from the foil where the ^3S state is excited, yields values considerably different from those found previously, which came from positions corresponding to only a fraction of a lifetime.[9,10] It must be stressed that the reason for the apparently non-exponential character of the decay curve at short times is as yet not understood; this question deserves further study.[9] It should also be noted that the experimental accuracy is not yet sufficient to provide a stringent test of the coefficients of the Z^{-2} and $(\alpha Z)^2$ terms.[28] It will be quite interesting to see how Eq. (3.20) holds up -- if and when the experimental accuracy reaches the 1% level.

As mentioned earlier, there is now available a much improved measurement of the lifetime for the $2^3S_1 \rightarrow 1^1S_0 + \gamma$ transition in helium itself,[14] viz., $(\tau^{exp.})^{-1} = 1.10 \times 10^{-4}$ sec.$^{-1} \pm 30\%$, giving

$$\tau^{exp.} = (.91^{+.40}_{-.20}) \times 10^4 \text{ sec.} \tag{3.21a}$$

This agrees quite well with

$$\tau^{th.} = .841 \times 10^4 \text{ sec.,} \tag{3.21b}$$

obtained from a six-parameter Hylleraas-type wavefunction.[29]

Thus, the theoretical and experimental values are in agreement over a range covering fourteen orders of magnitude. This is surely an impressive triumph for Q.E.D. in a domain rather different from that in which it has been tested heretofore.

When Feinberg and I worked on this problem we were gratified by the fact that, modulo the nonrelativistic wavefunctions, in leading order the answer for the decay rate could be given in essentially closed form. We were however disappointed by the fact that the electron-electron interaction showed so much discretion, appearing only implicitly, by way of the nonrelativistic wavefunctions. We now turn to a quite different and only recently discovered physical system, which appears to be a new kind of atom; the excited states of this "atom" can undergo radiative E1 and M1 decay. Feinberg and I were delighted to find that the M1 decays for this system involve interaction between the bound particles in a much more direct way than in the case of the He-like ions. Of course, quite apart from this feature, the new system has charms of its own.

IV.   CHARMONIUM AND ALL THAT[30]

A.   Introduction

Let me now turn to the developments which around 20 months ago electrified the community of elementary-particle physicists.   In November, 1974, the following discoveries were announced:

(i)   A relatively narrow peak in the e^+-e^- invariant mass-spectrum was observed by an MIT-BNL group[31] in the reaction

$$p + Be \rightarrow e^+ e^- + X \tag{4.1}$$

as shown in Fig. 6.   The peak mass was ~ 3.1 GeV and the width was less than the energy resolution ~ 20 MeV, "consistent with zero."

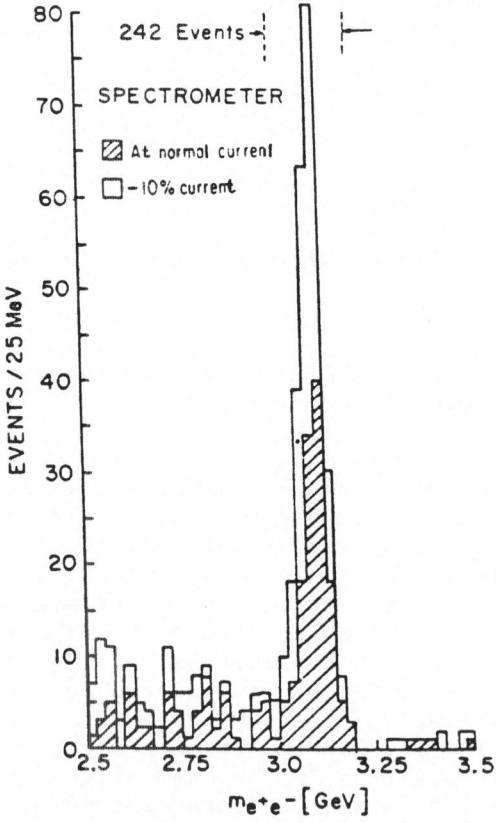

Fig. 6. Discovery of the J-particle: $e^+ e^-$ mass-spectrum observed in $p + Be \rightarrow e^+ e^- + X$ (from Ref. 31).

(ii) Extremely high and sharp peaks in the elastic and various inelastic e^+e^- cross-sections were observed by a SLAC group[32] in colliding beam experiments at the SPEAR facility. As shown in Fig. 7 these peaks were seen in the Bhabha scattering, $e^+e^- \to e^+e^-$, in $e^+ + e^- \to \mu^+ + \mu^-$, and in $e^+ + e^- \to$ hadrons. The value of the c.m. energy at the peak was $3.105 \pm .003$ GeV, with an upper limit of 1.3 MeV for the full width at half-maximum.

These discoveries were confirmed within a week, by three groups working at Frascati.[33]

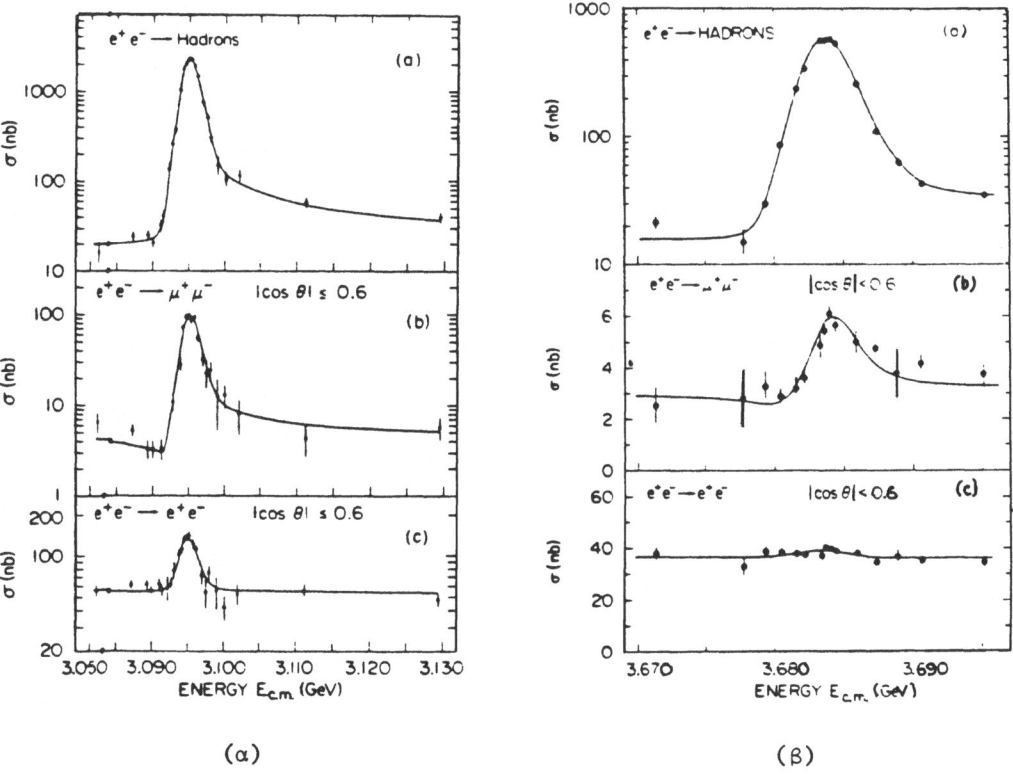

Fig. 7. (α) Evidence for the ψ; (β) evidence for the ψ'. (From Ref. 35.)

The natural interpretation of the observation at BNL was that it corresponded to the production of a new particle or resonance -- called J -- followed by its decay into e^+e^-:

$$p + Be \rightarrow J + X , \qquad J \rightarrow e^+e^- . \qquad (4.2)$$

Similarly, the SLAC experiments corresponded to the formation of a new particle -- called ψ -- followed by its decay into both leptonic and hadronic channels:

$$e^+e^- \rightarrow \psi \rightarrow e^+e^-, \mu^+\mu^-, \text{hadrons} \qquad (4.3)$$

Of course, because of the coincidence of the mass and the common e^+e^- decay channel, J and ψ were immediately regarded as the same particle, now often designated as J/ψ.

(iii) A week after the discovery of the ψ, the SLAC group found another narrow resonance, called ψ', at 3684 MeV.[34] Analysis of the final states showed an important decay channel for the ψ' to be

$$\psi' \rightarrow \psi + \text{hadrons.} \qquad (4.4)$$

There followed a period of intense experimental activity accompanied by equally intense theoretical analysis and speculation.[35] The remarkable aspect of the ψ was that unlike other states coupled to both leptons and hadrons, such as the ρ, ω and ϕ mesons, which have <u>total</u> widths in the 5 MeV -- 150 MeV range, but purely leptonic widths of the order of one to ten KeV, the ψ had a total width of the order of only .1 MeV although its leptonic width was "normal."

Analysis of the experiments showed that the ψ and ψ' could be regarded as <u>hadrons</u>, with the same quantum numbers as the photon, $J^{PC} = 1^{--}$, as one would naively guess. This assignment was confirmed by observation of interference just below resonance of the amplitudes $A(e^+e^- \rightarrow \gamma \rightarrow \mu^+\mu^-)$ and $A(e^+e^- \rightarrow \psi \rightarrow \mu^+\mu^-)$. Study of the hadronic final states showed consistency with an assignment $I = 0$ and $G = -1$ for the isospin and G-parity of both the ψ and ψ'.

B. The Three C's: Charm, Color, and Confinement

1. The first detailed dynamical models for the ψ and ψ' involved the notion of <u>charm</u>. What is charm? A good characterization was actually given long ago: "... it's a sort of bloom... if you have it, you don't need to have anything else; and if you don't have it, it doesn't much matter what else you have."[36]

A more technical definition was given a half-century later, around 1964. There were two routes to charm,[37] one via the strong interactions, the other via the weak interactions -- the front door and the back door, in terminology once used by B. d'Espagnat in discussing approaches to the symmetries of particle interactions.

Front door. The introduction of a new quantum number or "flavor," as the current jargon has it (which like the third component of isospin, I_3, and the strangeness S is conserved by both the strong and electromagnetic interactions) during the post-SU(3) period was motivated by relatively slender, more or less aesthetic considerations. Recall that in 1962-63 it was discovered that the hadrons could be classified into multiplets of the group SU(3), with the states in a multiplet labelled by the eigenvalues of I_3 and Y = B + S (B = baryon number), taken as two commuting generators of SU(3), and SU(3) a broken symmetry of the strong interactions. Now the mass spectrum of the vector mesons (ρ,ω,ϕ, K*...) suggested that some still higher strong-interaction symmetry might be valid and SU(4) was a natural candidate; but SU(4) has three commuting generators, two of which could be taken as I_3 and Y, leaving a third to be identified with some as yet undiscovered quantum number -- now called C, for "charm" -- which was zero for the known particles.

Back door. The weak-interaction route to charm came from considerations of lepton-hadron symmetry. Recall that after the discovery that ν_μ, the neutrino associated with the muon, was distinct from the electron-neutrino ν_e, one had four leptons: e^-, ν_e, μ^-, ν_μ. However, only three basic hadrons were needed to carry the conserved charges of baryon number B, electric charge number Q and strangeness S. In the eightfold way of Gell-Mann and Ne'eman, as distinct from the Sakata-model where these were taken as P, N, and Λ, the simplest possibility was that introduced by Gell-Mann and Zweig: three hypothetical spin-1/2 fermions q_1, q_2, q_3 called "quarks," each with baryon number 1/3 and with (q_1,q_2) an iso-spin doublet (u,d) of "up" and "down" quarks, with Z = 0 (Y = 1/3), and q_3 a "strange" quark "s", with S = -1 (Y = - 2/3). All known mesons and baryons could then be considered, at least on a formal level, as composites formed from the q_i and their anti-particles $\bar{q}_i$.[38] Thus either in the quark scheme, or in less economical schemes which avoided the introduction of fractional electric charge and baryon number, one was led to the idea of a fourth basic hadron in order to be able to express notions of symmetry between leptons and hadrons. In the modern version of the standard quark model, the fourth quark, usually designated by "c", is an isosinglet, with S = 0 and Q = 2/3 like the "u", but with charm quantum number C = +1. (The generalized Gell-Mann-Nishijima formula then reads Q = I_3 + Y/2 + C/2.)

The hypothesis of a fourth quark received a major boost after
the discovery of renormalizable, unified gauge theories of weak and
electromagnetic interactions, based on the Yang-Mills idea of
local gauge invariance of a field theory under a non-Abelian group
and the Higgs-Kibble mechanism for the spontaneous symmetry breaking
of the local symmetry.[39] This came from the observation that the
introduction of a fourth quark allows one to accomodate in an
elegant way, consistent with the ideas of unified gauge theories,
both (i) the presence of neutral currents as observed in strangeness-
nonchanging high-energy neutrino interactions, and (ii) the suppression of strangeness-changing neutral currents, inferred from the
smallness of the K_1°-K_2° mass difference and the low rate for,
e.g., $K_L^\circ \to \mu^+\mu^-$.[40]

2. We have talked at length about quarks. So where are they?
Another thread in the story must now be drawn in. Exploration in
the mid-sixties of the mass spectrum of the hadrons revealed the
existence of an approximate SU(6) invariance of the strong quark-
binding interactions corresponding to joint transformations of the
SU(3) indices and the spin indices associated with the non-relativistic SU(2) spin group. This could be made compatible with
Fermi-statistics for the quarks only if there were another hidden
degree of freedom, nowadays described as the introduction of a
three-valued "color" index for the quark field which can be used to
antisymmetrize the baryon wavefunctions.[41] In the so-called
standard version of unified gauge theories of weak, electromagnetic
and strong interactions, the Lagrangian is imagined to be invariant
not only under global SU(3) transformations of the color-index,
but also under local $SU(3)_{color}$ transformations. This implies that,
a la Yang-Mills, there is a set of eight massless vector mesons
in the theory. The exchange of these "color-gluons" between quarks
is imagined to provide not only binding, but <u>binding with a
vengeance</u>: Because of the infrared complexities of such theories
one is allowed to entertain the hope that somehow the associated
long-range forces will be such that only objects which are color-
neutral, i.e. singlets under $SU(3)_{color}$, are observable as real
particles, having asymptotic states associated with them in Hilbert
space, and presto!, both the color-triplet quarks and the color-
octet gluons are forever out of view, e.g., protected from leaving
tracks in a bubble chamber.[42,43] The masses of the quarks then become "effective masses" or parameters which can be determined only
indirectly by experiment. Although we cannot explore these issues
further here, we want to note, for use in the following section,
that the idea that the quarks are permanently confined inside
hadrons expresses itself in the modern quark-binding industry by
way of the assumption that the force between, say, a "q" and a
"$\bar{q}$" <u>grows</u> with their separation <u>r</u>, in such a way that it would
require infinite energy to break them up. Thus, for large <u>r</u>, one
assumes, typically, that $V_{q\bar{q}} \sim kr$ with $k > 0$.[44]

C. The Psions as $(c, \bar{c})$ Bound States

1. Although the ψ and ψ' could not themselves be charmed particles (more generally, could not possess a non-zero value for an additive quantum number conserved by strong and electromagnetic interactions) a number of authors advocated the viewpoint that they are made up of charmed quarks in the same sense that, e.g., the "classical" vector mesons are regarded as composites in the SU(3) quark model: $\phi \sim \bar{s}s$, $\omega \sim (\bar{u}u + \bar{d}d)/\sqrt{2}$, $\rho_0 \sim (\bar{u}u - \bar{d}d)/\sqrt{2}$, etc. In particular, in close analogy with the ϕ, it was suggested that $\psi \sim \bar{c}c$.[45] This picture has an immediate qualitative appeal: It related the narrow hadronic widths of the psions to the success of the IZ (Iizuka-Zweig) rule, which states that in the SU(3) quark model decay processes are strongly suppressed if they can only proceed by the annihilation of a $q\bar{q}$ pair present in an initial-state hadron or by the creation of a pair which forms part (or all) of a final-state hadron. This empirical rule accounts, e.g., for the fact that the ϕ, which is largely $\bar{s}s$, decays predominantly into $K\bar{K}$ rather than $\pi^+\pi^-\pi^\circ$, for which the phase space is much bigger; note that since the pions, unlike the K's, contain <u>no</u> s or $\bar{s}$ quarks one can only get rid of the initial $\bar{s}s$ pair by mutual annihilation. An extended IZ rule would then inhibit the transformation of psions into ordinary hadrons, which have no c-quarks in them. Provided that $c\bar{q}$ systems ("charmed" mesons) have masses above $m_{\psi'}/2 \sim 1.85$ GeV there would be no open decay channels consistent with the IZ rule. Of course, this does not by itself <u>explain</u> the narrow widths, but it does at least provide, one might say, a unification of two mysterious facts. On a more technical level, it was suggested that the validity of asymptotic freedom in non-Abelian unified gauge theories (a property which states that at short distances the quark interactions become weak and may be treated perturbatively) might in fact account for the validity of an extended IZ rule when heavy quarks ($m_c \sim$ few GeV) are involved.

The idea that $m_c \sim$ few GeV arose in part from the study of strangeness-changing second-order weak processes within the frame-work of unified gauge theories. For example if, on the one hand, the <u>c</u> were degenerate in mass with the ordinary quarks the suppression of such effects would be too strong. On the other hand, an estimate of the $K_L K_S$ mass difference, which is proportional to $G_F \alpha m_c^2$ in the standard model, shows that m_c cannot be too large. In addition, one would expect to have already seen charmed hadrons unless m_c exceeded m_q by at least 1 GeV or so. Thus, before the discovery of the psions one had arrived at the conclusion that, roughly, 1.5 GeV $\lesssim m_c \lesssim$ 5 GeV.[46]

Let us now consider the bound-state models and their implications in more detail.

2. The bound states of a particle anti-particle system such as $(c, \bar{c})$ can be classified by $(J)^{PC}$ where J is the total angular momentum, and P and C are the eigenvalues of parity and charge conjugation operators, respectively. For a state which in nonrelativistic spectroscopic notation is described by $^{2S+1}L_J$, P and C are related to the orbital and spin angular momentum quantum numbers L and S by

$$P = -(-1)^L, \qquad C = (-1)^{L+S} . \tag{4.5}$$

It follows from (4.5) that if the $(c, \bar{c})$ system has any triplet-S $(^3S_1)$ bound states, they will have

$$J^{PC} = 1^{--} , \tag{4.6}$$

that is, precisely the quantum numbers of the ψ and ψ'. The basic hypothesis of most models of the psions as bound states is therefore:

(a) ψ is the ground state of "orthocharmonium," i.e., can be identified with the lowest lying $^3S_1$ state of a bound $(c, \bar{c})$ system; ψ' is the first excited state, i.e., a radial excitation of ψ.

Tests of this hypothesis alone would be difficult to come by. But the addition of further hypotheses leads to a number of predictions which can be tested. These hypotheses all concern the dynamics of the binding.

(b_1) The energy levels of the bound $(c\bar{c})$ system may, in first approximation, be found by solving a nonrelativistic Schroedinger equation:

$$(\frac{\vec{p}^2}{2\mu} + V) \phi = W\phi, \tag{4.7}$$

where $\mu = m_c/2$ is the reduced mass, W is the binding energy, and $M = 2 m_c + W$ is the mass of the bound state with wavefunction ϕ.

(b_2) If the binding interaction V is written as the sum of a spin-independent part $V_{s.i.}$ and a spin-dependent part V_{spin},

$$V = V_{s.i.} + V_{spin} , \tag{4.8}$$

then

$$V_{s.i.} \gg V_{spin} . \tag{4.9}$$

(b_3) $V_{s.i.}$ is a local, central potential, popularly taken to be of the form[47]

$$V_{s.i.} = ar + d - \alpha_s/r , \tag{4.10}$$

with $\underline{a}$ and $\underline{d}$ parameters to be determined and α_s small.[48]

   3. Before discussing (b_3), we should emphasize that (b_1) and (b_2) alone already lead to certain qualitative predictions, regardless of the precise form of $V_{s.i.}$.

   One expects to find in the neighborhood of ψ and ψ' other ψ-like states as follows:

   (i) ψ and ψ' should have singlet-S state $(^1S_0)$ "hyperfine" partners, the ground and first excited states of "paracharmonium." The quantum numbers of these states, denoted as η_c and η_c' respectively, would be, 0^{-+}, according to Eq. (4.5).

   (ii) There should be bound orbital excitations of at least the ψ, the lowest-lying being P-states. In particular, one expects to see a multiplet of triplet-P states: $^3P_0$, $^3P_1$, and $^3P_2$, with quantum numbers 0^{++}, 1^{++} and 2^{++} respectively. We shall denote them by $\chi_J (J = 0, 1, 2)$.

   These expectations are summarized in Fig. 8. The η_c and η_c' states have been drawn relatively near to and below the ψ and ψ' states corresponding not only to the assumption (b_2) that the spin-spin forces are small, but also that they are attractive in singlet states, as in positronium. Since $C = +1$ for the η_c, the η_c' and the χ_J states, they will not show up as narrow resonances in e^+-e^- annihilation, assumed to proceed via a one-photon intermediate state $(C = -1)$. However, if these states exist, then new radiative decay modes become available to the ψ and ψ'. In particular, one may look for the E1 decays of the ψ':

$$\psi' \rightarrow \chi_J + \gamma \; . \tag{4.11}$$

Even if the photon is not directly observed, χ_J will itself have direct hadronic decays so that there should be a signal in those final hadronic channels that have $C = +1$. Furthermore, χ_J can itself undergo a radiative E1 transition to the ψ:

$$\chi_J \rightarrow \psi + \gamma \; . \tag{4.12}$$

So observation of two-photon events in the region of the ψ' mass might reveal the existence of the cascade process, (4.11) followed by (4.12).

   In addition, both the ψ' and the ψ should undergo "ordinary" radiative M1 transitions to their $^1S_0$ partners:

$$\psi' \rightarrow \eta_c' + \gamma \tag{4.13a}$$
$$\psi \rightarrow \eta_c + \gamma \tag{4.13b}$$

and, finally, the ψ' should be also able to go into η_c via a "relativistic" M1 transition:

$$\psi' \rightarrow \eta_c + \gamma \ . \tag{4.13c}$$

So at last we return to the subject discussed at length in the atomic case. Although the designation of the decays (4.13a), (4.13b), and (4.13c) as M1 transitions is essentially model-independent, relying only on the fact that we have $\Delta J = 1$, and <u>no</u> change of parity, it should be emphasized that the use of the adjectives "ordinary" and "relativistic" is justified only if the non-relativistic description, in which the ψ and ψ' radial wavefunctions are orthogonal and η_c and η_c' have the same radial wavefunction as ψ and ψ' respectively, is indeed a good first approximation.

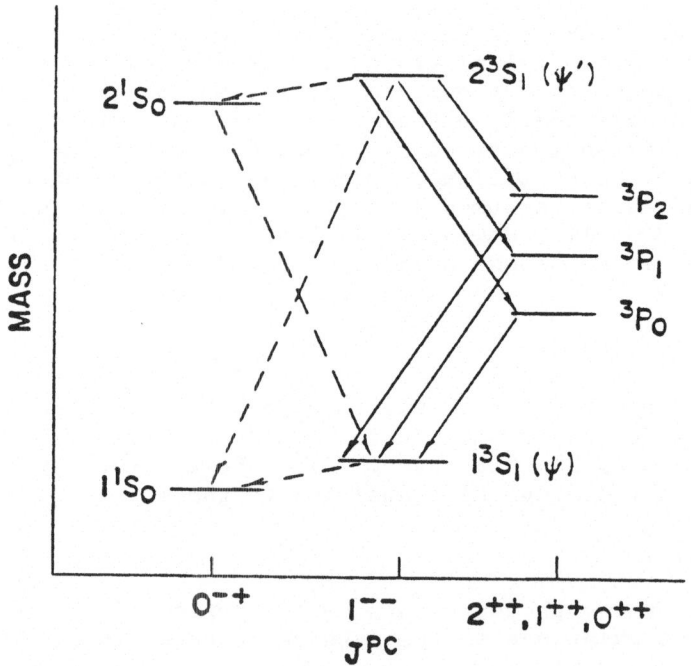

Fig. 8. Schematic partial spectrum of charmonium and radiative decays expected on the basis of hypotheses (b_1) and (b_2) of the text. Solid and dashed arrows indicate E1 and M1 decays, respectively. Possible $^1P_1$ and $^1D_2$ states between ψ and ψ', and states above ψ' are omitted.

How have these qualitative predictions fared in light of the intense experimental activity referred to above? After some initial disappointment, existence of the states of the predicted type, together with the existence of most of the predicted radiative decays has largely been confirmed.[49] In particular, there is now strong evidence for the existence of <u>four</u> states with masses in between that of the ψ and ψ', all with C = +1. Denoting these collectively with the symbol χ, one has a $\chi(3545)$, $\chi(3505)$, $\chi(3455)$ and a $\chi(3410)$. In more detail, from observation of hadronic final states, there is evidence for the processes

$$\psi' \to \chi + \gamma , \quad \chi \to \text{hadrons} \tag{4.14}$$

at $m_\chi \sim 3550$, 3500, and 3400 MeV. From observation of photon pairs there is evidence for

$$\psi' \to \chi + \gamma, \quad \chi \to \psi + \gamma \tag{4.15}$$

at $m_\chi \sim 3550$, 3500 and 3450. The process (4.14) has not been seen at $m_\chi \sim 3450$ and the process (4.15) has not been seen at $m_\chi \sim 3400$. In addition there is evidence from observation of three photons, for a process

$$\psi \to X + \gamma, \quad X \to \gamma + \gamma \tag{4.16}$$

with $m_X \sim 2850$ MeV.

Thus it seems very likely that at least five narrow C-even states have been identified in the range of interest. According to a recent analysis of Chanowitz and Gilman,[50] if one assumes that the four observed states between ψ and ψ' correspond in some order to the three $\chi_J = {}^3P_J$ and to $\eta_c' = 2{}^1S_0$ the most likely assignment compatible with all the data is

$$\chi(3410) \leftrightarrow \chi_0$$
$$\chi(3505) \leftrightarrow \chi_1 \tag{4.17a}$$
$$\chi(3545) \leftrightarrow \chi_2$$

and

$$\chi(3455) \leftrightarrow \eta_c' . \tag{4.17b}$$

The "X" (seen at DESY but not at SLAC) is the one and only candidate for the ground state of paracharmonium:

$$X(2850) \leftrightarrow \eta_c . \tag{4.17c}$$

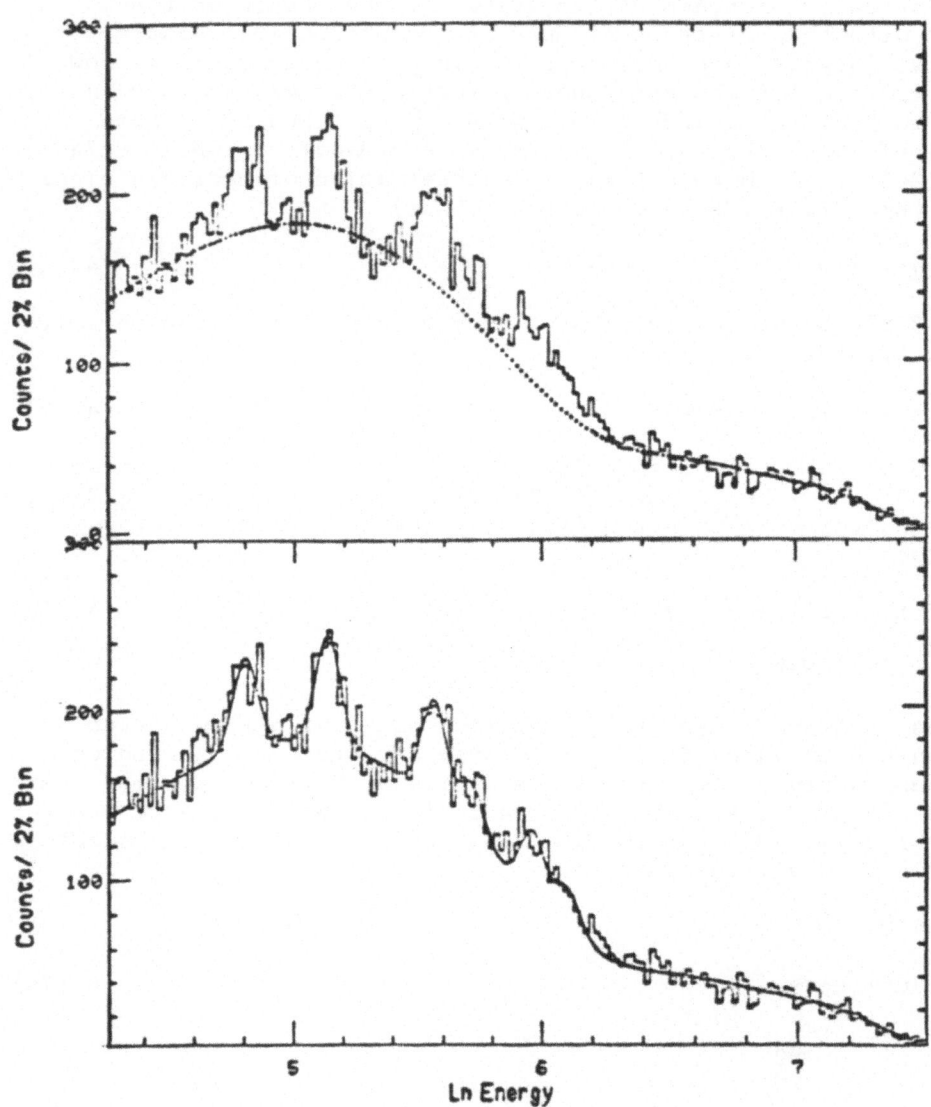

Fig. 9.   Inclusive γ-ray spectrum for ψ' → γ + anything. The
          abscissa is ℓnE$_γ$, with E$_γ$ in MeV. The dotted line
          represents a Monte Carlo calculation of the background;
          the solid line represents, a best fit, allowing for
          ψ' → γ + χ$_J$, χ$_J$ → γ + ψ. (From Ref. 51).

Thus, from a qualitative point of view, there is at present impressive agreement with the idea that the psions represent different states of a particle, anti-particle system, whose dynamics is perhaps simpler than one would expect.

Quite recently existence of the radiative decays has been confirmed by study of the inclusive γ-ray spectrum of the ψ',[51]

$$\psi' \to \gamma + \text{anything} . \tag{4.18}$$

The peaks seen in Fig. 9 at E_γ = 121, 169, 260 and at E'_γ = 448, 390 and 304 MeV are interpretable respectively as $\psi' \to \chi_J + \gamma$ (J = 2, 1, 0) followed by $\chi_J \to \psi + \gamma'$ (J = 2, 1, 0). The χ-masses are correspondingly 3561, 3511 and 3414 MeV, in good agreement with the values listed earlier (eq. 4.17a).

What about the quantitative aspects of (b_2) and (b_3)? First, the good news. If one neglect spin-dependent forces altogether and drops the presumably small r^{-1} term in $V_{s.i.}$ one may solve the resulting Schrödinger equation for the 1S and 2S states (exactly, as it turns out, in terms of Bessel functions of order 1/3). Since, in the nonrelativistic limit the amplitude for, e.g., $\psi \to e^+e^-$ may be expressed in terms of m_c and $\phi_{1S}(\vec{r} = 0)$, one may determine the three parameters m_c, $\underline{a}$, and $\underline{d}$ by fitting, say m_ψ, $m_{\psi'}$ and $\Gamma(\psi \to e^+ + e^-)$. (A typical result of such an exercise is $m_c \sim 1.6$ to 2 GeV.) The resulting prediction for $\Gamma(\psi' \to e^+e^-$ is within a factor of two of the experimental value of 2.2 keV. Not too bad, considering the simplicity of the model. Furthermore, the first excited P-states are predicted to be at about 2/3 of the way between ψ and ψ', i.e. at around 3500 GeV, right in the ball park of the χ-states. The inclusion of the r^{-1} term changes these numbers only slightly.

Now for the bad news: (i) If one uses the radial wavefunctions given by the model to compute the widths for the E1 decays of ψ' ($\psi' \to \chi_J + \gamma$) one gets numbers which are larger than the experimental ones by a factor of five to ten. (ii) If (b_2) holds, then the mass differences Δm between the ψ and η_c and between the ψ' and η'_c ought to be <u>small</u> compared to the $m_{\psi'} - m_\psi$ = 580 MeV. The identification of either X(2850) with η_c or χ(3450) with η'_c yields Δm's of the order of 250 MeV -- hardly a small fraction of 580 MeV. (iii) The hypothesis (b_2) arises in part from the idea that if the binding force comes from the exchange of massless vector gluons, then analogy with Q.E.D. suggests that V_{spin} is of order $(v/c)^2$ relative to $V_{s.i.}$.[52] If one imagines that V_{spin} is related to $V_{s.i.}$ in the same way that the non-relativistic spin-dependent inter-actions between two-electrons (coming from the reduction of the Breit operator) are related to the Coulomb interaction, then the splitting of the P-states must satisfy the bound[53]

$$(E(^3P_2) - E(^3P_o))/E(^3P_1) - E(^3P_o)) \geq 0.8$$

whereas the above assignment gives 0.4 for this ratio.

5. We cannot go into detail about the theoretical efforts that have been made to deal with these difficulties. We mention only two. First, the El widths have been shown by Eichten et al.[54] to be strongly affected by the coupling to other channels, above the threshold for production of charmed particles, which of course is not taken into account by the simple potential model; inclusion of this coupling reduces the widths considerably, although $\Gamma(\psi' \to \chi(3410) + \gamma)$ still exceeds the experimental width ($\sim$ 10 keV) by a factor of three or so. Further, very recently it has been suggested by Schnitzer[55] that if the quarks have a large anomalous "magnetic" moment ($\kappa \sim 1$) as far as their coupling to gluons is concerned, then the large ψ-η_c mass difference can be accounted for, and the problem of the P-state splittings can be somewhat alleviated without giving up the nonrelativistic picture.

We shall conclude by describing very briefly the extension of the theory of the Ml transitions to the psions.[56] The diagrams (b) and (c) of Fig. 5 are still relevant, except that the exchanged quanta represent gluons (or their "instantaneous" equivalent). For vector gluons, the potential to be inserted in a Dirac-like equation describing the $(c,\bar{c})$ bound state will have the form

$$V_v = X_v(r) + Y_v(r)\,\vec{\alpha}_1 \cdot \vec{\alpha}_2 + Z_v(r)\,\vec{\alpha}_1 \cdot \hat{r} \quad \vec{\alpha}_2 \cdot \hat{r}\,.$$

The matrix element for a relativistic Ml transition is then found to have the form

$$M = i < \chi_f \left| \frac{\vec{\sigma}_1 \cdot \vec{k}\ x\vec{\epsilon}}{2m_c} - \frac{\vec{\sigma}_2 \cdot \vec{k}\ x\ \vec{\epsilon}}{2m_c} \right| \chi_i > I_T$$

where χ_i and χ_f are the initial and final Pauli spin-wavefunctions and I_T is a sum of integrals over radial wavefunctions determined from the nonrelativistic Hamiltonian

$$H_{n.r.} = \vec{p}^2/m_c + X_v(r)\,.$$

The main point to be made is that I_T and hence the decay rate depends sensitively on the choice of Y_v and Z_v, relative to X_v. We consider two special cases: (α) $Y_v = Z_v = 0$. Then

$$I_T = < X_v/3m_c >(\zeta^2-2) + < r\ X_v'/6m_c > (\zeta^2-1)$$

where $\zeta = k/\Delta E$ is the ratio of the photon energy to the ψ'-ψ mass difference ΔE and the bracket indicates a matrix element taken between the $|2s>$ and $|1s>$ spatial wavefunctions. (β) Y_v and Z_v are determined by assuming that the propagator of transverse gluons has the form $(\delta_{ij} - q_i q_j / \vec{q}^2) \, F(\vec{q}^2)$, where $F(\vec{q}^2)$ is the Fourier transform of the confining potential $X_v(r)$. Then

$$I_T = I_T \big|_{case(\alpha)} + < r \, X_v'/6m > \pm < U_{ss}(\vec{r}) > / \Delta E$$

where $\vec{\sigma}_1 \cdot \vec{\sigma}_2 \, U_{ss}(\vec{r})$ is the induced spin-spin interaction $[U_{ss} = \nabla^2 X_v / 6m_c^2]$ and the "+" sign is for $2^3S_1 \to 1^1S_0$, the "−" sign for $2^1S_0 \to 1^3S_1$. For decays between states of the same principal quantum number, I_T is replaced by $J_T = <ns \,|\, -1 + (rX_v'/6m_c) \,| ns>$ for case (α) and $J_T = <ns \,|\, -1 + rX_v'/3m_c \,| ns>$ for case (β). The resulting widths are shown in Table II, together with a third case (γ) $V_x = X_s(r)\beta_1\beta_2$, corresponding to a potential arising from exchange of scalar gluons.[57]

It is clear from Table II that the relativistic M1 rates depend strongly on the Dirac covariant character of the potential and therefore afford a more sensitive probe of relativistic aspects of quark interactions than the E1 decays.

As an application of these results, consider the ordinary M1 decay $\psi \to X + \gamma$. Using the experimental value B.R. $(\psi \to X + \gamma)$ x B.R. $(X \to 2\gamma) \sim 2 \times 10^{-4}$ and a theoretical estimate[58] B.R. $(X \to 2\gamma)$ $\sim 10^{-3}$ we infer $\Gamma(\psi \to X + \gamma) \sim .2 \, \Gamma(\psi \to all) \sim 14$ keV, in (too) good agreement with the third column of Table II, for case (α) or (β). Again, the experimental upper limit of 4.5 keV for $\Gamma(\psi' \to X + \gamma)$ is compatible with the first column, for case (α) and (β). Further, the experimental upper limit of 10% for B.R. $(\psi' \to \eta_c' + \gamma)$ implies

Table II. M1 decay rates of charmonium S-states for different inter-
          action potentials: (α) spin-independent vector-type
          potential, (β) potential from transverse vector-gluon
          exchange, (γ) potential from scalar-gluon exchange.

Case	$R(2^3S \to 1^1S)$ (keV)	$R(2^1S \to 1^3S)$ (keV)	$R(1^3S \to 1^1S)$ (keV)	$R(2^3S \to 2^1S)$ (keV)
(α)	.007	1.0	15	11
(β)	1.3	.052	13	9
(γ)	7.3	.58	30	15

$\Gamma(\psi' \to \eta'_c + \gamma) < 23$ keV which is compatible with the last column. With regard to the second column, one knows only that B.R. $(\psi' \to \eta'_c + \gamma)$ x B.R. $(\eta'_c \to \psi + \gamma) \sim 1\%$. It we take the last column seriously then B.R. $(\psi' \to \eta'_c + \gamma) = 5\%$, and we infer B.R. $(\eta'_c \to \psi + \gamma) = 20\%$; from the second column we then infer $\Gamma(\eta'_c \to$ all$) \sim$ (5 to .1) keV -- this is about two orders of magnitude smaller than expected from the gluon-annihilation picture of hadronic decays of the psions: $c + \bar{c} \to$ color gluons $\to$ hadrons and, as emphasized in Ref. 50, poses a serious problem.

In summary, at present the evidence for the relativistic M1 transitions in charmonium is relatively weak: there are only a few events interpretable as $\psi' \to \chi(3455) + \gamma$, $\chi(3455) \to \psi + \gamma$, and $\psi' \to X + \gamma$ has not been seen. However, if one considers the 40-odd years it took before such transitions were seen in atoms or ions, then as far as psions are concerned, we have been in the desert for only a very short time. So let us be patient.

6. I should stress, in conclusion, that there are many other issues concerning the new particles which I have not been able even to touch on. These include the problem of a quantitative understanding of the hadronic decays, the considerable structure exhibited by the e^+e^- annihilation cross-section above the mass of the ψ', and most exciting of all, the possibility that charmed particles have finally been discovered. For these and related developments I must refer you to the reviews already quoted,[59] to the forthcoming proceedings of the 18th International High-Energy Physics Conference held last week at Tbilisi, and to recent and hopefully future issues of the New York Times.

Finally, I should remark that yet another area involving magnetic dipole transitions has become of intense interest very recently. I refer of course to the possibility of detecting parity-violation effects in atomic transitions;[60] such effects are presumably induced by the weak electron-nucleon interaction arising from the exchange of a neutral heavy boson. Its study, which could be of great significance in elucidating the nature of the weak neutral current, involves atomic, nuclear and particle physics, a welcome feature in this era of specialization. It would seem that M1 transitions of one type or another will continue to be of interest, for some time to come, to both atomic and particle physicists.

<u>Acknowledgements</u>

I would like to thank both G. Feinberg and G. A. Snow for several discussions helpful in the preparation of this report.

*Supported in part by the National Science Foundation, under Grant No. GP-41822X.

REFERENCES

1. P. A. M. Dirac, Proc. Roy. Soc. London A <u>114</u>, 243 (1927.
2. The radiative decays of the metastable states of the H and He
    sequences were previously reviewed from the point of view of
    theory by G. W. F. Drake, in <u>Proceedings of the Third Inter-</u>
    <u>national Conference on Atomic Physics</u>, Boulder, Colorado,
    1972, edited by S. J. Smith and G. K. Walters (Plenum Press,
    New York-London, 1973), p. 269; a detailed review of
    experimental techniques was given by R. Marrus, <u>ibid</u>, p. 291.
3. M. Goeppert-Mayer, Ann. Physik <u>9</u>, 273 (1931).
4. G. Breit and E. Teller, Astrophys. J. <u>91</u>, 215 (1940).
5. A. H. Gabriel and C. Jordan, Nature <u>221</u>, 947 (1969); Mon.
    Notices, Roy. Astron. Soc. <u>145</u>, 241 (1969). For more recent
    work see C. P. Bhalla, A. H. Gabriel, and L. P. Presnayakov,
    <u>ibid</u> <u>172</u>, 359 (1975).
6. H. K. Griem, Astrophys. J. <u>156</u>, L103 (1969); <u>161</u>, L555 (1970).
    A correct order-of-magnitude estimate of the M1 decay rate
    in helium was also made by G. Feinberg, who had independently
    noted that this rate had been underestimated in Ref. 4.
    (private communication, 1969; referred to by M. J. Rees and
    D. W. Sciama, Comments on Astrophysics and Space Physics <u>1</u>,
    35 (1969)).
7. R. Marrus and R. W. Schmieder, Phys. Lett. A <u>32</u>, 431 (1970);
    Phys. Rev. Lett. <u>25</u>, 1245 (1970).
8. R. Marrus and R. W. Schmieder, Phys. Rev. A <u>5</u>, 1160 (1972).
9. J. Bednar, C. L. Cocke, B. Curnutte, and R. Randall, Phys.
    Rev. A <u>11</u>, 460 (1975).
10. H. Gould and R. Marrus (private communication).
11. H. Gould, R. Marrus, and R. W. Schmieder, Phys. Rev. Lett.
    <u>31</u>, 504 (1973).
12. H. Gould, R. Marrus, and P. J. Mohr, Phys. Rev. Lett. <u>33</u>,
    676 (1974).
13. H. Gould and R. Marrus, in <u>Proceedings of the Fourth</u>
    <u>International Conference on Beam-Foil Spectroscopy</u>,
    <u>Gatlinburg, Tennessee, 1975</u>, edited by I. A. Sellin and
    D. J. Pegg (Plenum Press, New York-London, 1976), Vol. I.,
    p. 305.
14. H. W. Moos and J. R. Woodworth, Phys. Rev. Lett. <u>30</u>, 775
    (1973); Phys. Rev. A <u>12</u>, 2455 (1975).
15. G. Feinberg and J. Sucher, Phys. Rev. Lett. <u>26</u>, 681 (1971).
16. G.W.F. Drake, Phys. Rev. A <u>3</u>, 908 (1971).
17. W. R. Johnson, Phys. Rev. Lett. <u>29</u>, 1129 (1972).
18. See Ref. 16. The starting point of Drake's calculation is
    the time-honored two-electron Dirac-Briet Hamiltonian H_{D-B}
    [(3.7) without projection operators but with V_{12} con-
    taining the Breit operator], modified by the inclusion
    of the radiation field. Another discussion of the M1
    transitions, also based on H_{D-B}, is given by I. L. Beigman

and U. I. Safronova, Zh. Eksp. Teor. Fiz., $\underline{60}$, 2045 (1971) [Sov. Phys. JETP $\underline{33}$, 1102 (1971)]. As mentioned in footnote 4 of Ref. 15, H_{D-B} has difficulties, which it shares with H_{Dirac}, Eq. (3.12), quite apart from the restrictions on the use of the Breit operator (see Ref. 24). However, incorrect is an assertion in footnote 5 of Ref. 15 which stated that Drake's modification of H_{D-B} involved double-counting. This was the result of a misunderstanding of the notation in Ref. 16, as correctly pointed out in G. W. F. Drake, Phys. Rev. A $\underline{5}$, 1979 (1972).

19. D. L. Lin and G. Feinberg, Phys. Rev. A $\underline{10}$, 1425 (1974) and D. L. Lin, Columbia University Ph.D. dissertation, 1975 (unpublished).

20. For related work see R. Barbieri and J. Sucher, CERN preprint, (derivation from first principles of a gauge invariant formula for the radiative corrections to atomic decay rates) and G. W. F. Drake, Phys. Rev. A $\underline{9}$, 2799 (1974) (cancellation of the low-frequency contributions to the $\alpha \log Z\alpha$ terms).

21. E. Kelsey, University of Maryland Ph.D. dissertation, 1974 (unpublished); Annals of Physics, 1976 (to appear).

22. W. R. Johnson (private communication). I thank Dr. Johnson for providing me with this result; it is implicit in Lin's thesis (Ref. 19).

23. W. R. Johnson and C.-P. Lin, Phys. Rev. A $\underline{9}$, 1486 (1974); see also S. Feneuille and E. Koenig, C. Acad. Sci. B. $\underline{274}$ 46 (1972).

24. See G. E. Brown and D. G. Ravenhall, Proc. Roy. Soc. A $\underline{208}$, 552 (1951); J. Sucher "Relativistic Hartree-Fock Type Calculations: A Caveat," University of Maryland T.R. #77-018 (unpublished).

25. M. O'Carroll and J. Sucher, Phys. Rev. Lett. $\underline{21}$, 1143 (1968).

26. E. J. Kelsey and J. Sucher, Phys. Rev. A $\underline{11}$, 1829 (1975). The main tool is Schwinger's integral representation of the Coulomb Green's function. The first order correction arising from e^2/r_{12} was independently calculated by P. J. Mohr, who used similar techniques. The value of b_1 obtained by him is identical with ours (P. J. Mohr, private communication, 1975).

27. M. Anderson and F. Weinhold, Phys. Rev. A $\underline{11}$, 442 (1975).

28. However, inclusion of these terms improves the agreement between the theoretical value and the central experimental value for all the accurately measured lifetimes.

29. See Ref. 15. The wavefunction of Ref. 16 (not designed to be highly accurate for low Z) gives $\tau = .79 \times 10^4$ sec.

30. This part of our review is an extended version of what could be covered in the original talk. It is written primarily for the benefit of non-high-energy physicists. Accordingly the references emphasize review papers and are far from exhaustive.

31. J. J. Aubert et al., Phys. Rev. Lett. $\underline{33}$, 1404 (1974).

32. J. E. Augustin et al., Phys. Rev. Lett. 33, 1406 (1974).

33. G. Bacii et al., Phys. Rev. Lett. 33, 1408 (1974).

34. J. E. Augustin et al., Phys. Rev. Lett. 34, 764 (1974).

35. For authoritative reviews of the experimental and theoretical situation concerning the new particles as of August, 1975, see Proceedings of the 1975 International Symposium on Lepton and Photon Interactions at High Energies, edited by W. T. Kirk (Stanford Linear Accelerator Center, Stanford, 1975). (Referred to as 1975 SLAC Symposium, hereafter). Fig. 7 is taken from the review talk of S. C. C. Ting, and Fig. 8 from that of G. Abrams, p. 155 and p. 25, respectively, of these Proceedings.

36. Maggie, in James M. Barrie's play What Every Woman Knows (1908).

37. P. Tarjanne and V. Teplitz, Plys. Rev. Lett. 11, 447 (1963); Y. Hara, Phys. Rev. 134B, 701 (1964); M. Gell-Mann, Phys. Lett. 8, 214 (1964); B. J. Bjorken and S. L. Glashow, Phys. Lett. 11, 225 (1964), where the name "charm" was introduced; D. Amati et al., Nuovo Cim. 34, 1732 (1964).

38. For reviews of SU(3) and of the quark-model, including many of the original papers, see M. Gell-Mann and Y. Ne'eman, The Eightfold Way (W. A. Benjamin, Inc., New York, 1964) and J. J. J. Kokkedee, The Quark Model (W. A. Benjamin, Inc., New York, 1969). For more recent reviews, see H. J. Lipkin, Phys. Rep. 8C, 174 (1973) and N. P. Samios, M. Goldberg and B. T. Meadow, Rev. Mod. Phys. 46, 49 (1974).

39. For reviews of unified gauge theories see J. Bernstein, Rev. Mod. Phys. 46, 7 (1974); S. Weinberg, ibid, 225 (1974); E. S. Abers and B. W. Lee, Phys. Rep. 9, 1 (1973); M. A. B. Bég and A. Sirlin, Ann. Rev. Nuc. Sci. 34, 379 (1974).

40. It was first shown by S. Glashow, J. Iliopoulos and L. Maiani (Phys. Rev. D2, 1285 (1970)) that with the introduction of charm one could avoid violation, by higher-order (divergent) corrections, of the selection rules built into the phenomenological weak-interaction Lagrangian; the extension to the unified gauge theories was made by S. Weinberg (Phys. Rev. D5, 1412 (1972).

41. For an up-to-date review of color models of the hadrons see O. W. Greenberg and C. A. Nelson, U. of Md. Tech. Rep. #77-022 (to appear in Physics Reports).

42. For a review of the theoretical ideas behind "quark confinement" see R. Dashen in 1975 SLAC Symposium (op. cit.) p. 981.

43. A comprehensive scheme in which the basic hadrons have integer charges and are not confined has been developed by J. C. Pati and A. Salam. For reviews, see J. C. Pati, in Gauge Theories and Modern Field Theories, ed. R. Arnowitt and P. Nath (MIT Press, Cambridge, 1976); J. C. Pati and A. Salam, invited paper presented at "Neutrino Conference 1976," Aachen, June 1976.

44. For an alternative, purely algebraic approach to the narrow resonances, which is based on commutators of vector and axial charges within the framework of broken SU(4) symmetry, see S. Oneda, invited paper presented at the Joint International Symposium on Mathematical Physics, Mexico City, January 5-8, 1976; S. Oneda and E. Takasugi, U. of Md. T.R. # 76-139.

45. T. Appelquist and H. D. Politzer, Phys. Rev. Lett. 34, 43 (1975); A. De Rujula and S. L. Glashow, Phys. Rev. Lett. 34, 46 (1975); T. Appelquist et al., Phys. Rev. Lett. 34, 365 (1975); S. Borchardt et al., Phys. Rev. Lett. 34, 38 (1975).

46. For a review see M. K. Gaillard, B. W. Lee and J. L. Rosner, Rev. Mod. Phys. 47, 277 (1975) and B. W. Lee in 1975 SLAC Symposium, p. 635.

47. H. J. Schnitzer, Brandeis University report, 1974 (unpublished); E. Eichten et al., Phys. Rev. Lett. 34, 369 (1975); B. J. Harrington et al., Phys. Rev. Lett. 34, 706 (1975); J. Kogut and L. Susskind, Phys. Rev. Lett. 34, 767 (1975), Phys. Rev. D12, 2821 (1975); R. Barbieri et al., CERN report no. TH2036, 1975 (unpublished). Many further references on the bound state models may be found in J. S. Kang and H. J. Schnitzer, Phys. Rev. D12, 841 (1975); D12, 2791 (1975). An example of a quite different approach is that of N.-P. Chang and C. Nelson, Phys. Rev. Lett. 35, 1492 (1975).

48. The $-\alpha_s/r$ term represents a Coulomb-like attraction arising from gluon-exchange at short distances; α_s is expected to be small because of asymptotic freedom.

49. This evidence comes from experiments performed at DESY and at SLAC. See W. Braunschweig et al., Phys. Rev. Lett. 57B, 407 (1975); G. J. Feldman et al., Phys. Rev. Lett. 35, 821 (1975); V. Lüth, invited talk at the Washington meeting of the American Physical Society, April, 1976; and G. Goldhaber, invited talk at the Wisconsin International Conference on the Production of Particles with New Quantum Numbers, Madison, Wisc., April 1976 (unpublished); B. Wiik, rapporteur report at the 18th International High Energy Physics Conference (Tbilisi, U.S.S.R., July, 1976) (unpublished).

50. M. S. Chanowitz and F. J. Gilman, Phys. Lett. 63B 178 (1976).

51. D. H. Badtke et al. (Maryland-Pavia-Princeton-San Diego-SLAC-Stanford collaboration), preliminary report submitted to 18th International High Energy Physics Conference, Tbilisi, U.S.S.R., July, 1976. Fig. 9 is from the more recent final report; I thank D. H. Badtke, L. H. Jones and G. Zorn for informative communications concerning this work.

52. Explicit forms for the spin-dependent interactions arising from such assumptions may be found in H. J. Schnitzer, Phys. Rev. Lett. 35, 1540 (1975), Phys. Rev. D13, 74 (1976); J. Pumplin, W. Repko and A. Sato, Phys. Rev. Lett. 35,

1538 (1975); G. Feinberg and J. Sucher, Phys. Rev. Lett. <u>35</u>, 1740 (1975); R. Barbieri <u>et al</u>., Ref. 44; A. De Rujula <u>et al</u>., Phys. Rev. D<u>12</u>, 147 (1975).

53. H. J. Schnitzer, Phys. Rev. D<u>13</u>, 74 (1976).

54. E. Eichten, <u>et al</u>., Phys. Rev. Letters <u>36</u>, 500 (1976).

55. H. J. Schnitzer, Brandeis University report, 1976 (unpublished). Another possibility, that the large spin-dependent forces come from the exchange of axial-gluons, is considered in Ref. 56.

56. G. Feinberg and J. Sucher, Phys. Rev. Lett. <u>35</u>, 1740 (1975).

57. Table II was prepared by use of Eqs. (14) and (15) of Ref. 56, using updated values for the photon energies and the assignments (4.17b,c), viz. $k(2^3S \rightarrow 2^1S_0) = 222$ MeV, $k(2^3S_1 \rightarrow 1^1S_0) = 740$ MeV, $k(2^1S_0 \rightarrow 1^3S_1) = 338$ MeV, and $k(1^3S_1 \rightarrow 1^1S_0) = 238$ MeV. We take the occasion to note that: (i) The values presented in Table I of Ref. 56 for the case of a scalar-gluon were incorrect because of a numerical error. (ii) The denominator $m_1 + m_2$ should have read $2m_1$ in Eq. (2) and $4m_1 m_2/(m_1 + m_2)$ in Eq. (12); the formulas were used only for the equal-mass case, where there is no change.

58. T. Appelquist and H. D. Politzer, Phys. Rev. D<u>12</u>, 1404 (1975).

59. See also A. DeRujula, in <u>Gauge Theories and Modern Field Theories</u>, ed. R. Arnowitt and P. Nath (MIT Press, Cambridge, 1976), p. 190; R. H. Dalitz, in <u>Few Body Dynamics,</u> eds., A. N. Mitra, I. Slaus, V. S. Bhasin, and V. K. Gupta (North Holland Publishing Co., Amsterdam, 1976), p. 632.

60. See the invited talks by M. A. Bouchiat, N. Fortson, and P. G. H. Sandars presented at this conference, and that by I. B. Khriplovich, presented at the 18th International High Energy Physics Conference (Tbilisi, U.S.S.R., July, 1976).

THE ELECTRIC AND MAGNETIC DIPOLE MOMENTS OF THE NEUTRON

Norman F. Ramsey

Harvard University

Lyman Lab. of Physics, Cambridge, MA. 02138

I. INTRODUCTION

In 1950 Purcell and Ramsey[1] pointed out that the parity arguments then used to prove that particles and nuclei could not have electric dipole moments, must be based on an experimental rather than a theoretical basis. As a test of this assumption, Smith, Purcell and Ramsey[2] used a neutron beam magnetic resonance apparatus to search for a neutron electric dipole moment and concluded that such a moment divided by the proton charge (μ_e/e) was experimentally less than 5×10^{-20} cm. Later, from the work of Lee and Yang[3] and Wu, et al,[4] it became apparent that the parity assumption was indeed invalid, but Landau[5] and others pointed out that the parity argument against an electric dipole moment could be replaced by one based on time reversal invariance. However, Ramsey[6] emphasized that time reversal invariance like parity at an earlier time, was merely assumed and must rest on an experimental basis. In 1964 Christenson, Cronin, Fitch, and Turlay[7] discovered the CP violating mode in the decay of the K_L^0 meson into two charged pions, which strongly suggested a violation of time reversal symmetry.

Since then a number of theoretical predictions[8] have been made for nucleon electric dipole moments on the basis of theories developed to account for the K_L^0 decay. Although the different predictions cover a wide range of values, some were as large as 10^{-19} cm for μ_e/e and most predicted 10^{-22} cm or larger. Since most of the range of predicted values was accessible to experimental search, several different experiments to measure the neutron electric dipole moment were started by Baird, Dress, Miller and Ramsey,[9,10,13,14] by Nathan and Shull,[11] by Cohen, Lipworth, Silsbee, and Ramsey,[12] by Smith and Pendlebury,[15] and by Apostolescu, Ionescu, Bujor,

Mecterts and Petroscu.[16]   The successive limits of the different
experiments is given in Fig. 1.   The greatest sensitivity at each
time has been provided by the experiments of Dress, Miller, Ramsey
and Baird[9,10,13,14,15] and their most recently published experiment[13] provides the greatest sensitivity of any experiment so far
published.   This experiment was based on a neutron beam magnetic
resonance study of 80 m/sec neutrons and provided a limit
μ_e/e of 10^{-23} cm.   It became apparent at the end of this Oak Ridge
experiment that a further increase in precision would require a
high flux of neutrons at velocities of 100 m/sec or less.   For
this reason, the apparatus was moved to Grenoble to take advantage
of the cryogenic moderator at the Institute Laue-Langevin (ILL)
reactor.

The earlier apparatus with considerable modifications has now
been operating at the Grenoble reactor and has produced a lower
limit for the neutron electric dipole moment than any obtained
previously.   The experiments were done in collaboration with W. B.
Dress and P. D. Miller of Oak Ridge National Laboratory in the
United States, Paul Perrin of the CENG in Grenoble and Michael
Pendlebury of Sussex University, England.

II.   METHOD AND APPARATUS

The apparatus used in this experiment is essentially one to
measure with high precision the precessional frequency of the neutron spin in a weak magnetic field with a neutron beam magnetic
resonance apparatus similar to that used for measuring the magnetic
moment of the neutron.   A strong electrostatic field is then applied
successively parallel and antiparallel to the magnetic field H.
If the neutron had an electric dipole moment the torque due to this
dipole moment in the electric field would make the precessional
frequency of the neutron spin somewhat greater with the electric
field in one direction and somewhat less in the opposite.   By setting an experimental limit on the change in the precessional frequency, a limit is thereby set on the electric dipole moment of
the neutron.   The main requirements in the experiment are to achieve
a very high sensitivity and to eliminate spurious effects that might
either lead to a false apparent electric dipole moment or might ob-
scure an actual moment.

A schematic view of the apparatus is shown in Fig. 2.   The
neutron beam comes from the cryogenic moderator at the ILL reactor.
The neutrons are conducted from the moderator through a neutron
conducting tube of rectangular cross sections on whose surface
they are totally reflected at glancing angles of two degrees or less.
The use of such neutron conducting pipes, which becomes possible
with sufficiently slow neutrons, markedly enhances the intensity
by overcoming the normal diminution of beam intensity with the in-

NEUTRON EDM EXPERIMENTAL RESULTS

VALUE D (cm)	LABORATORY (year)	REFERENCE
$< 5 \times 10^{-20}$	ORNL (1951, 1957)	2
$(-2 \pm 3) \times 10^{-22}$	ORNL (1967)	9
$< 3 \times 10^{-22}$	ORNL (1968)	14
$(+2.4 \pm 3.9) \times 10^{-22}$	MIT-BNL (1967)	11
$< 1 \times 10^{-21}$	BNL (1969)	12
$< 1 \times 10^{-21}$	ALDERMASTON (1968)	15
$< 5 \times 10^{-23}$	ORNL (1969-1972)	10
$0.2 \pm 3.9 \times 10^{-22}$	ROMANIA (1970)	16
$< 1 \times 10^{-23}$	ORNL (1973)	13

Figure 1 -- Experimental results for the electric dipole moment of the neutron

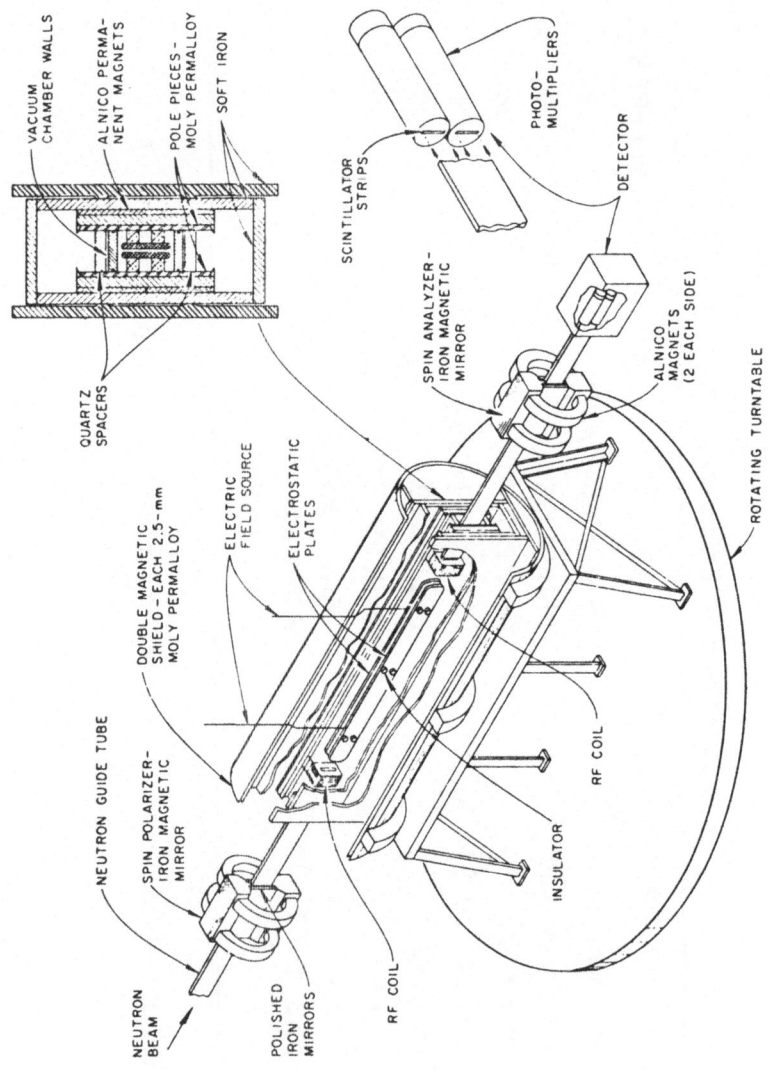

Figure 2 -- Experimental arrangement of the magnetic resonance spectrometer

verse square of the distance from the moderator. This gain of intensity is badly needed to compensate in part for the even greater loss of intensity by the selection of extremely slow neutrons.

As shown in Fig. 2, the neutron beam goes through a portion of the pipe in which the walls consist of magnetized iron. Depending upon the orientation of the neutron spin, there is either a positive or negative magnetic interaction between the neutrons and the magnetic induction of the walls in the magnetized region. The combination of this positive or negative mean magnetic interaction with the coherent forward scattering amplitude of the neutrons by the wall material leads to total reflection at the walls for neutrons of one spin orientation while the neutrons with opposite spins are not reflected by that portion of the pipe and instead penetrate through the walls and are lost. Consequently following the spin polarizing magnetic mirror, the neutrons are mostly polarized. The analyzing device to determine if there has been a change in the neutron spin orientation is a second spin analyzing magnetic mirror. If the neutron spin remains unaltered between the first and the second of these magnetic field regions, most of the neutrons will be transmitted by the second region. If, on the other hand, the neutrons have been reoriented by approximately 180 degrees between the two iron mirror sections, the neutrons whose orientation has changed will not be totally reflected in the second magnetic mirror with a consequent reduction in beam intensity. Therefore, if the oscillatory fields are all in phase, the minimum of detected beam intensity occurs at the precessional frequency of the neutron. On the other hand, as shown by the author,[17] if the oscillatory magnetic field is provided in two separate segments with a 90 degree phase shift between them, the shape of the resonance curve is that of a dispersion curve with the steepest portion of the slope at the spin precession frequency as shown in Fig. 3. If the frequency of the oscillator is set so that the detected neutron intensity is at the position of the steepest slope, the presence of a neutron electric dipole moment can be detected by successively reversing a strong electrostatic field. If there is an electric dipole moment the torque due to the electric field will increase the precessional frequency of the neutron for one orientation of the field and decrease it for the opposite. At a fixed frequency of the oscillator, this change in the precessional frequency of the neutron spin will then be detectable with high sensitivity as a change in the neutron beam intensity.

The electric field is applied over a length of 196 cm. and typically has a value of about 100 kV/cm. The static magnetic field was about 17G and the neutron beam was 89% polarized.

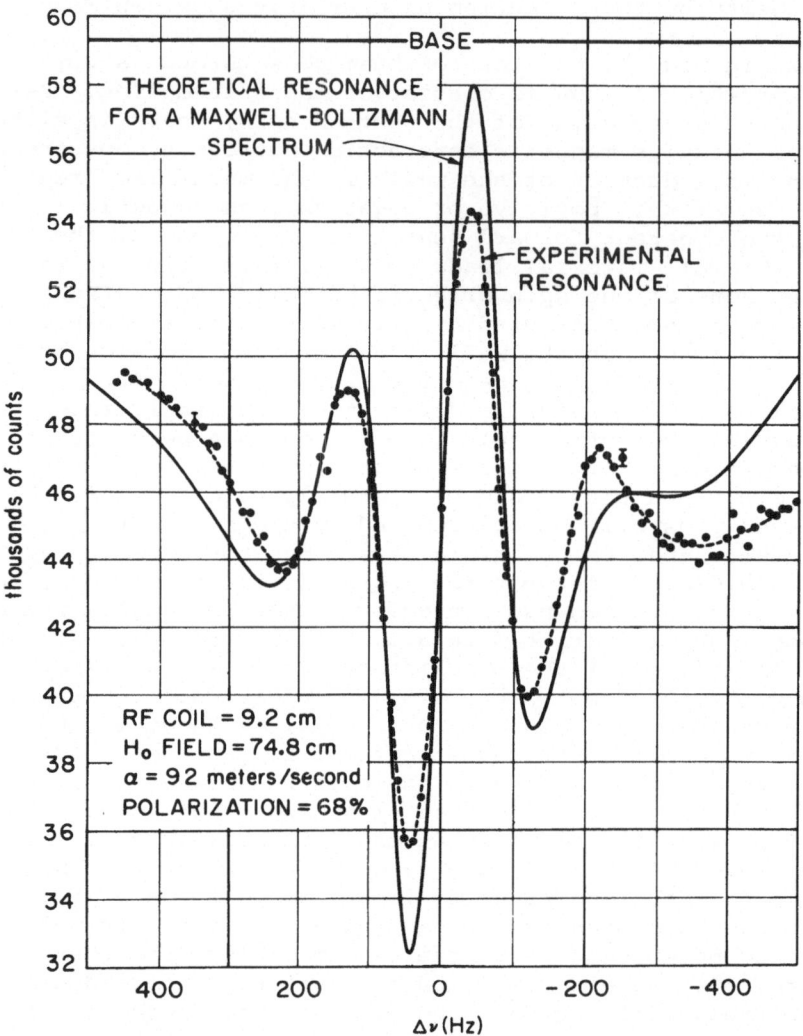

Figure 3 -- Typical magnetic resonance with a phase shift of π/2 between the two oscillatory fields. The calculated transition probability for a Maxwell-Boltzmann distribution characterized by a
temperature of 1°K is shown in the solid curve. The departure of
the experimental curve from the theoretical one when far from
resonance is to be expected from the known departure of the beam
velocity from a Maxwell-Boltzmann distribution.

Great care in the experiment must be taken to avoid spurious effects which could either simulate a non existent electric dipole moment or mask an existing one. Fortunately, a number of things can be done to eliminate or minimize such spurious effects. The relative phase of the two oscillatory fields, can be shifted from +90 degrees to -90 degrees in which case the slope of the curve at the resonance position is reversed with a consequent reversal of the effect of the electric field on the detected neutron beam intensity. This reversal in the electric dipole moment effect eliminates many possible spurious effects. The phase was reversed once per second. Fortunately, in addition, many of the possible spurious effects cancel themselves due to the parity or time reversal symmetry of the effect. For example, there can be an effect of the electric field upon the frequency due to the force from the electric field pulling the magnets together and thereby changing the magnetic field. However, this effect and many others go as E^2 and consequently cancel on subtracting of results with reversed electric fields. A check on the existence of such E^2 effect can also be obtained from observations at zero electric field. Likewise to detect magnetic effects from the field reversing mechanism, the leads to the source of potential are reversed at intervals. In addition, measurements are made when no potential is present but when the reversing switches are successively changed. The importance of such control measurements is illustrated by the fact that for some months there was a very small residual effect when the switches were reversed in the absence of any potential. This was eventually eliminated by moving the reversing switches still further from the apparatus and by increasing their magnetic shielding.

An important source of a spurious effect has been observed in recent runs of high sensitivity. Whenever there is a spark across the electric plates, the accompanying current produces a slight magnetic field which in turn produces a very small residual change in the permanent magnetic field due to the hysteresis of the iron. Even if the neutron counts during the period of the spark are excluded, the residual change in the permanent magnetic field can give a false result. This trouble, however, can be eliminated if the existence of sparks are recorded and if care is taken to assure that equal amounts of measurements with the fields in opposite directions are utilized in each interval between sparks.

One of the most bothersome spurious effects is that due to motion of the neutrons with a velocity $\vec{v}$ through the electric field $\vec{E}$ since such motion produces an effective magnetic field $\vec{E} \times \vec{v}/c$. This effective magnetic field can then interact with the known neutron magnetic moment to produce an added precession frequency which will look like that due to an electric dipole moment since it will reverse with the reversal of $\vec{E}$. This effect is drastically reduced

by making E parallel to $\vec{H}$. If exact parallelism could be obtained
the effect would be completely eliminated since this spurious magnetic field would be perpendicular to the initial magnetic field
with the result that the effect would go as E^2 instead of $\vec{E}$. However, due to residual magnetism of ferromagnetic materials and magnetic shields, one can never be absolutely certain as to the direction of the magnetic field with the result that $\vec{E}$ and $\vec{H}$ cannot be
made exactly parallel and the perpendicular component of $\vec{E}$ can produce an apparent electric dipole effect through the $\vec{E}$ x $\vec{v}$/c effective magnetic field. The existence of such an effect, however, can
be detected by changing the velocity of the neutrons since the
spurious effects should be proportional to the neutron velocity.
Consequently, all the data is analyzed in terms of an electric dipole moment and an apparent electric dipole proportional to the
neutron velocity. The neutron velocity is altered in either of two
ways. In some cases, the velocity is changed by changing the angle
of neutron reflection from mirrors and in all cases the measurements
are repeated many times with the direction of the neutrons through
the apparatus reversed. For this reason, as can be seen in Fig. 2,
the basic neutron resonance apparatus is fastened to a turn table
which can be rotated to have the neutrons pass through the apparatus
in opposite directions. The necessity for experiments at altered
velocity greatly increases the running time of the experiment since
the $\vec{E}$ x $\vec{v}$/c effect must be measured with equal precision to that
desired for the neutron electric dipole moment.

                            III.   RESULTS

     The results of the present phase of measurements at the Institute
Laue-Langevin are

$$\mu_E/e = (0.4 \pm 1.5) \times 10^{-24} \text{ cm.}$$

In other words, the neutron electric dipole moment, if it exists at
all, is less than 3 x 10^{-24} cm. To emphasize the smallness of this
result, I should emphasize to nuclear physicists that this is 10^{-24} cm
not cm^2; it corresponds to 10^{-48} cm^2. If the neutron were expanded to
the size of the earth this asymmetry would correspond to an incre-
mental height of 0.01 cm in the northern hemisphere.

     There have been numerous theoretical predictions as to the
value of the neutron electric dipole moment. All theories that
account for the CP violating decay of the K_L^o meson[7] predict non zero
values for the neutron electric dipole moment. The predictions of
these theories[18-32] are shown in Fig. 4. Each lettered block in the
figure corresponds to the prediction of a different theory. One
source of interest of the present experimental limit is that it provides extreme difficulties for many of the theoretical predictions

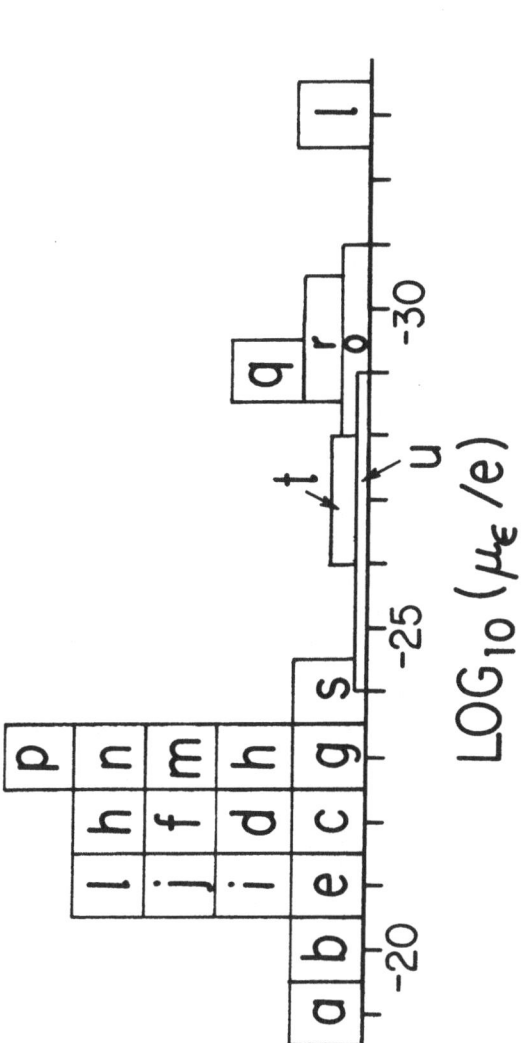

Figure 4 -- Theoretical predictions of the neutron electric dipole moment. Each lettered block corresponds to a different theory with the references to the different theories given by the corresponding letters in the references.[18-32] The basis of the different theories are indicated in square brackets in the references with EM indicating a theory which attributes the time reversal violation to the electromagnetic theory, W attributes it to the weak forces, MW to a milliweak force, SW to a new superweak force. Normally, the rectangle indicating each theory is a square spanning one decade; where the authors propose a wider spread, the rectangle is adjusted accordingly but with the same area as for other theories.

and significant difficulty for most of the theories except those
which attribute the time reversal asymmetric interaction to a new
super weak force.

IV.   NEUTRON ELECTRIC DIPOLE MOMENT EXPERIMENTS
WITH BOTTLED ULTRA-COLD NEUTRONS

The neutron electric dipole moment experiment so far described
depends upon the fact that neutrons at a velocity of 80 m/sec will
be totally reflected by many materials at glancing angles of
approximately 5 degrees. As the velocity of the neutrons diminish,
the glancing angle for total reflection increases until finally
at a velocity of 6 m/sec total reflection can be obtained even at
normal incidence on many surfaces. Under such circumstances, it
is possible in principle to store neutrons in an enclosed bottle.
For many years,[33,34,35] we have been anxious to do such experiments
which would provide for the neutrons many of the advantages of the
successive oscillatory field experiments with stored atoms[33,34]
including those with the hydrogen maser.[34] Up until recently, however, we have had no prospect of obtaining access to such ultra-cold
neutrons.

Zeldovitch,[36] Vladimirski[37,38] and the late Dr. F. L. Shapiro
have discussed bottled ultra-cold neutrons and Shapiro and his
collaborators[39] have shown that ultra-cold neutrons can be stored
in bottles for up to 20 seconds. Improvements in the techniques
with ultra-cold neutrons and storage bottles have been made by
Steyerl,[40] Ageron,[41] Lobashov,[42] Taran,[43,44] Pendlebury, Gollub and
Smith[45,46] and Miller, Dress and Ramsey[46,47] and others. Specific
experiments to use neutron bottles to measure the neutron electric
dipole moment have been discussed by Ramsey, Miller and
Dress,[33,35,46,47] by Taran[43,44] and by Pendlebury, Smith and
Gollub.[45,46]

The experiment at ILL now being prepared to use ultra-cold
bottled neutrons to set a limit to the neutron electric dipole
moment is a collaboration between J. Byrne, R. Gollub, J. M.
Pendlebury, K. F. Smith, N. F. Ramsey, W. B. Dress, P. D. Miller,
A. Steyerl, P. G. H. Sandars, P. Ageron and P. Perrin of the
University of Sussex, Harvard University, Oak Ridge National Laboratory, Munich, Oxford University, ILL and CENG. Ultra-cold neutrons
at approximately 6 meters per second will be led by a neutroncon-
ducting pipe into the apparatus shown in Fig. 5. The neutrons will
be stored in a cylinder approximately 15 centimeters in diameter
and 10 centimeters high with the top plates being metallic -- probably beryllium -- and the sides of the cylinder being of beryllia
insulator. The oscillatory field is applied to the admission and

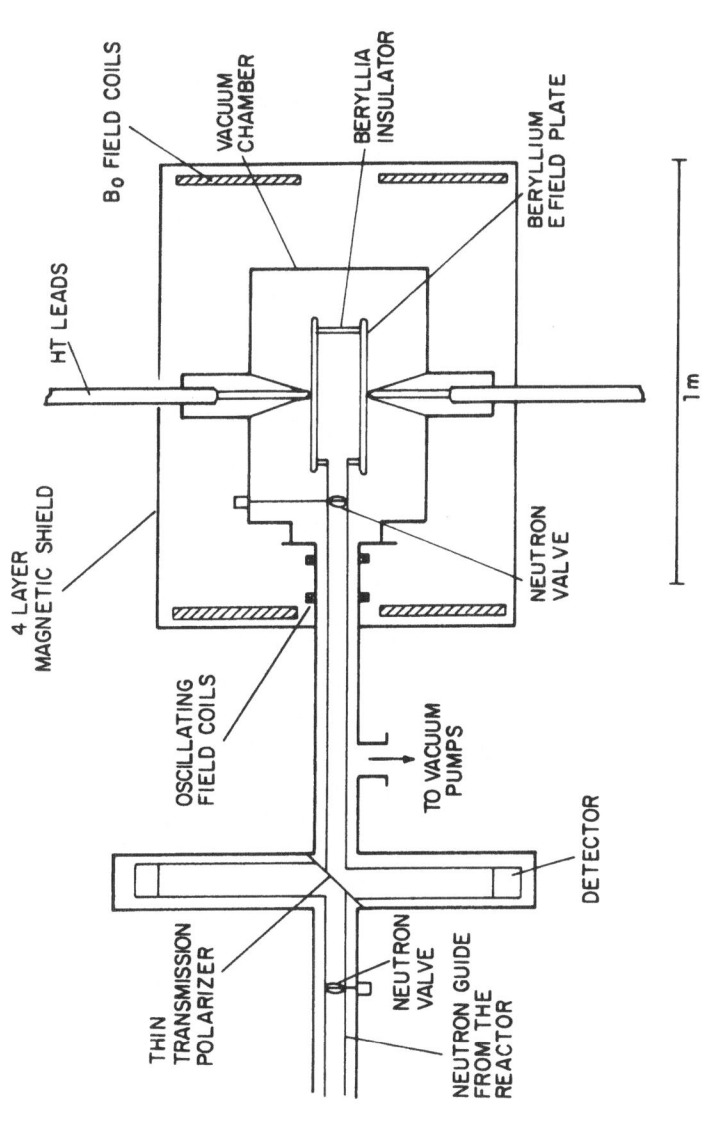

Figure 5 -- Schematic diagram of apparatus for determining the neutron magnetic resonance experiment with bottled neutrons.

exit tubes so the resonance can be observed by the previously described successive oscillatory field technique.[33] The resonance
will be observed in a similar fashion to our present neutron beam
experiment and observation will be at the steepest point of the
resonance curve. The change in beam intensity correlated with the
application of an electric field will then be examined to set a
limit to the neutron electric dipole moment.

   The use of stored ultra-cold neutrons possesses two particularly
important advantages. The resonance curve for 18 second storage
time of the neutron should be approximately 800 times narrower than
in the present experiment with a corresponding increase in sensitivity. Furthermore, as mentioned earlier, a large fraction of
running time in the present experiment must be devoted to eliminat-
ing the $\vec{E}$ x $\vec{v}$/c effect. Since it is the average value of $\vec{v}$ that is
important, this effect is drastically diminished when the neutrons
enter and leave by the same exit hole with an 18 second storage
time instead of passing through the apparatus at a velocity of
80 m/sec. As a result of the reduced effective magnetic field
from $\vec{E}$ x $\vec{v}$/c, it should also be possible to use a much weaker static
magnetic field with an accompanying reduction in the field stability
problem.

   Although the new experiment being planned will have the above
marked advantages, it must be recognized that it will still be an
extremely difficult one. The limit has by now been pushed to such
a low value that care must be taken to avoid all possible systematic
effects. Although some of these are intrinsically reduced in an
experiment with bottled neutrons, other serious problems will remain.
For example, problems due to stray magnetic fields (especially when
associated with reversals of the electric field) and to magnetic
field changes resulting from electrical sparks can be just as serious
with bottled neutrons as with neutron beams. These problems have
already caused much difficulty in the beam version of the experiment
and should be even more formidable in the bottled neutron experiment
which seeks to lower the limit for the neutron electric dipole moment
by a factor of 100 to 1000.

   The apparatus is planned to be capable of being operated in
either of two fashions. In one, pulsed ultra-cold neutrons will be
admitted for a few seconds and then stored with the neutron valve
closed for approximately 30 seconds before the valve is reopened so
the neutrons can escape past the oscillatory field for a second time.

   In the second mode of operation, the neutrons will continuously
be introduced and permitted continuously to bounce out of the neutron
bottle with a mean storage time of approximately 18 seconds. The
precision of the two methods of observation are comparable and the
two procedures should mutually compliment each other.

With an electric field of 30 kV/cm and a multilayer Mumetal or Moly-Permalloy magnetic shield, it should be possible to achieve a limit on the electric dipole moment of 10^{-26} cm. To go to a lower limit will probably require superconducting magnetic shields. These are currently contemplated but decisions on a subsequent phase of the experiment will not be taken until later. With superconducting shields and sufficiently long observation times, it should be possible to lower the limit to 10^{-27} cm. With a larger cell diameter and other improvements, sensitivity of the order of 10^{-28} cm might ultimately be reached.

V.   OTHER NEUTRON BEAM MAGNETIC RESONANCE EXPERIMENTS

Since it will take more than a year before the ultra-cold neutron beam can be available at ILL and before the apparatus for the bottled neutron experiment can be ready and since the new apparatus will be required to achieve a significant improvement in the present limit, the collaborators of the present experiment plan to modify the present apparatus so it can be used during the coming year markedly to improve the accuracy of the measurements of the neutron electric dipole moment. At present, the neutron electric dipole moment is the least accurately known of all the nucleon and lepton magnetic moments. The magnetic moments of the negative electron, the positive muon and the proton are all know to a fractional error less than 3×10^{-8} whereas the fractional error in the neutron moment is 1000 times greater or 3×10^{-5}. Although the present apparatus was designed with the neutron electric dipole moment exclusively in mind, by coincidence it turns out to be an appropriate design for measuring the magnetic moment of the neutron. Although the magnetic field is low and probably cannot easily be raised much above 800 gauss, this disadvantage is more than offset by the large magnetic gap which permits an accurate calibration of the magnetic field because of the smaller inhomogeneities which result from the increased gap. In the previous most accurate experiment[48] the precision of the result was primarily limited by this field inhomogeneity and the consequent difficulty in calibrating the magnetic field accurately.

The magnetic moment will be measured in an apparatus which is essentially the same as that now being used in the neutron electric dipole moment experiment. Permanent magnets, however, can be added to increase the magnetic field from 15 oerstead to 800 oerstead. The magnetic field can be calibrated in several alternative ways. One is by the use of a proton NMR magnetometer and another is by the use of a rubidium magnetometer. A still different alternate is to pump water at high speed through a high magnetic field storage region to polarize the protons and then to have the water pass through the neutron beam pipe at high velocity, with the resonance being observed by the separated oscillatory field method;[17] in this

case the second oscillatory field region has many of the characteristics of a volume filled with molecules in "super radiant" states.
The flowing water method has the advantage of a close similarity
between the averagings done by the neutrons and by the protons as
each are confined to the neutron pipe. It is anticipated that all
three methods will be used. The greatest possible care must be
devoted to assuring that the magnetic field at the time of the
proton calibration is the same as that during the measurements with
the neutron.

    With this technique, it appears that it should be relatively
easy to improve the accuracy of the measurement of the neutron
magnetic moment by at least a factor of 100 and hopefully by a
somewhat larger factor.

    There are two other interesting neutron beam experiments which
we would very much like to do as soon as we can fit in the time
without significant interference with the primary priority we
attach to the neutron electric dipole moment experiments. From
the experimental point of view, the two experiments are closely
related. Since both involve the measurement of a small parity violating reorientation of the neutron spin when it passes through
matter.

    Kabir, Karl and Obryk[49] have pointed out that when a neutron
passes through an optically active medium, (one whose constituents
are parity asymmetric as solutions of levulose), the transverse
component of the neutrons polarization should precess about the
direction of propagation about 10^{-5} radians in traversing a centimeter of a representative optically active medium. Since an electric
dipole moment of 10^{-24} cm provides a precession of only 0.3 x 10^{-5}
radians, there should be sufficient sensitivity in an adaptation of
the apparatus for that experiment to observe such a precession;
among the apparatus changes required in such an adaptation would be
the provision of a weak longitudinal magnetic field instead of a
weak transverse field. If sensitivity alone were the only requirement, the measurement should be relatively easy. Unfortunately,
the primary difficulty will arise from the existence of the neutron
magnetic moment and the necessity of demonstrating that there is no
small magnetic perturbation associated with insertion of the interacting material which produces a comparable or greater precession
This can be seen from the fact that a magnetic field of 3 x 10^{-8}
gauss would produce 10^{-5} radians of precession while a 100 m/sec neutron traversed the 1.9 meter length of our present apparatus. Some
benefit could be obtained by making the apparatus as short as possible,
but there is still a severe requirement for eliminating any change in
magnetic field associated with the change in the sample. A great
improvement in this problem can be obtained by simultaneously running neutron beams through the apparatus in opposite directions. The
parity violating effect depends on the direction of the neutron

velocity while the magnetic precession is independent of the neutron
direction so by simultaneously measuring the precession for opposite
directions of neutron motion the two effects can be distinguished.

A second experiment of even more fundamental interest is to
look for a similar parity violating precession of the neutron in
passing through a medium which is not optically active. Michel[50]
and Stodolsky[51] have pointed out that a precession of 1.4×10^{-6}
radians should occur if cold neutrons passed through at about 1
meter of say bismuth. The source of the parity violating rotation
in this case would not be the optical activity of the medium but
instead would be the parity violating weak interaction of the neutron. It would be of great interest in this way to observe directly
the weak interactions of neutrons and the parity violating character
of the weak interaction provides a unique signature through the
spin precession to distinguish this interaction from the strong
forces that usually dominate the interaction of neutrons with matter.
The problem of magnetic effects in this case is of course even
more severe than in the case of an optically active medium because
the precession angles from the desired effects are even smaller.
However, as in the previous experiment, great benefit could be obtained in distinguishing the precession due to the weak forces from
those due to magnetic fields by simultaneously making observations
on neutrons which pass through the apparatus in opposite directions.

Although the possibility of doing these experiments was first
discussed as a neutron beam magnetic resonance experiment, they can
also be done by Mezei's interesting neutron spin echo technique.[52]
To avoid unnecessary duplication, we have discussed a combined effort
with Mezei to observe these two interesting effects probably using
a modification of his neutron spin echo technique. The principle
problem is finding time to fit these experiments in to our respec-
tive programs, particularly in view of the primary priority attached
to further lowering the limit on the electric dipole moment of the
neutron.

REFERENCES

*This work was partially supported by the U.S. Energy Research and
Development Administration, The National Science Foundation, The
French Commissariat a l' Energie Atomic and the Institut Laue-
Langevin.

1.   E. M. Purcell and N. F. Ramsey, Phys. Rev. 78, 807 (1950).

2.   J. H. Smith, thesis, Harvard University, 1950 (unpublished);
     J. H. Smith, E. M. Purcell and N. F. Ramsey, Phys. Rev. 108,
     120 (1957).

3.   T. D. Lee and C. N. Yang, Phys. Rev. 105, 1671 (1957).

4.   C. S. Wu, E. Ambler, R. W. Hayward, D. D. Hoppes and R. P. Hudson, Phys. Rev. 105, 1413 (1957).

5.   L. Landau, Zh. Eksperim. i Teov. Fiz. 32, 405 (1957 [translation: Soviet Physics - JETP 5, 336 (1971)]).

6.   N. F. Ramsey, Phys. Rev. 109, 225 (1958).

7.   J. H. Christenson, J. W. Cronin, V. L. Fitch, and R. Turlay, Phys. Rev. Letters 13, 138 (1964).

8.   References to many of the theoretical predictions for nuclear electric dipole moments are given in references 12-21 in reference 9 below.

9.   P. D. Miller, W. B. Dress, J. K. Baird and N. F. Ramsey, Phys. Rev. Letters 19, 381 (1967).

10.   J. K. Baird, P. D. Miller, W. B. Dress and N. F. Ramsey, Phys. Rev. 179, 1285 (1969).

11.   C. Shull and R. Nathan, Phys. Rev. 19, 384 (1967).

12.   V. W. Cohen, E. Lipworth, R. Nathan, N. F. Ramsey and H. B. Silsbee, Phys. Rev. 177, 1942 (1969).

13.   W. B. Dress, P. D. Miller and N. F. Ramsey, Phys. Rev. D7, 3147 (1973).

14.   W. B. Dress, J. K. Baird, P. D. Miller and N. F. Ramsey, Phys. Rev. 170, 1200 (1968).

15.   K. Smith and M. Pendlebury, private communication (1968).

16.   Apostolescu, Ionescu, Ionescu-Bujor, Meiterts and Petroscu, Rev. Romaine Phys. 15, 343 (1970).

17.   N. F. Ramsey, Molecular Beams, Oxford University Press (1956).

18.a. G. Feinberg, Phys. Rev. B140, 1402 (1965) [EM1].

19.b. F. Salzman and G. Salzman, Phys. Rev. Let. 15, 91 (1965) [EM1].

20.c. G. Barton and E. D. White, Phys. Rev. 184, 1660 (1969) [EM2].

21.d. D. J. Broadhurst, Nucl. Phys. B20, 603 (1970) [EM2].

22.e.   Babu and Suzuki, Phys. Rev. 162, 1359 (1967) [MWΔS=0].

23.f.   N. T. Meister and T. K. Rhada, Phys. Rev. B135, 769 (1964)
        [MWΔS=0].

24.g.   G. R. Gourishankar, Can. Journ. Phys. 46, 1843 (1968)
        [MWΔS=1].

25.h.   P. McNamee and J. C. Pati, Phys. Rev. 178, 2273 (1969)
        [MWΔS=0,1].

26.i.   K. Nishijima and L. J. Swank, Nuc. Phys. B3, 565 (1967)
        [MWΔS=0].

27.j.   K. Nishijima, Prog. Theor. Phys. 41, 739 (1969) [MWΔS=0].

28.k.   D. G. Boulware, Nuovo Cimento 40A, 1041 (1965) [MWΔS=0].

29.l.   L. Wolfenstein, Phys. Rev. Let. 13, 562 (1964) [SWΔS=2].

30.m.   A. Pais and J. Primack, Phys. Rev. D8, 625 and 3036 (1973)
        [MW].

31.n.   T. D. Lee, Phys. Rev. D8, 1226 (1973) and Physics Reports 9,
        143 (1974) [MW].

32.o.   L. B. Okun, Comments on Nuclear and Particle Physics III,
        135 (1969) [SW].

32.p.   R. N. Mohapatra, Phys. Rev. D6, 2026 (1972) [MW].

32.q.   J. Frenkel and M. E. Ebel, Complex Cabbibo Angle and CP
        Violation in a Class of Gauge Theories, Univ. of Wisconsin
        Preprint and reference 32.r. below [MW].

32.r.   L. Wolfenstein, Nuclear Physics B77, 375 (1974) [SW].

32.s.   S. Weinberg, private communication [MW]. The predicted value
        of this gauge theory which attributes CP non-conservation
        purely to the exchange of Higgs bosons is $\mu_E/e = 2.3 \times 10^{-24}$ cm.

32.u.   R. N. Mohapatra and J. C. Pati, Phys. Rev. D11, 569 (1975)
        [MW].

32.t.   S. Pakvasa and S. F. Tuan, Phys. Rev. Lett. 34, 553 (1975)
        [MW].

33.   N. F. Ramsey, Rev. Sci. Inst. 28, 57 (1957).

34. D. Kleppner, N. Ramsey, P. Fijilstadt and M. Goldenberg,
     Phys. Rev. Lett. 1, 232 (1958) and 8, 361 (1960).

35. N. F. Ramsey, Proposal to ILL for Neutron Electric Dipole
     Moment Experiment (1969).

36. Y. B. Zeldovitch, Zh. Eksp. Zero. Fiz. 36, 1952 (1959) [JETP
     9, 1389 (1952)].

37. V. Vladimirskii, JETP 12, 740 (1961).

38. A. Droshkevich, JETP 16, 56 (1963).

39. V. I. Luschikov, Yu N. Pokotilosky, A. V. Stvelkov and F. L.
     Shapiro, Zh. Eksp. Teov. Fiz., Pis mo v Redaktsiya 9, 40 (1969)
     Sov. Phys. - JETP Lett. 9, 23 (1969).

40. A. Steyerl and W. B. Trustedt, Private Communication, 1973.

41. P. Ageron, J. M. Astrus and J. Verdier, Preliminary Project
     of an Ultra-Cold Neutron Source at the High Flux Reactor,
     Institute Laue-Langevin, Grenoble (unpublished).

42. V. M. Lobastov, G. D. Porsev and A. B. Serebrov, Preprint
     No. 37, Konstantinov Institute of Nuclear Physics of the
     Academy of Science of the USSR, Leningrad, 1973.

43. Elementarnaya Teoria Metoda Electricheskova Kipolnova Momenta
     Neitrona C. Pomoshyn UXN. I Protochini.

44. Variant. Yu V. Taran, Dubna, Preprint P3-7147, Elementarnaya
     Teoris Metoda Opredeleria Electrischeskova Dipolnova Momenta
     Neitrona C. Pomoshyn UXN. II Nakopitelnu Variant, Yu V. Taran,
     Dubna Preprint P3-71479.

45. R. Gollub and J. M. Pendlebury, Contemp Phys. 13, 519 (1972).

46. J. Byrne, R. Gollub, J. M. Pendlebury and K. F. Smith in
     collaboration with N. F. Ramsey, W. B. Dress, P. D. Miller,
     A. Steyerl, P. G. H. Sandars, P. Ageron and P. Perrin, A
     Proposal to Search for the Electric Dipole Moment of the
     Neutron Using Bottled Neutrons, Institute Laue-Langevin (1974)
     and Rutherford Laboratory (1975).

47. P. D. Miller, Measurements of the Electric Dipole Moment of the
     Neutron. P. D. Miller, Second International School on Neutron
     Physics, Alushta, Crimea, USSR (1974).

48. V. W. Cohen, N. R. Corngold, and N. F. Ramsey, Phys. Rev. 104,
     283 (1958).

49. P. K. Kabir, G. Karl and E. Obryk, Phys. Rev. D10, 1471 (1974).

50. F. C. Michel, Phys. Rev. 133, B329 (1964).

51. L. Stodolsky, Max Planck Institute, MPI-PAE/PTh9 (1974).

52. F. Mezei, Z. Physik 255, 146 (1972).

SPECTROSCOPY OF HIGHLY IONIZED ATOMS PRODUCED BY A LOW-INDUCTANCE VACUUM SPARK

U. Feldman[*] and G.A. Doschek

Naval Research Laboratory
E.O. Hulburt Center for Space Research
Washington, D.C. 20375

INTRODUCTION

The late fifties and early sixties saw a trend of growing interest in the spectroscopy of highly ionized atoms. This new trend was motivated by advances in solar physics and in high temperature plasma physics. Powerful new sources were developed to generate high temperature plasmas, e.g., the NRL Pharos,[1] Harwell's Zeta[2], and a variety of θ-pinches.[3] High power lasers were also used to produce highly ionized atoms.[4] Common to all these new sources was their complexity and high cost, precluding them from ordinary laboratories.

The need for a simple but powerful source was felt. A search into the known traditional spectroscopic sources led to the development of the low inductance vacuum spark.[5] This light source was particularly successful in achieving a wide range of desirable temperatures.[6,7,8,9] To the best of my knowledge, the low inductance spark is still capable of achieving a temperature as high as any other known laboratory light source, and at the same time is also the simplest and least expensive of them all.

LOW INDUCTANCE SPARK

For a pair of electrodes in vacuum, the breakdown voltage for a separation of about 1 mm is 50 kV or more. It is possible to lower the breakdown voltage in a variety of ways. One such way[10] is to provide an insulating surface between the electrodes along which breakdown can occur. This is the so-called sliding spark.[11]Low voltage breakdown across gaps as large as a few

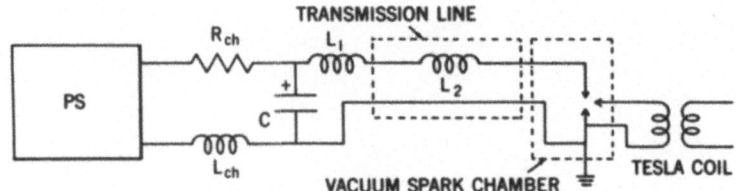

Figure 1   Schematic diagram of the discharge circuit where PS is the power supply R_{ch} and L_{ch} are the resistance and inductance of the charging circuit respectively. C is the capacitor, L_1 is the inductance of the capacitor, and L_2 is the inductance of the transmission line (Ref. 5).

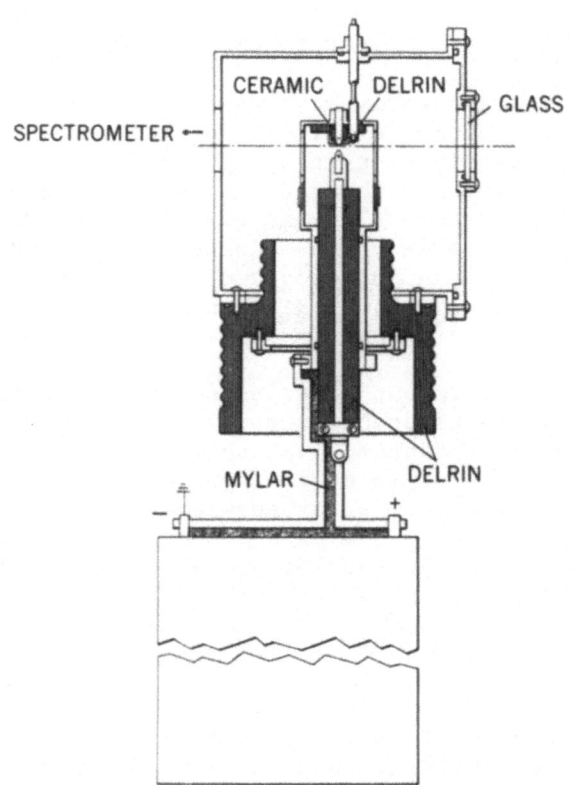

Figure 2   Side view schematic of the spark chamber and capacitor (Ref. 5).

centimeters is possible with this method. Another method is to
inject charge carriers in the electrode gap by other means[12]
in order to initiate a discharge. The three electrode arrangement
(Figure 1,2) combines these two ideas. A third electrode is
separated from the ground electrode of the spark gap by a ceramic
insulator. At a predetermined time in the high voltage charging
cycle, a sliding spark is produced along the ceramic, consequently
triggering the breakdown of the main spark across the gap between
the main electrodes. The nature of the triggering mechanism is
probably not too critical to the spectral characteristics of the
source. The critical factors are the geometric configuration of
the electrodes, the electrical properties of the main discharge
circuit, and the material of the anode. The ground electrode and
the trigger mechanism usually do not produce intense spectral
lines. The discharge circuit consists of a $C = 14-30$ µF capacitor
charged to $V = 10-20$ kV. The total inductance is from $L = 23-300$ nH
and the current $I = 100-300$ kA (see Table 1).

TABLE 1

V(kV)	C(µF)	L(nH)	I(kA)	Spectra Observed	References
19	14	160	176	He Like Ni	5
14	15	294	100	He Like Cu	7
16	15	80	220	He Like Cu	9
10	30	23.6	356	He Like Cu	14

The properties of the low inductance spark are very complex.
During the several cycles of the R-L-C discharge (5-10 µsec/cycle),
the characteristics of the spark change dramatically. Since we
are interested primarily in the highest temperature regions, it
is sufficient to consider only the first quarter cycle of the
discharge.[6,7]   Figure 3 is a typical oscillogram showing the
variation of electric current with time (dI/dt) and the X-ray
spectral signals recorded by scintillation detectors.

As can be seen, the X-ray emission takes place during the
first quarter cycle, and each X-ray pulse signal is approximately

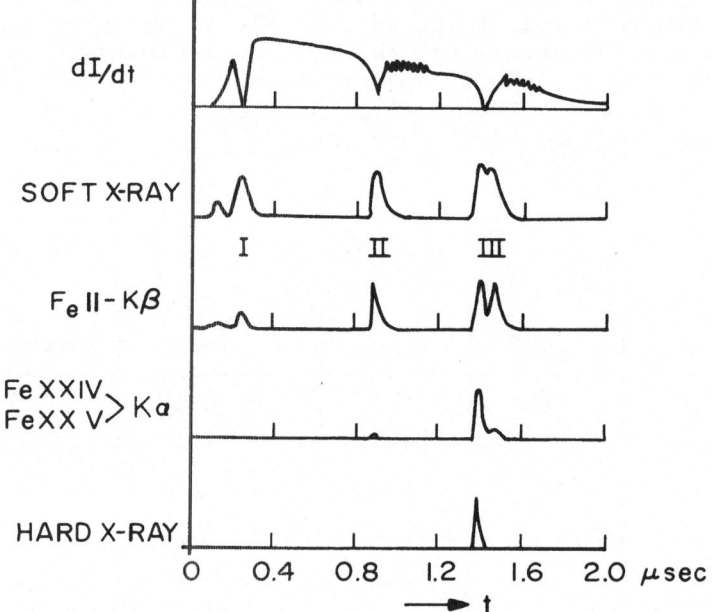

Figure 3   Oscillograms showing the dI/dt waveform and the X-ray
            spectral signals from the scintillation detectors
            (Ref. 7).

coincident with dips in the dI/dt trace. Figures 4 and 5 show
both coarse spatial resolution pinhole photographs and high
resolution slit photographs of two different single spark plasmas.[13]
There are three slit photographs for each plasma; each photograph
was obtained with a different filter between the spark and the
film. Three filters of Fe, Mn, and Cr were used as a Ross filter
system. The pinhole photographs taken through a Be window show
intense ball-like emission features that are seen to be resolved
into multiple images in the high-resolution photos. The pinhole
photographs also show faint extended emission that is too weak to
be seen in the high resolution slit photographs. The emission
cloud farthest to the left in the pinhole photographs may be line
emission and bremsstrahlung from the anode. This emission corresponds to the soft X-ray peak in Figure 3.

     The simplest appearing plasma is shown in Figure 4.
In the slit images, one intense, compact feature is present,
along with a much weaker and extended image. These weak and
extended features are shortly demonstrated to be closely associated with the intense sources. The distance scales in Figures
4 and 5 give the dimensions through the plasma. The width of the
intense feature in Figure 4 is ~20 µm. Twenty microns is the
combined resolution of the imaging slit and the densitometer
slit. Thus, at least in the Z-direction the size of the intense
pinched region could be considerably less than 20 µm. Even the
separation of the less intense broad feature from the compact
source is small, about 120 µm. The size of the broad feature is
about 100 µm along the anode-cathode axis in the manganese filter
tracing.

     Information of the temperature distribution within the
plasmas in Figures 4 and 5 may be obtained by comparing the
filter tracings. The manganese filter passes radiation from the
cooler ions, Fe II- ~Fe XXII, but absorbs the radiation from
~Fe XXII-Fe XXV. The chromium filter absorbs radiation from all
the ions, and the iron filter transmits all the line radiation.
Therefore, the features that appear in the manganese tracing but
not in the chromium tracing are regions of the plasma that
primarily emit line radiation, and this line radiation must arise
from the cooler ions, Fe II- ~Fe XXI. The features in the
tracing due to plasma that primarily emits ~Fe XXII-Fe XXV line
radiation have nearly the same intensity in both the manganese
and chromium tracings and are more intense in the iron filter
tracing than in the manganese and chromium tracings. This radiation
corresponds to the X-ray peaks under No. III in Figure 3. In
addition to the line radiation, all three filters also transmit
continuum radiation, due to both free-free and free-bound radiation
processes. Using the scale in Figure 4, the intense feature near
zero microns is seen to emit line radiation primarily due to
~Fe XXII-Fe XXV, and we therefore interpret the feature as a hot

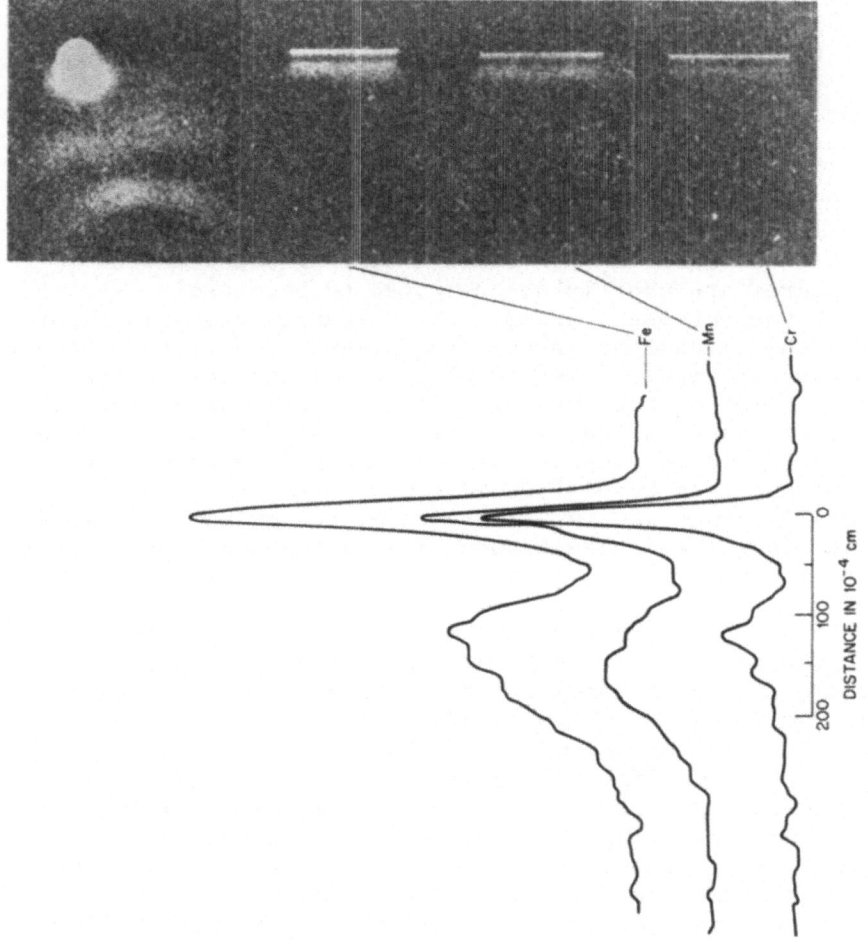

Figure 4   Pinhole slit photographs of a vacuum spark plasma and densitometer tracings of the slit photographs. The ordinate is in photographic density units (Ref. 13).

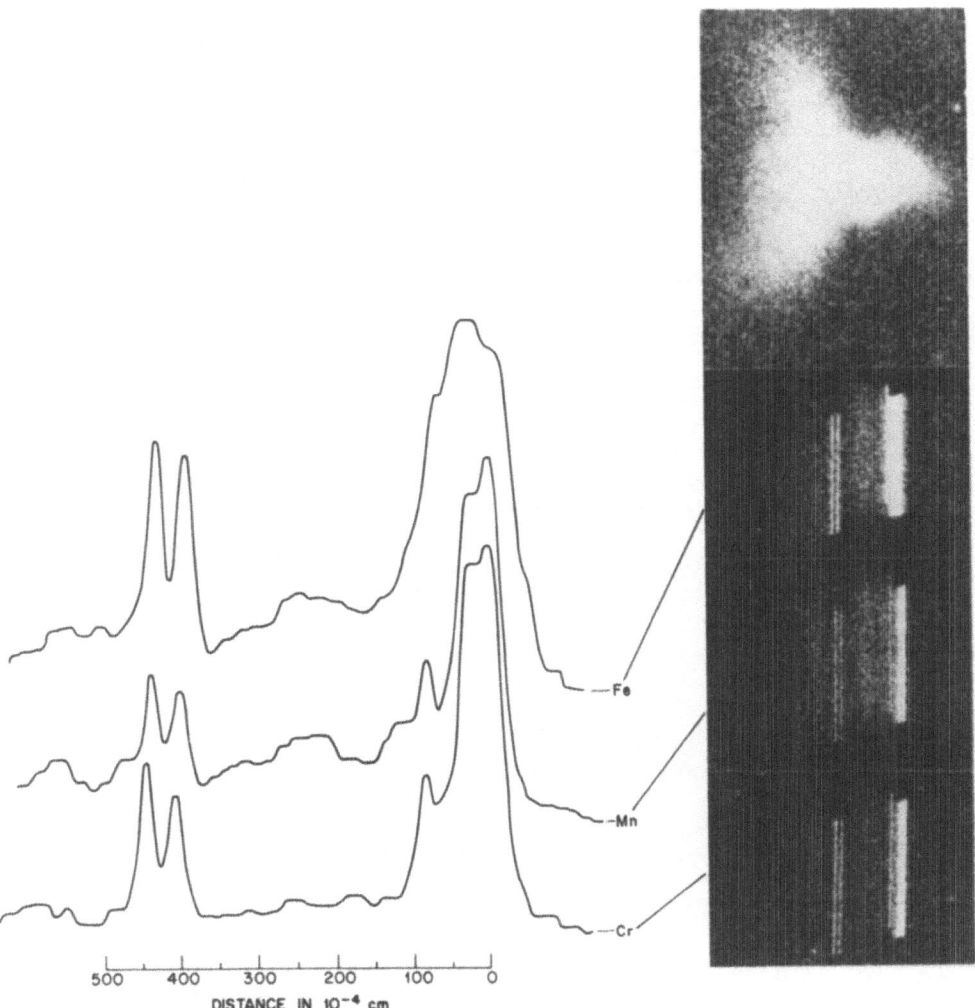

Figure 5   Pinhole slit photographs of a vacuum spark plasma and
            densitometer tracings of the slit photographs. The
            ordinate is in photographic density units (Ref. 13).

plasma source. In the manganese tracing cool plasma sources that
emit Fe II- Fe XXI are found on the left-hand shoulder of the
intense feature, and in the broad feature between about 140 µm
and 250 µm. There is also a small hot component between ~100 µm
and ~140 µm.

    As can be seen from Fig. 3, the X-rays are emitted at three
different times. The events from which they originate correspond
to the different phases in the spark. The X-ray emission in
phase II originates in the relatively large volume of plasma
corresponding to the middle cloud in the pinhole photograph of
Figures 4 and 5. This radiation consists of low energy X-ray
lines and continuum. The electron density in the very compact region
of Phase III has a temperature of 2 keV or higher and electron
densities of 10^{21} cm^{-3}. The size of the compact region is on the
order of 10 µm or less, and its lifetime is on the order of
10^{-9} sec.

    The less compact region of Phase III has an electron temperature
of about 1 keV and probably a lower electron density than the
compact region. These two regions are responsible for making the
source so valuable for spectroscopy.

    One major disadvantage of the source is its non-repro-
ducibility. The geometrical arrangement is very critical.
Although all operating conditions may appear normal, a high
density and temperature region may not always be produced. Different investigators have their own schemes for operating the
source, with greater or lesser degrees of success.
However, we are not aware of any method that guarantees a high
temperature plasma everytime the source is discharged.

    It should be mentioned that once good working conditions
are achieved, an increase in the capacitor size or in the voltage
does not lead to more intense or hotter sparks. On the contrary,
increasing the voltage or capacitor size usually lowers the performance. However, occasionally a very large X-ray pulse is
generated by increasing the energy in the capacitor by about a
factor of 5. This pulse is strong enough to produce an X-ray
picture of the spark chamber.

                        EMISSION LINE SPECTROSCOPY

    A review of the ionization and excitation properties of the
low inductance vacuum spark (LIVS) is given below according to
types of transitions.

1s - np

In the past decade spectra of the H I and He I isoelectronic sequences and associated satellite lines due to transitions from doubly excited configurations have been studied extensively. The study was motivated mainly by observations of some of these types of lines in solar spectra of active regions and solar flares.[15,16,17] It was shown[18,19] that the ratio of the satellite line intensities to the resonance line intensity of the heliumlike ions may be used to determine the electron temperature of the plasma and the departure of the plasma from ionization equilibrium.

By now the spectrum of the He I-like sequence and its satellite lines is known for elements as far as $Cu^{6,7,8,9,14,20}$ and the H I-like sequence and its satellite lines are known as far as Fe. For the higher Z ions the only transitions detected are from n = 2,3. The knowledge of the high end of the sequence comes from the LIVS. For the lighter elements the spectra were obtained using a high voltage spark, the LIVS and laser-produced plasmas. It seems that the LIVS in its present electrical and geometrical configuration will not be able to achieve spectra of significantly higher members of these sequences, e.g., Mo^{14}. Fig. 6 is a high dispersion Ti spectra from a LIVS. Figure 7 shows Fe spectra from a LIVS and from a solar flare.

$n\ell - n'\ell'$

The LIVS is a convenient source for studying the spectra of transitions for which $n' > n$. In such cases even though the light source emits spectra from lower degrees of ionization than those under investigation, the different types of spectra are still well-separated in wavelength. A great deal of our knowledge of transitions for which the ground configuration is of the form $2s^1 2p^k$ was achieved with the aid of the LIVS. Although in many cases the lower members of the isoelectronic sequences were produced with the regular vacuum spark, the higher members were generated with the LIVS. The study of transitions of the types $2\ell - 2\ell'$ is not convenient to perform with the LIVS. One main reason is the multi-thermal properties of the source. The wavelength region occupied by these transitions overlays the region occupied by transitions of the type $3\ell - n\ell'$, produced by lower degrees of ionization. Because of this circumstance, it is more convenient to study the $\Delta n = 0$ transitions for high degrees of ionization using spectra of laser-produced plasmas, or spectra from another light source of more defined temperature. The temperatures found in laser-produced plasmas cover a much smaller range than in the LIVS. However, the lines may be studied using a LIVS, particularly if the spectral resolution is high. In

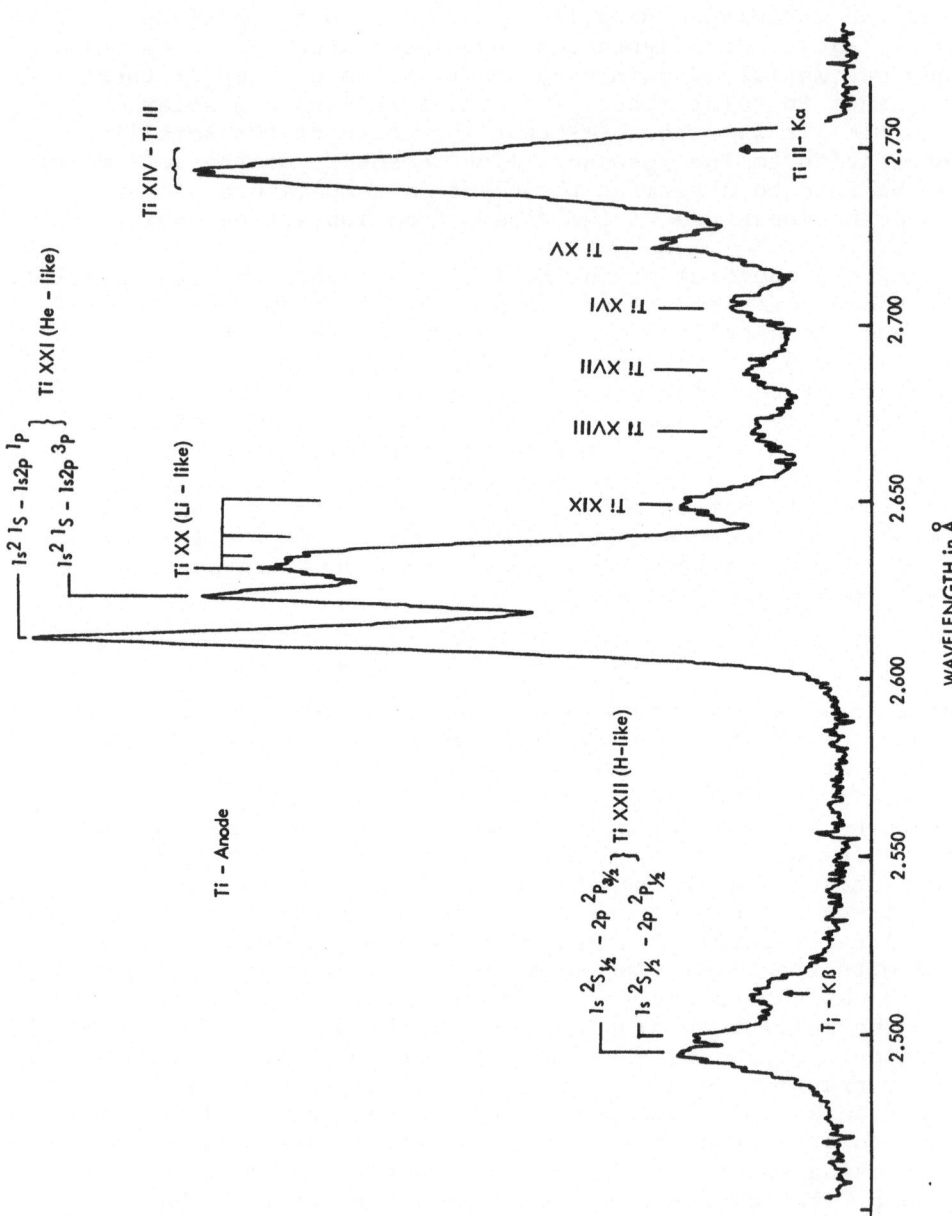

Figure 6   Microdensitometer scan of X-ray emission spectra
obtained from a titanium electrode (Ref. 7).

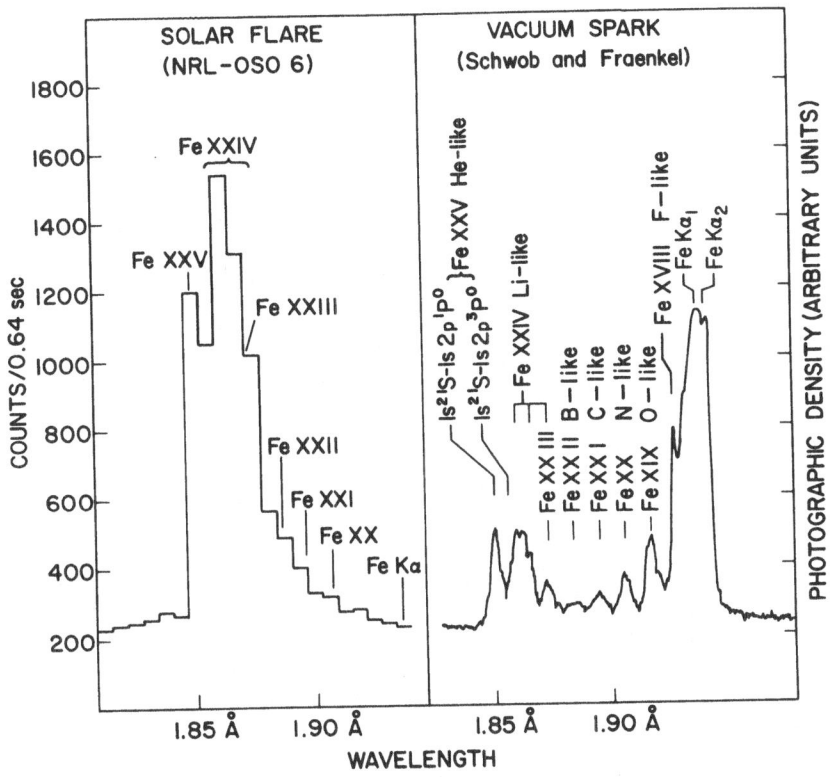

Figure 7   A comparison of a solar flare iron line spectrum with the iron spectrum produced by a LIVS (Ref. 13).

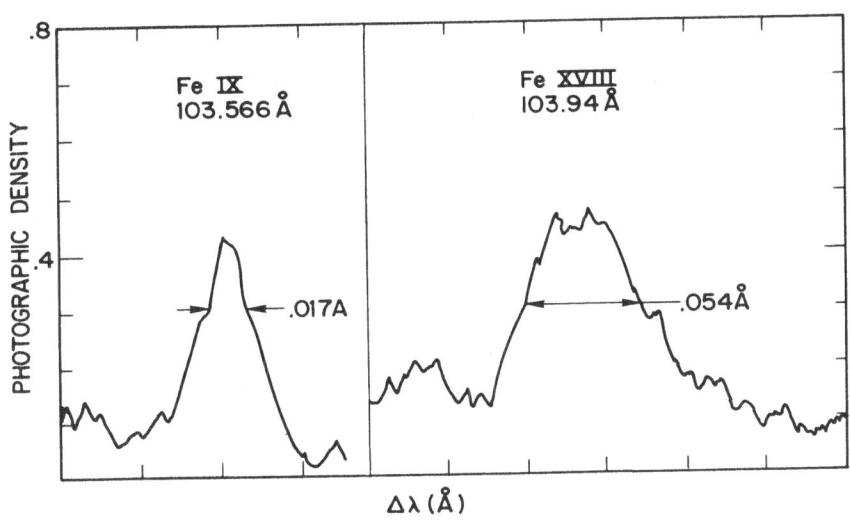

Figure 8   Profiles of Fe IX and Fe XVIII lines from a LIVS.

spectrographs with very high resolution, the origins of the lines according to temperature will be obvious from their widths ($\Delta\lambda \propto \lambda\sqrt{T}$).

For ions with n=3 ground state configurations, the situation is similar. The $\Delta n = 0$ transitions can be investigated more conveniently in laser-produced plasmas.

In the above mentioned type of sequences very little is known beyond Ni, because the primary motivation for the investigation had been the study of astrophysical plasmas, particularly the Sun. (Ni is the heaviest element that exists in reasonable abundance in these plasmas.) Research in heavier elements has only begun recently as a result of interest in high temperature plasma machines for the fusion program, such as the tokamak.

Emission Line Profiles in LIVS Spectra

The profiles and widths of the emission lines in LIVS spectra give information on the ion temperature, non-thermal motions, and magnetic field stengths in the plasmas. In the 100-300 Å region, lines from a broad range of ionization stages are emitted. For example, emission from an ion such as Fe IX arises in cold, low density clouds. Emission from ions such as Fe XVIII arises in substantially hotter regions while emission from Fe XXIV might be expected to occur in the highest temperature pinches. Figure 8 shows profiles of the Fe IX 103.566 Å line and the Fe XVIII 103.94 Å line. These profiles show structure which is probably real. The width of the Fe IX line corresponds to a temperature of 0.26 keV, which is about a factor of two higher than in ionization equilibrium. The width of the Fe XVIII line corresponds to a temperature of 2.6 keV, which is about a factor of five times greater than the ionization equilibrium temperature (see Figure 9).[21,22,23] The only manner in which the line widths can be interpreted as pure thermal Doppler broadening is to assume that the plasma is in a state of transient ionization.

However, the observed structure in the profiles suggests that broadening mechanisms other than thermal Doppler broadening are also important. Non-thermal streaming motions or plasma turbulence is a possible broadening mechanism. If the line widths are interpreted as a combination of thermal and non-thermal Doppler broadening, the non-thermal velocities are ~21 km-s^{-1} for Fe IX and ~94 km-s^{-1} for Fe XVIII.

A second plausible broadening mechanism are the intense magnetic fields generated by the discharge current. The predicted strengths of these fields are in the megagauss range for the pinched regions of the plasma. Fig. 10 shows the full widths at

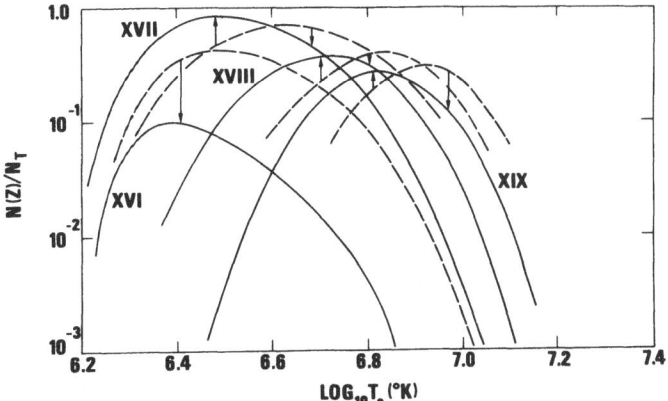

Figure 9   The ionization equilibrium of Fe XVI-Fe XIX.---results
from Ref. 21,22. ——— results from Ref. 23 (see Ref. 23).

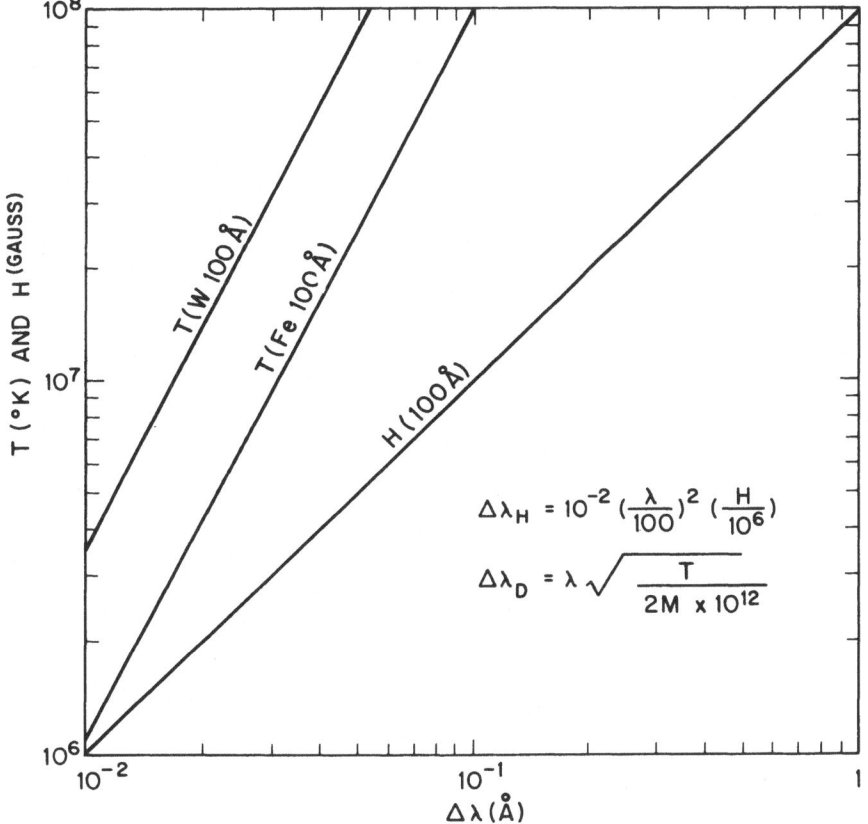

$$\Delta\lambda_H = 10^{-2} \left(\frac{\lambda}{100}\right)^2 \left(\frac{H}{10^6}\right)$$

$$\Delta\lambda_D = \lambda \sqrt{\frac{T}{2M \times 10^{12}}}$$

Figure 10   The full width at half maximum of lines near 100 Å due
to both thermal Doppler broadening and broadening induced
by the Paschen-Back effect.

half maximum of lines near 100 Å due to both thermal doppler broadening and broadening induced by the Paschen–Back effect. In terms of the Paschen–Back effect, the width of the Fe XVIII line corresponds to magnetic fields of about 5 megagauss. These fields do not seem unreasonably large, as can be seen from the equation B_θ = I/5r, where I is the current (I = 150,000 A) and r is the radius of the pinched region (r = 5.0×10^{-3} cm, see Figs. 4 and 5).

Finally, Stark broadening does not appear to be a likely broadening mechanism because the levels involved are not degenerate, and therefore not subject to first order Stark broadening.

Although ions similar in ionization potential to the Fe XVIII ion are characteristic of fairly high temperatures (~0.5 keV), these ions would have negligibly small abundances in the ~5 keV plasma in the most intense pinched regions assuming ionization equilibrium. We therefore looked carefully for the two lines of Fe XXIV at 192 Å and 255 Å, first identified in solar flare spectra.[24] The lines are due to the transitions, 2s $^2S_{1/2}$ – 2p $^2P_{3/2}$ and 2s $^2S_{1/2}$ – 2p $^2P_{1/2}$. Fe XXIV reaches peak abundance near a temperature of 1.7 keV, but has a substantial abundance at significantly higher temperatures, due to dielectronic recombination of Fe XXV. No trace of the Fe XXIV lines was found. A possible explanation for the absence of these lines is that the highest temperature pinches are simply too hot for the Fe XXIV ion to exist in significant abundance, while lower temperature pinches are too cold to form Fe XXIV. The Fe XXIV emission observed in X-ray spectra is not inconsistent with this view, because this emission is produced primarily by dielectronic recombination of Fe XXV. In contrast, the Fe XXIV lines near 200 Å are produced mostly by electron impact excitation.

Another possible explanation for the absence of the Fe XXIV lines is that they are formed in the high temperature pinched regions, but are broadened by very high magnetic fields. If the fields were no stronger than those calculated for Fe XVIII, the Fe XXIV lines would be ~0.2 Å wide, because the broadening scales as λ^2. However, if magnetic fields are indeed responsible for the broadening, they would likely be much stronger in the high temperature pinched region. Assuming r ~ 50 μm for Fe XVIII and r ~ 5 μm for Fe XXIV, the fields are expected to be ~10 times stronger for Fe XXIV. In this case, the Fe XXIV lines would be ~2 Å wide. They would be smeared out over the strong continuum radiation, and hence they would not be detectable.

To determine the relative importance of Doppler and Paschen–Back broadening, one can measure the widths of lines of a significantly heavier ion than Fe XVIII. The heavy ion should have an ionization potential nearly the same as the ionization potential of Fe XVIII. If the line broadening is mainly Doppler, then the width of the heavy ion line should be significantly less than the Fe XVIII line, while if the broadening is induced magnetically, the width should be the same as the Fe XVIII line (Figure 10). It is also possible to find an ion present in the high temperature and magnetic field region that has lines in two different wavelength regions, e.g., 100 Å and 350 Å. The relative widths of the lines will determine the origin of the broadening because Doppler broadening scales as λ and broadening by magnetic fields scales as λ^2.

Absorption Lines

In addition to producing a complex emission line spectrum characterized by emission lines from an extremely broad range of ionization stages, the LIVS also produces an intense continuum and an absorption line spectrum. The continuum is probably produced by free-free and free-bound emission. The intensity of the continuum is roughly proportional to $N_e^2 \Delta V$, where N_e is the electron density and ΔV is the emitting volume. During a pinch, if the total number of electrons is assumed constant for simplicity, the continuum intensity will increase linearly with N_e, and N_e will be proportional to $1/r^2$ where r is the plasma radius. Therefore, it seems likely that most of the observed continuum is produced in the intense, high temperature pinch. The continuum radiation passes through a cooler and less dense plasma cloud between the pinched region and the observer. Thus, absorption lines are also observed in LIVS spectra.

The plasma cloud, or clouds, that give rise to the absorption spectra, appear to be formed by material ejected from either the anode, cathode, or the insulator that separates the cathode from the trigger electrode. Evidence supporting this conjecture was found in three spectra obtained with the following anode-cathode combinations: Fe+Fe, Fe+Fe, and W+Fe. In the W+Fe spectrum, in which absorption lines of W VII were identified (Figure 11A), the absorption cloud must have originated at the anode. In one of the Fe+Fe spectra, strong absorption lines of Al IV were found (Figure 11B). Because aluminum is not present in the electrode material, but is present in the insulator ceramic, in this case the plasma cloud had its origin in material ejected from the insulator. On the other hand, in another Fe+Fe spectrum absorption lines of either Fe VI or Fe VII were strong (Figure 11C), indicating ejection from the anode or cathode.

From Doppler shift displacements of the absorption lines, the absorbing layer is moving at speeds of from 20 to 90 km sec^{-1} toward the observer. At 60 km sec^{-1}, the plasma could move 5 mm

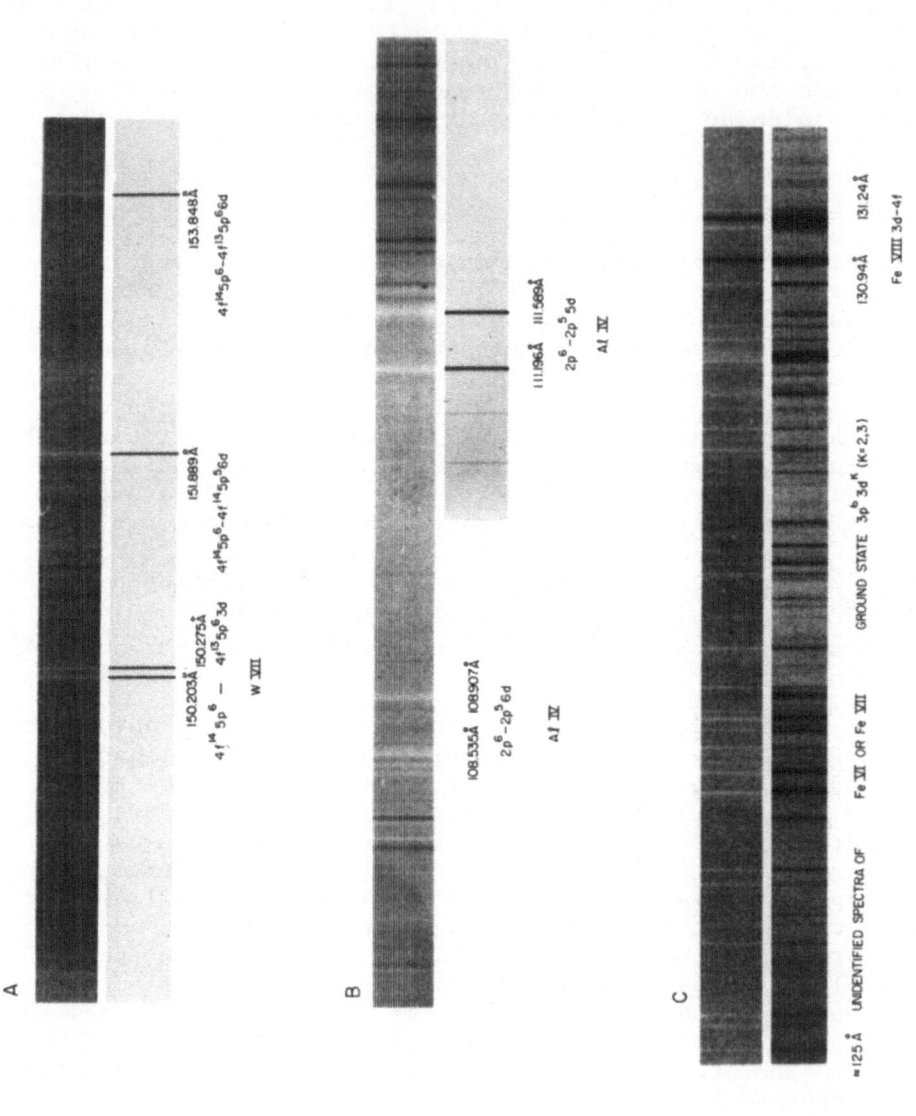

Figure 11 A) W VII lines, B) Al IV lines, C) Fe lines. The absorption lines were produced by the LIVS and the emission with a sliding spark.

in 83 ns. Five millimeters is about the cathode-anode separation. The lifetime of the very dense pinched plasma is quite short, on the order of several nsec. Therefore, it seems that the absorbing layer was produced before the dense pinches occurred and the absorption spectra were produced in a short time or times of several nsec durations.

The absorption spectra are very intense in ions with a closed shell, or with only one electron outside of the closed shell. The ground states of Al IV and W VII are $2p^6\ {}^1S_0$, and $5p^6\ {}^1S_0$, respectively. However, strong absorption spectra from more complex ground configurations are also observed (see the spectra of Fe VI or Fe VII in Fig. 11C). Only intermediate degrees of ionization are found and the lines are quite narrow. This indicates that the temperature in the absorption cloud is quite low, i.e., $T_e < 100$ eV. From the nature of the phenomenon, it is also likely that the electron density in the cloud is low, certainly much less than in the high temperature pinched regions. The observation that lines from ions with closed shells or with one electron outside of a closed shell are very intense has a simple explanation in terms of the ionization balance in the plasma. Inspection of ionization equilibrium curves for the elements[21,22,23] shows that ions with outer closed shells, and ions with only one electron outside of closed shells, exist abundantly over very broad temperature ranges. Although the calculations assume ionization balance, a similar result would be obtained under transient conditions.

The existence of absorption clouds in LIVS spectra gives rise to an entirely new technique for measuring relative f-values for ions of intermediate degrees of ionization. There are few measured f-values for such ions. The absorption spectra appear from a very small region during the first quarter cycle of the discharge. Therefore it is possible to obtain a spectrum free of contamination from emission lines of the same degree of ionization. Assuming that the emission lines arise from a large volume and over a long period of time, a pinhole in front of the slit and a fast shutter ($\sim 10^{-6}$ sec) could eliminate most of the cold emission lines.

ACKNOWLEDGEMENTS

We would like to thank A. Crooker, R.C. Elton, H.R. Griem, T.N. Lee and J. Reader for helpful discussions. We are grateful to the researchers of the Spectroscopy Section of the National Bureau of Standards for providing us with the opportunity to use their 10 m grazing incidence spectrograph.

FOOTNOTES AND REFERENCES

*On leave of absence from Tel-Aviv University Tel-Aviv, Israel.

1. R.C. Elton, E. Hintz and M. Swartz, Proc. Seventh International Conf. Phenomena Ionized Gases (Belgrade, 1965) 3, 190 (1966).

2. B.C. Fawcett, A.H. Gabriel, W.G. Griffin, B.B. Jones, and R. Wilson, Nature 200, 1303 (1963).

3. G.A. Sawyer, A.J. Bearden, I. Henins, F.C. Jahoda, and F.L. Ribe, Phys. Rev. 131, 1891 (1963).

4. B.C. Fawcett, A.H. Gabriel, and P.A.H. Saunders, Proc. Phys. Soc. (London) 90, 863 (1967).

5. U. Feldman, M. Swartz, and L. Cohen, Rev. Sci. Instr. 38, 1372 (1967).

6. L. Cohen, U. Feldman, M. Swartz, and J.H. Underwood, J.O.S.A. 58, 843 (1968).

7. T.N. Lie, and R.C. Elton, Phys. Rev. A 3, 865 (1971).

8. R.C. Elton, and T.N. Lee, Space Sci. Rev. 13, 747 (1972).

9. Schwob, J.L., and B.S. Fraenkel, Space Sci. Rev. 13, 589 (1972).

10. G. Balloffet, Ann. Phys. (Paris) 5, 1243 (1960).

11. B. Vodar and N. Astoin, Nature 166, 1029 (1950).

12. T.N. Lee, Astrophys. J. 190, 467 (1974).

13. U. Feldman, S. Goldsmith, J.L. Schwob and G.A. Doschek, Astrophys. J. 201, 225 (1975).

14. W.A. Cilliers, R.U. Datla, and H.R. Griem, Phys. Rev. A 12, 1408 (1975).

15. A.B.C. Walker, Jr., and H.R. Rugge, Astrophys. J. 164, 181 (1971).

16. W.M. Neupert, Solar Phys. 18, 474 (1971).

17. G.A. Doschek, Space Science Reviews 13, 765 (1972).

18. A.H. Gabriel, Monthly Notices Roy. Astron. Soc. 145, 241 (1972).

19. C.P. Bhalla, A.H. Gabriel, and L.P. Presnyakov, Monthly Notices
    Roy. Astron. Soc. 172, 359 (1976).

20. M. Klapisch, J.L. Schwob, B.S. Fraenkel and J. Oreg, private
    communication (1975).

21. C. Jordan, Monthly Notices Roy. Astron. Soc. 142, 501 (1969).

22. C. Jordan, Monthly Notices Roy. Astron. Soc. 148, 17 (1970).

23. V.L. Jacobs, J. Davis, P.C. Kepple, and M. Blaha, to be
    published in the Astrophys. J. (1976).

24. W.M. Neupert, Philos. Trans. R. Soc. London, A 270, 143
    (1971).

25. A. Crooker, private communication (1976).

INFLUENCE OF ELECTRON CAPTURE ON X-RAY PRODUCTION IN HEAVY-ION COLLISIONS

Hans D. Betz, F. Bell, and E. Spindler

Sektion Physik, Universität München, 8046 Garching,

West-Germany

SUMMARY

Recent investigations are summarized which show that electron capture and other quenching processes can play an important role for the x-ray production in ion-atom collisions. In particular, electron capture affects the distribution of excited projectile states inside solids and causes nonproportional target-thickness dependence of projectile x-ray yields. This may give rise to difficulties in the measurement of projectile x-ray cross sections, but allows determination of lifetimes of projectile states. New results are presented on the oscillatory behavior of projectile x-ray yields as a function of target species and on the spectral shape of radiative electron capture.

INTRODUCTION

Passage of projectile ions through solid targets leads to complex collision processes which have been studied with increasing efforts in the past years. The most easily observed effects of ion-atom collisions are characteristic x-rays emitted from projectiles and target atoms. Measurements of such x-rays can obviously be utilized to deduce information about excitation processes in ion-atom encounters. In the past, most investigators have interpreted x-ray spectra as a direct consequence of Coulomb- and quasimolecular ionization mechanisms[1] and one was generally surprised that not all experimental data could be understood in these terms. A major reason for these apparent discrepancies is that in ion-atom collisions excitation of electrons proceeds not only into continuum

493

states but also into bound states. Inner-shell vacancy production
in heavy targets by light projectiles can be well described by
excitation of the inner-shell electrons into the continuum since
(i) close lying bound states of the target atom are occupied and
(ii) transfer to bound projectile states proceeds with extremely
small cross sections. The situation changes when projectile- and
target species don't differ drastically in nuclear charges Z_1 and
Z_2, respectively, and when the collision velocity, v, is large
enough to allow extensive stripping of the collision partners.
Then, collisional redistribution of bound electrons becomes important and can compete with ionization into the continuum. In
very general terms these additional effects can be called collisional quenching; in most cases, however, it is meaningful to interpret them as electron capture processes, i.e. electron transfer
either from the target to the projectile or vice versa. In the
following, we will consider capture processes only, but it should
be understood that more complex quenching effects may occur.

     Among the many consequences of electron capture on the x-ray
production in ion-atom collisions we limit ourselves to the discussion of projectile x-rays in foil targets and radiative electron
capture. This includes equilibrium excitation of ions inside solids,
target-thickness effects, lifetime determination, and evaluation
techniques of x-ray cross sections.

EQUILIBRIUM EXCITATION OF PROJECTILES INSIDE FOIL TARGETS

     Charge- and excitation states of fast ions passing through
matter fluctuate due to collisional excitation, loss, and capture
of electrons, and deexcitation via Auger- and radiative effects.
When the lifetime, τ, of excited states is short compared to the
average time, t_c, between two successive collisions, excited
ions will have returned to the ground state before another encounter occurs. In solids, however, the condition $t_c \ll \tau$ will often
be fulfilled, i.e. the collisions follow each other so rapidly
that a certain build-up of excitation results. It must then be
expected that a substantial fraction of excited states is no longer destroyed by Auger- and radiative decays but by collisional
quenching processes such as, for example, capture of target
electrons into empty projectile states. As a consequence, the
number of observed x-rays will be smaller compared to what one
would expect on the basis of primary vacancy production. Since a
great variety of charge- and excitation states are involved, a
detailed description of the various competing processes, on the
basis of coupled rate equations, is extremely difficult and will
not be attempted here. Instead, we describe an approximation
which turned out useful for practical purposes.

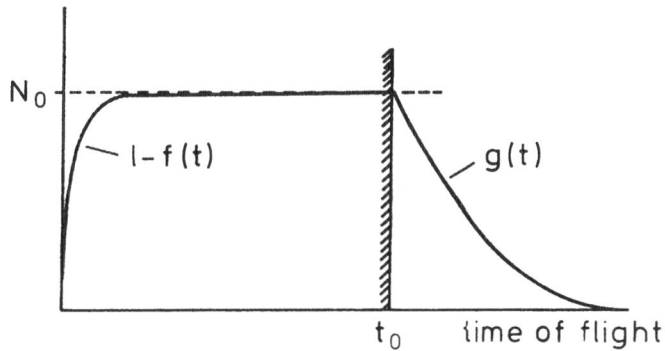

Fig. 1. Build-up and decay of an excited projectile state inside
and behind a target, schematically.

Fig. 1 illustrates build-up and decay of any one excited state
of a projectile beam moving through and exiting from a target.
Here, path lengths are expressed in terms of flight times, t. We
characterize build-up and decay by functions $f(t)$ and $g(t)$, respectively, and normalize $f(o) = g(o) = 1$ and $f(\infty) = g(\infty) = 0$. The time
scale for $g(t)$ is displaced relative to $f(t)$ by the total foil
transit time, t_o. It becomes obvious that a certain equilibrium
excitation will be reached, provided that the foil is not too thin:
then, a certain fraction, N_o, of the beam will be found in the
particular excitation state. In general, we obtain for the excited
state fractions

$$N^i(t) = N_o\{1 - f(t)\}$$

$$N^a(t) = N_o\{1 - f(t_o)\}g(t)$$

where index i and a refer to inside and outside the foil, respectively. It can be shown that for $t \gg t_o$ the total x-ray yield per
ion, Y, from decays of the state considered is given by[2]

$$Y(t_o) = \{t_o + <\tau> - <\tau_i> \} N_o/\tau_R, \qquad (1)$$

where τ_R is the radiative lifetime of the state. $<\tau>$ and $<\tau_i>$
represent certain averages of τ and the effective lifetime, τ_i,
of the state inside the foil, as defined by

$$\langle \tau \rangle = \int_0^\infty t g'(t)dt \; ; \; \langle \tau_i \rangle = \int_0^\infty t f'(t)dt \; ,$$

where the prime denotes the derivative with respect to t.

This formal but quite general description becomes more transparent when only ground state and a single excited state are considered[3]

$$Y(t_0) \simeq \{t_0 + \tau - \tau_i\} N_0/\tau_R \; ; \qquad\qquad (2)$$

in this case we have the simple relation

$$N_0 = \sigma_v/\sigma_T \; ,$$

where σ_v denotes the vacancy production cross section and $\sigma_T = \sigma_v + \sigma_c + (nv\tau)^{-1}$ with electron capture cross section σ_c and target density n (atoms/cm^3).

Under equilibrium conditions inside the foil, the rate of producing an excited projectile state (σ_v) is balanced by the rates to destroy this excited state via Auger- and x-ray decays ($\{nv\tau\}^{-1}$) and collisional quenching (σ_c). Thus, the vacuum lifetime τ of a vacancy reduces inside the foil to $\tau_i = \{nv\sigma_T\}^{-1}$. When the ions emerge from the foil, the fraction N_0 carries a vacancy which then decays according to lifetime τ.

The above description of projectile excitation which was suggested and tested in refs. 2-4 has been found useful by many other investigators for similar and other collision systems[5-8]. We may note that the condition $t_c \ll \tau$ is often valid in solids where t_c is extremely short, but it may also be fulfilled in gaseous targets provided that metastable states with sufficiently long lifetimes are present[6].

Finally, we point out that in the absence of any collisional quenching processes, and when vacancy production is slow compared to the normal decay, $\sigma_v \ll (nv\tau)^{-1}$, one gets $\sigma_T \simeq (nv\tau)^{-1}$ and Eq.(1) reduces to the well-known standard expression

$$Y(t_0) \simeq \omega \sigma_v x_0 \; , \qquad\qquad (3)$$

where ω is the fluorescent yield and x_0 the target thickness in atoms/cm^2.

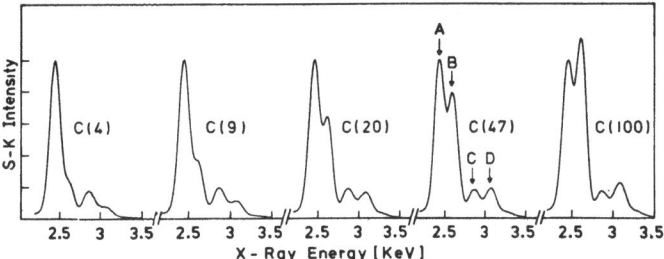

Fig. 2.   Relative intensities of K x-ray satellites of 95-MeV
sulphur ions in carbon, observed with 150 μm carbon
absorber.   In each case, the target thickness is indicated in units of μg/cm^2 (ref.9).

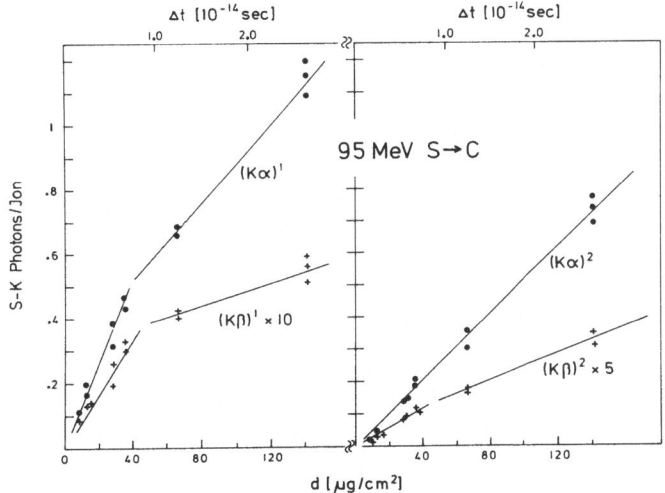

Fig. 3.   Absolute yield of four resolved K x-ray lines of 95-MeV
sulphur in carbon, as a function of target thickness in
μg/cm^2 (ref.3).

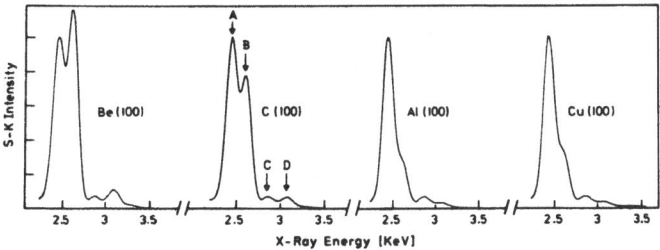

Fig. 4.   Relative intensities of K x-ray satellites of 95-MeV sulphur
ions in various targets of thickness 100 μg/cm^2 (ref.3).

TARGET THICKNESS DEPENDENCE OF PROJECTILE X-RAY YIELDS

In the past, one generally assumed that total x-ray yields are proportional to the target thickness as is obvious from Eq.(3), except when σ_v is so large that the fraction of excited ions in the beam is no longer small compared to 1. The necessary and sufficient condition for observing nonproportional yields can be expressed by[3,9]

$$\sigma_v + \sigma_c \gtrsim (nv\tau)^{-1} \quad , \quad (\sigma_T x_o \gg 1) \tag{4}$$

in which case Eqs.(1) or (2) become useful. In the following, we briefly comment on some experimental results and consequences related to this nonproportional behavior of projectile x-ray yields.

Fig. 2 shows K x-ray spectra of 95-MeV sulphur ions in carbon targets of various thickness, measured with a Si(Li) detector. Two $K\alpha$ (A,B) and two $K\beta$ (C,D) lines are resolved; the centroid energies are quite independent of target thickness, and lines B,D are hypersatellites to lines A,C. It can be seen that the intensity ratios of the lines vary strongly with target thickness. The four yield curves are shown in Fig. 3 (A-D correspond to $K\alpha^1$, $K\alpha^2$, $K\beta^1$, $K\beta^2$, respectively). Satellites A and C show pronounced nonproportional behavior as is described by Eqs.(2) and (4). Thus, the varying intensity ratio is readily understood. We emphasize that these ratios will still change even when the target thickness is beyond the one needed to achieve equilibrium excitation.

Another consequence of electron capture is demonstrated in Fig. 4. The intensity ratios of lines A-D also depend on the target species. Our data[10] shows that the ratio of lines A and B, for example, oscillates with Z_2. An explanation can be based on the formalism from the preceding section: it turns out[10] that the ratio in question is largely determined by the ratio σ_v/σ_c and detailed analysis of these cross sections involved reveals indeed an oscillatory trend $\sigma_c(Z_2)$. The latter effect will be treated later on.

Capture cross sections are smaller in light targets as compared to heavy ones. This effect becomes obvious from comparison of Figs. 3 and 5. In the latter case, σ_T is dominated by σ_c and is so large that pre-equilibrium thicknesses could not be realized experimentally if targets without backing are used.

Since a Si(Li) detector can not resolve individual multiplet states, high-resolution experiments have been performed in order to check whether our description of excited state populations is also valid for pure transitions. Fig. 6 reproduces the larger part of a sulphur $K\alpha$ spectrum, along with the growth curves for several transitions. Resolved lines from Fig. 6 are identified as follows:

1 : $(1s2p^2)^4P-(1s^22p)^2P$ 4 : $(1s2p)^1P-(1s^2)^1S$
2 : $(1s2p^2)^2D,^2P-(1s^22p)^2P$ 5 : $(2p^2)^3P-(1s2p)^3P$
3 : $(1s2p)^3P-(1s^2)^1S$ 6 : $(2p)^2P-(1s)^2S$.

Transitions number 1 and 3 show differential metastability, i.e.
there is almost no radiative decay inside the target and the nonproportional behavior is an extreme one.

Yield curves were also obtained for different collision
velocities. Fig. 7 presents growth curves for sulphur on alumi -
num in the energy range from 25- to 110 MeV. It can be seen that
no major change occurs between 55- and 110 MeV, whereas the nonproportional yield effect seems to fade away for low collision
energies. This indicates that condition Eq.(4) is no longer fulfilled for small v.

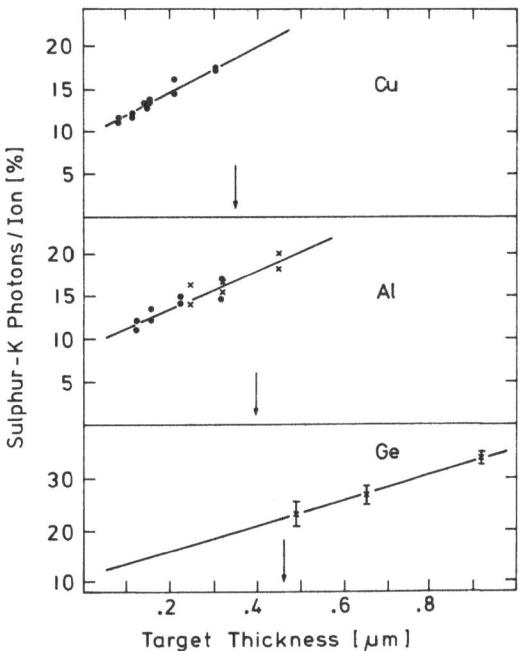

Fig. 5. Absolute K x-ray yield of 110-MeV sulphur in Cu, Al, and
         Ge, as a function of target thickness in units of
         μm (ref.9).

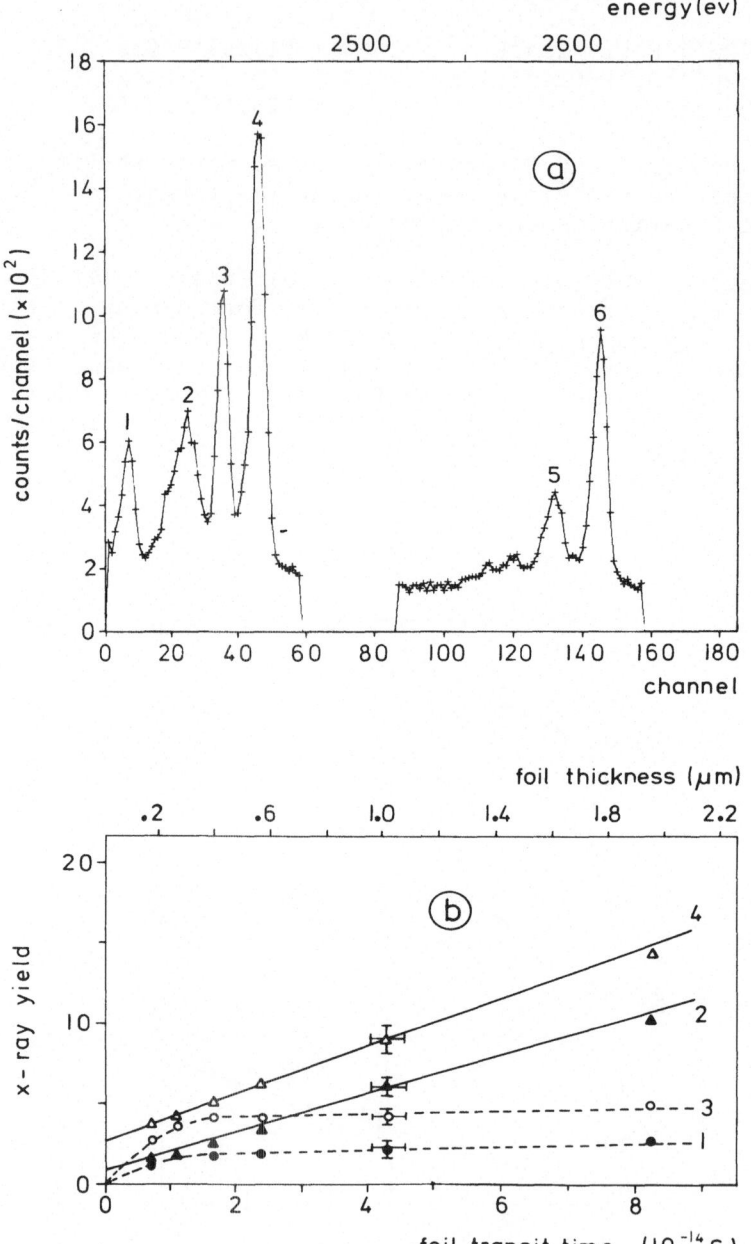

Fig.6. a) K x-ray spectrum of 92-MeV sulphur in 150 μg/cm² carbon;
b) multiplet x-ray yields as a function of target thickness.
See text for line identification (ref.11).

LIFETIME DETERMINATION OF PROJECTILE STATES

It becomes evident from Eqs.(1) or (2) that measurement of nonproportional x-ray yield curves allows determination of lifetimes τ of projectile states. Such procedures have been carried out for various sulphur states[3,4,11] and we will not repeat the details here. Some general remarks, however, are in order and may illucidate advantages and difficulties associated with this new technique.

The method can be utilized whenever nonproportional yield curves are measureable, i.e. condition Eq.(4) must be fulfilled. It is not necessary, however, to have detailed knowledge of how the excited state is produced or destroyed; thus, the actual quenching processes need not be understood. Eq.(1) is valid for a particular excitation state and does not depend on the assumption of a two-state system. When the nonproportional effect is a strong one, $\sigma_v + \sigma_c \gg (nv\tau)^{-1}$, one can neglect $\langle \tau_i \rangle$ and Eq.(1) becomes very simple. Evaluation of τ from a given yield curve is then straightforward. Lifetimes of individual multiplet states can be obtained provided that the desired transition is resolved. The target can be chosen to optimize experimental conditions.

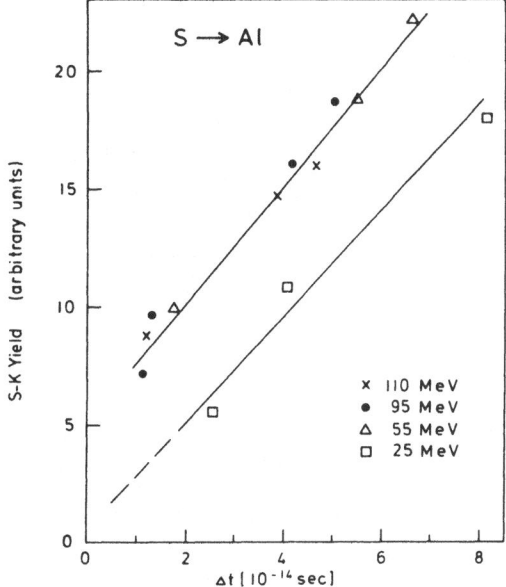

Fig. 7. K x-ray yields of sulphur ions passing through aluminum, as a function of target thickness expressed in transit times; parameter is the energy of the sulphur ions.

Target thicknesses must be known on an absolute scale.
Effects of energy loss of the ions and absorption of the x-rays in
the target should be kept small enough to be readily corrected for.
X-ray yields must not be measured absolutely, but runs with different target thicknesses must be normalized to each other. Only
those lifetimes can be extracted which are of the order of transit
times of ions through the foil ($\tau \simeq t_o$). Cascade deexcitation of
excited ions behind the foil are incorporated in Eq.(1) but not
in Eq.(2) and may present problems in the analysis.

In few-electron systems, lifetimes of particular states can
be calculated theoretically with excellent accuracy. This allows
tests of the experimental lifetime technique[11]. It seems feasible
to apply the technique also to more complex states where calculations are cumbersome. Finally, total lifetimes τ_i may be measured
to give information on the magnitude of $\sigma_v + \sigma_c$.

PROJECTILE X-RAY CROSS SECTIONS IN FOIL TARGETS

Occurrence of x-ray yields which are nonproportional to target thickness makes it difficult to extract vacancy production
cross sections. This will be illustrated by means of an example.

In an early MIT-BNL collaboration numerous x-ray production
cross sections have been measured[12] and some of that data has
been analyzed as a function of Z_2 on the simple-minded basis of
Eq.(3). Results[13] are reproduced in Fig. 8. The most striking
feature is a pronounced oscillatory trend $\sigma(Z_2)$ which was not
understood for a long time and has been explained only very
recently[2]. Effects of quasimolecular level matching[1] can hardly
account for the experimental findings because the collision
velocities involved are simply too large; furthermore, the
oscillations seem to increase with v, whereas quasimolecular
phenomena are expected to fade with increasing v.

It is now clear that in the cases Fig. 8 one deals with
pronounced nonproportional yields and analysis can not be based
on Eq.(3). Instead, Eq.(2) is more appropriate and one then
realizes that the experiment is sensitive to the quantity N_o, whereas σ_v can not be extracted directly. In fact, it can be shown here
that σ_c is large enough to approximate Eq.(2) by $Y \simeq (t_o + \tau)\tau_R^{-1} \sigma_v/\sigma_c$,
i.e. the experimental yield reflects the <u>ratio</u> between the
cross sections for vacancy production and electron capture into
these vacancies, and not at all σ_v alone.

In the case of 120-MeV chlorine we have evaluated the
relevant cross sections: σ_v can be described as Coulomb ionization
and varies smoothly with Z_2 (Fig.9). In the case of σ_c we use a

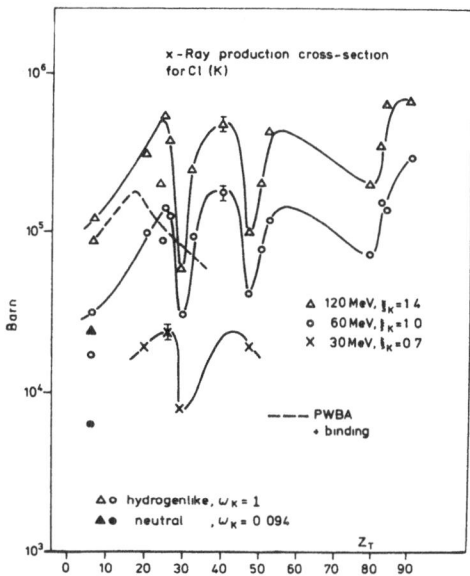

Fig. 8. Production cross sections for chlorine K x-rays, as a function of target species; parameter is the energy of the chlorine ions (ref.13).

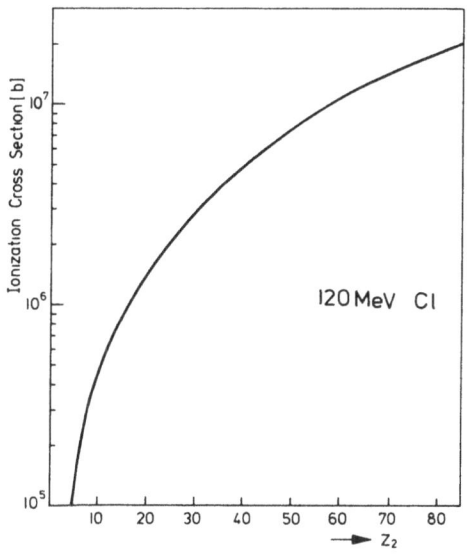

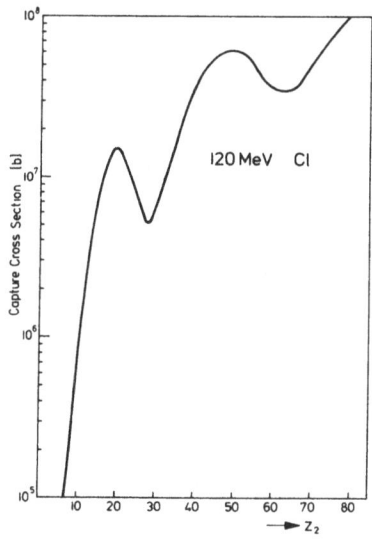

Fig. 9. K shell vacancy production cross section for 120-MeV chlorine ions, as a function of target species (ref.2)

Fig.10. Cross section for electron capture into the K shell of 120-MeV chlorine ions, as a function of target species (ref.2)

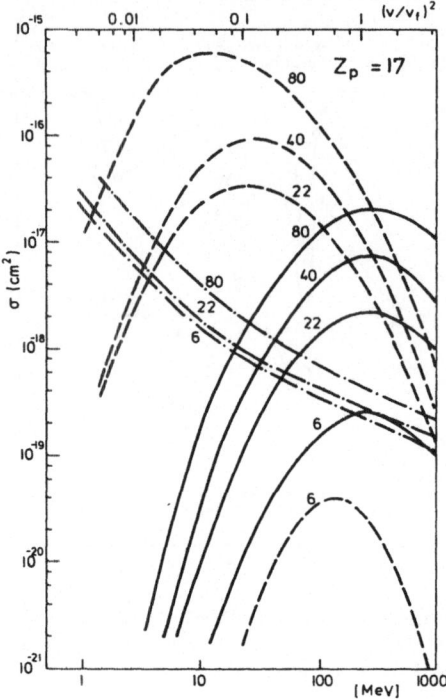

Fig. 11. Cross sections for chlorine, as a function of chlorine
energy; parameter is the target species:── σ_v; ── ── σ_c;
── · ── $\{nv\tau\}^{-1}$ (ref.2).

scaled Brinkman-Kramers formalism[14] and obtain an oscillatory
behavior as is shown in Fig. 10. The resulting theoretical ratio
σ_v/σ_c explains the experimental oscillations satisfactorily.

Fig. 11 demonstrates the relative importance of the cross
sections σ_v, σ_c, and $\{nv\tau\}^{-1}$ for chlorine ions as a function
of beam energy. It becomes obvious that condition (4) is well
fulfilled in a large velocity range and for most targets except
the lightest ones. Values of σ_v and σ_c for very low energies
should not be taken too seriously since quasimolecular effects
will then become important and both Coulomb ionization and
Brinkman-Kramers formalisms must be modified. Then, electron
capture or direct charge exchange competes with quasimolecular
effects[15] and the collision processes become more complex.

The example above demonstrates the problems of measuring
meaningful projectile x-ray cross sections in solids. On the
one hand, when vacancy production cross sections are desired,

one must check for the absence of nonproportional yield effects; use of gaseous targets is, of course, more straightforward and eliminates most of the difficulties. On the other hand, study of yields from foils provide direct and useful information on yet poorly studied collision processes inside solids.

RADIATIVE ELECTRON CAPTURE (REC)

A partial review on REC has been given recently[14]. Additional results have been reported on REC line widths[16], cross sections[17,18] and angular distributions[17]. As regards REC production cross sections and angular distributions, it appears that the data can be understood in its most important aspects using existing theoretical descriptions[19,21]. In the following, therefore, we comment only on the spectral line shape which is determined by the momentum distribution of the captured target electrons.

Some investigators[22,16] have reproduced experimental shapes at and near the REC maximum by theoretical techniques. In essence, they used hydrogen-like atomic wave functions for the target electrons with adequate Slater screening to obtain correct REC peak widths. It has been argued, however, that such a procedure is improved by using Hartree-Fock wave functions which make it unnecessary to choose appropriate Slater-type screening constants[14]. Nevertheless, the results of these more rigorous calculations give substantially poorer agreement[14]: theoretical REC distribution widths turn out to be significantly smaller than experimental ones (compare Figs. 12 and 13). The reason for this discrepancy had not been fully understood.

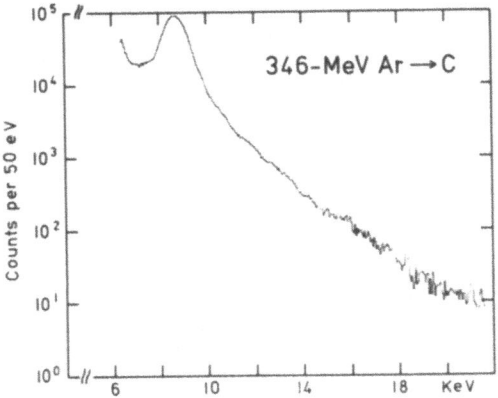

Fig. 12. Experimental REC profile, corrected for absorber effects.
          Characteristic Ar lines are suppressed.

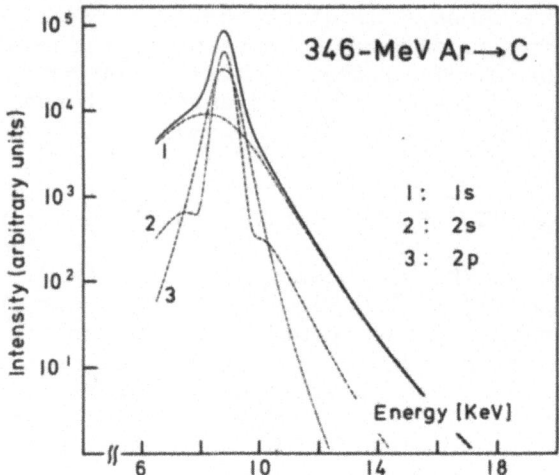

Fig. 13. Theoretical REC profile calculated with free atom Hartree-
         Fock wave functions.

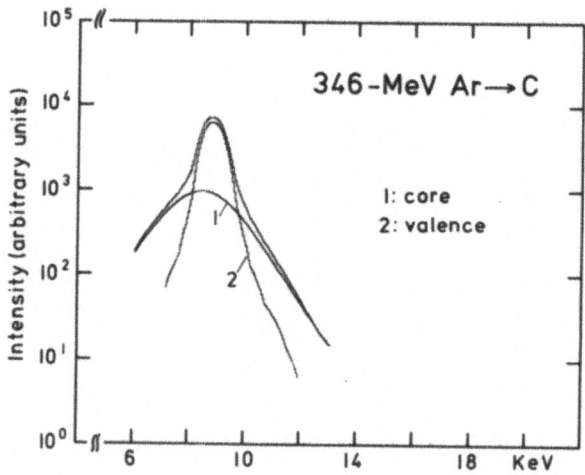

Fig. 14. Theoretical REC profile; contributions of carbon valence
         electrons are taken from experimental Compton profiles.

Our most recent analysis of the situation resulted in the following. Let us consider the above example of 346-MeV argon ions passing through carbon foils with thicknesses of the order of $\sim$100 μg/cm^2 (Fig.12). The velocity distribution of the valence electrons differs in solid state targets and in free target atoms. Thus, atomic wave functions may not be justified when solids are used. To be more consistent, we have taken experimentally determined Compton profiles for the valence electrons of carbon[23] and calculated the corresponding REC profile; the 1s core electrons were treated as before, namely by HF functions for isolated atoms. Results are given in Fig.14. It turns out that the main peak of the experimental REC distribution is now reproduced with good accuracy and the former discrepancies in the half-widths are no longer present. Contributions of the C-1s electrons account for part of the high energy tail, but for REC intensities below $\sim$1 % of the maximum intensity deviations still occur and increase for higher x-ray energies. It is not yet clear whether this is a result of the break-down of the impulse approximation on which calculations are based[21], or whether effects of other origins have yet to be identified such as sorts of bremsstrahlung[22,24]. We hope that these remaining questions can be clarified and that new measurements in both gaseous and solid targets can be used to decide whether the analysis from above is meaningful and can be reversed. If the latter holds true, REC measurements could be used to determine experimental Compton profiles.

REFERENCES

1. J.D. Garcia, R.J. Fortner, and T.M. Kavanagh, Rev. Mod. Phys. 45,111(1973).
2. F. Bell, H.-D. Betz, H. Panke, W. Stehling, and E. Spindler, submitted to J. Phys. B.
3. H.-D. Betz, F. Bell, H. Panke, G. Kalkoffen, M. Welz, and D. Evers, Phys. Rev. Lett. 33,807(1974).
4. H. Panke, F. Bell, H.-D. Betz, and W. Stehling, Nucl. Instr. Meth. 132,25(1976).
5. F. Hopkins, Phys. Rev. Lett. 35,270(1975).
6. D.L. Matthews, R.J. Fortner, and G. Bissinger, Phys. Rev. Lett. 36,664(1976).
7. T.J. Gray, P. Richard, K.A. Jamison, J.M. Holl, and R.K. Gardner, reprint May 1976 (Kansas State University).
8. F. Hopkins, J. Sokolov, and A. Little, preprint June 1976 (State University of New York, Stony Brook).
9. H.-D. Betz, F. Bell, H. Panke, G. Kalkoffen, M. Welz, and D. Evers, Phys. Lett. 49A,133(1974).
10. H. Panke, thesis 1976 (University of Munich).

11. H. Panke, F. Bell, H.-D. Betz, W. Stehling, E. Spindler, and R. Laubert, Phys. Lett. 53A,457(1975).

12. H.W. Schnopper, A.R. Sohval, H.-D. Betz, J.P. Delvaille, K. Kalata, K.W. Jones, and H.E. Wegner, Proc. Conf. on Inner Shell Ionization Phenomena and Future Applications, Ed. R.W. Fink, Atlanta (1972), USAEC Technical Inf. CONF-720404, p.1348.

13. F. Bell and H.-D. Betz, in Atomic Collisions in Solids, Eds. S. Datz, B.R. Appleton, and C.D. Moak, Plenum Press 1975, p.397. (Proc. Conf. Atomic Collisions in Solids, Gatlinburg Tenn., 1973).

14. H.-D. Betz, M. Kleber, E. Spindler, F. Bell, H. Panke, and W. Stehling, Conf. on the Physics of Electronic and Atomic Collisions, Seattle, 1975, Eds. J.S. Risley and R. Geballe, University of Washington Press, p.520.

15. J.H. McGuire, Phys. Lett. 48A,387(1974).

16. A.R. Sohval, J.P. Delvaille, K. Kalata, K. Kirby-Docken, and H.W. Schnopper. J. Phys. B 9,L25(1976).

17. J. Lindskog, J. Pihl, R. Sjödin, A. Marelius, K. Sharma, R. Hallin, and P. Lindner, preprint April 1976 (University of Uppsala).

18. J.A. Tanis, B. L. Doyle, W.W. Jacobs, and S.M. Shafroth, preprint 1976 (University of North Carolina).

19. H.A. Bethe and E.E. Salpeter, Encyclopedia of physics, vol.35, Ed. S. Flügge (Springer, Berlin 1957) p.408.

20. J.S. Briggs and K. Dettmann, Phys. Rev. Lett. 33,1123(1974).

21. M. Kleber and D.H. Jakubassa, Nucl. Phys. A252,152(1975).

22. P. Kienle, M. Kleber, B. Povh, R.M. Diamond, F.S. Stephens, E. Grosse, M.R. Maier, and D. Proetel, Phys. Rev. Lett. 31,1099(1973).

23. W.A. Reed, P. Eisenberger, K.C. Pandey, and L.C. Snyder, Phys. Rev. B10,1507(1974).

24. D.H. Jakubassa and M. Kleber, Z. Phys. A273,29(1975).

ION-INDUCED CONTINUUM X-RAY EMISSION

F.W. Saris and Th.P. Hoogkamer

FOM-Institute for Atomic and Molecular Physics

Kruislaan 407, Amsterdam/Wgm., The Netherlands

ABSTRACT

This paper reviews continuum X-ray emission produced by ion-impact on gaseous and solid targets. Molecular orbital X-rays are considered in detail in order to establish the circumstances under which they can be distinguished from radiative electron capture and from various kinds of bremsstrahlung. The level of understanding of continuum X-ray production appears to parallel that of ion-induced inner-shell ionization in general.

1. INTRODUCTION

In the very first paper on heavy-ion induced X-ray emission, written in 1934 by Coates[1]) from the Radiation Lab., the author notes: "In no case do the curves show any evidence for continuous radiation". This was an important remark, for the bremsstrahlung continuum was known to dominate electron-induced X-ray spectra. Yet, after many years of investigating characteristic X-ray emission, the ion-induced X-ray researchers have recently focussed on the continuum radiation in their spectra, which has become visible through the availability of energy dispersive X-ray detectors. The motivation for this new interest is two-fold:
1) Investigating continuum radiation appears to be of fundamental importance for the understanding of atomic interactions in ion-atom collisions; 2) As background radiation underlying characteristic X-ray signals, continuum X rays limit the detection sensitivity in analytical applications such as trace element analysis.

The purpose of this paper is to review the properties of ion-induced continuum X-ray spectra and thus show how progress in this field parallels that of ion-induced inner-shell ionization and decay. First we will outline some characteristics of molecular orbital X-ray emission, radiative electron capture and various kinds of bremsstrahlung. Then we will discuss under what experimental conditions these features have been identified unambiguously. This will lead to some conclusions and possible extensions.

2. GENERAL FEATURES AND EXCITATION MECHANISMS

2.1. Molecular Orbital X rays

In a slow collision of two atoms the nuclei may approach each other so closely that the atomic electrons rearrange adiabatically to form quasi-molecular states in the two-center Coulomb field. These quasi-molecular states will change continuously as a result of the varying internuclear distance during the collision. In addition the nuclear motion will impose considerable excitation on the electrons in the transient molecular levels. Generally this will result in excitation or ionization of the projectile and target atom when separated. Sometimes, however, the excited quasi-molecule may decay during the collision if the lifetime of the excited molecular state is not too different from the collision time. Inner-shell vacancy lifetimes are of the order of 10^{-15} sec, only a factor ten or hundred longer than most collision times. It is well known that inner-shell vacancies are copiously produced in heavy-ion-atom collisions even at moderate energies. It is to be expected, therefore, that in many cases vacancies are present in the inner-molecular levels and their decay during the collision may result in X-ray spectra characteristic of these quasi-molecular states. As the internuclear distance changes continuously, the transition energy will vary also, giving rise to continuous X-ray spectra instead of characteristic lines.

It seems inappropriate, however, to use the name non-characteristic radiation (NCR) instead of MO-X rays, as is done by some authors. The continuum spectral distribution is very characteristic for the molecular states involved in the collision.

Before we go into a detailed discussion of MO-X rays we will briefly mention some other mechanisms for continuum X-ray production.

2.2. REC

At high projectile velocities typically equal to or higher
than the orbiting velocities of inner electrons, continuum X-ray
emission due to radiative electron capture will occur. This radiation is emitted when electrons of the target are captured into
vacancies in the fast moving projectile. It will show a maximum
photon intensity at an X-ray energy which is higher than the binding energy of the electrons final state, due to the relative
motion of projectile and target electron. The shift will be equal
to $(m_e/M)(E_p/A_p)$, m_e and M are the electron and proton mass, E_p
the projectile energy and A_p its mass number. The width of the REC
continuum is proportional to the projectile velocity and the mean
velocity of the target outer-electrons. For light materials REC is
seen as a relatively narrow peak and therefore readily identified.
For heavy targets a peak may not be recognizable due to contributions from many outer-shell electrons. Since REC has been reviewed
at this conference already[2], there is no reason to dwell more on
it here. Recently, Briggs and Dettmann[3,4] have formulated a uniform theory for MO-X ray emission and REC.

2.3. Various kinds of bremsstrahlung

It is not always appreciated that in any particle-induced
ionizing event three different kinds of bremsstrahlung may occur.
1) The well-known projectile bremsstrahlung or NNB (for nucleus-
nucleus bremsstrahlung) results from the deflection of the incoming
particle in the field of the target nucleus. 2) If in such a collision an electron is ionized then it is being accelerated, a process
that is accompanied by radiation (RI for radiative ionization,
sometimes also called inner-bremsstrahlung). 3) The secondary electron when going through a dense medium at some kinetic energy will
be deflected by target nuclei, thus causing secondary electron
bremsstrahlung (SEB).

3. QUANTITATIVE ANALYSIS OF ION-EXCITED CONTINUUM SPECTRA

In this section we will try to identify different contributions from the different mechanisms and compare the experimental
spectral distributions to those generated theoretically, in order
to establish the level of understanding in this field of research.
Before doing this we will briefly outline some experimental parameters.

A typical experimental set-up for these studies will consist
of a vacuum chamber in which the ion beam enters to hit the target
(gaseous or solid) viewed by an X-ray detector at 90°. Some arrange-

ments allow the detector also to look at the scattering center under different angles. Frequently one is able to insert filters in
front of the detector window in order to reduce pile-up effects
from low-energy signals. The beam current is measured on target,
or via a scattered particle detector calibrated with a Faraday cup.
Usually a beam current integrator gates the detector electronics
in order to obtain the number of X rays emitted per incoming ion.

Of course in Coates' pioneering work with MeV Hg^+ ions continuum X rays must have been present also. In those days the detector resolution and efficiency were not sufficient to separate the
weak continuum signals from the noise level or from the tails of
characteristic lines. With Si(Li) or Ge(Li) detectors this is not
a problem any longer, although one should be aware of two kinds of
background signals associated with their use. In experiments with
high energy light ions one will produce copious amounts of γ-rays
due to interactions with the target matrix or imbedded light impurities. The efficiency of the very thin Si(Li) detectors is too
little to detect γ-rays of several tenths of keV or more, but
Compton scattering can cause an almost flat background in the recorded spectra. A second effect typical for energy dispersive
systems is charge loss due to incomplete collection of the photoionization avalanche near the edge of the active area in the detector[5]). This causes a background on the low energy side of characteristic lines of at least 1 count/channel per 1000 counts in the
peak channel of the characteristic line. The high energy side of
a characteristic spectrum drops sharply (Gaussian fits are usually
applicable) and is therefore a region more suited to look for the
continuum X-ray signals.

For convenience we will divide the discussion of the spectra
into different subsections for the different projectile velocities
and projectile-target combinations used.

3.1. $Z_p \ll Z_t$, $v_p \lesssim v_K$

Fig. 1 shows measured cross sections for continuum X-ray
emission versus X-ray energy (2 – 50 keV) in collisions of 2 MeV H^+
on C and Al. At low energies the spectrum is dominated by SEB since
secondary electrons are produced with high probability up to an
energy $T_m = 4 (m/M)(E_p/A_p)$ beyond this X-ray energy the spectra drop
steeply, for only tightly bound electrons can acquire more energy
than T_m. This is reflected in the spectra of fig. 1: the SEB is
more intense and extends to higher X-ray energies in Al than in C.
Folkman[6,7], who recorded these cross sections, also gave a quantitative theoretical description using the binary encounter theory

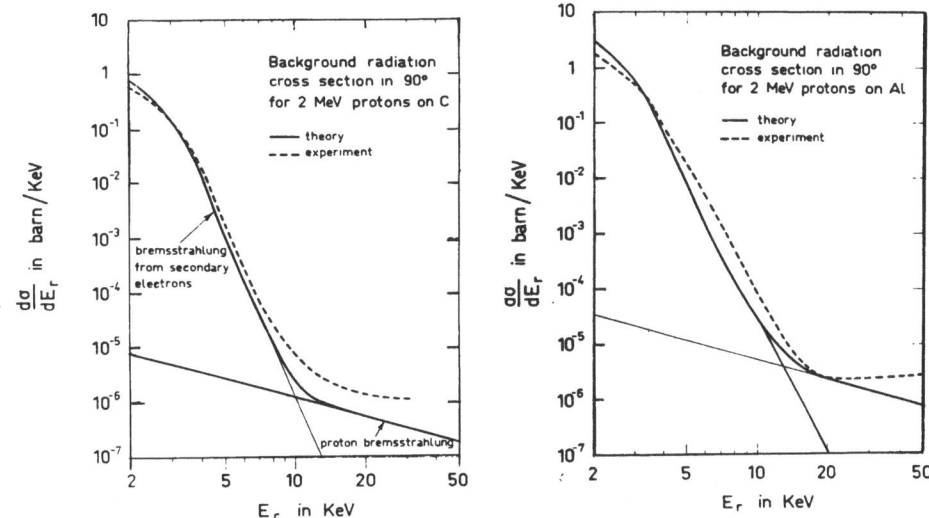

Fig. 1. Experimental and theoretical background radiation
cross sections for a thin sample (ref. 6).

and scaling law. Very good agreement is obtained, see also Sohval
et al.[8]). Above 10 keV X-ray energy a less steep background appears
which can be accounted for by dipole NNB or proton bremsstrahlung.
Compton scattering from γ rays probably causes the deviation of experiment and theory above 30 keV.

The angular distribution of secondary electron bremsstrahlung
can be expected to be non-isotropic. For high secondary electron
energies the distribution is strongly forward peaked also
after some slowing down in the target material. Electron bremsstrahlung distributions are peaked away from the beam axis[10]).
Although both angular distributions will be smeared out, a mean
increase of the SEB yield at 90° and decrease at forward and backward angles is to be expected. This has recently been confirmed by
measurements of Ishii et al.[11]) for 1.5 MeV and 4 MeV H$^+$ → Al.
A clear 90° peaking of the X-ray distribution is seen for the highest X-ray energies, approaching isotropy at lower energies. In
conclusion, when using proton-induced X-ray emission for trace
element analysis, positioning the X-ray detector at a backward
angle instead of the usual 90° geometry will improve signal to
background ratios.

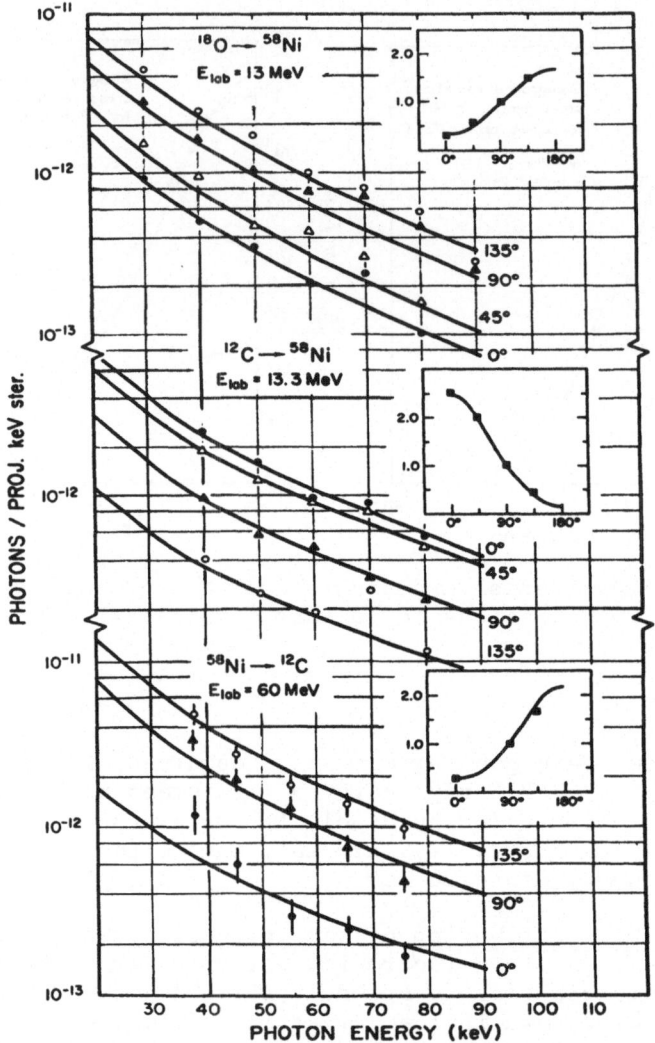

Fig. 2. Comparisons of calculated NBB absolute thick-target
yields with X-ray intensities as a function of
photon energy at four angles of observation. Each
energy point represents an integration over a 10-keV
photon energy interval. The statistical errors
shown on the bottom graph are typical. The insets
(vertical scale relative) show a comparison of the
calculated angular distribution (solid lines) with
data (squares) for the photon energy interval
40 – 60 KeV (ref. 12).

The production of nucleus-nucleus bremsstrahlung at high X-ray
energies in fast asymmetric collision systems has been investigated
in detail recently by Trautvetter et al.[12]. They compare measured
X-ray continuum spectra for $O + Ni$, $C + Ni$ and $Ni + C$ with calculations of dipole and quadrupole bremsstrahlung[13], see fig. 2.
Theory predicts interference between dipole and quadrupole NNB to
depend critically on the ratio of Z/A for target and projectile.
Indeed, as is shown in fig. 2, a small change in this ratio leads
to a spectacular variation in the anisotropy of the emitted radiation. Also the absolute intensity measured is in good agreement
with the calculated NNB yields. Trautvetter et al. also note that
NNB becomes vanishingly small for symmetric collision systems only
for medium or small atomic numbers.

3.2. $Z_p \simeq Z_t$, $v_p \ll v_K$

Under these very adiabatic conditions, in particular for gas
phase collisions, all REC and bremsstrahlung processes are many
orders of magnitude below measured X-ray intensities. So in this

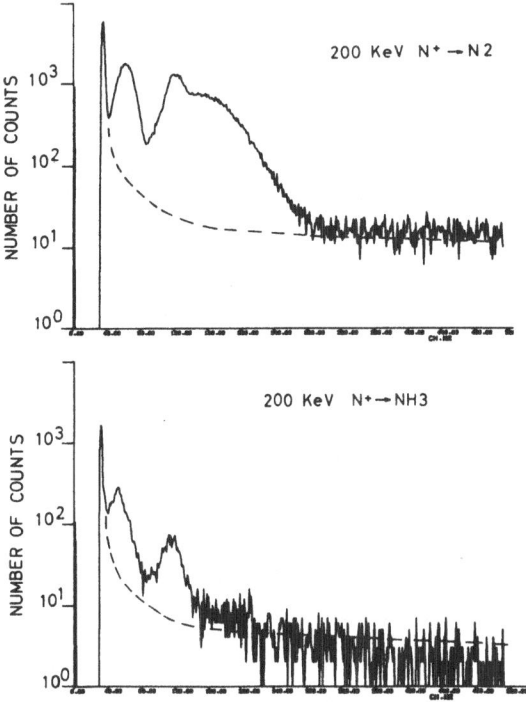

Fig. 3. The X-ray spectra for 200 keV N^+ on N_2 and NH_3.
        The $N^+ \to N_2$ spectrum shows in addition to the K_α
        and $K_{\alpha\alpha}$ line a broad continuum extending up to
        about 2 keV X-ray energy (ref. 16).

subsection we will mainly deal with MO-X rays, first K shells will
be discussed then higher molecular orbitals.

In addition to characteristic lines a broad X-ray continuum
is observed during 200 keV N^+ impact on N_2, see fig. 3 top. One
can identify the first line as $N-K_\alpha$, the second as a $N-K_{\alpha\alpha}$. The
latter being due to double K-shell vacancies, which can decay via
a 2-electron-1-photon transition[14]. Note that $N-K_\alpha$ is attennua-
ted by a factor 10^{-4} compared to $N-K_{\alpha\alpha}$ due to the low transmission
of the detector window at lower X-ray energies. The broad X-ray
band is interpreted as the radiative decay of vacancies in the
quasi-molecular $1s\sigma$-state.

The production mechanism for this spectrum can best be explained with the aid of the energy level diagram of fig. 4,[15]
showing how the atomic levels of the colliding N atoms correlate
via the MO-levels to the united atom (Si) levels when the internuclear distance goes to zero. N-K vacancies can be produced via
the rotational coupling of the $2p\sigma$ and $2p\pi$ molecular orbitals.
In a single collision 1 as well as 2 2p vacancies can be transferred simultaneously thus the $N-K_\alpha$ and $N-K_{\alpha\alpha}$ line of fig. 3 are
mainly due to a single collision of the N projectile with one of
the N atoms in the N_2 molecule.

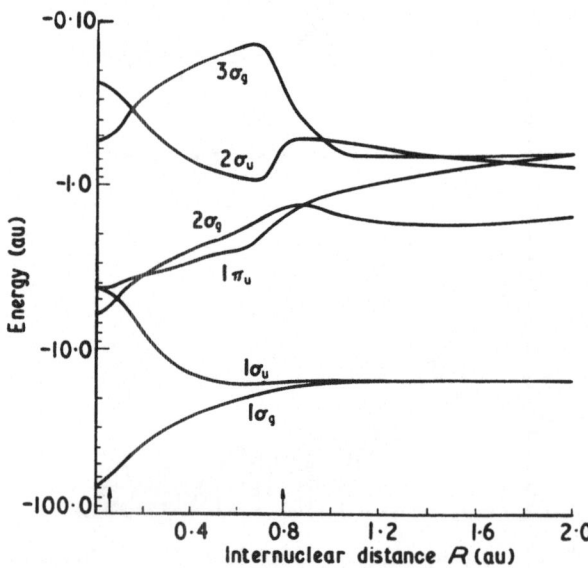

Fig. 4. The molecular orbital diagram for the N_2 system
        (ref. 15).

This collision model, however, forbids vacancies to be present in the $1s\sigma$ MO under single collision conditions at low velocities. Therefore a double collision is suggested for the production of the continuum X-ray band of fig. 3, top. In the first collision N-1s vacancies are produced, half of them are carried away by the projectile which may collide with the second atom of the same molecule before the vacancy decays. During the collision the vacancy can follow the $1s\sigma$ MO and radiatively decay there. The continuum X-ray spectrum of fig. 3, top, extends from the $N-K_\alpha$ line to the transition energy at distance of closest approach.

Clear evidence for this double collision model has been obtained by comparing the spectrum of fig. 3, top, to that for 200 keV $N^+ + NH_3$ in fig. 3, bottom. Here only single collisions occur, the N-H collisions not being effective for N-K vacancy production. Indeed on top of the electronic noise of the detector only $N-K_\alpha$ and $N-K_{\alpha\alpha}$ lines are observed. It is to be expected, however, that at high impact energies $1s\sigma$ vacancies should be produced directly in a single collision. The decay of such vacancies during the same collision should lead again to a continuum X-ray spectrum, the properties of which will be discussed along with theoretical calculations of radiative ionization in the next sections.

In fig. 5a it is shown that excellent agreement between measured and calculated spectral distributions can be obtained using the two collision model for 50 keV $N^+ \rightarrow N_2$,[16]). The calculations are based on theoretical MO-energy levels and estimated radiative rates. The probability of the projectile carrying a 1s vacancy into the second collision is accounted for by taking the ratio of the K-shell ionization cross section divided by $4\pi R^2$ (R, being the internuclear distance in a N_2 molecule). Note that in fig. 5 the spectrum has been corrected for window absorption effects and the detector resolution has been accounted for in the calculated spectral distribution.

When the projectile energy is increased to 140 keV the MO-X ray intensity increases but for 200 keV the absolute intensity is the same as at 140 keV. The calculated spectra follow this trend but have to be multiplied by a constant factor (2.5) in order to bring them in overlap with the experimental curves, see fig. 5b,c.

SEB from fast electrons, created in a collision with the first N atom and deflected in the field of the second N atom of the same molecule would sharply increase with projectile energy. The observed energy dependence does not show this and we believe that the discrepancy of a factor 2.5 between measured and calculated absolute MO-X ray yields is not surprising given the simplifying assumptions made.

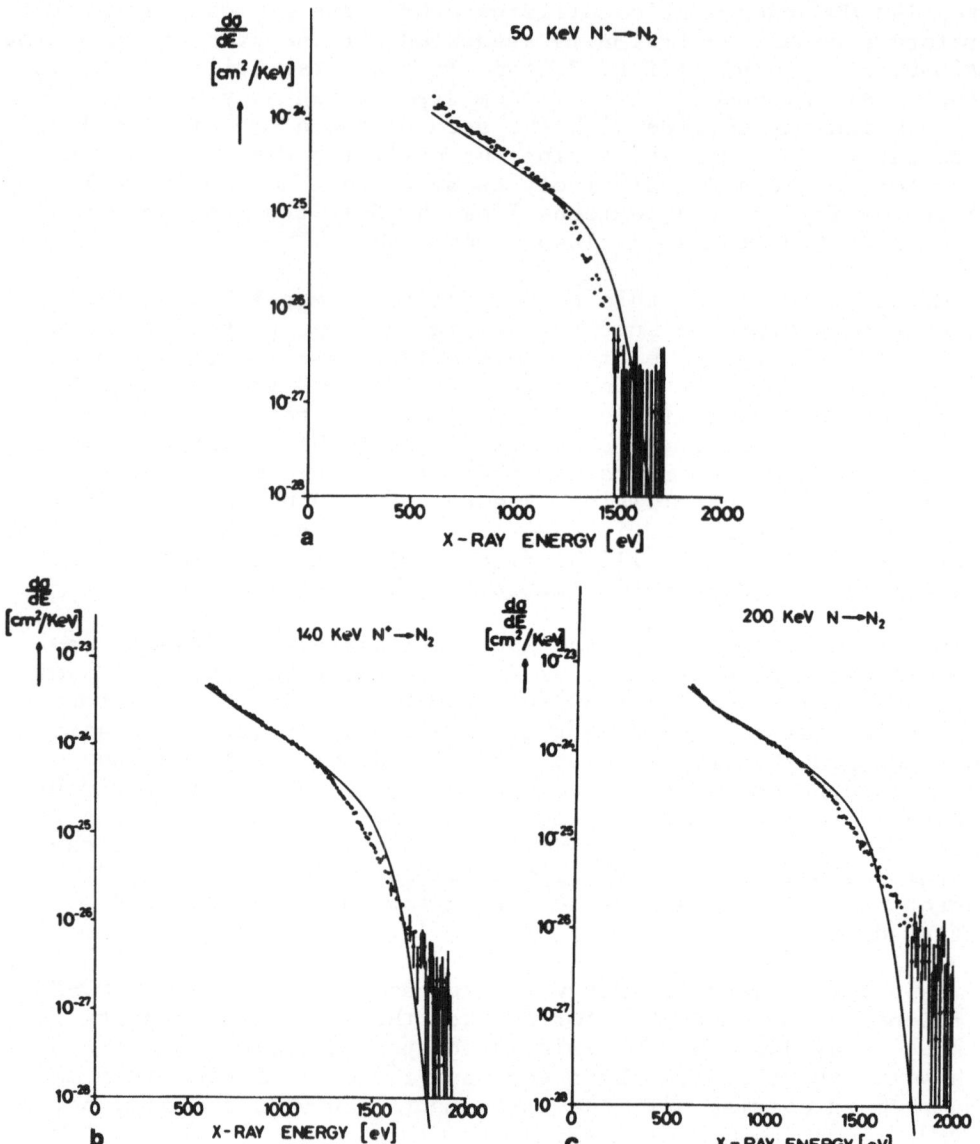

Fig. 5. Comparison between experimental (o) and theoretical
        (solid line) cross sections for MO–X ray production
        in collisions of:
        a)   50 keV $N^+ \to N_2$;   b) 140 keV $N^+ \to N_2$ (theory $\times 2.5$)
        c) 200 keV $N^+ \to N_2$ (theory $\times 2.5$) (ref. 16).

The change in shape of the experimental curve near the united atom limit, which leads to a deviation between calculated and measured yields, is to be ascribed to a dynamic broadening (Heisenberg broadening) not yet included in the quasi-static calculation. The collision time becomes too short to allow for an accurate measurement of the sharply varying $2p\pi$, $2p\sigma$ to $1s\sigma$ transition energy. In this case a full Fourier analysis would be required, ref. 17, see also Anholt, ref. 18.

Very similar experimental results have been obtained for C^+, impact on graphite by MacDonald et al.[19] in the energy region 50 - 300 keV and Laubert et al.[20] for 30 keV - 2.5 MeV.

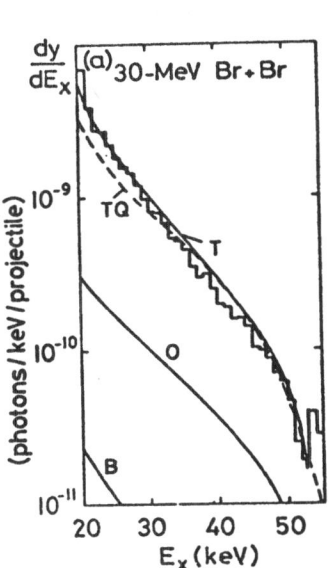

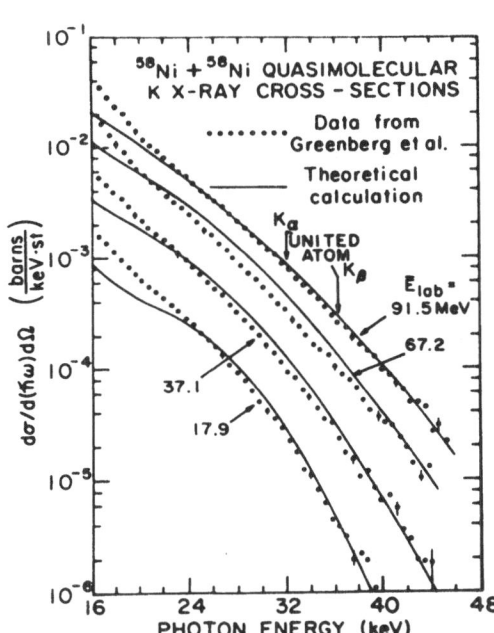

Fig. 6. K-MO X ray spectrum from 30 MeV Br on KBr. The curve labeled T is the two-collision quasi-static calculation and TQ the two-collision quantum-mechanical calculation (ref. 22, 29).

Fig. 7. Continuum X-ray spectra produced in Ni + Ni at various bombarding energies. The curves are absolute two-collision quantum-mechanical calculations (ref. 29).

MO-K X ray spectra have been investigated also for much heavier systems. It is important to note that v/v_K is almost identical for 200 keV N + N, 17 MeV Ni + Ni,[21]), 30 MeV Br + Br,[22,23]),
42 MeV Kr + Kr,[24]), 67 MeV Nb + Nb,[24,25,26]), 115 MeV La + La,[27])
and 144 MeV Bi + Bi,[24]), which have all been investigated. Again a
two-collision mechanism, in the solid targets, is responsible for
the broad MO-X ray continua observed. Meyerhof[22,28]) showed that
the simple theoretical two-collision approach was applicable for
Br + Br, see fig.6. He also estimated a one collision MO contribution
and nucleus-nucleus bremsstrahlung. Müller[29]) did a full dynamical
calculation for this system as well as for Ni + Ni, see fig.7, and
finds excellent agreement with the experiments of the Yale group.

For the heaviest collision systems the Dubna group discovered
two components of MO-X rays, c_1 and c_2 in fig. 8. c_2 is ascribed
to the radiative decay of vacancies in the 1sσ MO due to a double
collision identical to Br + Br and Ni + Ni. The double collision
mechanism was substantiated by a Kr + Kr gas collision experiment
of the Dubna group, see fig. 9, where the c_2 continuum is absent.
This figure also proves that c_1 can be produced in a single collision. In order to understand this we need to look at the correlation diagram of these heavy atom collisions, see fig. 10. Through
coupling with higher MO's, vacancies will be transferred into the
2pσ MO. As is seen in the diagram, the 2pσ goes through a minimum
and a radiative transition, c_1, will give rise to photons harder
than K_α which is indeed observed[30]). The 2pσ vacancies can be for

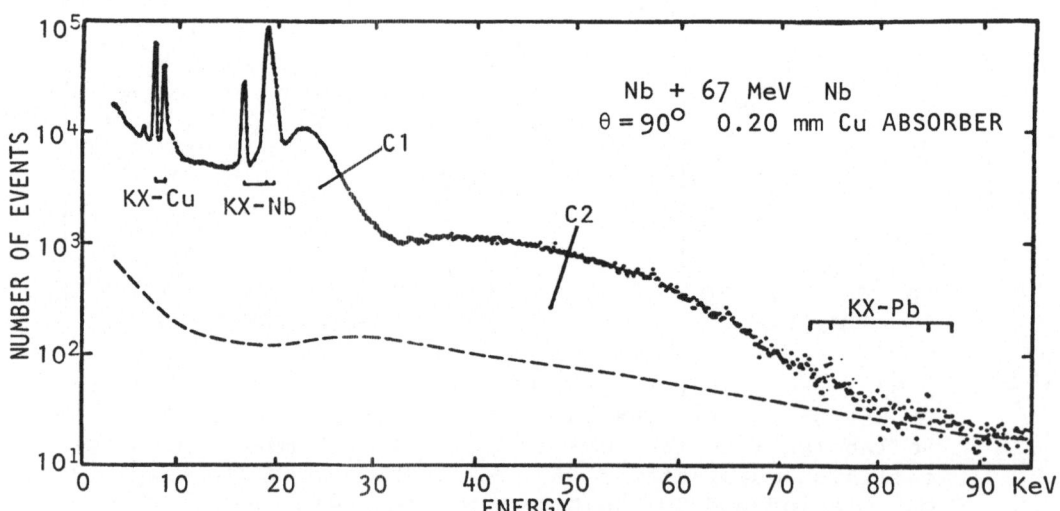

Fig. 8. The X-ray spectrum measured by bombarding 67 MeV
Nb ions on a Nb solid target (ref. 24).

med in double collisions but apparently also in single collisions. For lighter collision systems the minimum in the $2p\sigma$ MO is much shallower and c_1 cannot be identified.

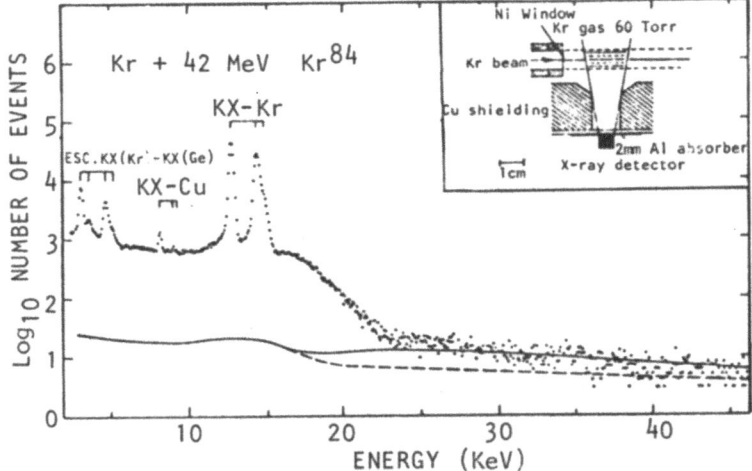

Fig. 9. The X-ray spectrum measured by bombarding 42 MeV Kr on a Kr gas target (ref. 24).

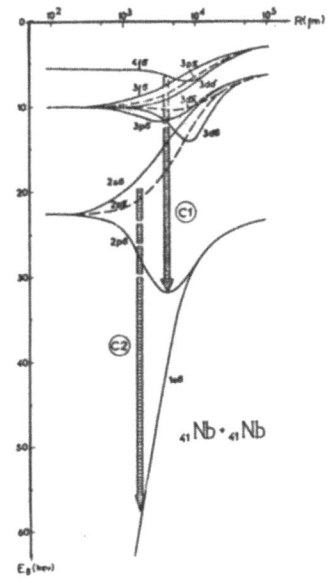

Fig. 10. Molecular level diagram for the Nb + Nb system (ref. 30).

The field of MO-X rays was initiated five years ago by the
observation of broad band X-ray emission during Ar impact of
solids and its interpretation as the radiative decay of projectile
L-shell vacancies during collisions with target atoms[31]. Since
then many experiments have been done to substantiate this speculation (ref. 32). Systematics in the high-energy end of the spectra
as function of target atomic number and projectile atomic number
and energy have been found (ref. 33,34,35), as well as some evidence for recoil effects when using solid targets (ref. 36).

Recently we have found some interesting effects in gas phase
experiments involving L-shell MO-X rays which we like to report
here. Fig. 11 shows the spectra emitted in collisions of 300 keV
Ar on Cl_2 and on Ar. In $Ar + Cl_2$ collisions one sees a broad band
plus the Cl K_α X-ray; in $Ar + Ar$ a broad band is observed also but
it does not extend as far as in $Ar + Cl_2$ and Ar K_α X-rays are
absent.

These features can be identified via the energy level diagram
of fig. 12, taken from recent work of Eichler et al.[37]. In close
collisions, vacancies present in the δ orbital will be transferred
to the 3σ MO via $3d\delta$, $3d\pi$, $3d\sigma$ rotational coupling[38]. According to
Eichler et al., they will pass through the avoided crossing and go
through the minimum in the 2σ MO. Radiative transitions in the
minimum, similar to the c_1 above, will give rise to X rays near

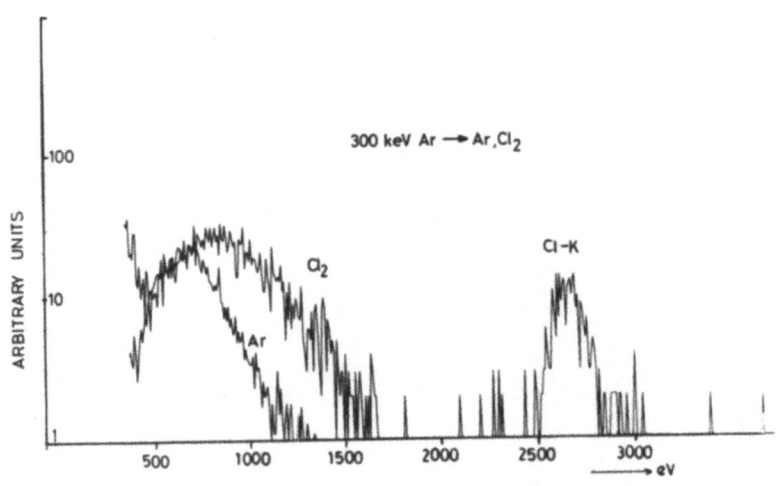

Fig. 11. The X-ray spectra for 300 keV $Ar^{2+} \rightarrow$ Ar and Cl_2.

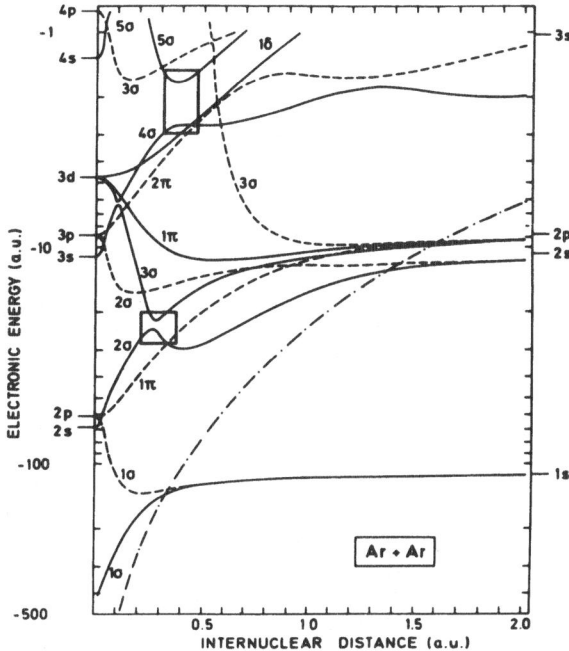

Fig. 12. Correlation diagram for the system Ar + Ar. The full curves indicate "gerade", the dashed ones "ungerade" states (ref. 37).

30 a.u. equivalent to ~800 eV which is consistent with the observations for Ar + Ar in fig. 11. Barat and Lichten[39] ignored this crossing and in their diagram the vacancy would only follow the 3σ MO into the 2p level of the separated atoms. The maximum photon energy then could never be higher than 20 a.u. or ~ 500 eV, which is clearly at variance with the measurements.

In the Ar – Cl$_2$ case a second collision can occur, just as in the N$_2$ molecule[40]. Vacancies in the Ar – L shell which follow the 2pπ MO in the second collision, can decay there to give rise to higher energy MO–X rays than in the Ar + Ar collisions. In close collisions the vacancies are rotationally coupled to the 2pσ MO to give Cl–K X rays, clearly in agreement with the measured spectra.

In the last three years a systematic investigation has been made of X-ray continua at about 8 keV, emitted in I – Au collisions and similarly heavy targets[41]. A peaked structure appears in the spectrum that was speculatively interpreted as the superheavy (Z = 132) quasi-atomic M shell X ray, resulting from radiative decay of projectile L-shell vacancies during I–Au collisions[42,43].

Gas-target measurements, however, show evidence that the broad
peak at 8 keV is produced in a single collision at threshold impact energy of ~ 2 MeV for which the distance of closest approach
is still larger than the united atom M-shell radius[44,45]. The
many electron relativistic I – Au energy level diagram of Rosen
et al.[46] allows an explanation of the peak structure in terms of
radiative transitions $4d - 3p_{3/2}$ in the quasi-molecule. For internuclear distances R between 0.12 and 0.18 a.u. the transition
energy is flat $E_x = 7.5 - 8$ keV. The $3p_{3/2}$ vacancies are produced
near the distance of closest approach (through coupling to higher
orbitals or to the continuum) and a small fraction, given by the
collision time versus the radiative lifetime, will decay during
the collision upon separation of the two nuclei. The rest of the
vacancies appear as characteristic lines emitted by the separated
atoms. This seems to be consistent with experimental findings.

So far these investigations of L-shell and M-shell MO X rays
reveal many interesting qualitative features of ion-atom collisions. Quantitative descriptions such as for the K-shell MO X ray
spectra have as yet not been possible. This to be attributed to
the multitude of radiative transitions, the lack of knowledge on
the distribution of initial excited states involved and their
fluorescence yields.

3.3. Near symmetric collision systems at high velocities

Mainly in search for REC and single collision MO-X rays many
groups have studied near symmetric collision systems at high velocities. Yet, the experimental situation becomes very complex since
contributions from bremsstrahlung processes are probably large.
Therefore controversies are raging over the interpretation of
continuum X ray emission in such systems as S impact on Ne and Al
by the Munich group[47,48], Si impact on Al and SiH_4 in Tennessee[49],
Br, Cl,O and C impact on Au at McMaster[50] and O impact on Zr and
Au at Stanford[51].

From the similarity in the S + Ne and S + Al spectra, the Munich
group concludes that single collision MO-X rays dominate. They are
surprisingly successful in parametrizing the high energy X-ray
tails using a simple expression for the dynamic broadening effect.
The Tennessee group finds, however, that the intensities, when
properly corrected for vacancy sharing effects, are not similar
and make the statement that their gas-target results would not be
inconsistent with SEB effects. The continuum X rays in the spectra
from McMaster and Stanford are claimed not to be due to SEB and
Anholt, in search for an explanation, has done calculations for
Radiative Ionization. The shape of the calculated spectra is well
reproduced. The absolute intensity, however, is low by an order of
magnitude.

In view of this we want to comment on the use of the binary encounter approximation in SEB and RI calculations for these heavy-ion atom collisions. One should take into account the quasi-molecular states that are formed during the collisions. This leads to a considerable alteration of the binding energies[52]. In addition, the inner-electron, since it cannot be distinguished to belong to one of the two nuclei, is accelerated through momentum transfer from both nuclei in relative motion during the collision. A quantitative description of this direct Coulomb ionization process under quasi-molecular conditions is not yet available, although a possible theoretical approach was formulated by Briggs[53]. For RI and SEB induced by heavy ions one needs the differential cross sections, i.e. the ionization probability as function of impact parameter and the secondary electron energy distribution. As long as these are not available an unambiguous identification of the X-ray continuum produced in the above collision systems remains unwarranted.

In spite of these difficulties we have looked for single collision MO-X ray emission in 1.5 MeV $N^+ + NH_3$ collisions[54]. In section 3.2. it was shown how K-shell MO X rays were observed in collisions of $N^+ + N_2$ and not in $N^+ + NH_3$ at an impact energy of 200 keV and below. Fig. 13 shows the X-ray spectra for these two collision systems at an impact energy of 1.5 MeV. In contrast to fig. 3 we now observe the same broad continuum, the intensity in $N^+ + N_2$ being slightly higher than in $N^+ + NH_3$. The production mechanism for this continuum may be explained with the aid of fig. 14,[55]. In a violent collision between two N atoms a vacancy may be produced in the $1s\sigma$ MO. During the same collision this vacancy can decay thus giving rise to a broad continuum in the X-ray region between $N - K_\alpha$ and $Si - K_\alpha$. In the 1.5 MeV $N^+ + N_2$ collision the double scattering mechanism which was discussed earlier, probably adds to the single-collision contribution to give the small difference in the relative intensities of the X-ray continuum and the characteristic lines in fig. 13. Note that the $N - H$ collisions cannot account for a double collision effect in $N^+ + NH_3$ since the $N - K$ shell ionization cross section in 1.5 MeV $N^+ + H$ is almost two orders of magnitude smaller than the cross section in $N^+ + N$ collisions.

A theoretical description of these single collision MO-X rays, which is not yet available, will have to include radiative ionization of the 2p levels also. As is shown in fig. 14 the final state of single collision MO-X rays is identical to that for radiative ionization of the 2p level: a photon, an electron in the continuum and a vacancy in either $2p\pi$ or $2p\sigma$. In addition we want to note that it is perhaps not so surprising that the high energy tails of many MO-X ray spectra are readily parametrized using the Heisenberg uncertainty principle. If the collision is sufficiently diabatic to yield direct coupling of the $1s\sigma$ MO to the continuum then

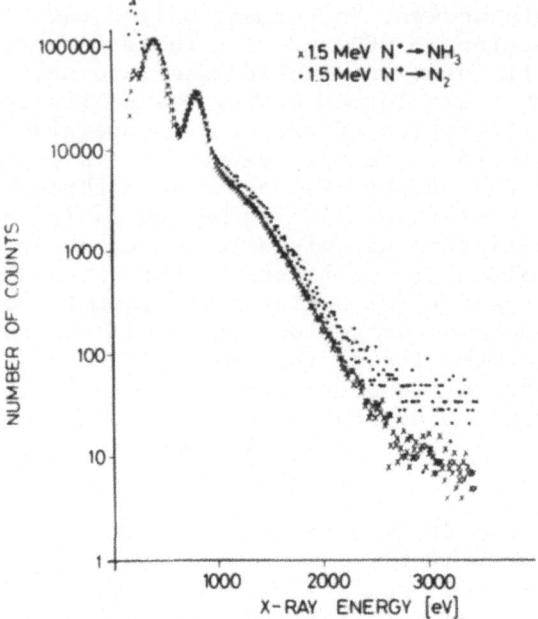

Fig. 13. The X-ray spectra for 1.5 MeV $N^+ \rightarrow N_2$ and NH_3
(ref. 54).

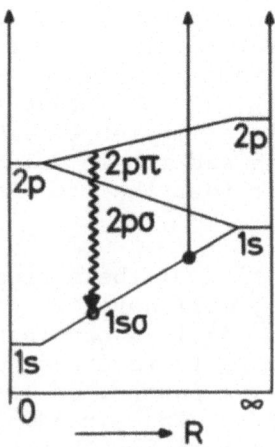

Fig. 14. Schematic diagram showing one-collision MO-X ray
production.

it should not surprise that little information is left of the adiabatic MO-energy levels and dynamic broadening effects dominate.
Since we find this to be the case for 1.5 MeV $N^+ + NH_3$ it should
hold also for any equivalent system with $v_p/v_K \geq 0.3$,[56]).

Few data are available for $v_p/v_K > 1$. Under these conditions
molecular orbitals are certainly ill-defined and X-ray continuum
spectra are explained in terms of REC and RI[57,58]). Kienle et al.
[57]) showed this convincingly for 140 MeV Ne + Ne and He, see fig.
15. The peak is due to radiative capture of target electrons directly into the K-shell of the fully stripped projectile. The high
energy tail however can only be accounted for if radiative
ionization of target K-shell electrons is included. A more detailed
theoretical description of this process was published later by
Kleber et al.[59]).

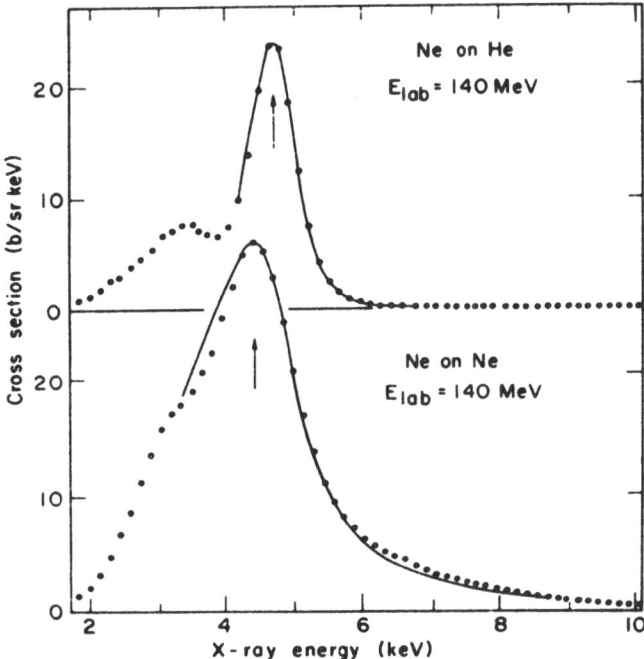

Fig. 15. Cross sections for X-ray production at $\Theta = 90^\circ$ in
collisions between 140 MeV Ne on He and Ne. The
experimental results are shown as points, and the
solid curve is the theoretical shape. Note the
greater importance of radiative ionization compared
with radiative electron capture for the Ne target
(ref. 57).

3.4. Doppler-shift and anisotropic continuum X-ray emission

Several authors have noticed that it is important to determine the Doppler shift of continuum X-rays from heavy-ion collisions. The Doppler velocity characterizes the radiating system independent of the production mechanism of the X-ray continuum, and
corrections must be made to experimentally observed anisotropies
before comparison to theoretically predicted center of mass anisotropy can be verified[60,61,62]. In many cases one has found that
the anisotropy of continuum X-ray emission can be made symmetric
about $\Theta_{lab} = 90^o$, when assuming a Doppler velocity equal to the
center of mass velocity. This clearly is sufficient to prove that
the radiation is emitted during the ion-atom collision and secondary electron bremsstrahlung can be excluded as radiation source.
It cannot be automatically concluded that the origin is of quasi-
molecular nature since NNB and RI will also be subject to the
center of mass Doppler shift.

It has been suggested[63,64] that the quasi-molecular system
under the influence of rotation of the internuclear axis would
radiate, leading to a pronounced asymmetry effect at an X-ray
energy near the united atom limit. At the recent Freiburg meeting[4]
the theoretical basis for these induced transitions has been
severely criticized. Experimentalists, on the other hand, find
distinct asymmetries that seem to rise towards the high energy
limit of the X-ray continua. It seems very likely that in ion-atom
collisions radiation from certain spectroscopic terms are preferred
over others. This will undoubtedly lead to a pronounced asymmetry
in MO-X ray emission not due to the rotation of the internuclear
axis but simply as a result of collision-induced atomic (molecular)
alignment[62,65,66].

Wölfli et al.[67,68] found intensity fluctuations in the anisotropy of continuum X rays emitted in Ca + Ca, Fe + Fe and Ni + Ni
collisions at 25 and 40 MeV, see fig. 16. They put forward an
interpretation as quasimolecular transitions in a rotating two-
center system[69], although it is not yet clear how the quantiza-
tion comes about, see also ref.[64]. Perhaps interference effects
such as considered by Trautvetter[12] or by Lichten[70] play a role
also.

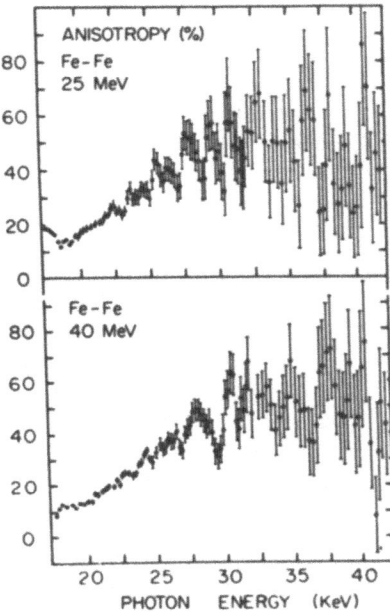

Fig. 16. Fine structure of the X-ray anisotropy $I(90^\circ)/$
$I(30^\circ) - 1$ observed in Fe - Fe collisions at two
different beam energies. In both cases periodic
intensity fluctuations appear if the evaluation
is performed with sufficiently good energy resolution (ref. 67).

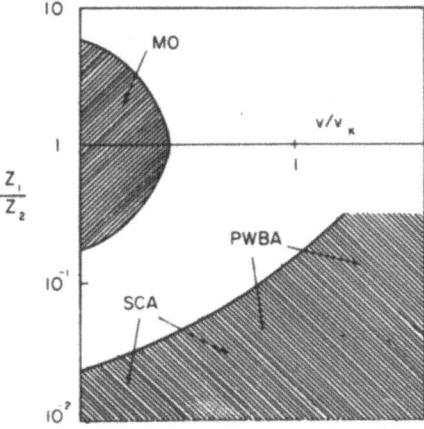

Fig. 17. Approximate regions of validity for various models
connected with inner-shell vacancy production in
ion-atom collisions (ref. 9).

4. CONCLUSIONS AND FUTURE EXTENSIONS

The picture that appears in summarizing the field of ion-induced X-ray continuum emission is very similar to that of ion-induced inner-shell processes in general (see fig. 17, ref. 9).

For Z_1/Z_2 small, inner-shell ionization is dominated by direct Coulomb ionization and various theoretical approximations describe the characteristic X-ray yield adequately. This is also true for continuum X-rays, where for solid targets SEB and NNB can be calculated successfully as well as RI for gaseous media.

In the other extreme, when Z_1/Z_2 is near unity and v/v_K small, then the molecular model applies and relatively intense continuum X-ray emission can be identified unambiguously as double collision MO-X rays. Good quantitative agreement between calculated spectral distributions and measurements is obtained for K-shell MO X rays. The multitude of possible transitions and the lack of knowledge of their fluorescence yields hinders quantitative description of L and M shell MO X rays. However, two kinds of continua c_1 and c_2 have been distinguished which may be considered as pictorial representations of the molecular states involved in heavy ion-atom collisions.

The region in-between of fig. 17 is still no-man's land. An unambiguous identification of the collision processes involved must wait for a theoretical description of direct coupling of quasi-molecular states to higher empty states in or near the continuum. One would hope that differential measurements such as pioneered by Schmidt-Böcking et al.[71,72] and Burch et al.[73] will guide these theoretical investigations.

The field of ion-induced continuum X-ray production has been in a most propitious situation with theoretical stimulation and guidance from the Frankfurt group. Not only did they produce the first ab initio calculations of relativistic wave functions in heavy and super-heavy molecules[74], the nuclear dipole and quadrupole bremsstrahlung[13] calculations and MO-X ray distributions[29] but in particular did they predict a most interesting new phenomenon[75,76], see fig. 18. In this diagram it is shown that the $1s\sigma$ MO in very heavy systems may dive into the negative energy continuum, for U + U at R = 35 fm. Should a vacancy exist in the $1s\sigma$ MO then the breakdown of the vacuum state under influence of the extremely high electric field is expected. Electrons from the negative continuum can fill the vacancy thereby creating positrons.

Estimates of the probability for vacancy formation in the $1s\sigma$ MO at zero impact parameter U + U collisions at 1500 MeV range from 10^{-6} to 10^{-1}. The Frankfurt group recently claims that at

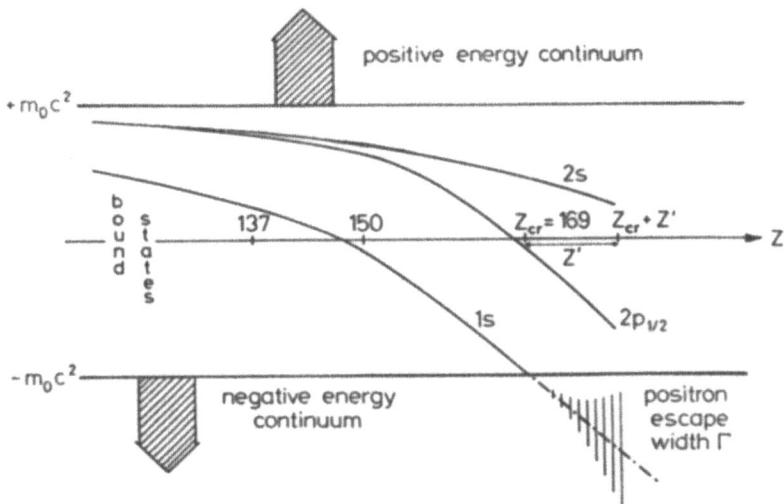

Fig. 18. The solutions of the single particle Dirac
equation. The positive and negative continuum
bound states (within the gap) are indicated.
Beyond Z critical the bound 1s-state dives into
the negative energy continuum. If there was a
K-hole present a positron emission could occur.

small internuclear distances relativistic effects dominate and may
lead to ionization probabilities as high as 20%,[64],[63]). This obviously gives great encouragement for the proposed experiments with
super-heavy molecules, see also ref. 77.

In addition to research possibilities in transient super-
heavy molecules one may speculate about another possible extension
of MO-X ray investigation which seem just as appealing. It has been
suggested that stimulated emission in the X-ray wavelength region
might be obtained via controlled charge exchange processes in intense beams or plasmas of highly stripped ions[78],[79]). In the present review we have demonstrated how excited ions can be made to
radiate at several times the transition energy of the isolated
particle, when the excited state decays in the transient molecule
during collision with another heavy particle. The control on the
X-ray transition energy is in the projectile target combination
and internuclear distance. We speculate that detailed investigations of these parameters may bring them under sufficient command
as to allow not only X-ray lasing in dense and hot plasmas of
heavy particles, but also tuning throughout a considerable wavelength region.

ACKNOWLEDGEMENTS

The authors should like to acknowledge the stimulating interest of J. Kistemaker and the skillful technical assistance of S.
Doorn. In addition F.W. Saris wishes to thank the Solid State
Science Branch, Chalk River Nuclear Labs., Canada, for their collaboration, financial support and warm hospitality, during various
stages of the work described in this paper.

This work is part of the research program of the Stichting
voor Fundamenteel Onderzoek der Materie (Foundation for Fundamental Research on Matter) and was made possible by financial support
from the Nederlandse Organisatie voor Zuiver-Wetenschappelijk Onderzoek (Netherlands Organization for the Advancement of Pure
Research).

REFERENCES

1) W.M. Coates, Phys.Rev. 46 (1934) 542.
2) H.-D. Betz, Invited talk at FICAP, July 26-30 (1976), Berkeley, U.S.A. To be published in FICAP proceedings, Eds. R. Marrus et al. (Plenum Press).
3) J.S. Briggs, K. Dettmann, Phys.Rev.Lett. 33 (1974) 1123; and private communication.
4) K. Dettmann, Proc. 2nd Int.Conf.on Inner-Shell Ionization Phenomena, Freiburg (1976), Invited Papers. Eds. W. Mehlhorn and R. Brenn, p. 57.
5) J.F. Chemin, I.V. Mitchell, F.W. Saris, J.Appl.Phys. 45 (1974) 532.
6) F. Folkmann, C. Gaarde, T. Huus, K. Kemp, Nucl.Instr.Meth. 116 (1974) 487.
7) F. Folkmann, J.Phys. E: Sci.Instr. 8 (1975) 429.
8) A.R. Sohval, J.P. Delvaille, K. Kalata, H.W. Schnopper, J.Phys.B: Atom.Mol.Phys. 8 (1975) L426.
9) D.H. Madison, E. Merzbacher, Atomic Inner-shell Processes, Ed. B. Crasemann (Academic Press, 1975, New York) p.2.
10) H.K. Tseng, R.H. Pratt, Phys.Rev. A3 (1971) 100.
11) K. Ishii, S. Morita, H. Tawara, Phys.Rev. A13 (1976) 131.
12) H.P. Trautvetter, J.S. Greenberg, P. Vincent, Phys.Rev.Lett. 37 (1976) 202.
13) J. Reinhardt, G. Soff, W. Greiner, Z.Phys. A276 (1976) 285.
14) Th.P. Hoogkamer, P. Woerlee, F.W. Saris, M. Gavrila, J.Phys.B: Atom.Molec.Phys. 9 (1976) L145.
15) J.S. Briggs, M.R. Hayns, J.Phys.B: Atom.Mol.Phys. 6 (1973) 514.
16) W.E. Meyerhof, Th.P. Hoogkamer, F.W. Saris, Proc.FICAP, Book of Abstracts, Eds. R. Marrus et al. (1976), p. 56; Th.P. Hoogkamer et al. (1976) to be published.
17) J.H. Macek, J.S. Briggs, J.Phys.B: Atom.Molec.Phys. 7 (1974) 1312.
18) R. Anholt, J.Phys.B: Atom.Molec.Phys. 9 (1976) L249.
19) J.R. MacDonald, M.D. Brown, T. Chiao, Phys.Rev.Lett. 30 (1973) 471.
20) R. Laubert, Proc.5th Int.Conf.on Atomic Collisions in Solids, U.S.A., 1973. Eds. S. Datz, B.R. Appleton and C.D. Moak (Plenum Press, 1975, New York) p. 389.
21) J.S. Greenberg, C.K. Davis, P. Vincent, Phys.Rev.Lett. 33 (1974) 473.
22) W.E. Meyerhof, T.K. Saylor, S.M. Lazarus, A. Little, B.B. Triplett, L.F. Chase, R. Anholt, Phys.Rev.Lett. 30 (1973) 1279; Phys.Rev.Lett. 32 (1974) 502.
23) C.K. Davis, J.S. Greenberg, Phys.Rev.Lett. 32 (1974) 1215.
24) K.H. Kaun, W. Frank, P. Manfrass, Proc.2nd Int.Conf.on Inner-Shell Ionization Phenomena, Freiburg (1976). Invited Papers. Eds. W. Mehlhorn and R. Brenn, p. 68.

25) P. Gippner, K.H. Kaun, H. Sodan, F. Stary, W. Schulze, Yu.P. Tretyakov, Phys.Rev.Lett. 52B (1974) 183.

26) W. Frank, P. Gippner, K.H. Kaun, P. Manfrass, Yu.P. Tretyakov, Z.Physik A277 (1976) 333.

27) W. Frank, P. Gippner, K.H. Kaun, H. Sodan, Yu.P. Tretyakov, Phys.Lett. 59B (1975) 41.

28) W.E. Meyerhof, to be published in Science (1976).

29) B. Müller, Invited Lectures, Review Papers and Progress Reports of the IX Int.Conf.on the Physics of Electronic and Atomic Collisions, Seattle, 1975, Eds. J.S. Risley and R. Geballe (University of Washington Press, 1976) p. 481.

30) H.K. Heinig, H.U. Jäger, H. Richter, H. Woittennek, Phys.Lett. 60B (1976) 249.

31) F.W. Saris, W.F. van der Weg, H. Tawara, R. Laubert, Phys.Rev. Lett. 28 (1972) 717.

32) F.W. Saris, F.J. de Heer, Atomic Physics, vol. 4, Eds. G. zu Putlitz, E.W. Weber and A. Winnacker (New York and London, Plenum Press, 1975) p. 287.

33) G. Bissinger, L.C. Feldman, Phys.Rev.Lett. 33 (1974) 1.

34) J.A. Cairns, A.D. Marwick, J. Macek, J.S. Briggs, Phys.Rev. Lett. 32 (1974) 509.

35) J.A. Cairns, L.C. Feldman, New Uses of Ion Accelerators, Ed. J.F. Ziegler, Plenum Press, New York and London, 1975) p.431.

36) F.W. Saris, Proc.5th Int.Conf.on Atomic Collisions in Solids, U.S.A., Eds. S. Datz, B.R. Appleton, C.D. Moak,(Plenum Press, New York, 1975) p. 343.

37) J. Eichler, U. Wille, B. Fastrup, K. Taulbjerg, to be published in Phys.Rev. (1976).

38) B.G. Schmid, J.D. Garcia, J.Phys.B: Atom.Molec.Phys. 9 (1976) L219.

39) M. Barat, W. Lichten, Phys.Rev. A6 (1972) 211.

40) F.W. Saris, C. Foster, A. Langenberg, J. van Eck, J.Phys.B: Atom.Molec.Phys. 7 (1974) 1494.

41) P. Armbruster, Proc.2nd Int.Conf.on Inner-Shell Ionization Phenomena, Freiburg, Invited Papers, Eds. W. Mehlhorn and R. Brenn (1976) p. 21.

42) P.H. Mokler, S. Hagmann, P. Armbruster, G. Kraft, H.J. Stein, K. Rashid, B. Fricke, Atomic Physics, Vol. 4, Eds. G. zu Putlitz, E.W. Weber and A. Winnacker (New York and London, Plenum Press, 1975) p. 301.

43) F.C. Jundt, H. Kubo, H.E. Gove, Phys.Rev. A10 ,(1974) 1053.

44) H.O. Lutz, W.R. McMurray, R. Pretorius, I.J. van Heerden, R.J. van Reenen, B. Fricke, J.Phys.B: Atom.Molec.Phys. 9 (1976) L157.

45) Ch. Heitz, P. Armbruster, W. Enders, F. Folkmann, S. Hagmann, G. Kraft, P.H. Mokler, Proc.2nd Int.Conf.on Inner Shell Ionization Phenomena, Freiburg (1976), Book of Abstracts, p. 15.

46) A. Rosén, D. Ellis, B. Fricke, T. Morovic, Proc.2nd Int.Conf.on Inner Shell Ionization Phenomena, Freiburg (1976), Book of Abstracts, p. 18, and to be published.

[47] H.-D. Betz, F. Bell, H. Panke, W. Stehling, E. Spindler, M. Kleber, Phys.Rev.Lett. 34 (1975) 1256.

[48] F. Bell, H.-D. Betz, H. Panke, E. Spindler, W. Stehling, M. Kleber, Phys.Rev.Lett. 35 (1975) 841.

[49] R. Laubert, R.S. Peterson, J.P. Forester, K.-H. Liao, P.M. Griffin, H. Hayden, S.B. Elston, D.J. Pegg, R.S. Thoe, I.A. Sellin, Phys.Rev.Lett. 36 (1976) 1574.

[50] W.R. Stott, J.C. Waddington, Phys.Lett. 56A (1976) 258.

[51] R. Anholt, T.K. Saylor, Phys.Lett. 56A (1976) 455.

[52] C. Foster, Th.P. Hoogkamer, P. Woerlee, F.W. Saris, J.Phys.B: Atom.Molec.Phys. 9 (1976) 1943.

[53] J.S. Briggs, J.Phys.B: Atom.Molec.Phys. 8 (1975) L485.

[54] F.W. Saris, W. Lennard, I.V. Mitchell, F. Brown, T.P. Hoogkamer, to be published (1976).

[55] W.R. Thorson, J.H. Choi, Abstracts IX Int.Conf.on the Physics of Electronic and Atomic Collisions, Seattle (University of Washington Press, 1975) p. 298.

[56] H.-D. Betz, F. Bell, E. Spindler, M. Kleber, Proc.2nd Int.Conf. on Inner Shell Ionization Phenomena, Freiburg (1976) Book of Abstracts, p. 11.

[57] P. Kienle, M. Kleber, B. Povh, R.M. Diamond, F.S. Stephens, E.H. Grosse, M.R. Maier, D. Proetel, Phys.Rev.Lett. 31 (1973) 1099.

[58] H.W. Schnopper, John P. Delvaille, Proc.5th Int.Conf.on Atomic Collisions in Solids, U.S.A. (1973), Eds. S. Datz, B.R. Appleton, C.D. Moak (Plenum Press, New York, 1975) p. 481.

[59] D.H. Jakubassa, M. Kleber, Z.Physik A273 (1975) 29.

[60] W.E. Meyerhof, T.K. Saylor, R. Anholt, Phys.Rev. A12 (1975) 2641.

[61] F. Folkmann, P. Armbruster, S. Hagmann, G. Kraft, P.H. Mokler, H.J. Stein, Z.Physik A276 (1976) 15.

[62] R.S. Thoe, I.A. Sellin, P.M. Griffin, K.-H. Liao, D.J. Pegg, R.S. Peterson, to be published (1976).

[63] B. Müller, W. Greiner, Phys.Rev.Lett. 33 (1974) 469.

[64] W. Betz, W. Greiner, M. Gros, B. Müller, J. Reinhardt, Proc.2nd Int.Conf.on Inner Shell Ionization Phenomena, Freiburg (1976), Invited Papers, Eds. W. Mehlhorn and R. Brenn, p. 79.

[65] R.S. Thoe, I.A. Sellin, M.D. Brown, J.P. Forester, P.M. Griffin, D.J. Pegg, R.S. Peterson, Phys.Rev.Lett. 34 (1975) 64.

[66] R.K. Smith, W. Greiner, to be published (1976).

[67] W.Wölfli, Ch. Stoller, G. Bonani, M. Stöckli, M. Suter, W. Däppen, Phys.Rev.Lett. 36 (1976) 309.

[68] Ch. Stoller, W. Wölfli, G. Bonani, M. Stöckli, M. Suter, Proc. 2nd Int.Conf.on Inner Shell Ionization Phenomena, Freiburg (1976), Book of Abstracts, p. 1.

[69] W. Däppen, Proc.2nd Int.Conf.on Inner Shell Ionization Phenomena, Freiburg (1976), Book of Abstracts, p. 4.

[70] W. Lichten, Phys.Rev. A9 (1974) 1458.

[71] H. Smidt-Böcking, I. Tserruya, R. Schulé, R. Schuch, K. Bethge, Nucl.Instr.Meth. 132 (1976) 489.

[72] I. Tserruya, H. Schmidt-Böcking, R. Schulé, K. Bethge, R. Schuch, H.J. Specht, Phys.Rev.Lett. 36 (1976) 1451.

[73] D. Burch, W.B. Ingalls, H. Wieman, R. van den Bosch, Abstracts IX Int.Conf.on the Physics of Electronic and Atomic Collisions, Seattle (University of Washington Press, 1975) p. 306.

[74] B. Müller, Thesis, University of Frankfurt am Main (1973).

[75] B. Müller, H. Peitz, J. Rafelski, W. Greiner, Phys.Rev.Lett. 28 (1972) 1235.

[76] R.K. Smith, H. Peitz, B. Müller, W. Greiner, Phys.Rev.Lett. 32 (1974) 554.

[77] J. Rafelski, B. Müller, Phys.Rev.Lett. 36 (1976) 517.

[78] M.O. Scully, W.H. Louisell, W.B. McKnight, Optics Comm. 9 (1973) 246.

[79] J.F. Seeley, M.O. Scully, W.H. Louisell, IX Int.Conf.on Quantum Electronics, Amsterdam, June 14-18 (1976), Abstracts, p. 99.

THE IONIZATION OF INNER SHELLS OF ATOMS TAKING

ACCOUNT OF OUTER SHELL REARRANGEMENT

M. Ya. Amusia
Ioffe Physico-Technical Institute
Academy of Sciences of the USSR
Leningrad

1.   Introductory Remarks

The application of the general many-body theory and methods formulated with its help, in particular, the so-called random phase approximation with exchange (RPAE) /1-4/ and the many-body perturbation theory (MBPT) /5/, allows us to achieve a rather satisfactory description of ionization processes for many outer and intermediate shells in a number of atoms. It appears that the effect of inter-action among all electrons of a given subshell is rather significant and ionization has to be considered as a many-electron process, because all electrons of a subshell take part in it. The study of the collective nature of the ionization process in the outer and intermediate shells is very important for the theory of many-elec-tron atoms. A significant result obtained in the investigation of the influence of interelectron interaction on the ionization process is the observation of its important role /3,6,7/. Especially strong interelectron correlations manifest themselves in the action of a many-electron shell upon the few-electron shell. The study of photo-ionization of outer ns^2 noble gas atom subshells demonstrates that the corresponding cross section is strongly affected by the outer np^6 subshell, which acts as a powerful screen, completely modifying the form of the cross section curve /3/. For $5s^2$ subshells in Xe, Cs, and Ba it appears that not only the effect of the outer np^6 but also that of the inner $4d^{10}$ subshell is prominent. The calculations were performed by the RPAE method. They proved that the joint action of $5p^6$ and $4d^{10}$ completely alters the form of the $5s^2$ subshell cross section /3/. This permits us to speak of the complete collectiviza-tion of few-electron shells. The measurements /8,9/ confirmed the rather high accuracy of the RPAE predictions.

But there are some experimental data that cannot be understood within the framework of this approximation /10/. Because the deviations exceed the experimental errors, it is expedient to conclude that it is necessary to go outside the frame of RPAE. A number of papers have been devoted to this question, i.e., to the generalization of RPAE in the theory of a high-density electron gas /11/. However, even for such an idealized model no consistent generalization of RPAE exists that would take into account next-order corrections in the characteristic small parameter of the high-density electron gas, $\alpha = 1/P_0$, where P_0 is the Fermi momentum, connected with the electron density by the relation $\rho = P_0^3/3\pi^2$.* However, even if the corresponding approximation in the theory of homogenous gases were to exist, its applicability to the electron shells of atoms is not self-evident. This is due to the strong inhomogeneity of electron density in atoms and the good energy separation of electron shells. Physically it seems rather probable that if in the inner or intermediate shell ionization process a slow electron is removed, the outer shells are able to rearrange, and the outgoing electron feels this rearrangement. The latter, being an outlet beyond the RPAE framework, is manifested in variations of both the remaining ion energy and the field in which the outgoing electron moves. We shall show later that the effect of the rearrangement** essentially changes the ionization probability near the threshold, as a result of which noticeably better agreement with experiment is achieved.

2.   The RPAE Method

In order to discuss the possibilities of generalizing RPAE, let us elucidate its content. This method is set forth at length in /3/. The physical content of RPAE is in that along with the direct reaction of atomic electrons to an external perturbation (single-electron approximation), it takes into account the variation of the self-consistent field which results under the action of this external perturbation. The variation of the self-consistent field in its turn affects all atomic electrons, including the one under consideration.

The RPAE equations may be derived by generalization of the Hartree-Fock (HF) approximation to systems with variable external fields /3/. Another method is based on the summation of Feynmann diagrams. Instead of the Coulomb interaction, the effective interaction Γ is introduced, which is presented diagrammatically in Fig. 1, where a line with an arrow to the right denotes an electron in

---

*We use the atomic system of units, $e = \hbar = m = 1$, in this report; the energy is given in Rydbergs.

**The effect of this rearrangement was discussed also in /4b/.

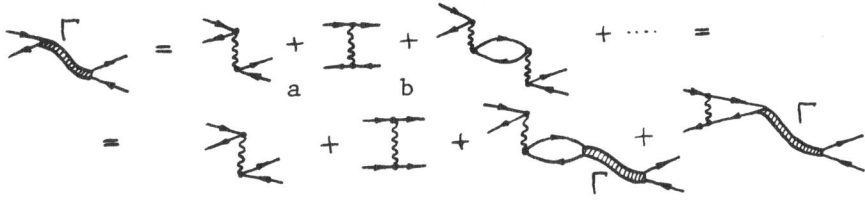

Fig. 1

an excited state in the HF approximation, a line with an arrow to the left a vacancy or a hole in the same approximation, and a vertical wavy line a Coulomb interaction matrix element. The quantity Γ satisfies the equation

$$\langle n_1 n_3 | \Gamma(\omega) | n_2 n_4 \rangle = \langle n_1 n_3 | U | n_2 n_4 \rangle -$$

$$- \left(\sum_{\substack{n_5 \leqslant F \\ n_6 > F}} - \sum_{\substack{n_5 > F \\ n_6 \leqslant F}} \right) \frac{\langle n_1 n_6 | U | n_2 n_6 \rangle \langle n_5 n_3 | \Gamma(\omega) | n_6 n_4 \rangle}{\omega - E_{n_5} + E_{n_6} + i\delta(1 - 2n_{n_5})} \qquad (1)$$

where the condition $n_5 \leqslant f$ indicates summation over occupied and $n_6 > f$ summation over vacant states; n_{n_5} is the Fermi step

$$n_{n_5} = \begin{cases} 1 & n_5 \leqslant F \\ 0 & n_5 > F \end{cases} \qquad (2)$$

The matrix elements $\langle n_1 n_3 | u | n_2 n_4 \rangle$ are expressed in terms of the Coulomb elements $\langle n_1 n_3 | v | n_2 n_4 \rangle$:

$$\langle n_1 n_3 | U | n_2 n_4 \rangle = \langle n_1 n_3 | V | n_2 n_4 \rangle - \langle n_1 n_3 | V | n_2 n_4 \rangle \qquad (3)$$

The amplitude of the atomic electron interaction with a weak external field A (either electromagnetic or created by the incoming electron) is determined by the expression

$$\langle n_1 | A | n_2 \rangle + \left(\sum_{\substack{n_3 \leqslant F \\ n_4 > F}} - \sum_{\substack{n_3 > F \\ n_4 \leqslant F}} \right) \frac{\langle n_4 | A | n_3 \rangle \langle n_4 n_2 | \Gamma(\omega) | n_3 n_1 \rangle}{\omega - E_{n_4} + E_{n_3} + i\delta(1 - 2n_{n_4})} \qquad (4)$$

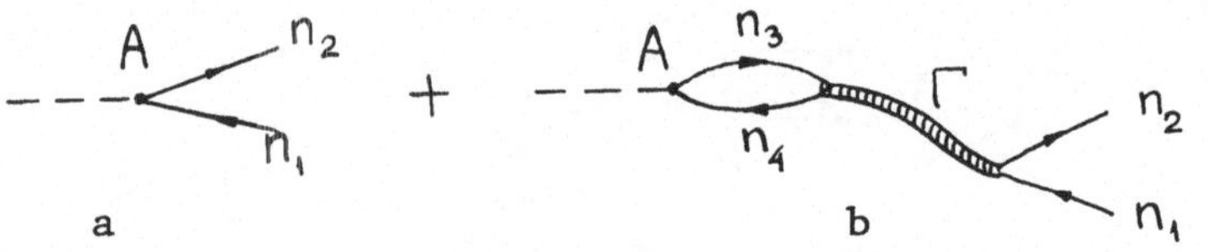

Fig. 2

where $\omega = E_{n_2} - E_{n_1}$. The amplitude of interaction with the external field is presented[1] diagrammatically in Fig. 2, where the dotted line denotes the external field.

The first term in (4), depicted in Fig. 2a, describes the direct interaction of an electron with the external field, whereas the second, depicted in Fig. 2b, describes the effect of the excitation (virtual or real) of other electrons of the given atom.

The photoionization cross section in the single-electron approximation is determined by the formulas

$$\sigma_\gamma^r = \sum_{\substack{n_2 > F \\ n_1 \leqslant F}} \frac{4\pi^2\omega}{c} |\langle n_2 | z | n_1 \rangle|^2 , \qquad (5)$$

$$\sigma_\gamma^\nabla = \sum_{n_2 > F, n_1 \leqslant F} \frac{4\pi^2}{\omega c} |\langle n_2 | P_z | n_1 \rangle|^2 . \qquad (6)$$

Here z is the dipole moment operator, P_z the z-component of the momentum operator, and $\omega = E_{n_2} - E_{n_1}$. Taking account of correlations within the RPAE framework requires that the sum (4) be substituted for $\langle n_2 | z | n_1 \rangle$ in (5) or $\langle n_2 | P_z | n_1 \rangle$ in (6), with $\langle n_2 | z | n_1 \rangle$ or $\langle n_2 | P_z | n_1 \rangle$ respectively used as the matrix element of the external field A. The fast-electron inelastic scattering cross section is expressed via generalized oscillator strength densities (GOSD) $\partial f(q)/\partial\omega$, which are connected with the e^{iqr} operator matrix element by the relation

$$\frac{\partial f(q,\omega)}{\partial\omega} = \frac{\omega}{q^2} \sum_{\substack{n_2 > F \\ n_1 \leqslant F}} |\langle n_2 | e^{i\vec{q}\vec{r}} | n_1 \rangle|^2 , \qquad (7)$$

where q is the momentum transferred to the atom by the fast electron, and $\omega = E_{n_2} - E_{n_1}$.

One of the attributes of RPAE is the satisfaction of the same sum rules which are valid for exact wave functions. Thus, for example, the following relation holds:

$$S \equiv \sum_{\substack{n_2 > F \\ n_1 \leqslant F}} f_{n_2 n_1} + \sum_{n_1 \leqslant F} \frac{1}{2\pi^2 \alpha} \int_{I_{n_1}}^{\infty} \sigma_\gamma(\omega) \, d\omega = N. \tag{8}$$

Here f is the dipole photoexcitation oscillator strength, α the fine structure constant, I_{n_1} the ionization potential of an electron of the n_1 shell, and N the number of atomic electrons.

Analogously, for the density of oscillator strength the following sum rule is fulfilled:

$$S_q \equiv \sum_{n_1 \leqslant F} \int_{I_{n_1}}^{\infty} \partial f(q, \omega) \big/ \partial \omega \, d\omega = N. \tag{9}$$

Within the RPAE framework the results of photoionization cross section and oscillator strength calculations in the length and velocity forms (with r and ∇ operators) coincide.

It is essential for what follows to note that a broad class of RPAE diagrams may be accounted for by a proper choice of the self-consistent field in which the electron leaving the atom moves. We have in mind all time-forward diagrams which belong to the same hole state (see Fig. 3).

The effect of the sequence of diagrams in Fig. 3 is equivalent to considering the outgoing electron as moving in the field of an ion with a hole i instead of that of a neutral atom. For many-electron shells the main contribution comes from correlations of the given shell electrons. In RPAE the effect of such correlations is accounted for by using (1) and (4) with limitation in the sums over $n \leqslant f$ to one term, which belongs to the ionized shell. Therefore, if correlations within one subshell are considered, their main part may be accounted for by the proper choice of the field in which the outgoing electron moves. In a number of cases taking account of intrashell correlations permits us to achieve rather satisfactory agreement with experiment, e.g., for outer Ne, Ar, Kr, and Xe shells, for the $4d^{10}$ Xe shell, and for some others.

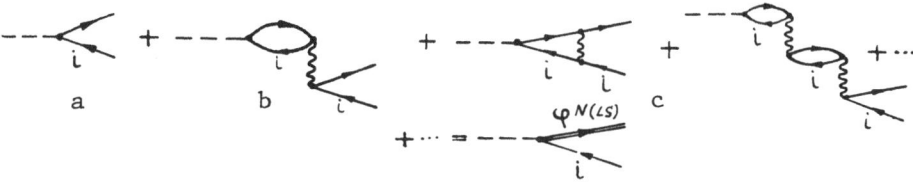

Fig. 3

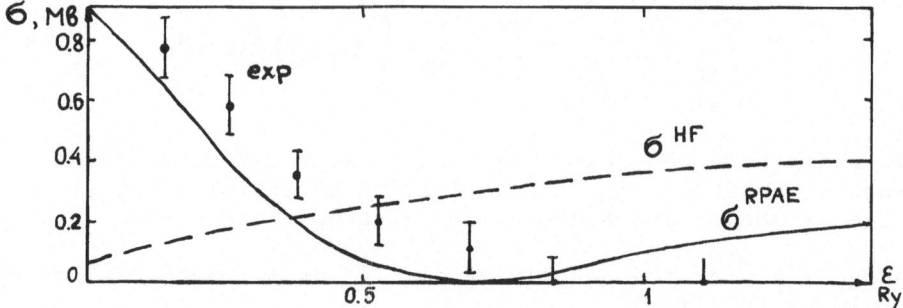

Fig. 4.   Photoionization cross section of the $3s^2$ subshell in Ar
          taking account of intershell correlations. Experimental
          data from /8/.

     In intershell correlations the main role is played by time-
forward diagrams, which corresponds to transitions from different
hole states.   In RPAE the role of electron correlations within two
shells is accounted for by limiting the sums over n ⩽ f in (1) and
(4) to two terms, which corresponds to two interacting subshells.

     Calculations performed within the RPAE framework demonstrate
that the outer ns^2 subshells of noble gas atoms are effectively
screened by the np^6 subshell /3/.   As a result the ns^2 subshell
photoionization cross section is modified qualitatively, which is
illustrated in Fig. 4.   It appears that there are cases in which in
spite of good energy separation of shells the interaction even among
three shells is essential /3/.   Thus, the $5s^2$ subshell in Xe, Cs,
and Ba is under strong action of $5p^6$ and $4d^{10}$, two many-electron
shells.   The results of calculations taking the influence of $5p^6$
and $4d^{10}$ on $5s^2$ into account are given in Fig. 5.   As is seen from
this figure, in which the single electron cross section for Xe is

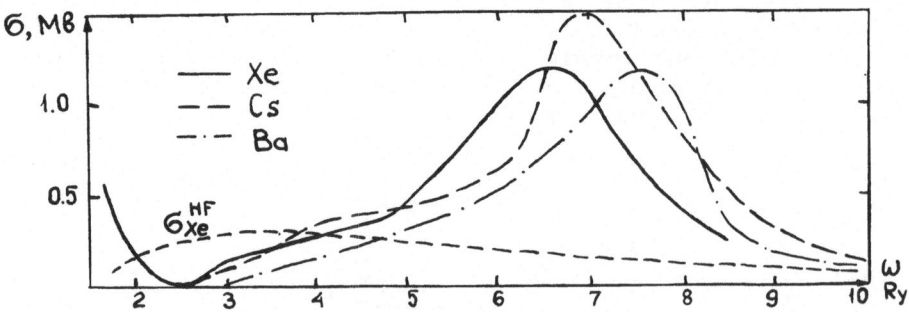

Fig. 5.   Photoionization cross section of the $5s^2$ subshell in Xe,
          Cs, and Ba, taking account of intershell correlations.

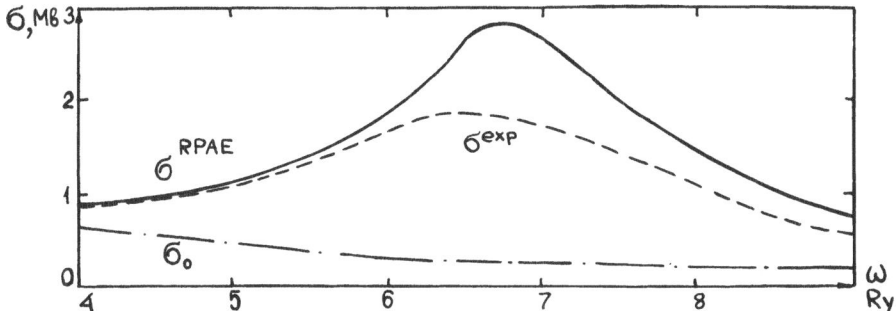

Fig. 6. Summary photoionization cross section of the $5p^6$ and $5s^2$
         shells in Xe in the vicinity of the $4d^{10}$ ionization
         threshold. Experimental data from /12/.

also presented, inclusion of intershell correlations modifies the
$5s^2$ cross section so strongly that it is possible to speak of complete collectivization of the $5s^2$ subshell.

An impressive collective effect is the discovery of a maximum
in the photoionization cross section of the $5p^6$ and $5s^2$ subshells
in the vicinity of the $4d^{10}$ threshold. The summary photoionization
cross section of these shells, one of which is a many-electron shell,
is given in Fig. 6. The examples given above relate to the photoionization cross section. But in inelastic electron scattering, in
GOSD calculations the use of RPAE also permits us to achieve rather
good results. Thus, taking account of only intrashell correlations
leads to satisfactory agreement with experiment in description of
inelastic scattering of electrons on Ar and Xe, as illustrated by
Fig. 7. Up to rather large values of q, the effect of the $4d^{10}$
shell on $5s^2$ in Xe is noticeable (see Fig. 8).

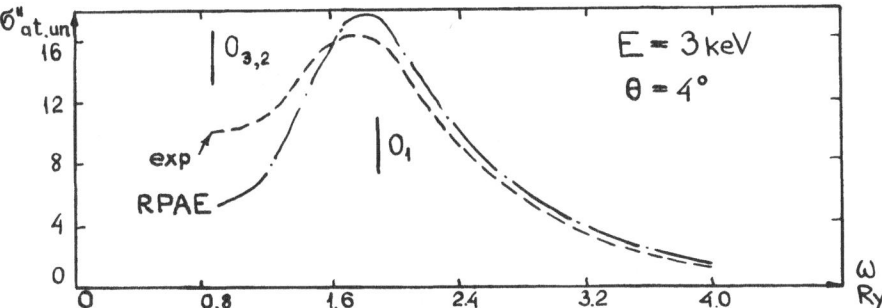

Fig. 7. Cross section of fast-electron inelastic scattering on Xe.
         Experimental data from /13/.

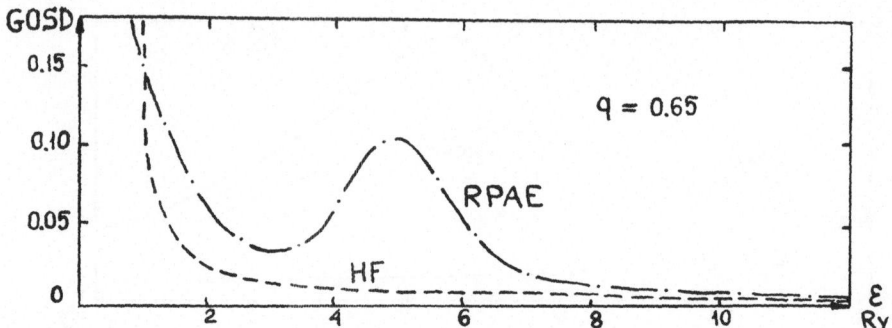

Fig. 8. Generalized oscillator strength (GOS) of the $5s^2$ subshell
         in Xe taking account of $4d^{10}$.

The differential inelastic electron scattering cross section
which describes $5s^2$ and $5p^6$ subshell ionization has a broad maximum
in the vicinity of $4d^{10}$, which is quite noticeable for scattering
angles up to $3-5°$.

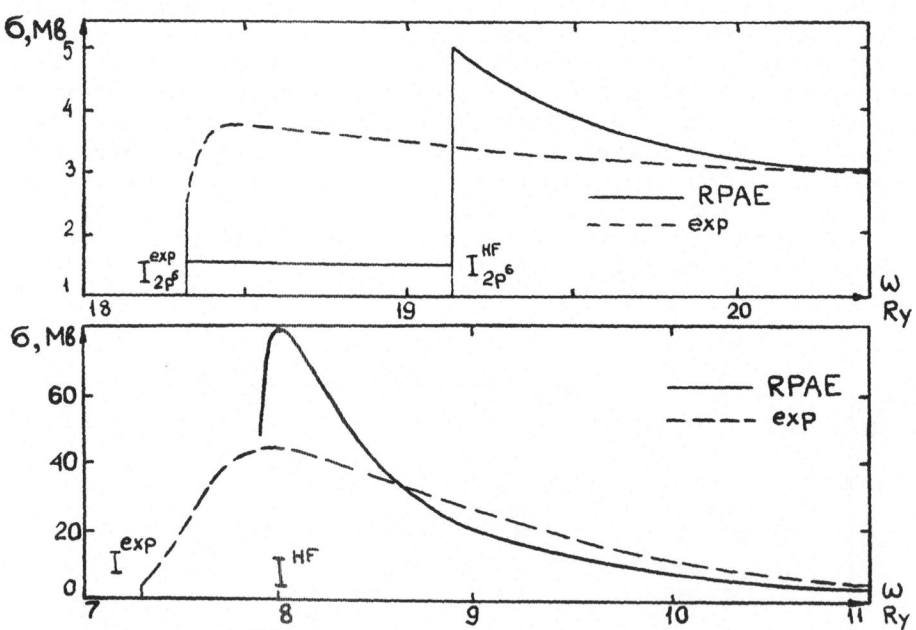

Fig. 9. Photoionization cross section of the $2p^6$ shell in Ar (top),
         and the $4d^{10}$ shell in Ba. Experimental data for Ar /14/,
         for Ba /15/.

There are a number of other results and predictions of the
RPAE method that agree with experiment. On the other hand, there
also exist definite deviations that are beyond the limits of experimental error. These are, first, a significant difference between
the ionization potential and its HF value, and, second, a noticeable difference between the RPAE and experimental cross sections
just above threshold. As an example we give the photoionization
cross section for the $2p^6$ shell of Ar and the $4d^{10}$ shell of Ba
(Fig. 9).

In the RPAE method the HF hole energy should be used as the
ionization potential. In Fig. 9 the cross sections -- theoretical
RPAE and experimental -- are given with their corresponding thresholds. It is evident that the RPAE cross sections in the vicinity
of the thresholds are much larger than the experimental ones. For
this reason a simple substitution of the experimental threshold for
the HF threshold in (1) and (4) does not lead to agreement with
observed data.

3.   Limitations of the RPAE Method

Both of these effects -- the difference between HF threshold and
experimental ionization potential and the too large cross section
near threshold -- may be removed by generalization of RPAE. At
first, let us consider the shift of threshold as compared to its HF
value. The shift of hole energy may be achieved by summation of
diagrams, as shown in Fig. 10. All these diagrams describe the
process of **hole** propagation, in which for part of the time the hole
exists as an exciatation of electron shells -- two holes and one
electron (Fig. 10a) or even more complicated configurations
(Fig. 10b,c).

Making use of the notations of Fig. 10A and summing the sequence
of Fig. 10B, we find for the hole Green function the expression
$G_i(E) = [E + (E_i + \Sigma_{ii})]^{-1}$. The combination $E_i + \Sigma_{ii}$ must be equal

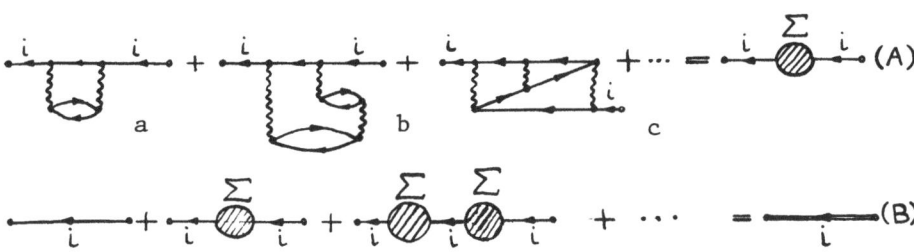

Fig. 10

to the experimentally determined ionization energy, if Σ_{ii} takes into account all possible processes, not only those presented in Fig. 10a, but processes of any complexity. But as is shown by concrete calculations, rather satisfactory agreement with experiment may be reached if one limits oneself to the contribution to Σ of the term given in Fig. 10a.

Strictly speaking, the ionization potential of an atom is defined as the difference between the total energies of the atom and the corresponding ion. This quantity does not coincide with the HF hole energy since the ion is rearranged in the process of removal of the electron, in consequence of the state variation of all other atomic electrons, and therefore the ion energy is altered.

It is possible to say in this sense that the diagonal matrix element of operator Σ between hole states i, which in many-body theory is called the self-energy part of the single-particle Green function, is equal to the energy of rearrangement resulting from the removal of the electron from the state i.

The use of experimental values of ionization threshold in the single-electron approximation shifts the cross section curve as a whole. Since for intermediate and inner shells $\Sigma_{ii}>0$, the experimentally observed values I_{exp} are less than I^{HF}, and σ_γ^r -- according to (5) -- is decreased near threshold, whereas σ_γ^∇ increases. Usually $\sigma_\gamma^r(HF)>\sigma_\gamma^\nabla(HF)$ and therefore the use of experimental thresholds in calculations makes σ_γ^r and σ_γ^∇ closer. The substitution of experimental (or corrected theoretical) thresholds in RPAE* equations, in which the main part of "time-forward" diagrams is already taken into account by the choice of self-consistent field, leads to increasing of the role of correlations near threshold and, as a consequence, to decrease of the σ_γ^r cross section in this domain. As a result $\sigma_{RPAE}^{r(exp)}$ (with experimental threshold) may be smaller than $\sigma_{RPAE}^{\nabla(exp)}$. For outer shells of some atoms (for example, alkaline earths) this is not the case: $\Sigma_{ii}<0$ and the difference between σ_γ^r and σ_γ^∇ increases. The use of experimental thresholds (or thresholds theoretically corrected as compared with the HF approximation) in accounting for intershell correlations leads in general to an increase of the correlation's role, since as a rule the shells are experimentally closer to each other than follows from calculations in the HF approximation.

The shift of the threshold as compared to the HF value is not unique, nor is it the most essential manifestation of the atom's rearrangement in the process of ionization. If the removed electron energy is small, the ion has enough time to rearrange. As a result,

---

*This question was considered rather at length in /4/.

not only the direct interaction of the outgoing electron with the
hole (Fig. 3c) appears to be essential, but also a more complicated
process in which the hole and the electron interact not directly
but through the virtual excitation of atomic shells which leads to
the screening of the hole. In the diagrammatical language this means
that processes analogous to those in Fig. 11 should be taken into account. Let us try to understand qualitatively to what effects taking
account of these diagrams leads. As a result of accounting for the
contribution of the diagrams in Fig. 3c, the outgoing electron
"feels" the field of hole i. Consequently, the photoionization
cross section, which is equal to zero if the outgoing electron is
considered as moving in the neutral atomic field, becomes finite if
the field of hole i is taken into account. Because the sign of the
contribution of the diagram in Fig. 11a is opposite to that of Fig.
3c, the hole field is weakened and the cross section at threshold
decreases. For considerable electron energies the leading role is
played by the electron-hole interaction not at large distances
(where the main contribution comes from the diagrams of the type of
Fig. 3c) but at intermediate and small distances. Besides, the
role of screening decreases. Therefore, in taking account of diagrams like those of Fig. 11, it is expected that the cross section
will decrease at threshold and not change far from threshold. As
a result the maximum at threshold, if it exists in single-particle
or RPAE calculations, has to be shifted to the higher-energy side.
It is possible to assert also that the maxima at the inner shell
thresholds must not only be shifted, but become noticeably broader.
In order to clarify the reason for this, consider the photoionization
near the inner-shell threshold purely qualitatively. Let us limit
ourselves to the matrix element diagonal in the hole state (see Fig.
3). Let us trace what singularities appear in summing up these diagrams for energies of quanta in the vicinity of the inner-shell ionization potential. To the first order of interaction U, we have

$$\sum_{n'>F} \langle n'|d|i \rangle \, \frac{1}{E'-E_i-\omega} \, \langle in'|U|iE \rangle, \qquad (10)$$

$$\omega = E - E_i$$

Fig. 11

Replacing summation in (10) by integration over the continuum spectrum and neglecting the dependence of the matrix element of U on E
near the ionization threshold, we obtain from (10)

$$\langle 0|d|i\rangle\langle i0|U|i0\rangle\ln E \tag{11}$$

For $E \to 0$ higher-order terms of perturbation theory in U also diverge,
so that to each power of the matrix element of interaction corresponds a power of the logarithm of the ratio of E divided by I_i, the
potential of shell i. Collecting only the divergent terms in all
orders, we obtain

$$\langle E|D|i\rangle = \frac{\langle E|d|i\rangle}{1 + \langle i0|U|i0\rangle\ln\frac{I}{E}} \tag{12}$$

In the vicinity of the threshold, if in matrix element (3) the first
term dominates, then $\langle i0|U|i0\rangle > 0$ and (12) has no singularities. If
the second term dominates, which corresponds to the diagrams of
Figs. 1b and 3c, a pole appears in (12) near the threshold and the
photoionization cross section abruptly increases. For the outer
shell the first term in (3) dominates, $\langle i0|U|i0\rangle > 0$, and $\langle E|D|i\rangle$
has no singularities near threshold. For the inner shells the
predominance of the second term is obvious, $\langle i0|U|i0\rangle < 0$, and the
ionization cross section should increase compared to the single-
particle value. Because the sequence of diagrams in Fig. 3 is
taken into account by the choice of self-consistent field, calculations with such single-electron wave functions should for deep
shells lead to an anomaly -- a considerable increase of the transition probability near threshold, which was indeed observed near
the L-shell thresholds of Ar (Fig. 9) and Kr. Taking account of
the diagrams of Fig. 11 is equivalent to reducing the second term
in U [see (3)], in consequence of which the pole in (12) may not
appear even for inner shells and the sharp maximum does not arise.

Apart from ion rearrangement processes, taking account of outgoing electron wave function variations due to ion polarization

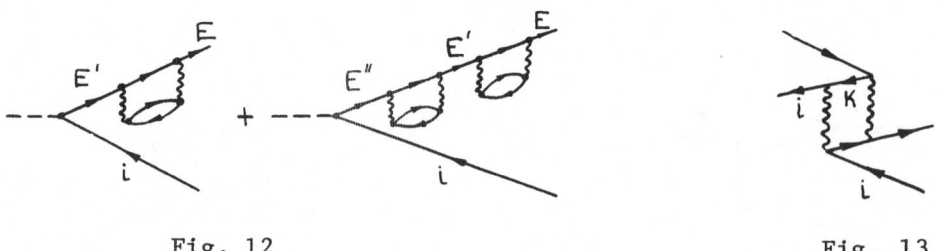

Fig. 12   Fig. 13

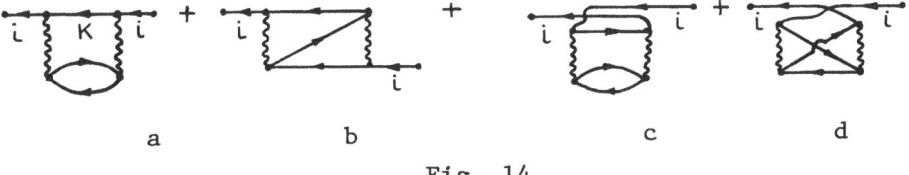

Fig. 14

provoked by electron motion is also outside the RPAE framework. As a result of polarization the self-consistent field in which the outgoing electron moves is altered. In the diagrammatic language this corresponds to taking account of processes similar to that given in Fig. 12. The particle-hole interaction is also altered -- apart from the matrix element of Fig. 1a, more complicated processes of particle-hole interaction must be taken into consideration, such as those given in Fig. 13.

To take all diagrams outside the scope of RPAE properly into account is impossible. Therefore one of the most important problems in the development of the theory of many-electron atoms is to find the next best approximation after RPAE, which takes account of real physical processes beyond the scope of RPAE and at the same time is comparatively simple to use.

4. Corrections to the Ionization Potential

The simplest method of determining the ionization potential is to calculate the diagrams of second order in the interelectron interaction, as given in Fig. 14 /3/. The calculation of their contributions gives values which are in reasonable agreement with experimental data, as illustrated by Table 1, where the quantity Σ_{ii} is denoted as ΔI_i^{RPAE}. The corrections to the ionization potential which are obtained by subtraction of the total HF energy

TABLE 1

Atom	Level	$\Delta I_i^{exp} = I_i^{exp} + E_i$	$\Delta I_i^{HF} = \varepsilon_i - \varepsilon_0 + E_i$	ΔI_i^{RPAE}
Ar	$2p^6$	0.76	0.85	0.80
	$3s^2$	0.40	0.13	0.45
	$3p^6$	0.01	0.10	0.026
Xe	$4d^{10}$	0.54	–	0.60
	$5s^2$	0.16	–	0.21
	$5p^6$	-0.04	–	-0.06

of the ion, ε_i, from that of the neutral atom, ε_o, are also given
in the table. For the inner shells these figures are rather close
to the experimental data, whereas for the outer shells they differ
strongly. This is not accidental. The calculations demonstrate
that for the inner shells the main contribution to ΔI comes from
the so-called monopole polarizability of the outer shells /3/; in
diagram language, we have Fig. 14a, in which the intermediate state
state k coincides with i. As for the outer shells, all diagrams
are important, and the contributions of Fig. 14a and c are often
close in magnitude but of opposite sign.

Let us consider the shift of the threshold against the HF
value for the inner shell. In determining the energy of hole i
it is necessary to take into account that the action of the ith
electron on itself should be excluded. In the lowest order of the
Coulomb interelectron interaction this is achieved by taking along
with the Hartree also the Fock exchange term into account. However,
in the higher orders the absence of an electron in a definite state
i, because of the self-consistency procedure, influences the wave
functions of all outer electrons. This, in turn, reflects on the
wave function and energy of hole i itself. The effect of hole i
on other wave functions of the atom and consequently on the hole
state itself is described by the diagram in Fig. 14a. Let us
elucidate this statement because it will be very important for us
in further considerations. The Hartree diagrams in the first two
orders of perturbation theory, which contribute to the self-energy
of hole i, are depicted in Fig. 15a and b. It is important to
note that in the loops the summation is performed over all hole
states, including the ith. Since there is no electron on the i
level, it is necessary to exclude its action on state i, i.e.,
self-action, by eliminating the i term in the sums over hole states.
The elimination of this contribution, which, let it be noted, violates the Pauli principle, is achieved by taking exchange terms
into account. Thus, the term with i in the loop in Fig. 15a is
equal in magnitude and of opposite sign to the contribution of the
intermediate state i of Fig. 15c. The contribution of state i in

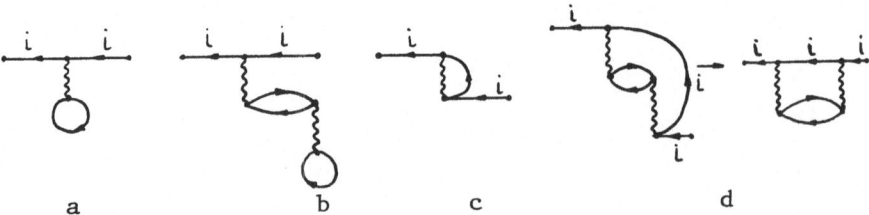

a   b   c   d

Fig. 15

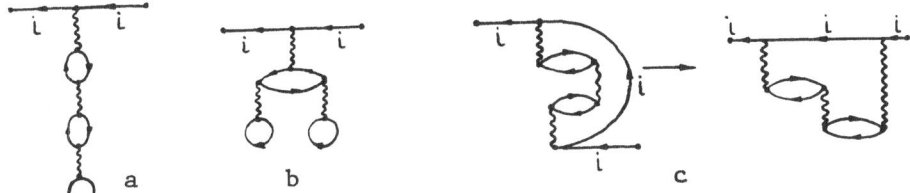

Fig. 16

the hole loop of Fig. 15b is compensated by that of Fig. 15d; they
are equal in magnitude and of opposite sign. Fig. 15d may be reduced to the form of Fig. 14a with k = i by deformation only, without destroying its topological structure.

Therefore, the shift of hole position determined by the contribution of Fig. 14a with k = i is tantamount to taking account
of ion rearrangement, its relaxation due to the creation of a vacancy. In the third order there are Hartree diagrams of two kinds
given in Fig. 16a and b. It may be shown, that contributions from
Fig. 16a and b in the loops of which term with hole state i occur
are canceled by taking into account the contributions of Fig. 16c
and Fig. 17.

Considering the higher-order Hartree diagrams, we may con-
vince ourselves that ion relaxation may be taken into account with
the help of the diagram in Fig. 18, where Γ is determined for the
initial neutral atom according to (1) without exchange terms. If
the exchange terms are also taken into account, the inner-shell
ionization potential shift is determined by the diagram in Fig. 18,
where Γ [see (1)] now includes exchange terms.

5.   Variation of the Outgoing Electron Wave Function

Apart from threshold shift, the ion rearrangement manifests
itself in the variation of the outgoing electron wave function. As
a result of the rearrangement, the latter moves not in the "frozen"

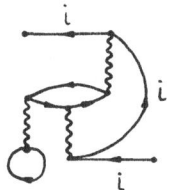
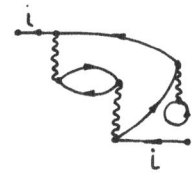
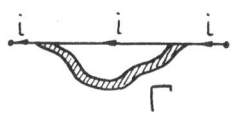

Fig. 17   Fig. 18

hole field, but in the altered screened one. The fact that the
outgoing electron moves in the "frozen" hole i field may be taken
into account in two ways -- either by eliminating the i term in the
sum over occupied states in the HF equations for the outgoing electron
or by summation of the diagonal in the hole state time-forward diagrams of the RPAE method. In the first case the equation for the
outgoing electron is of the form

$$-\Delta \phi_{\varepsilon}(r) + \sum_{j \neq i \leqslant F} \int \phi_{j}^{*}(r') \frac{dr'}{|r-r'|} \left[\phi_{j}(r') \phi_{\varepsilon}(r) - \phi_{j}(r) \phi_{\varepsilon}(r') \right] - \tag{13}$$
$$- \frac{Z}{r} \phi_{\varepsilon}(r) = E \phi_{\varepsilon}(r) ,$$

where ϕ_j is given by the system

$$-\Delta \phi_{K}(r) + \sum_{j \leqslant F} \int \phi_{j}^{*}(r') \frac{dr'}{|r-r'|} \left[\phi_{j}(r') \phi_{K}(r) - \phi_{j}(r) \phi_{K}(r') \right] - \tag{14}$$
$$- \frac{Z}{r} \phi_{K}(r) = E_{K} \phi_{K}(r) \quad , \quad K \leqslant F.$$

In the second, the diagrams given in Fig. 3 have to be taken into
account. The two methods are equivalent, because each term of
sequence (3) is equal in magnitude and of opposite sign to the
contributions of the diagrams in Fig. 19, where the index i near
the loop or internal line indicates that only the ith term of the
sum over states k ≤ f has been taken into account.

However the presence of hole i is not fully taken into account
in (13), because in (14) the variation of wave functions ϕ_j with
$j \neq i$ is neglected. Let us show to what taking account of the ϕ_j
variation corresponds. Just as was done in Sec. 4, we take into
account the existence of hole i in the second- and third-order Hartree diagrams which describe the self-energy of the removed electron (Fig. 20a, b, c). The contribution of Fig. 20a', b', c',
which takes account of the interaction of the removing electron
with hole i, is equal but of opposite sign to the term with the ith
hole state in the loops of Fig. 20a, b, c. But Fig. 20a', b', c'
takes into account the fact that the removing electron feels the
screened field of hole i. The main contribution to the screening
comes from outer-shell excitations, because their polarizability is
the greatest. Therefore it may be said that the diagrams of

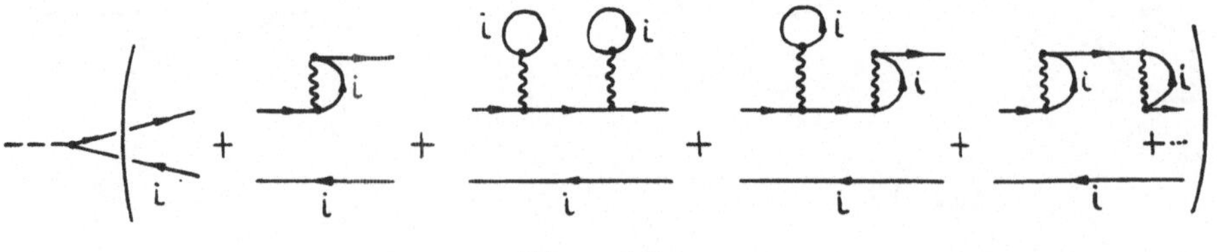

Fig. 19

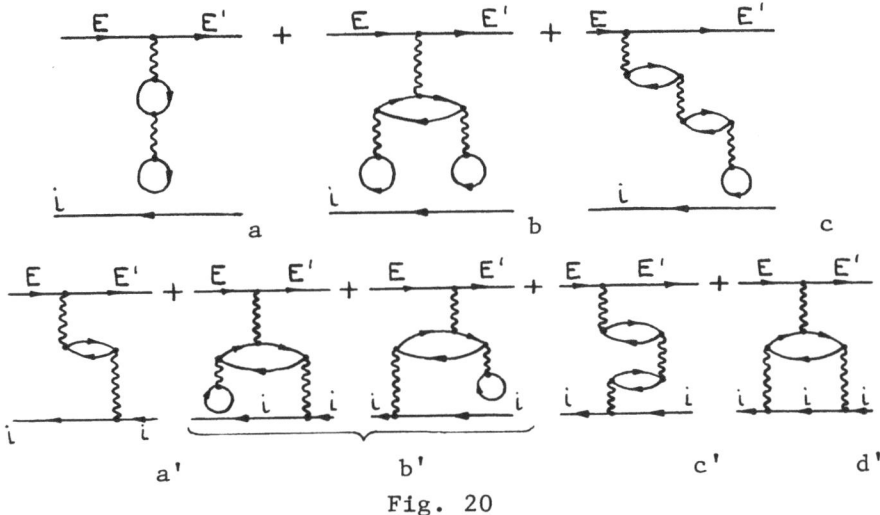

Fig. 20

Fig. 20a', b', c', d', take into account the variation of field
created by hole i under the action of the outer shells. The contribution of Fig. 20a' compensates one term in the contribution of
Fig. 20a, that with state i in the lower hole loop. But the latter
together with the Coulomb potential describes the self-consistent
field which acts upon an electron in the k-state. In Fig. 20a
this is the field of a neutral atom. Therefore, the contribution
of Fig. 20a takes account of the variation of the field in which the
kth electron moves in the presence of hole i, as compared with that
of a neutral atom. The situation in higher orders is the same.

As a result we conclude, that the variation of field which
acts upon the outgoing electron due to the presence of hole i may
be taken into account either by eliminating the term k = i in sums
over k ≤ f in loop diagrams, or by adding the contribution of Fig.
20a', b', c', d' and all similar diagrams in higher orders of perturbation theory. The latter diagrams describe the hole field
variation, that is, its screening.

Taking account not only of Hartree diagrams (Fig. 20) but also
of exchange diagrams makes the consideration much more complicated
but does not alter the conclusion that Fig. 20a', b', c', d' etc.
(together with exchange terms) compensates corresponding terms in
the HF diagrams which describe the self-consistent field acting
upon the outgoing electron. Thus the diagrams of Fig. 20a', b',
c', d' etc. may be accounted for by proper modification of the HF
equation describing the outgoing electron motion.

In (13) the absence of an electron in state i is accounted
for in the lowest order of perturbation theory in interelectron

interactions if the ϕ_k are determined by (14). To take into account the contribution of Fig. 20a', b', c' and similar diagrams in higher orders, the variation of the kth electron wave function under the action of the ith hole must be taken into account. This is achieved by solving the HF equation without the term j = i:

$$- \Delta \phi_k(r) + \sum_{j \neq i \leq F} \int \phi_j^*(r') \frac{dr'}{|r-r'|} \left[\phi_j(r')\phi_k(r) - \phi_j(r)\phi_k(r') \right] -$$
$$- \frac{Z}{r} \phi_k(r) = E_k \phi_k(r). \qquad (15)$$

Just as the solution of the HF equations (13) and (14) permits us to take account of the sequence of diagrams in Fig. 3, the solution of system (13) and (15) includes the much more complicated sequence of diagrams of Fig. 20a', b', c', d' and all similar diagrams in higher orders.

The effect of rearrangement is essential if the electron leaves the atom sufficiently slowly, for a time exceeding the period of rearrangment τ. The quantity τ can easily be estimated,

$$\tau \sim \frac{1}{\Delta E} = \left[E_i - (\mathcal{E}_o - \mathcal{E}_i) \right]^{-1}, \qquad (16)$$

where the notation is the same as in Table 1. The electron leaves the atom for a time T ~ $(E)^{-\frac{1}{2}}$, E being its energy. Comparing T and τ, we obtain the upper bound on E, for which the effect of rearrangement is still essential:

$$E < \left[E_i - (\mathcal{E}_o - \mathcal{E}_i) \right]^2 / 2 . \qquad (17)$$

For higher energies the rearrangement has no time to proceed and the outgoing electron moves in the field of a "frozen" hole. The system (13) and (15) takes into account the interaction of a particle and a hole which appear as a result of external field action. It is meaningful to consider in this sense that rearrangement is an example of postcollision interaction /16/.

6.   Generalization of the RPAE Method - GRPAE

In this report as a generalization of RPAE -- GRPAE -- we assume a method of calculating of effective interaction with the help of (1) in which, however, the wave functions of excited states and the continuum spectrum are determined by the solution of (13) and (15) and the differences between hole energies and their HF values are taken into account. Hole wave functions are determined by (14).

The cross section was obtained from (5)-(7), and the amplitude from (4), where the shifting of the hole energy and the effect of

rearrangement on excited-state wave functions were taken into
account, just as in calculating Γ.

If in the sums over $n \leqslant f$ in (1) and (4) only a single term
is considered, we will speak of the GRPAE correlation in one transition (or one shell). If in the sums two terms (i and k) are
taken into account, we will speak of two-transition or intershell
GRPAE correlation.

Using functions which are solutions of (13) and (15) is equivalent to assuming that the rearrangment has time to proceed not
only in the final but in intermediate virtual states. This assumption is quite meaningful if both interacting shells are inner shells.
If one of them is an outer shell, its rearrangement is neglected.
As was demonstrated in Sec. 4, only rearrangements due to hole
creation in inner or intermediate shells may be taken into account
by choosing the self-consistent field acting upon the outgoing electron. As for the outer shell, the contribution of "time-backward"
diagrams (see Fig. 14b), neglected in GRPAE, is essential in calculating the ionization potential. Therefore, the proposed method
cannot improve either the threshold position or the ionization cross
section for outer shells as compared to values given by RPAE. Taking account of rearrangement in GRPAE means that, apart from Figs.
1 and 2, the contribution of Fig. 18 is included, which leads to
variation of the hole energy as compared with its HF value, and
electron-hole screened interaction within the particle-hole loop is
taken into account. The corresponding diagrams are given in Fig.
21. Our experience in calculations by GRPAE permits us to formulate
the following criterion of its applicability: if $E_i - (\varepsilon_o - \varepsilon_i)$ is
close to ΔI_i^{exp}, the method is effective.

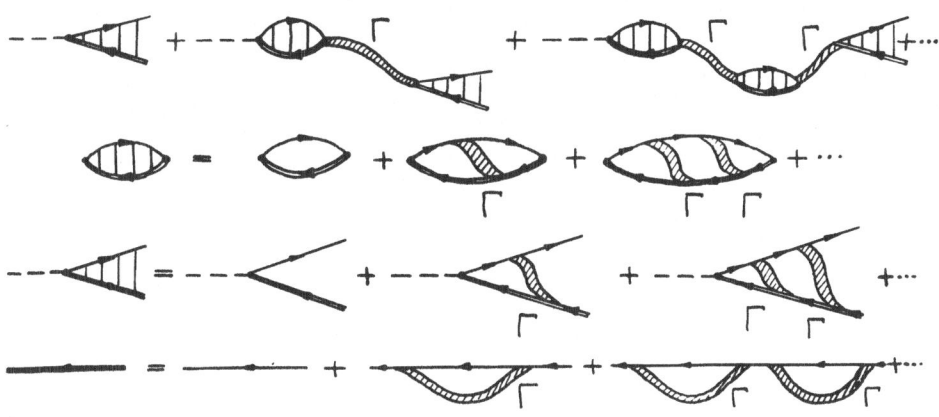

Fig. 21

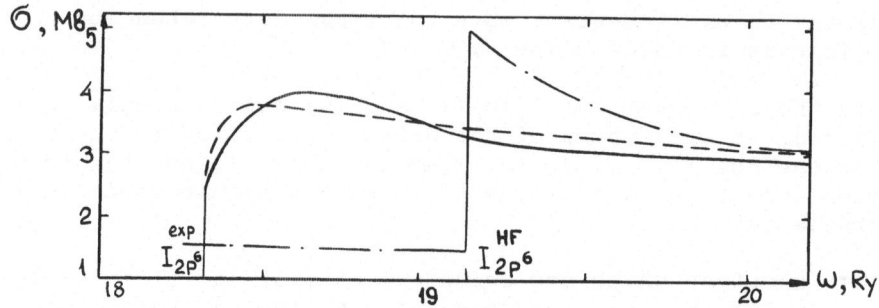

Fig. 22. Photoionization cross section of the $2p^6$ subshell in Ar. − − − experiment /14/, — · — RPAE, ——— GRPAE. The same notation is used in all subsequent figures.

7. Results of Calculations by the GRPAE Method

We shall give here the results of photoionization cross section and generalized oscillator strength calculations obtained with the help of (4), (5), and (7) by solving equation (1) with excited-state wave functions determined from (13) and (15). This method was first applied to the study of $2p^6$ subshell photoionization of Ar /10/, where without account of rearrangement there was a narrow threshold peak −− contrary to the experimental data /14/. By taking account of rearrangement this narrow peak was eliminated and much better agreement with experiment was achieved (Fig. 22).

An analogous consideration was applied to generalized oscillator strength, and qualitatively the same picture proved to be correct −− the oscillator strengths of dipole, monopole, and quadrupole transitions decrease significantly at threshold, and with increase of outgoing electron energy the effect of rearrangement becomes insignificant.* In Fig. 23 the results of generalized oscillator strength calculations for the $2p^6$ subshell of Ar are presented for monopole, dipole, and quadrupole transitions. The role of RPAE correlations within the $2p^6$ shell for dipole and quadrupole transitions is rather small, the effect of rearrangement being significant. In the monopole transition both effects are important (Fig. 23a). The effect of rearrangement was considered for the $4d^{10}$ shell of La taken as an example /17/. In the investigation of photoionization of metallic La /18/ it appeared that the cross section has a strong narrow maximum situated above the $4d^{10}$ ionization threshold, the position of the latter being determined from electron spectroscopic data /19/. In order to explain this peculiarity it was assumed /20/ that the maximum is a manifestation of a discrete $4d \to 4f$ excitation level in trivalent La^{+++} falling into its own continuum spectrum, in consequence of which the level at-

---

*Obtained together with S. A. Sheinerman and S. I. Sheftel.

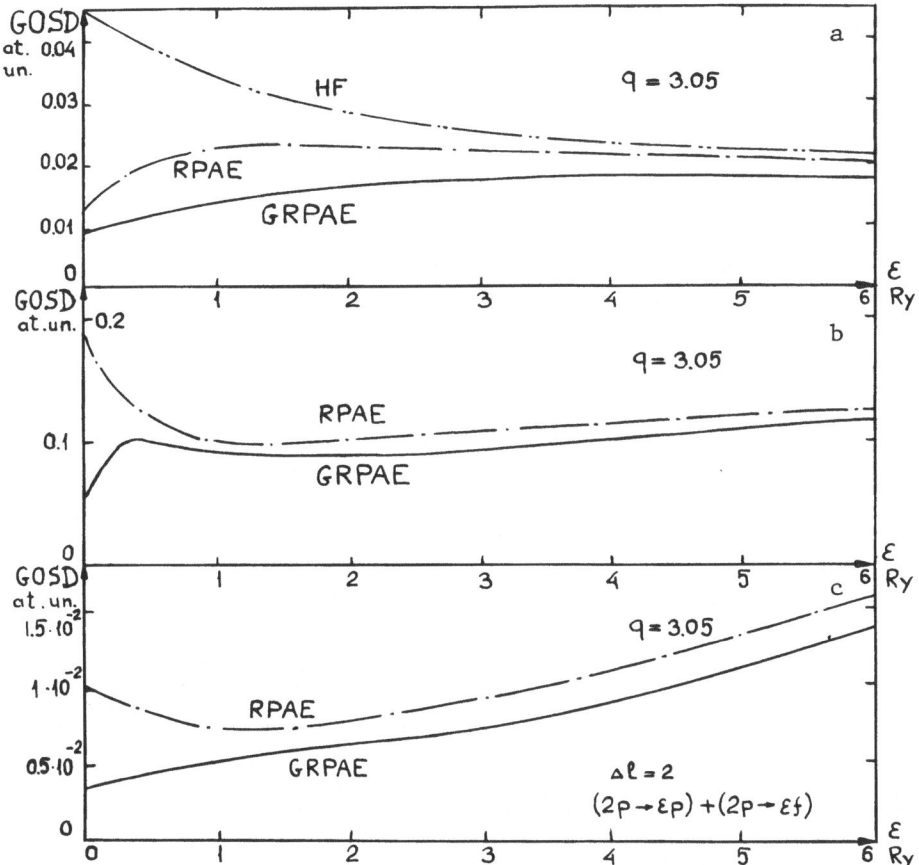

Fig 23. GOSD's of the 2p^6 subshell in Ar: (a) monopole, (b) dipole,
         (c) quadrupole transitions.

tains noticeable width and pronounced asymmetry -- as is usual for
autoionizing states. We shall not discuss here the problem of
photoabsorption by metallic La. As to atomic La, the RPAE calculation gives for the 4d^{10} shell a sharp maximum at threshold (see
Fig. 24). Taking account of rearrangement leads to its shifting
from threshold to the higher-energy side, but it is placed still at
a lower energy than on the experimental curve /18/. It was interesting to trace the variations in the relative role played by rearrangement with increase in nuclear charge. As an object we choose
the 4d^{10} shell in Xe, Cs, and Ba.*

_____

*It was investigated together with V. K. Ivanov.

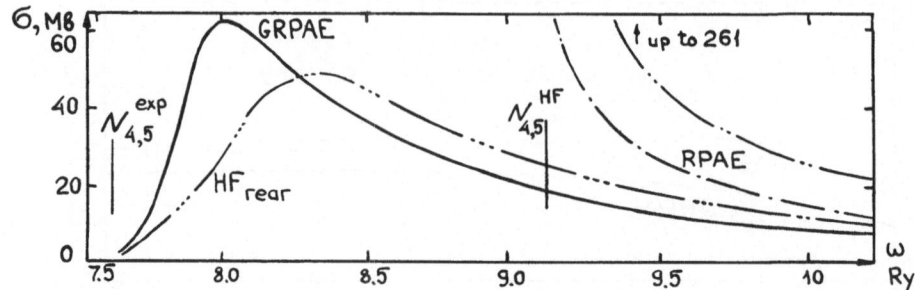

Fig. 24. Photoionization cross section of the $4d^{10}$ shell in La.

The general tendency proved to be the same as in the cases
discussed above -- the cross section maximum shifted to the higher-
energy side and its magnitude at threshold became smaller. The
results of the calculations are shown in Fig. 25. The curves
demonstrate that the relative role of rearrangement increases from
Xe to La. The role of the RPAE correlations changes in the same
direction: they are significant in Xe and become even more important for Ba and La.

The calculations in all cases mentioned above, just as in
the next section, were performed with r- and ∇-operators. An
important check of the quality of the wave functions used and of
the internal consistency of the accepted approximation is the coincidence of cross sections calculated with r- and ∇-operators by
(5) and (6) and fulfillment of the sum rules (8) and (9). For RPAE
it has been demonstrated /3/ that the rigorous sum rules are fulfilled and the r- and ∇-operators are equivalent. It is quite
possible that GRPAE has no such properties.

Our RPAE calculations contain some purely computational error,
which shows up in a small difference between results with r- and
∇-operators -- about 3-5%. It was interesting to learn what the
difference is between results of calculations with r- and ∇-opera-
tors in GRPAE, and whether it exceeds purely computational error.
It appeared that by shifting thresholds and taking rearrangement
into account in GRPAE, coincidence of r and ∇ results is achieved,
and the sum rule is fulfilled with an error not exceeding 3-5% and
rarely 10%.

Consider the intershell correlations in GRPAE. As an example
let us study the effect of the $4d^{10}$ shell in Xe, Cs, and Ba upon
the partial cross section of $5s^2$ and $5p^6$ subshell ionization.

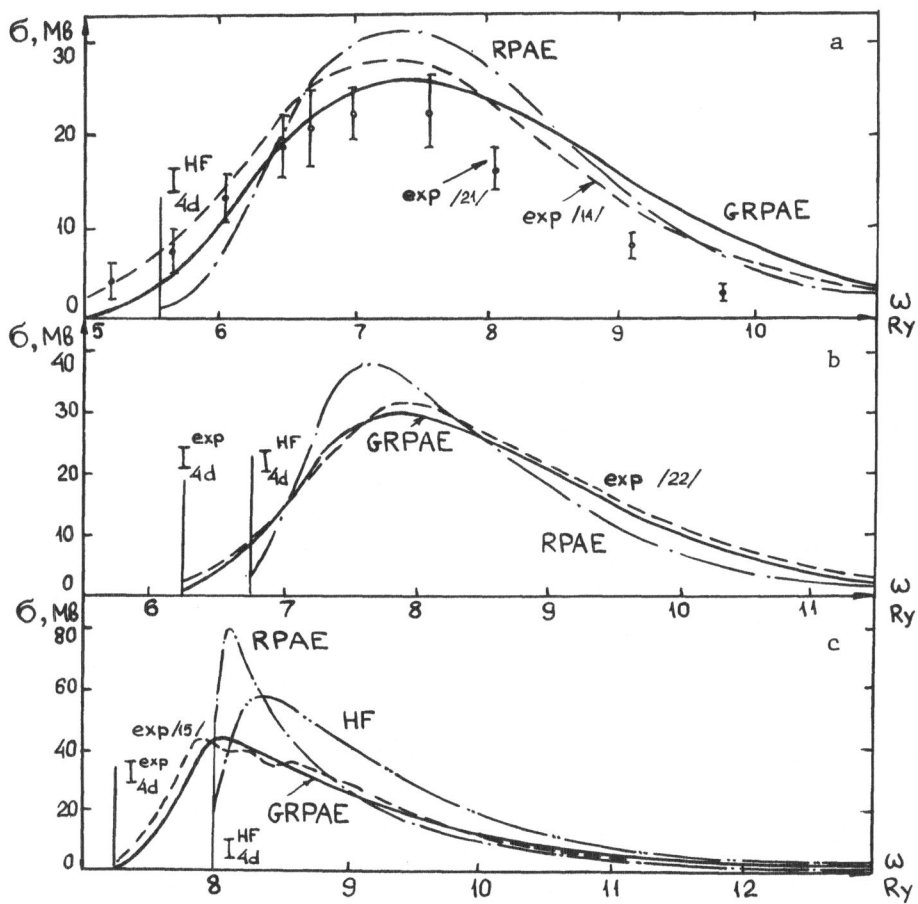

Fig. 25. Photoionization cross section of the $4d^{10}$ shell in (a) Xe, exp /21,14/; (b) Cs, exp /22/; (c) Ba /23/, exp /15^a/. RPAE data from /3/.

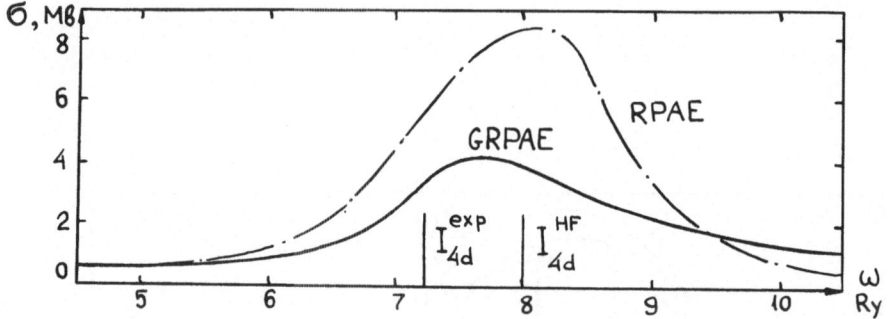

Fig. 26. Photoionization cross section of the $5s^2$ and $5p^6$
subshells in Ba.

The effect of $4d^{10}$ is essential even at the $5s^2$ ionization
threshold. But it is most noticeable in the vicinity of the $4d^{10}$
threshold, where under the influence of $4d^{10}$ a maximum of collective nature in the $5s^2$ ionization cross section appears. The results of GRPAE and RPAE calculations practically coincide in the
vicinity of the $5s^2$ threshold, whereas near $4d^{10}$ the collective
maximum due to rearrangement becomes lower and broader, the tendency increasing from Xe to Ba (see Fig. 26). The latter is a
result of the fact that for Ba the $4d^{10}$ ionization cross section
in RPAE has a sharp narrow maximum at threshold and for Xe a
broad maximum far removed from threshold.

Consider the effect of electron shell rearrangement upon the
angular distribution given by the formula

$$\frac{d\sigma}{d\Omega} = \frac{\sigma_\gamma}{4\pi}\left[1 - \frac{\beta(\omega)}{2}P_2(\cos\theta)\right], \tag{18}$$

where P_2 is the Legendre polynomial, and $\beta(\omega)$ the anisotropy parameter,
expressed in matrix elements of the dipole operator /3/, calculated
in this paper in GRPAE, and the elastic scattering phases of the
ionized electron in the ion field. The effect of rearrangement on
the phases was also taken into account.

The values of $\beta(\omega)$ given in this report reflect to what extent the difference between GRPAE and RPAE tells on the amplitudes
of the transitions $\ell \rightarrow \ell + 1$ and $\ell \rightarrow \ell - 1$ and on the elastic
scattering phases, where ℓ is the partial wave angular momentum.
The results of the calculations are presented in Fig. 27. As is
seen from Fig. 27, the difference between the RPAE and GRPAE results is rather characteristic -- the dependence of $\beta(\omega)$ on ω for
$4d^{10}$ is more stretched out in GRPAE than in RPAE. The effect of

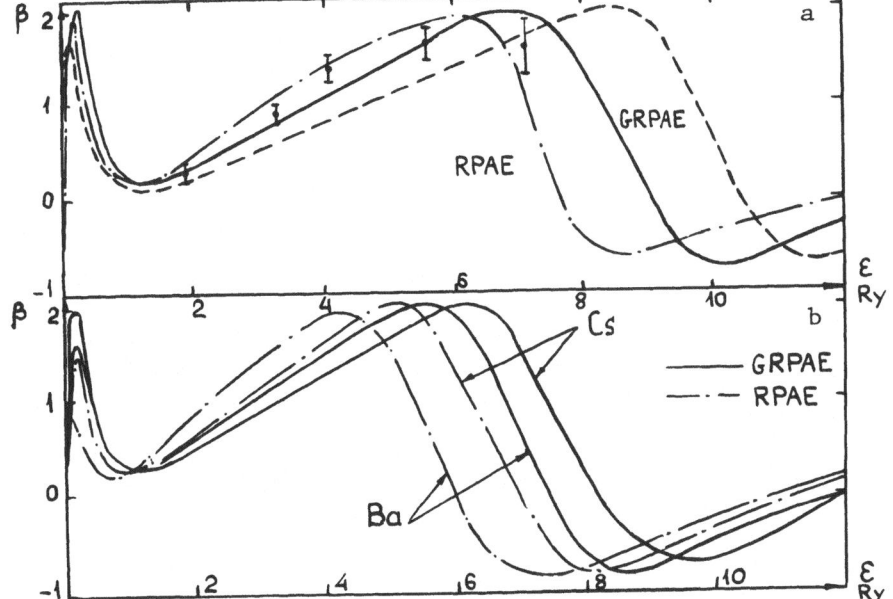

Fig. 27.   Anisotropy parameter $\beta(\omega)$ for the $4d^{10}$ subshell in
           (a) Xe, exp /28/ and (b) Ba and Cs; $- - -$ HF$_\Gamma$ /24/.

$4d^{10}$ strongly affects the angular distribution of the outer shells,
for example, $5p^6$ in Xe, as illustrated by Fig. 28.

8.   GRPAE Restrictions and Further Generalization of RPAE

Though GRPAE permits us to account for some many-electron correlations outside of the RPAE frame, it has some restrictions.   The
principal one is that rearrangement in GRPAE is considered a com-

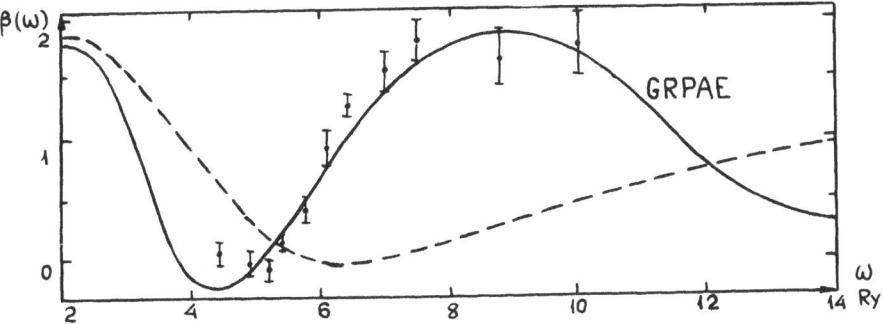

Fig. 28.   Anisotropy parameter $\beta(\omega)$ for $5p^6$ in Xe; $- - -$ without
           effect of $4d^{10}$, exp /28/.

pleted process. In fact it proceeds in time, the fast electrons
feeling the "frozen" core field and the slow ones the rearranged
field. This is why the rearrangement has to be switched on gradually depending on electron energy.

GRPAE permits us to sum the diagrams given in Fig. 21, the
rearrangement being taken into account in "time-forward" diagrams
like those presented in Fig. 20. Corresponding "time-backward"
diagrams are neglected.

GRPAE does not take into account the polarization potential
acting on outgoing electrons (Fig. 12). This does not lead to
significant errors for noble gas atoms, because the polarizability
of their ions and consequently the polarization potential acting on
outgoing electrons is small. In other atoms, especially the alkaline earths, neglecting the polarization of the remaining ion by
the removed electron must lead to noticeable mistakes.

The problem of going beyond RPAE in atomic physics has recently
been discussed from a rather general point of view by Fano and Chang
/25/.

To generalize RPAE without these restrictions it is necessary
to solve a system of equations much more complicated than (1), which
permits taking account of the difference between the best single-
electron and HF atomic field, the variation of the vertex of interaction with the external field, and the difference between the
electron-hole interaction and the purely Coulombic one due to
virtual atomic excitations -- all in "time-forward" and "time-
backward" diagrams.

If we do find a proper approximation making RPAE more accurate,
which is far from simple, the numerical solution of such a system of
equations will be an extremely difficult task. In principle a semi-
phenomenological generalization of RPA which flows from the Landau
theory of Fermi liquids and from the "theory of finite Fermi systems"
by Migdal /26,27/ is possible. The idea is to use the RPAE equations, but with the difference that for electrons and holes not the
pure HF self-consistent field but that with a phenomenological correction W is used, which, together with the HF potential, permits
us to describe elastic scattering on atom and ionization potential
in accordance with experimental data. In (1) (U + w) has to be
substituted for U, w being an additional phenomenological electron-
electron potential. The values W and w must be connected because
W is an addition to the self-consistent field due to w, the addition to the interelectron interaction.

The main difficulty in this method lies in choosing the simplest radial dependence of w and determining the parameters in such
a way that they are as equal as possible for all but the lightest

atoms. We don't know whether this program, widely used in nuclear
theory, can be realized for electron shells of atoms.

9. Concluding Remarks

The investigation of outer- and intermediate-shell ionization
by photons and electrons demonstrates the collective character of
these processes and the possibility of describing them by RPAE.

GRPAE, the generalization of RPAE proposed in this report,
permits us to take account of a considerable part of the electron
correlations which lie outside the RPAE frame. As is demonstrated
by the results of GRPAE calculations, the ionization near the inner-
shell threshold is accompanied by rearrangement of the outer shell.
The agreement with experiment that has been achieved testifies to
the fact that rearrangement is the main process which has to be
taken into account apart from RPAE correlations in describing inner-
and intermediate-shell ionization. Further investigation of ionization processes in the vicinity of these shell thresholds will assist
in the development of the theory of many-electron atoms.

The author expresses his gratitude to V. K. Ivanov, S. I.
Sheftel, and S. A. Sheinerman, with whom many results presented
in this report were obtained.

References

1. R. L. Altick and A. E. Glassgold, Phys. Rev., 133 (1964) A632.

2. M. Ya. Amusia, N. A. Cherepkov, and S. I. Sheftel, Phys. Lett.,
   24A (1967) 541.

3. M. Ya. Amusia and N. A. Cherepkov, Case Studies in Atomic
   Physics, 5, 2, 47-179 (1975).

4. G. Weindin, (a) J. Phys. B: Atom. Molec. Phys. 4 (1971) 1080;
   (b) in: Vacuum Ultraviolet Radiation Physics, Pergamon, Vieweg,
   1974, p. 225.

5. H. P. Kelly, Phys. Rev. 131 (1963) 684; 136 (1964) B 896.

6. C. D. Lin, Phys. Rev. A9 (1974) 171.

7. P. G. Burke and K. T. Taylor, J. Phys. B: Atom. Molec. Phys.,
   8 (1975) 2620.

8. J. A. R. Samson and J. L. Gardner, Phys. Rev. Lett., 33 (1974) 671.

9. R. G. Houlgate, J. B. West, K. Codling, and G. V. Marr,
   J. Phys. B: Atom. Molec. Phys., 7, 17, L470 (1974).

10. M. Ya. Amusia, V. K. Ivanov, and N. A. Cherepkov, V All-Union
     Conference on Physics of Electronic and Atomic Collisions,
     Book of Abstracts, Ushgorod, 1972, p. 131.

11. J. Hubbard, Proc. Roy. Soc., A243, 336 (1958), K. S. Singwi,
     M. P. Tosi, R. H. Land, and A. Sjolander, Phys. Rev. B1,
     (1970) 1044.

12. M. J. Van der Wiel and G. R. Wight, Phys. Lett., 54A (1975) 83.

13. V. V. Afrosimov, Yu. S. Gordeev, V. M. Lavrov, and
     S. G. Shchemelinin, Zh. Eksp. Teor. Fiz. 55 (1968) 1569.

14. R. Haensel, G. Keitel, P. Schreiber, and C. Kunz, Phys. Rev.,
     188 (1969) 1375.

15. (a) P. Rabe, K. Radler, and H.-W. Wolf, in: Vacuum Ultraviolet
     Radiation Physics, Pergamon, Vieweg, 1974, p. 247.
     (b) J. B. Connerade, D. Tracy, M. W. D. Mansfield, and
     K. Thimm, ibid., p. 243.

16. G. C. King, F. H. Read, and R. C. Bradford, J. Phys. B: Atom.
     Molec. Phys. 8, 13 (1975) 2210.

17. M. Ya. Amusia and S. I. Sheftel, Phys. Lett., 55A (1976) 469.

18. T. M. Zimkina and S. A. Gribovskii, J. Physique (France) 32
     (1971) C4-282.

19. W. Lotz, J. Opt. Soc. Am., 60, 2 (1970) 206.

20. J. L. Dehmer and A. F. Starace, Phys. Rev. B5 (1972) 1792.

21. J. B. West, P. R. Woodruff, K. Codling, and R. G. Houlgate,
     J. Phys. B: Atom. Molec. Phys. 9 (1976) 407.

22. H. Petersen, K. Radler, B. Sonntag, and R. Haensel, Preprint
     DESY-SR 74/14, 1974, Hamburg.

23. M. Ya. Amusia, V. K. Ivanov, and L. V. Chernysheva, IX ICPEAC,
     Abstracts of papers, Seattle, 1975, p. 1135.

24. D. J. Kennedy and S. T. Manson, Phys. Rev. A5 (1972) 227.

25. T. N. Chang and U. Fano, Phys. Rev. A13, 1 (1976) 263.

26. L. D. Landau, Zh. Eksp. Teor. Fiz., 30 (1956) 1058; 32
     (1957) 59.

27. A. B. Migdal, Teoria konechnikh Fermi-sistem i svoistva
     Atomnikh Yader (in Russian), "Nauka," Moscow 1965.

28. G. V. Marr, in: Photoelectron Emission: proceedings of the
     Darsbury Study Weekend, 6-7 March, 1976, DL/SRF/R8, p. 81.

SUBJECT INDEX

ABMR, 201
Absorption Beats, 183
Absorption Spectroscopy, 487
ALCATOR, 396
Alignment, 240
  predissociation, 176
Analyzing Power, 328
Angular Distribution
  radiation, 240
Anisotropy
  in atomic collisions, 283-291
Anomalous Moment
  electron, 39-41
  muon, 41-43
Anomalous Fluorescence, 186, 194
Anti-Protononium, 63
Asymmetry, 242
Atomic Beam Detection, 211
Atomic Beam Magnetic
  resonance, 201
  resonance, apparatus, 206
Atomic Collisions
  anisotropy and time dep.,
    283-291
Atomic Data Importance
  to thermonuclear reactor,
    383, 384
Autoionization, 315
  atomic, 168

Beam Deposition Instability,
  385, 401
Beam Foil Excitation
  anisotropy from, 283-286
  quantum beats, 283-286
Binding Energies, 314

Bound States
  of pi-mu atom, 95
Bremsstrahlung, 511
  inner, 511
  nucleus-nucleus, 511, 514
  power loss in H.T. plasma,
    383, 385
  secondary electron, 511-513

Charge Exchange, 287-289
  inj. neut. on H.T. protons,
    386
Charge State Production, 494
Charm, 434
Charmonium, 63, 432
Chemi-Ionization, 346
Circular Dichroism, 13
Coherent Pumping, 179
Coincidence Measurements
  electron, photon, 286-287
Collective Emission Effects, 196
Collisional Broadening, 157
Collisional Excitation, 494
  power loss in H.T. plasma,
    383, 385
Collisional Ionization
  power loss in H.T. plasma,
    383, 385
Color, 434
Configuration Interaction, 313
Confinement, 434
Continuum X-Ray Emission, 528
Corona Equilibrium, 379, 383
Corona Model, 379, 383
Correlation Coefficients, 242

Coulomb Collisions
  in tokamak plasma, 385
CPT Theorem, 81
Cross Section
  $O^{5+} + H \rightarrow O^{4+}$, 386, 388–389
  $O^{5+} + H_2 \rightarrow O^{4+}$, 386, 388–389
  $Fe^{10+} + H \rightarrow Fe^{9+}$, 386, 388–389
  $H + O \rightarrow H^+$, 386, 388–389
  $H + H^+ \rightarrow 2H^+$, 386, 388–389
  $e + H \rightarrow H^+$, 386, 388–389

Decay
  predissociative, 168
  radiative, 168, 175, 177
Decay Rates
  positronium annihilation,
    108–112
Detection
  atomic beam, 211
  radioactive, 207
Dielectronic Recombination,
    399–400, 402
  power loss in H.T. plasma,
    383, 385
Dipole Polarizability, 300
Dissociation Limit, 170
Doppler Broadening, 174, 186
Double Collision Model, 517
Dye Laser
  pulsed, 171

Effective Magnetic Field, 459
Elastic Scattering, 363
Electric Dipole Moment
  neutron, 453, 460
  theory, 461
Electron
  anomalous moment, 39–41
Electrons
  polarized, 325
Electron-Atom Scattering,
    293, 301
Electron-Molecule Scattering,
    293
Electron Attachment Rates, 271
Electron Capture, 493
Electron Correlation, 313
Electron Correlation Effects,
    541–542

Electron Excitation
  of He(2, ^{1}P), 286–287
Electron Scattering
  from opt. pumped states, 289
Electron Shakeup Spectra, 321
Electron Temperature
  in tokamak plasma, 377–378, 383
Emission
  spontaneous, 167
Emission Spectroscopy, 480
Energy Levels
  iodine, 169
Exchange Excitation, 361
Exotic Atoms, 63
  high Z muonic atoms, 55–57
  muonic helium, 53–55
  pi-mu, 95
  positronium, 103–124
E2 mixing, 69

Fano Effect, 333–334
Faraday Effect, 18
Faraday Rotation, 23, 27
Fe^{N+} Abundances in
  tokamak plasma, 402–403
Fe^{N+} K Alpha X-Rays from
  tokamak plasma, 403–404
Fe^{N+} Resonance Lines, 402
Fermi-Segre Formula, 6
Fine Structure
  helium, 38–39
  muonic helium, 53–55
  positronium ground state, 43–44
  positronium n=2, 115–120
Fine Structure Constant
  helium fine structure, 39
  Josephson effect, 37–38
  summary derived values, 46
Fine Structure Splittings, 151
Fluorescence, 168, 176
Fluorescence Decay, 174
Fluorescent Anomaly, 186
Forbidden Decays
  in positronium annihilation
    108–112
Forbidden Transitions, 415, 437
Free-Free Radiation
  power loss in H.T. plasma,
    383, 385

Free Induction Decay, 180
Fusion Power, 391–392
Fusion Yield, 401

Gallium Arsenide, 350
Gauge Theories, 1
g-2
    electron, 39–41
    muon, 41–43

Hadronic Atoms, 63
Hamiltonian
    rotational, 170
    spin orbit, 170
Heating
    ohmic of tokamak plasma, 376
Helium
    fine structure, 38–39
    nuclear charge radius, 54–55
Helium-3 Ion
    hyperfine structure, 46–47
Highly Excited Atoms, 269
Highly Ionized Atoms, 473
Hydrogen-Like Spectra, 481
Hyperfine Pumping, 211
Hyperfine Structure, 201, 207
        239
    in hadronic atoms, 72
    muonium, 44–45
    of stored ions, 125–145
    positronium, 105–107
    2S ^{3}He$^+$, 130–145
    ^{25}Na, 222
    ^{3}He$^+$, 46–47

Impurities
    in high temperature plasmas,
        375
Incoherent Pumping, 179
Inelastic electron
    scattering, 543–544
Inner-Shell Ionization, 509–510
Inner-Shell Vacancy
    production, 494
Interference, 175, 177
Iodine, 169
Ionization
    of pi-mu atom, 96
    two photon, 339, 343

Ionization Potential, 549
Ionization Processes, 537
Ions
    behavior in H plasma, 375–390
    C and O in H.T. plasmas, 378
    Fe ions in tokamak plasma,
        383–384
    imp. in H.T. plasma, 375–390
    line rad. from H.T. plasmas,
        378
    Mo ions in tokamak plasma,
        381, 383
    multicharged Au, W, 378
    multicharged O, 375, 380, 382
    multicharged Mo, 375, 381
    O ions in tokamak plasma,
        380, 382
Ion-Atom Collisions, 287–289
Ion Density and Temperature
    profile in TFTR plasma,
        386–387
Ion Storage
    for RF spectroscopy, 126–130
Ion Traps, 127, 128, 132–136
Iron Spectrum, 477–480
Isolde, 202, 239
Isotope Production
    targets, 203–204
    yield, 205
Isotope Separator, 207
Isotope Separation, 175, 228
Isotope Shift, 155, 239, 251
    shape, 252
    sodium isotopes, 221–225
    volume, 252

Josephson Effect
    determination of alpha, 37–38

Kaonic Atoms, 65
K Zero Long
    decay into pi-mu atom, 95

Lamb Shift
    pi-mu atom, 96
    summary of experiments, 53
    theory, 47–52
Laser Action
    mirrorless, 196

Laser Excitation, 211, 263
Laser Spectroscopy, 167,
    215-225, 258-259, 262
  time resolved, 179
Lawson Criterion, 392
LEED, 354
Lifetime, 168
  hyperfine, 174
  rovibronic, 171
Lifetime Measurement, 501
Limiter, 376-377
Line Broadening, 486
Line Profiles, 484
Line Radiation
  high temperature plasmas, 378
  spatial dist. tokamak plasma,
    382
  time dep. in tokamak plasma,
    380-381, 383
Liquid Drop Model, 252

Magnet
  four-pole, 204
  six-pole, 204
Magnetic Dipole
  transitions, 415
Magnetic Dipole Moment
  neutron, 453, 465
Magnetic Dipole Moments, 201
Magnetic Moment
  of antiproton, 83
  of sigma minus, 87
  ^{23}Na nucleus, 225
Masses
  of fundamental particles, 75
Mean Free Path
  inj. neut. in H.T. plasma, 386
Metastable Atomic States, 209
Microwave Spectroscopy
  positronium fine structure,
    115-120
Modulation Transients, 180
Molecular Spectroscopy, 161
Molecular Orbital X-Rays,
    509-510
Moller Scattering, 356
Mott Scattering, 362
Multiphoton Spectroscopy, 147
Muonic Atoms, 228

Muonic Helium, 53-55
Muonic High Z Atoms, 55-57
Muonium, 63
  hyperfine structure, 44-45
Muon, 45
  anomalous moment, 41-43
  in pi-mu atom, 95

Negative Electron Affinity, 350
Neutral Currents, 1, 23, 27
Neutral Vector Boson, 2
Neutrino-Electron Scattering, 3
Neutron
  electric dipole moment,
    453, 460
  magnetic dipole moment,
    453, 465
  spin echo, 467
Neutron Bottle, 462-463
Neutrons
  ultra-cold, 462
Nuclear Charge Radius, 251-252,
    254
Nuclear Compressibility, 253
Nuclear Excitation, 227
Nuclear Magnetic Resonance, 243
Nuclear Moments, 201, 209, 245
Nuclear Phase Transition, 253
Nuclear Polarization
  muonic helium, 37
Nuclear Shapes, 209, 257
Nuclear Spin, 208, 261-262

Ohmic Heating
  of tokamak plasma, 376
Optical Detection, 243
Optical Nutation, 180
Optical Pumping, 212, 239
  hyperfine, 210
  Na atomic beam, 215-225
Orientation
  by beam foil excitation,
    285-286
  by electron impact, 286-287
  by ionic impact, 287-289
  predissociation, 175-176
ORMAK, 378-379
Oscillator Strengths, 541
Osmium-189, 231

Parity, 453, 466
  non-conservation, 326
Parity Non-Conservation, 27
Parity Violation, 1, 23,
    361, 365
PEGGY, 335-339, 361
Photoelectron Spectroscopy, 313
Photoionization, 301, 540
Photon Echo, 180
Pion
  charge radius, 96
  in pi-mu atom, 95
Plasma
  heating by energetic H, D,
    375, 385
  next gen. tokamak param., 378
  present gen. tokamak param.,
    377
Plasmas
  high temperature magnetic
  confined, 375-390
Plasma Current Density
  measurement, 410
Plasma Diagnostics, 391-414
Plasma Heating,
  by adiabatic compression, 396
  by energetic neutral beam, 394
Plasma Impurities, 397
  diagnostic value, 401
  identification in tokamak, 410
  scattering of laser light, 410
Plasma Instability
  beam deposition, 401
Plasma Ion Temperature
  measurement, 406-408
Plasma Power Loss
  bremsstrahlung, 399-400
  by collisional ionization,
    383, 385
  by collision excitation,
    383, 385
  by radiative recombination
    383, 385
  charge exchange, 379-380
  dielectronic recombination,
    383, 385, 399-400, 402
  due to Fe in H plasma,
    383, 385
  free-free radiation, 383, 385

  radiative, 399-400
Plasma Radiation, 401
PLT (Princeton Large Torus), 396
Polarization, 240
  degree, 328
Polarized Electron Beams
  applications, 355
Polarized Electron-Electron
  scattering, 356
Polarized Electron Source
  field emission, 347
  LEED, 354
  low energy Mott scattering,
    332
  NEA, 350-351
  optically pumped, 344
  photoionization by pol. light,
    333
  photoionization by pol. at.
    beam, 335-339
  photo emission, 349-350
  resonant 2 photon ionization,
    343
Polarized Electrons, 325
  characteristics, 329-330
  figure of merit, 329
  source characteristics, 331
Polarized Hydrogen, 363
Polarized Neutrons, 457
Positive Electron Affinity, 351
Positronium, 44, 63
  annihilation rates, 108-112
  fine structure n=2, 115-120
  fine structure, 43-44
  formation of, 104
  hyperfine structure, 105-107
  production of n=2, 113-114
Predissociation
  gyroscopic, 169, 172, 175, 177
  hyperfine, 169, 174, 175, 177
  magnetic, 175, 177
  molecular, 167
  quantum interference, 173
  tuning, 175
Pressure Shifts, 157
Proton
  charge radius, 50-52
  form factors, 50-52
  gyromagnetic ratio, 38

Psions, 415

Quadrupole Moment, 225
Quantum Beats, 179-180
Quantum Defect, 8
Quantum Electrodynamics, 415
   atomic physics tests, 37-62
Quarks, 416, 437
Quenching
   magnetic, 175

Radiative Recombination
   power loss in H.T. plasma,
      383, 385
Radiative Electron Capture,
      505, 511
Radioactive Detection, 207
RADOP, 240
Raman Beats, 180
Random Phase Approximation,
      537-538
Rate
   predissociation, 175
Rearrangement After
   ionization, 554
RF Spectroscopy
   of 2S ^{3}He$^+$, 125-145
Ross Filter, 477
Rotational Constant B, 170
Rydberg Atoms, 181
   collisions, 271-272
   core polarizabilities, 273
   electron attachment, 271
   field ionization, 275
   fine structures, 273
   stark spectroscopy, 274
   threshold ionization, 275
Rydberg Levels, 153
Rydberg States, 269
R-Matrix Theory, 293

Scattering, 326
   deep inelastic, 359
   longitudin. pol. electrons,
      359
   Moller, 356
   Mott, 326
Slow Neutron Beam, 454
Spark Sources, 473, 480

Spectrometer
   quantum beat, 181
Spectroscopy
   high resolution laser, 215-225
Spin
   density matrix, 328
   Pauli operator, 327
   polarization, 327
Spin Exchange, 363
Spontaneous Emission
   amplified, 196
State
   dissociative, 167
   rovibronic, 167
States
   ortho, 173
   para, 174
Stimulated Emission Beats, 183
Stimulated Quantum Beats, 185
ST, 396
Superradiance, 179-180, 187,
      194, 198
Superweak Force, 462
Symmetric Collisions, 524
Szilard-Chalmers Method, 228

TFR
   toroidal fusion reactor, 396
TFTR
   toroidal fusion test reactor,
      391, 393, 396
Thomson Scattering
   in tokamak plasma, 410-411
Threshold
   superradiance, 188, 190
Time Constant, 168
Time Reversal, 9
Time Reversal Symmetry, 453
Tokamak Machines, 376
   FTR, 380-383
   JET, 377-378
   JT-60, 377-378
   PLT, 386
   TFTR, 377-378, 386-387
   T-20, 377-378
Tokamak Plasma, 376-377
   next generation parameters, 378
   present generation parameters,
      377

Tokamak Reactors
  ALCATOR, 396
  ATC, 394, 397
  diagnostics, 391-414
  ORMAK, 394
  PLT, 396
  ST, 396, 401-402
  TFR, 396
  TFTR, 391, 393, 395-396,
      399
  T-4, 397

UV Spectroscopy
  of tokamak impurities,
      401-402

Vacancies
  production, 520, 523
  radiative decay, 520
Vacuum Polarization
  in hadronic atoms, 74

Vacuum UV Radiation
  from tokamak plasma,
      401-402
Van Der Waals Coefficient, 301

Weak Interactions, 2
Weak Neutral Currents, 361
Weinberg-Salam Model, 2

X-Ray Continuum
  from tokamak plasma, 404-405
X-Ray Fluctuations
  from tokamak plasmas, 404
X-Ray Radiation
  from tokamak plasma, 401-404
X-Ray Yields, 498, 502

Zeeman Effect, 155, 207
Zeff
  imp. conc. in H.T. plasma,
      386